INDUSTRIAL
AND
LOCOMOTIVE
OF
SWITZERLAND

Compiled by Richard A. Bowen

INDUSTRIAL RAILWAY SOCIETY

Published by the INDUSTRIAL RAILWAY SOCIETY
at 24 Dulverton Road, Melton Mowbray, Leicestershire, LE13 0SF

© INDUSTRIAL RAILWAY SOCIETY 2009

ISBN 978 1 901556 59 9   (hardbound)

ISBN 978 1 901556 60 5   (softbound)

British Library Cataloguing-in-Publication Data
A catalogue record for this book is available from the British Library

Visit the Society at www.irsociety.co.uk

Sadly, shortly after delivering the manuscript to the Society, the author passed away peacefully in his sleep at his home in Germany. He was 70 years of age. The Society is grateful to his widow, Pat, and his children, for their help in bringing the manuscript to completion. Richard was the acknowledged expert on Swiss Industrial Railways and other than changing the contact details on page 10 we have made no further changes to his text.

IAN BENDALL   IRS CHAIRMAN, LEICESTER
                                                                                                                                             SEPTEMBER 2009

Front cover photo:
HOLCIM (Schweiz) AG, Untervaz (GR23). *On 15 March 1991 MaK 700041 of 1980 (type DE501 C) leaves the works to reach the SBB exchange sidings. In order to so do it must cross the R. Rhein on a road/rail bridge which it is approaching here. The branch can also carry traffic to and from the RhB exchange sidings, but these must use the right hand pair of rails seen in the photograph.*
Rear cover photo:
Internationale Rheinregulierung (IRR) (SG34). *"JUNO" a Motor Rail 32/42HP locomotive (10091 of 1950) stands with a train of stone used for strengthening the sides of the channels of the canalised R. Rhein. It was seen here near Koblach [A] on 11 April 1988*

Produced for the IRS by Print Rite, Witney, Oxon. 01993 881662

All rights reserved. No part of this publication may be reproduced, stored in a retrieval system or transmitted in any form or by any means without prior permission in writing from the Industrial Railway Society. Within the UK, exceptions are allowed in respect of any fair dealing for the purpose of private research or private study or criticism or review as permitted under the Copyright, Designs and Patents Act 1988.

List of Contents

Introduction		CH	4
Conventions		CH	9
Railway Company Initials		CH	10
Abbreviations for Railway Equipment Dealers		CH	14
Other Abbreviations		CH	15
Cantons			
AG	Aargau	CH	16
AI + AR	Appenzell (Appenzell Innerrhoden + Ausserrhoden)	CH	56
BE	Bern	CH	58
BL	Basel Landschaft (Basel Land)	CH	97
BS	Basel Stadt	CH	115
FL	Fürstentum Liechtenstein	CH	127
FR	Fribourg	CH	128
GE	Genève	CH	133
GL	Glarus	CH	140
GR	Graubünden	CH	143
JU	Jura	CH	160
LU	Luzern	CH	166
NE	Neuchâtel	CH	180
NW + OW	Unterwalden (Nidwalden + Obwalden)	CH	184
SG	Sankt Gallen	CH	188
SH	Schaffhausen	CH	219
SO	Solothurn	CH	222
SZ	Schwyz	CH	238
TG	Thurgau	CH	247
TI	Ticino	CH	256
UR	Uri	CH	278
VD	Vaud	CH	285
VS	Valais	CH	309
ZG	Zug	CH	337
ZH	Zürich	CH	339
Miscellaneous		CH	407
Locomotive Index		CH	421
Location Index		CH	483
Colour Section		CH	497

INTRODUCTION

The Industrial Railway Society (formerly the Birmingham Locomotive Club - Industrial Locomotive Information Section) has established a series of Handbooks concerned primarily with the locomotives owned by industrial concerns. The author presented the first booklet for Switzerland in the same series (reference code CH) in 1971. This was in English with a synopsis in French. Since then the VRS under the guidance of M. Sébastien Jarne has produced an edition of "Locomotives Industrielles de Suisse" in 1994. This goes much deeper into the constructional details of the locomotives than is IRS practice, indicating, for instance, the number of cylinders or motors installed. It also follows the normal continental practice of indicating when the site was last visited and which locomotives were present at that time. This particular booklet, written in French, is restricted to those systems connected to the public network, which may be of various gauges.

Despite the existence of these booklets there have been requests for an updated English language CH, which will, hopefully, be met with this publication. In the meantime there is a great deal of information available on the Internet in the form of builders' lists and other very detailed databases. These should become more correct and complete as time goes by and it would have almost certainly been possible to provide such a tool which enhanced (rather than competed with) the existing schemes. Some of these sites are listed in the references and readers are invited to use them to discover additional facts - many of which will have come to light since this Handbook closed for publication.

Using the information assembled principally by Jens Merte and published by him, later editions of Herr A. Moser's classic book on main-line Swiss steam have been enhanced by Eisenbahn Amateur with the addition of appendices covering industrial steam locomotives. This information is not free from errors. Some of the input to Jens Merte's work is hand-written and not always easy to read and place names are not always rendered as accurately as one might hope. Where this Handbook differs from the information presented by Eisenbahn Amateur (particularly by omitting locomotives) this is normally deliberate. The difference is often based on detailed research carried out by M. Jarne. Sometimes, but not always, an explanation is to be found in the text.

One advantage of the Internet is to make available many records that could previously only be found (if at all) by hours of dusty searching. M. Jarne has carried just such a search of local government authorities' records, VHS records and those elsewhere to amass a great deal of information about erstwhile narrow gauge railway systems not connected to the public railways. The coverage by Canton is not consistent and so far only the tip of the iceberg has been explored. Nevertheless, the results are made available here. The major effort therefore in preparing this Handbook has been the reconciliation of variant data (conflicting sometimes) and preparation of appropriate notes when a clean resolution could not be found. One consequence was such a dramatic increase in the size of the work that initially it was proposed to restrict the publication to those firms that existed after an arbitrary date (1965). The IRS, however requested that the coverage be made as exhaustive as possible and further requested that their established practice of listing preserved locomotives be followed. The latter is a very volatile field where, in the author's opinion, a web-site can offer a more up-to-date picture than this Handbook can achieve.

In consequence, it is intended to list as many as possible of the locomotives almost certainly owned at any time by industrial firms in Switzerland, and to give details of some of the contractors' and public works systems. Locomotives working on public railways are excluded, with the exception of some heritage stock. Brief details are given of preserved or heritage locomotives. Additionally, a few systems operated without locomotives that personally appeal to the author are also included.

For the purpose of this Handbook the term locomotives includes motor coaches. The Handbook does not, in general, concern itself with tramcars, though preservation schemes are sometimes mentioned – chiefly to indicate that they do not have locomotives. Multiple units are not given coverage.

Items *in italics* require confirmation and should not be taken as reliable.

Layout of this Handbook

- Cantons

The Handbook is firstly divided into Cantons in alphabetical order of their two-letter code in general use with a miscellaneous section for those items that do not fit into the scheme. Within each section there is a list of concerns covered. These are arranged by alphabetical order of the principle name of the current or last operator to use locomotives. The alternative of ordering by location offers some advantages, but assumes a better than average knowledge of Swiss geography. A location index is provided in this Handbook to give some assistance in this field.

There are a few instances where the locality name is repeated or has a very similar alternative in more than one canton. The conventional way to distinguish them as used by the postal and other authorities is to follow the locality name by the Canton code as appropriate - for example, Münchwilen AG / Münchwilen TG and Sisseln AG / Siselen BE. This practice has been followed where applicable using the repertory of Swiss postal codes as a guide.

- Concerns

For every concern, there is a three or four-part header, covering up to three lines.

The first item is the latest known variation of the concern's title and the entries for each Canton are arranged by alphabetical order of this title. Following IRS convention any initial SA, AG or Christian name is disregarded in defining the order. In general the title is given in upper case; some exceptions are made if it is known that the firm uses lower case on its letterheads, visiting cards and advertisements.

The second item in the title is a location. Generally, this is the locality (Gemeinde or town) in which the site lies. However, sometimes it is the name of the railway station to which the siding leading to the concern is attached (as in the reports made to the state authorities).

Each concern has been given an increasing identification number (with some missing numbers) within the Canton and this forms the third item in the header (examples are SH2, ZH112).

The three broad classes recognised by the IRS are indicated as the (optional) fourth item in the header line as follows:

Industrial concern	(blank)
Contractors, public works and railway services	C
Museums and preservation	M

As a guide to an enthusiast trying to determine what he may or may not see in a particular area, the title line of concerns currently (2008-9) in possession of locomotives is given in **bold** script, while archive ones are in normal script. There are borderline cases where the firm is still active, has a railway siding in use but no locomotives. These are classed as active and an appropriate explanation given.

Concern names continually change; these changes are recorded in the Handelsregister. The principle names involved are listed in this Handbook in reverse order from the one in the header and immediately below it. Names used before or after the use of locomotives may also be given in order to provide a way of relating concerns to such names.

If possible an explanation of the concern's business is given. The location index indicates the Canton of locations with unfamiliar names. The local spelling of place names, without translation into English, has been used but English spellings of place names may be used for places outside Europe, particularly for those countries not using a Roman alphabet. Place names outside Switzerland are followed by the country identification letter [e.g. F for France] in square brackets and are as used for example in road vehicle registration identification.

In general the name of a railway station indicates the locality it serves; not that in which it lies. (Schwyz station is located in Seewen SZ, for example).

Local government changes mean than localities have merged or changed their names during the period under review.

In railway circles the name of the station to which the private siding is connected or to which goods have been sent for collection is frequently used when referring to a concern. The station name is also useful for rail travellers. In this Handbook either may be used and when there is a difference a cross-reference made in the text and the location index.

- Locomotives

These are grouped by gauge for each concern. The first line indicates the gauge, the location on the 1:50,000 maps of the Landeskarte der Schweiz and the date when this section was prepared - or seriously revised.

The first part of map reference is always the three-digit map number of the appropriate 1:50,000 Landeskarte followed by a separator "$".

The second part gives the approximate location in one of two forms:

(a) the classical form at a resolution of hectometres (west-east/south-north), e.g. 4940/1194

(b) a form for use with map.search.ch with a resolution of one metre, e.g. 494000,119450

The two are directly related:

to convert from classical to map.search format replace the "/" by "00," and add "00" at the end

to convert from map.search to classical delete digits 5, 6, 11 and 12 and replace the "," by a "/".

Field 1 - Inscription

Where possible the latest observed (or photographed) inscription on the locomotive is given.

A name or inscription in the form '---' indicates that it is not carried, but referred to in some way.

Especially for ex-main line stock, the inscription often includes the Swiss classification. Unfortunately this has become far from constant over time and varies from company to company; some companies stayed with a frozen version, some followed main line practice (including the changes) and some had their own local variations. There is tendency for preservation web-sites to give the classification that the author thinks that the locomotive would have if it were to be registered now - disregarding what is painted on the locomotive.

Basically the system consists of a string of letters and two digits; the first upper case letter indicates speed for locomotives or type of accommodation for motor coaches or trams. The following lower case letter indicates propulsion mechanism, while the first digit gives the number of driven axles and the last the total number of axles. Driving axles with fixed rack pinions are indicated by Roman numerals.

Details can found in, for instance, the reference book by Herr A. Moser or the Platform 5 publication; but a detailed study is not relevant to the industrial field.

Recently the UIC code of a three-digit class code, a three-digit number and a check bit has been imposed by the SBB and the OFT for any locomotive which might venture onto the public network. The introduction is patchy and examples of duplication already exist. A very brief and incomplete explanation of the full 12-digit code follows.

Given a vehicle with the number 9180 6 <u>110</u> <u>434</u>-8 then the first four digits define the responsible organisation and the type of vehicle, the next digit is a wild digit (see later), the next three (underlined) the class, the final three (underlined) the running number. Separated by a hyphen is the check digit. This is formed by multiplying each digit of the six digit group by either 1 (the odd positions) or 2 (the even positions), then adding the digits (not the values) of the products together. If the sum is a multiple of ten then the check digit is zero - else it is the difference from the sum to the next highest multiple of ten. The wild digit is chosen so that the same calculation applied to the 11 digit number leads to the same check digit.

Field 2 - Wheel arrangement in Whyte convention as modified by the IRS

The Whyte system of wheel classification is generally used but when driving wheels are connected by chains or gearboxes rather than visible rods they are, generally, shown as 4w (four-wheeled), 6w (six-wheeled) or whatever the case is. The IRS code 2-2w-0 becomes 1A, etc.

The following additional codes may be used:

- T   Side water tank or similar tank positioned externally and fastened to the frames
- WT   Well Tank - a tank located between the frames under the boiler
Continental records do not make the distinction between WT and T. Most industrial narrow gauge locomotives were built with a well tank for the sake of stability. The code WT has been added when this is a known feature of this class or can be determined from photographs or drawings. Consequently, most of the narrow gauge contract or public works locomotives shown as T may in fact be WT
- F   Fireless steam locomotive
- Tm   Enclosed locomotive for use on street tramways
- VB   Vertical-boilered steam locomotive
- ic   inside cylinders
- xc   x number of cylinders when more than two
- cc   compound (and xcc)
- CA   Compressed air locomotive
- D   Diesel locomotive (transmission unknown)
- P   Petrol locomotive (transmission unknown)
- B   Battery locomotive
- E   Locomotive with electrical transmission
- H   Locomotive with hydraulic or hydrostatic transmission
- M   Locomotive with mechanical transmission
- W   Locomotive taking power from an overhead wire
- RC   (Rail Car) A passenger-carrying motive power unit
- RT   Rack Tank locomotive

The Whyte system does include locomotives fitted with rack gear. In most cases, a pseudo Whyte code is given that ignores the rack pinions. In some cases the locomotive has no adhesion capability and relies purely on the rack-equipment. For these, there is no valid Whyte code; conventionally a code is given that approximates what the eye sees.

Field 3 - Wheel arrangement in simplified UIC code

For the sake of saving a few characters per line the otherwise very useful brackets and other separators have been omitted.

Some of the codes used that may not be obvious are:

- a   Akku (battery)
- b   Benzin (petrol)
- el   electrical supply via a trailing cable
- g   DC electrical supply, normally low tension
- h   Heissdampf (superheated steam)
- n   Nassdampf (saturated steam)
- p   Pressluft (compressed air)
- v   verbund (compound)
- w   AC electrical supply, normally high tension
- k   Kastenlok (enclosed locomotive, usually for tram systems)

The final digit (optional) indicates the number of motors or cylinders.

Field 4 - Builder's name (short form)

Very often this consists of one word taken from the registered title that enables the full name to be identified. Sometimes an abbreviation is used in place of a word. In the earlier IRS pocket books very short codes were used in order to save space. With the page size used in this

Handbook such short codes are not necessary. Indeed as a number of the names are not familiar to everyone reading this Handbook they might lead to confusion. Where the IRS normally uses a short abbreviation which is not employed in this Handbook the conventional IRS abbreviation is indicated in the locomotive index, which also contains some indications of the different VRS and IRS conventions.

Field 5 - Builder's type (optional)

Very often these contain spaces and, especially in hand written scripts, these have not remained consistent. Where possible a consistent rendering is followed here. It may not always correspond to the version used in some other enthusiast publications.

Field 6 - Builder's serial number

In most cases these form a continuous sequence. Some cases exist of builders having separate series per class and some (Jenbacher Werke, in particular) have a system whereby the recorded number and the number on the plate affixed are related, but not the same. Some builders record an order or manufacturing number in addition to the locomotive serial number, and steam locomotives were forever exchanging boilers, which originally may or may not have had the same number as the frame. See footnotes for more details.

Field 7 - Year

Where possible this is the year given on the builder's plate affixed to the locomotive. Failing that, it is the date recorded in the published lists for the particular builder. However, there are other dates such as: ordering date, completion date, delivery date, acceptance date and date the locomotive was put into operation. The owner or operator may quote any of these without qualification. The footnotes often give more details.

A date given as 1930(38) indicates that the locomotive was built in 1930 and rebuilt in 1938; if a second line is used it indicates a major rebuild involving a wheel arrangement or prime mover change.

Field 8 - Provenance

A letter code in the form (x) refers to footnote (x).

In the provenance footnotes IRS convention is followed (as requested) in that any date is given after the change. The "/" symbol is used to separate the dd/mm/yyyy part of the date where leading blank fields are omitted. At least one "/" is retained in the date.

The footnotes are allocated (a) to (z) and then (aa) to (az) etc. within an entry, but may not always be used. Also the sequence may restart at a gauge change if that results in lists which are easier to update. Somewhat rarely one letter appears to have been omitted; this almost always represents the result of an update that removed a locomotive from a list.

The initials used for various railway companies are explained in a separate table.

Field 9 - Disposal

Either a number code iis given {(9) - referring to footnote (9)} or a very short abbreviation code is used for scrapping, store or the like.

IRS convention is to assume in the absence of any disposal symbol that the locomotive may still be present and conversely if all locomotives have a disposal entry then rail activity has ceased. However, that is not always completely clear. If it is known that the whole entry refers to an "archive" operation, then the concern's heading is in normal (non-emphasised) script.

In a disposal footnote the normal continental practice of putting the date before any change of ownership or location is followed.

The numbers used correspond to the letters used in the provenance - (d) and (4), (aa) and (27) for instance.

Further Reading

Some of the useful books, magazines and web-sites dealing with the Swiss industrial scene in general are listed below; all have been used (with thanks) in compiling this Handbook.

Der Dampfbetrieb der Schweizerischen Eisenbahnen 1847-2006, A. Moser, Eisenbahn Amateur, 2006
Eisenbahnatlas Schweiz, Schweers+Wall (latest edition 2004)
Géographie des chemins de fer d'Europe, H. Lartilleux, Tome II, Premier volume, Suisse-Italie, Libraire Chaix, 1951
Krauss-Lokomotiven, Bernhard Schmeiser, Slezak, Wien 1977 (ISBN 3-900134-36-7)
Lieferverzeichnis der Firma S. Kronenberg, Sébastien Jarne, VRS, 2008
Lokomotiven der Schweiz, P. Willen, Orell Füssli Verlag, 1970 (Normal-Spur Triebfahrzeuge) and 1972 (Schmalspur Triebfahrzeuge) or later
O&K Dampflokomotiven, Lieferverzeichnis 1892-1945, Bude, Fricke & Murray, 1977 (ISBN 3-921894-00-X)
Swiss Railways, Locomotives and Railcars, Chris Appleby and Paul Russenberger, Platform 5, 1991
The Essential Guide to Swiss Heritage and Tourist Railways; Mervyn Jones, The Oakwood Press, 2007
Verzeichnis des Rollmaterials der Schweizerischen Privatbahnen/État du matériel roulant des chemins de fer suisse privés, Eidgenössischesamt für Verkehr
Verzeichnis der Kleinmotorfahrzeuge/État des petits véhicules moteurs, Schweizerische Bundesbahnen (401.6)
Verzeichnis der Triebfahrzeuge und Steuerwagen/État des véhicules-moteurs et voitures de commande, Schweizerische Bundesbahnen (401.1)
L'Escarbille, Lausanne
Eisenbahn Amateur, Zürich (EA)
Lokomotivfabriken, J. Merte (CD, latest edition 2008)
VST Revue, Büchler & Co. A.G.
www.bahn-express.de/
www.dampflok.at/
www.deutsche-kleinloks.de/
map.search.ch/
www.privat-bahn.de/Privatbahnen.html
www.rail.lu/
www.rangierdiesel.de/
www.rollmaterial.ch/online/abkuerzungen_d.php
www.seak.ch/lokgalerie_werk.htm
zefix.admin.ch

For a non-exhaustive selection of Web-sites concerned with only one location, see within the text.

Conventions

This list includes some translations the proper nouns that are generally left in the original language in the text.

Bauunternehmung or Bauunternehmer /Entrepreneur or Entreprise de construction (building contractor)
Tiefbau / génie civil (civil engineering)
Hochbau / bâtiment (structural engineering)
Lac Léman (Le Léman) / Genfersee (Lake of Geneva)
Öffentliche Bauarbeiten / travaux publics (public works)
Rhein / R. Rhine
Rhône / R. Rhône
Schotterwerk / carrière de ballast (ballast quarry)
Genève / Genf (Geneva)
ayyyy   in or after the year yyyy
byyyy   up to and including the year yyyy
cyyyy   about the year cccc (identical to ca. yyyy)

1xxx   a year not greater than 1999
19xx   a year between 1900 and 1999 (both inclusive)
xxxx   a year that may be either before or after 2000
19xx-yy
     a year that lies somewhere between 19xx and 19yy (both inclusive)
brickworks
     the singular noun is used for works making one or both of bricks and tiles
works   the singular noun "works" is used to mean "manufacturing plant"

Acknowledgements

The author wishes to express his sincere thanks to all that have assisted in getting together this collection of facts. My especial thanks go to Sébastien Jarne who has made his records available in computer readable format. Martin Baumann has assisted with details of preserved locomotives and Pat Bowen has provided proof reading, fact finding and constructive comments.

Photographs

The majority of the photographs are by the author. The exceptions are individual acknowledged gratefully by the author. The copyright remains the property of the photographer.

A number of photographs are incorporated in the text and within the section to which they apply. For technical printing reasons these are in greyscale form. However, a colour section has been added at the end; within this the order is generally by alphabetical order of builders.

Corrections and Additions

The author was aware that the information in this book was incomplete. He was, no doubt, responsible for some unintentional errors and errors of interpretation. He also knew that only one character wrong in a line can spoil the whole entry for any one of the approximately 3,500 locomotives in this Handbook. Advice of even the smallest correction would be welcomed by the Society either by email to swiss@irsociety.co.uk or by letter to:

Mr I.R. Bendall,
46 Orson Drive
Wigston
Leicestershire
LE18 2EJ

Compilation Date: 16 June 2009

Railway Company Initials

The following initials are used in this Handbook: for a complete list and history see elsewhere.

AB	Appenzeller Bahn
	Appenzeller Bahnen (from 1/1/1988)
ASm	Aare Seeland mobil
BA	Biasca-Acquarossa
BB	Bödelibahn
BBÖ	Österreichische Bundesbahnen [A]
BC	Blonay-Chamby
BDB	Bremgarten-Dietikon-Bahn
BFD	Brig Furka Disentis
BLS	Bern-Lötschberg-Simplon (15.12.1997 merged into BLS Lötschbergbahn)
BM	Bellinzona-Mesocco
BN	Bern-Neuchâtel (15.12.1997 merged into BLS Lötschbergbahn)
BOB	Berner Oberland Bahnen
BRB	Brienz-Rothorn-Bahn
BTB	Birsigtalbahn (Basel)
BTB	Bergdorf-Thun Bahn (1942 merged with EB to become EBT)
BT	Bodensee Toggenburg (merged with SOB 1.1.2001)
BTI	Biel-Täuffelen-Ins Bahn
BVB	Bex Villars Bretaye
CEV	Chemins de fer Électriques Veveysans
CFF	Chemins de fer fédéraux
CFTT	Chemin de fer Touristique du Tarn [F]
CGN	Compagnie Générale de Navigation
CGTE	Compagnie Genevoise des Tramways Électriques
ChA	Chur-Arosa
CJ	Chemins de fer du Jura
CMN	Chemin de fer des Montagnes Neuchâteloises
CP	Caminhos de Ferro Portugueses [P]
Crossrail	Crossrail AG, Wiler
DB	Deutsche Bundesbahn (till 31.12.1993)
	Deutsche Bahn AG (from 1.1.1994)
DFB	Dampfbahn Furka Bergstrecke
DSF	Draisinen Sammlung Fricktal, Laufenburg
DR	Deutsche Reichsbahn (1937 till the end of the war)
DR (DDR)	Deutsche Reichsbahn der DDR (DDR period till 31.12.1993)
DRG	Deutsche Reichsbahn Gesellschaft (1921-1937)
DVZO	Dampfbahn Verein Zürcher Oberland, Bauma
EB	Emmental Bahn (1942 merged with Bergdorf-Thun Bahn (BTB) to become EBT)
EBT	Emmental-Burgdorf Thun (formed 1942 from Bergdorf-Thun Bahn (BTB) and EB)
EH	Eisenbahn und Häfen, Duisburg [D]
ET	Euskotrenbideak (E)
FART	Ferrovie autolinee regionali ticinesi (from 1961)
FFS	Ferrovie federali svizzere
FMA	Chemin de fer Fribourg–Morat–Anet (Anet = Ins)
FRT	Ferrovie regionali ticinesi (Centovallibahn) (1923-61)
FW	Frauenfeld-Wil
GB	Gotthardbahn
GBS	Gürbetal-Schwarzenberg-Bahn (15.12.1997 merged into BLS Lötschbergbahn)
GFM	Gruyère-Fribourg-Morat
GTB	Gürbetalbahn (1901-43, then GBS)
GV	Genève-Veyrier
HBNPC	Houillères du Bassin du Nord et du Pas-de-Calais [F]
JS	Chemin de fer Jura-Simplon
KLB	Kriens Luzern Bahn
LEB	Lausanne-Echallens-Bercher
LJB	Langenthal Jura Bahn
LMB	Langenthal-Melchnau-Bahn
LPB	Ferrovia Locarno–Ponte Brolla–Bignasco (Maggiatalbahn)

MAV	Magyar Államvasutak (Hungarian State Railways) [H]
MCM	Monthey-Champéry-Morgins
MGB	Matterhorn Gotthard Bahn
MOB	Montreux-Oberland-Bernois
MThB	Mittelthurgaubahn
MVR	Transports Montreux-Vevey-Riviera
NIAG	Niederrheinische Verkehrsbetriebe AG [D]
NOB	Nord-Ost Bahn
NStCM	Chemin de fer Nyon-St. Cergue-Morez
OeBB	Oensingen Balsthal Bahn
ÖBB	Österreichischen Bundesbahnen [A] (see also BBÖ)
OG	Olot-Gerona [E]
OJB	Oberaargau-Jura-Bahnen
PKP	Polskie Koleje Państwowe (Polish State Railways) [PL]
PV	Chemin de fer Pont-Vallorbe
Rail4Chem	BASF AG/Bertschi AG/Hoyer/VTG AG
RBS	Regionalverkehr Bern - Solothurn
RdB	Régional des Brenets
RhB	Rhätische Bahn
RHB	Rorschach-Heiden-Bahn
	Rorschach-Heiden-Bergbahn (from 2006 part of the Appenzeller Bahnen group)
RhStB	Rheintalische Strassenbahn
RhV	Rheintaler Verkehrsbetriebe
RM	Regionalverkehr Mittelland AG, Burgdorf (created 1997 from EBT+SMB+VHB; 27.6.2006 merged into BLS)
RBW	Rheinische Braunkohlenwerke, Hürth [D]
RPB	Chemin de fer Régional Porrentruy-Bonfol
RSG	Régional Saignelégier-Glovelier
RVT	Régional Val-de-Travers
SAR	South African Railways [SA]
SBB	Schweizerische Bundesbahnen
SCB	Schweizerische Central Bahn
SEZ	Spiez-Erlenbach-Zweisimmen-Bahn (15.12.1997 merged into BLS Lötschbergbahn)
SeTB	Seetalbahn (Lenzburg-Emmenbrücke, 1922 merged into SBB)
SG	Sissach-Gelterkinden-Bahn
SGA	St. Gallen Gais Appenzell Bahn (part of the Appenzeller Bahnen group from 1.1.1988)
SiTB	Sihltalbahn
SMB	Solothurn-Münster-Bahn
SNB	Solothurn-Niederbipp-Bahn
SNCF	Société Nationale des Chemins de fer Français [F]
SOB	Schweizerische Südostbahn (from 1.1.2001 Schweizerische Südostbahn AG)
SSIF	Società Subalpina per Imprese Ferroviaria
	Società Subalpina di Imprese Ferroviarie SpA [I]
StEB	Stansstad-Engelberg-Bahn
ST	Sursee-Triengen
STB	Sensetalbahn (Flammat-Laupen to SBB 2001, Laupen-Gummenen closed 16.6.2003)
SZB	Solothurn-Zollikofen-Bern-Bahn (1922-84)
SZU	Sihltal Zürich Uetliberg
TB	Trogenerbahn (part of the Appenzeller Bahnen group from 2006)
TEL	Tramvie elettriche Lugano
TEM	Tram elettrici mendrisiensi
TF	Tramways de Fribourg
TPC	Transports Publics du Chablais
TRN	Transports Régionaux Neuchâtelois
TSB	Thunerseebahn
UeBB	Uerikon Bauma Bahn
VBSch	Verkehrsbetriebe Schaffhausen
VBZ	Verkehrsbetriebe Zürich
VHB	Vereinigte Huttwil Bahnen
VNJ	Varde-Nørre Nebel Jernbaneselskab, Varde [DK]
VVT	Vapeur Val-de-Travers, St. Sulpice NE
WMB	Wetzikon-Meilen-Bahn
VSB	Vereinigten Schweizer Bahnen
VBW	Vereinigte Bern-Worb Bahnen

WB	Waldenburger Bahn
WM	Wohlen-Meisterschwanden
WT	Worblentalbahn (1913-26)
WTB	Wynentalbahn
ZT	Nebenbahn Zell-Todtnau [D]

Abbreviations for Railway Equipment Dealers

Ammann	Ulrich Ammann Baumaschinen AG, Langenthal
Asper	Victor Asper Maschinenbau, Seestrasse 205, 8700 Küsnacht ZH (also built draisines)
(none)	Basler Lagerhausgesellschaft, Basel (an importing and forwarding concern)
BEMO	Beton- und Monierbau Gesellschaft mbH, Bernhard-Höfel-Strasse 11, 6020 Innsbruck [A] (also Germany and Sweden)
Cogefar	COGEFAR spa, Milano (I)
Cogefer	[CO.GE.FER] Cogefer S.p.a. Loc. Prato Risacco, snc - 00065 Fiano Romano [I]
Desbrugères	Établissements J. Desbrugères, Noyon, Oise [F]
ETRA	Eisenbahn Transportmittel AG, Angererstrasse 6, Zürich
IPE	I.P.E. Locomotori s.r.l., Via Ticino, 5 - 37060 Pradelle Nogarole Rocca [I]
Joly	Joly, Lausanne
A. Koppel	A. Koppel AG, Berlin [D]
Layritz	Elizabeth Layritz GmbH, Nonnenwaldstraße 5, Penzberg [D]
LSB	Lokservice Burkhardt AG, Weierstrasse 50a, Rüti ZH
A. Maffei	A. Maffei, Zürich (probably an agent of J.A. Maffei, München [D])
Marti	Marti AG, Bern and Winterthur
METRAG	AG für Mechanisierung im Gleisbau und Traktionsmittel, Uznach and elsewhere (subsidiary of SERSA)
MF	Mainische Feldbahnen, Schwerte [D] (later also Hattingen [D])
NEWAG	NEWAG GmbH & Co KG, Ripshorster Strasse 321, 46117 Oberhausen [D]
OnRail	OnRail Gesellschaft für Eisenbahnausrüstung und Zubehör mbH, Mettmann [D]
OR	see OnRail
PACTON	Eisenbahnservice, Spezialtransporte, Radevormwald [D]
Ralfo	Nuova Ralfo S.R.L., 50, Via Dell' Industria, 23854 Olginate [I]
railimpex	railimpex (Johannes Scheurich GmbH), Rebenstr. 9A, Mannheim [D] (in Käfertal)
RUBAG	Rollmaterial und Baumaschinen AG; Zürich, later Basel
SAD	SA Diesel, Uckange [F]
Scheurich	see railimpex
Stauffer	Stauffer Schienen- und Spezialfahrzeuge, Langdorfstrasse 16, 8500 Frauenfeld
	Stauffer Schienen- und Spezialfahrzeuge, Südstrasse 1, 8952 Schlieren (till ca. 2007)
SECO	SECO Rail, 78130 Les Mureaux [F]
SERSA	SERSA AG, Oberglatterstrasse 15, Rümlang
Shunter	Shunter BV, Albert Plesmanweg 87, 3088 GC Rotterdam [NL] (workshops)
Tafag	Tafag AG, Chräbelstrasse 3, Goldau
Unirail	Unirail, Recke [D]
WW	Wander-Wendel (an importer associated with MBA, Dübendorf)
WBB	Westdeutscher Bahnbedarf, Hattingen [D]
	Westdeutscher Bahn- und Baubedarf Horst Scholtz GmbH, Hattingen [D]
Würgler	Hans F Würgler Ing. Bureau, Zürich-Albisrieden

Other abbreviations (non-exhaustive list)

AG	Aktiengesellschaft
	The older form of A.G. is commonly found; it is rendered as AG in the lists
AGIP	Azienda Generale Italiana Petroli
ArGe	Arbeitsgemeinschaft / association (consortium)
AMP	Armee Motorfahrzeugparks
ARBED	Aciéries Réunies Burbach-Eich-Dudelange
	Vereinigte Hüttenwerke Burbach-Eich-Dudelingen
AVM	Armee-Verpflegungsmagazin / centre de distribution de ravitaillement militaire (military supply depot)
Aw	Ausbesserungswerk (railway workshop)
AZT	Arbeitsgemeinschaft Tunnel Zürich-Thalwil
BABHE	Bundesamt für Betriebe des Heeres (from 1.1.2004 Logistikbasis der Armee (LBA))
BAV	Bundesamt für Verkehr (formerly EAV)
BHS	Betriebsgesellschaft Historischer Schienenfahrzeuge (later Swisstrain)
BTA	Bundestankanlagen
Bw	Bahnbetriebswerk (German railway locomotive depot)
Carbura (SA)	the Swiss central service for the importation of combustibles and liquid fuels
CIBA	Chemische Industrie Basel
DDR	Deutsche Demokratische Republik
DrL	Druck Luft (compressed air)
dsm	dismantled
EA	Eisenbahn Amateur
EAV	Eidgenössisches Amt für Verkehr (later BAV)
EOS	Énergie Ouest Suisse
FKG	Finanzkontrollgesetz (?)
FWF	Schweizerische Verein der Feld- und Werkbahnfreunde, Otelfingen
HFB	Heeresfeldbahnen [D]
+GF+	Georg Fischer, Schaffhausen
GmbH	Gesellschaft mit beschränktes Haftung (limited liability company)
IRR	Internationale Rheinregulierung
km	kilomètre(s) / Kilometer(n) (kilometre(s))
KMW	Kombinatie Middelplaat Westerschelde v.o.f [NL]
LBA	Logistikbasis der Armee (formed 1.1.2004 from Bundesamt für Betriebe des Heeres (BABHE))
ÖAMAG	ÖAMAG-Magnesitbruch, Tagebau Inschlagalm [A]
ÖMV	Österreichische Mineralöl Verwaltung [A]
OFT	Office Fédérale des Transport (see BAV)
OOU	out of use
playgrd	playground
Platformwagen	/ wagon plat.
	In this Handbook the term is used to indicate a rail-borne motorised flat truck. The term was used by BBC, Jung, Kronenberg, Moyse and others during the last century
RAG	Ruhrkohle AG [D]
RTC	Rail traffic ceased (but locomotives present at time of the last visit)
RWE	Rheinisch-Westfälische Elektrizitätswerke [D]
SA	Société Anonyme
	The older form of S.A. is commonly found; it is rendered as SA in the lists
SGRK	St. Gallische Rheinkorrektion
Swisstrain	Vereinigung Swisstrain Schienenfahrzeuge
TSO	Travaux du Sud Ouest, Chemin du Corps de Garde, 77501 Chelles [F]
Vanoli	see C. Vanoli AG, Samstagern
VHS	Verkehrshaus der Schweiz
VRS	Verein Rollmaterialverzeichnis Schweiz
ZIMEYSA	Zone Industrielle Meyrin-Satigny
scr	scrapped
s/s	sold or scrapped; covering all cases where the locomotive is no longer with the given operator and its fate is not known
str	stored
wdn	withdrawn

Aargau   AG

AARGAUISCHE TORFGESELLSCHAFT, Muri AG   AG001
later: Torfwerk August Meyer, Muri (from 1927)
A peat extraction scheme created in 1915.

Gauge: 750? mm   Locn: ?   Date: 03.2002

| | 0-4-0T | Bn2t | Heilbronn II | 69 | 1878 | (a) | (1) |

(a)   new to Rommel & Schoch, Stuttgart-Heslach [D] (800 mm); purchased from St. Gallische Rheinkorrektion, Rheineck /1918.
(1)   1922 sold to RUBAG.

ALSTOM, Birr   AG002
formerly: Alstom (from 2001)
            ABB Alsthom (2000, Alstom applies from 1/1/2001)
            ASEA Brown Boveri (ABB) (1989-1999)
            Brown Boveri & Cie (BBC) (1891-1988)

In BBC days the site was referred to as Werk Birr and was connected to Birrfeld station, renamed Lupfig from 1994. There have been temporary loans of locomotives between BBC, Baden and this location. Since 1989 shunting has been performed by a road/rail device.

Gauge: 1435 mm   Locn: 215 $ 6587/2542   Date: 05.1988

| Em 2/2 | FRITZ | 4wDE | Bde2 | SIG,BBC | ? ,6088 | 1960 | new | (1) |
| | | 4wDM R/R | Bdm | unknown R/R | ? | 19xx | | |

(1)   1989 sold to SERSA.

ARBEITSGEMEINSCHAFT STAUWERKBAU KLINGNAU, Klingnau
                                                              AG003 C
A construction consortium (including Th. Bertschinger, Rothpelz & Lienhard and Ph. Holzmann) was formed to construct the power station at Klingnau (1931-1935). A different allocation of the same sources and destinations could be assigned to the first two locomotives in the list on the basis of the available information.

Gauge: 600 mm   Locn: 215 $ 6592/2718   Date: 03.2002

	0-4-0T	Bn2t	O&K 40 PS	1106	1903	(a)	(1)
	0-4-0T	Bn2t	O&K 40 PS	1588	1905	(b)	s/s
	0-4-0T	Bn2t	O&K 40 PS	5225	1912	(b)	(3)

(a)   new to C. Meister, Metz [F]; ex ? /1932.
(b)   new to Ph. Holzmann & Cie, Bauunternehmung (no site specified); ex ? /1932.
(1)   1933-39 sold to Kies AG Bollenberg, Tuggen.
(3)   1933 removed from boiler register.

ARGE BAUSTELLE N3, Birrfeld   AG004 C
After assisting in the building of the N3 motorway through the area, the locomotive was hired or loaned to the SBB for use while dismantling the catenaries over the siding to Chemie Reichhold during bridge construction.

Gauge: 1435 mm			Locn: ?				Date: 01.2008	
	6	CROCODILE DUNDEE						
		0-6-0DH	Cdh	KHD MS430 C	57204	1961	(a)	(1)

(a) hired from Asper /1991.
(1) 1993 returned to Asper.

ASEA BROWN BOVERI (ABB), Baden   AG005

formerly: Brown, Boveri & Cie (BBC) (1891-1988)

The Baden factory site was rail connected in 1891 and completed in 1892. The firm has since grown into an international concern, active in the electromechanical field. The presence of a number of Platformwagen in the locomotive fleet indicates that rail was used to move equipment around the site. Since 1989 rail traffic has been handled by the SBB. There have been temporary loans of locomotives between BBC, Birr and this location.

Gauge: 1435 mm				Locn: 215 $ 6652/2592				Date: 12.1990	
			0-4-0TVB	Bn2t	Cockerill II	2104	1898	new	(1)
	51		0-6-0WT	Cn2t	O&K 300 PS	3069	1908	new	(2)
	1		1ABE	1Aa1	BBC Transportwagen	845	1912	new	s/s
			1ABE	1Aa1	BBC Transportwagen	1008	1914	(d)	s/s
	3		4wBE	Ba2	Jäger,BBC Akku-Fahrz.	? ,1010	c1914	(e)	(5)
	5		1ABE	1Aa1	BBC Transportwagen	860	1917	(f)	s/s
			1ABE	1Aa1	BBC Akku-Fahrz.	1425	1920	new	s/s
			1ABE	1Aa1	BBC Akku-Fahrz.	1426	1920	new	s/s
	3		0-4-0F ic	Bf2	SLM	3299	1929	new	(9)
Tm		FERDI	0-4-0DE	Bde1	SLM,BBC Tm^{III}	4112,5459	1953	new	(10)
Tm		MAX	0-4-0DE	Bde1	SLM,BBC Tm^{III}	4129,6003	1954	new	(11)
Em 2/2		EMIL	4wDE	Bde2	SIG,BBC	? ,6089	1959	new	(12)
Em 2/2		WALTER	4wDE	Bde2	SIG,BBC	? ,6405	1964	new	(13)

(d) new; also quoted as of year 1915.
(e) new; may have been "10" initially; the exact building date is in doubt, BBC quote 1915, Jäger 1916.
(f) new; may have been of year 1918.
(1) 1917 sold to Elektrochemie Werke Laufenburg, Laufenburg.
(2) 1908-10 hired to Seetalbahn (SeTB) "51"; 1914 sold to ?, Bruxelles [B].
(5) 1966 to BBC, Münchenstein; then WM.
(9) 1956 sold to Papierfabrik Perlen, Perlen "4".
(10) also used at BBC, Münchenstein; BBC, Oerlikon; BBC, Sécheron; 1989 sold to Imprägnieranstalt AG (IZAG), Zofingen.
(11) 19xx to ABB, Münchenstein.
(12) 1972 to BBC, Oerlikon.
(13) sometimes used at BBC, Birr; 1990 to ABB, Oerlikon.

AUTOMOBIL- UND MOTOREN AG (AMAG)   AG006

Birrfeld

A loading and parking area for the main plant that is located at Schinznach Bad. A locomotive from Schinznach Bad was stationed here between 1973 and 1977-80. It is thought to be the one listed - to be clarified. In 1994 the location was renamed Lupfig.

Gauge: 1435 mm			Locn: 215 $ 6585/2554				Date: 09.1993	
		4wPM	Bbm	RACO 40 PS	982	1931	(a)	(1)

Schinznach Bad

In 1948 this concern created a site for the import of motorcars: a subsidiary site at Birrfeld was added later. The long private siding was removed in 1996. The RACO locomotive may have been rebuilt or replaced, and it may or may not have been the locomotive lent to Werk Birrfeld from 1973 till the period 1977-80 (confirmation required).

One of SBB's Austro-Daimler petrol engine shunting locomotives "Tm 884" came to AMAG in 1945 and has since been preserved and repatriated to Austria.

Also present on 2 May 1980 was an early ex-SBB tractor Tm 818 (RACO 982)

Gauge: 1435 mm Locn: 215 $ 6550/2555 Date: 02.1996

		4wPM	Bbm	Austro-Daimler	?	1930	(b)	(2)
		4wPM	Bbm	Robel	?	1931	(c)	s/s
Tm		4wPM	Bbm	RACO 40 PS	982	1931	(d)	(4)

(a) ex AMAG, Werk Schinznach Bad /1973.
(b) ex SBB "Tm 884" /1945.
(c) new to SBB "51"; shortly after "887"; later "898"; ex SBB "4359/59" /1958.
(d) ex SBB "Tm 818" 3/1962.
(1) 1977-80 to AMAG, Werk Schinznach Bad.
(2) 1995 to private preservation (Bahnhof Effingen), 1996 to Feld- und Industriebahnmuseum, Freiland [A].
(4) 1971 either s/s and another locomotive acquired or rebuilt. It may have been the locomotive sent in 1973 to AMAG, Birrfeld; if so 1977-80 returned; 1992 scrapped.

BATA SCHUH AG, Möhlin   AG007

The following locomotive was hired, probably for use as a stationary boiler.

Gauge: 1000 mm Locn: ? Date: 02.2008

| 0-4-0T | Bn2t | Jung 50 PS | 920 | 1905 | (a) | (1) |

(a) hired from Robert Aebi & Cie AG /1936.
(1) returned to Robert Aebi & Cie AG.

BAUGESELLSCHAFT FRICK, Frick   AG008 C

A public works concern participating in the construction of public railways around the Bötzberg.

Gauge: 750 mm Locn: ? Date: 04.2002

| 0-4-0T | Bn2t | SLM | 67 | 1875 | new | s/s |

BENKLER AG, Villmergen   AG009 C

The firm was founded in 1931: it performs track maintenance and was integrated into the SERSA Group in 2000. The workshops are connected to the SBB north of Wohlen SBB station; there is mixed-gauge track on site. Address: Nordstrasse 1, 5612 Villmergen.

Gauge: 1435 mm Locn: 225 $ 6611/2454 Date: 08.2006

		4wPM	Bbm	Breuer	?	19xx	(a)	(1)
	23	4wDH	Bdh	KHD A6M420 R	26099	1940	(b)	s/s
Tm 2/2	1 Martha	4wD	Bd	Steck	01	1993	new	
TM 237 897-4 Madeleine Em 2/2 102								
		4wDE	Bde	SIG,BBC	?,7639	1968	(d)	
Tmr 10Inv 95-0004		4wD	Bd	MATISA	?	19xx	(e)	

Gauge: 1000 mm

	23	4wDH	Bdh	KHD KG125 BS	57687	1963	(f)	
	24	4wDH	Bdh	KHD KG125 BS	57688	1963	(g)	
Tmf 2/2	25	4wDH	Bdh	Gmeinder 300 B	5553	1977	(h)	(8)
Tmf 2/2	26	4wDH	Bdh	Gmeinder 300 B	5586	1980	(i)	(9)
	Walter	4wD	Bd	Steck	?	1992	(j)	

(a) ex WM /1972.
(b) new to Marinewerft, Wilhelmshaven [D] "23"; ex VBW.
(d) a building date of 1966 is also quoted; ex WM (BDWM) "Em 2/2 102" /1994; additional number "TM 237 897-4" added /1999.
(e) rebuilt from track machine to a locomotive by /2003.
(f) purchased from Halbergerhütte, Brebach [D] "23", later "10" /1996; also described as type KG230BS – error or upgrade?

(g)	purchased from Halbergerhütte, Brebach [D] "24", later "11" /1996; also described as type KG230BS – error or upgrade?
(h)	new as type KG125 BS; purchased from Halbergerhütte, Brebach [D] "25" /1994 and overhauled by the RhB before use, also bears plate on cab-side plate "Béa P-21262".
(i)	new as type KG125 BS; purchased from Halbergerhütte, Brebach [D] "26" /1994 and overhauled by the RhB before use, also bears plate on cab-side plate "Elsbeth P-23968".
(j)	either this is a new construction or the standard gauge locomotive re-gauged and renamed - clarification required.
(1)	1974 scrapped before being used.
(8)	after 11/2005 to Stauffer Schienen- und Spezialfahrzeuge, Schlieren; 12/2007 to MOB "Tm 5".
(9)	before 5/2005 to Stauffer Schienen- und Spezialfahrzeuge, Schlieren; 12/2007 to MOB "Tm 6".

BERGWERKSILO, Herznach   AG010 M

The surface buildings of Jurabergwerke AG have been taken over by Hohl AG and the owner (U. Hohl) has built a small railway in the grounds and is responsible for a mining display in the Hauptstrasse.

Gauge: 500 mm   Locn: 214 $ 6453/2583 + 6454/2578   Date: 10.2007

4wDM	Bdm	O&K(M) RL 1a	?	1932	(a)	display
4wDM	Bdm	home-made	-	-	(b)	
4wDM	Bdm	O&K(M) RL 1a	4003	1930	(c)	
4wDM	Bdm	O&K(M) RL 1a	4439	1931	(d)	

(a)   ex Jean Bollini, Baulmes, JU (600 mm) /1995.
(b)   constructed from a skip.
(c)   supplied by MBA, Zürich to Tonwerke Oberentfelden, Kölliken (later Keller AG); ex Dachziegelwerk Frick, Frick (or Stahlton AG) by /1995.
(d)   ex Stahlton AG, Frick by /2007.

THEODOR BERTSCHINGER, Lenzburg & Baden   AG011 C

A construction firm founded in 1868 and closed in 1975.

Gauge: 780 mm   Locn: ?   Date: 04.2002

	0-4-0T	Bn2t	Krauss-M IV f	71	1869	(a)	s/s
	0-4-0T	Bn2t	Krauss-S IV o	425	1874	(b)	(2)
	0-4-0T	Bn2t	SLM	575	1889	(c)	(3)

Gauge: 750 mm   Locn: ?   Date: 04.2002

7	0-4-0T	Bn2t	Borsig	7842	1911	(d)	(4)
SAN PAOLO	0-4-0T	Bn2t	O&K 50 PS	4347	1911	(d)	(4)
	0-4-0T	Bn2t	O&K 70 PS	4374	1911	(d)	(4)

Gauge: 600 mm   Locn: ?   Date: 04.2002

0-4-0T	Bn2t	O&K 20 PS	373	1900	(g)	(7)
0-4-0T	Bn2t	O&K 30 PS	2213	1906	(h)	(8)
4wDM	Bdm	O&K LD 16	4685	1933	(i)	s/s
4wDM	Bdm	O&K RL 1c	11426	1939	(j)	s/s
4wBE	Ba	Oehler G5055	?	1955	(k)	s/s

Gauge: ? mm   Locn: ?   Date: 11.2003

4wBE	Ba	Oehler G3050	832	1950	(k)	s/s
4wBE	Ba	Oehler G3050	868	1950	(k)	s/s
4wBE	Ba	Oehler G3050	1025	1955	(k)	s/s
4wBE	Ba	Oehler G3050	1100	1957	(k)	s/s
4wBE	Ba	Oehler G3050	1101	1957	(k)	s/s
4wBE	Ba	Oehler G3050	1215	1960	(k)	s/s

(a)   new to Braunschweigische Bahn D] (785 mm); used for enlarging the canal at Holderbank /1901.
(b)   purchased from Lehmann & von Arx, Wangen an der Aare /1901, used for enlarging the canal at Holderbank.

(c) purchased from Pümpin & Cie, Bern /1889-1903.
(d) hired from Robert Aebi & Co, Zürich after /1931.
(g) new to Siegel & Sohn, Landau [D], but the Strasbourg [F] branch; imported /1909; used on a contract for modifying the course of the R. Thur at Wattwil /1909; in the Gossau SG depot /1911.
(h) new to Guyot; imported /1912; purchased from Robert Aebi & Cie, Zürich /1925.
(i) purchased via O&K, Zürich; other sources give the type as LD 3.
(j) purchased via O&K, Lager Zürich.
(k) new to Th. Bertschinger, Baden.
(2) after 1911 to O&K, Zürich.
(3) 19xx sold; after 1911 scrapped.
(4) either returned to Robert Aebi & Co, Zürich or s/s.
(7) 1911 hired to O&K, Zürich; 1918 hired and then sold to Société Suisse pour l'exploitation de la tourbe, Bavois via Fritz Marti, Bern.
(8) 1941 removed from boiler register.

BÖZENEGG-ERIWIS BAHN, Schinznach Dorf　　　AG012 M

A private operation has been started using the railway, but not the staithes or the locomotive of the Zürcher Ziegeleien AG.

Gauge: 600 mm　　　　　Locn: 214 $ 6522/2564　　　　　　　　Date: 04.2009

| | 4wDM | Bdm | Diema DS30 | 2054 | 1957 | (a) |

(a) ex AG Conrad Zschokke, Näfels /2006 via Tafag.

BUSS & CIE, BASEL/PRATTELN, Augst-Wyhlen　　　AG013 C

formerly:　Buss & Cie AG, Basel/Pratteln, Gesellschaft für Eisenkonstruktion, Wasser- und Eisenbahnbau (since 1901)
　　　　　Albert Buss, Basel, Werkstatt für Bau- und Kunstschlosserei (1884-1901)

A contractor involved in the construction of the hydroelectric power station at Augst-Wyhlen in 1910.

Gauge: 900 mm　　　　　Locn: 214 $ 6203/2552　　　　　　　　Date: 03.2002

	0-4-0T	Bn2t	Borsig	7442	1910	(a)	sold
HANSA	0-4-0T	Bn2t	Hanomag	6743	1913	(b)	s/s
	0-4-0T	Bn2t	Hanomag	6744	1913	(b)	s/s
GERTRUD	0-4-0T	Bn2t	Hanomag	6745	1913	(b)	s/s
	0-4-0T	Bn2t	Henschel Helfmann	9690	1909	new	s/s

(a) purchased new via Max Strauss, Karlsruhe [D].
(b) purchased new via Robert Aebi & Cie AG, Zürich.

ANGELO CASTELLI, Rheinfelden　　　AG014 C

A public works concern located in Basel. Between 1888 and 1890 it constructed part of the main line between Schöpfheim and Säckingen, including the bridge over the R. Wehra in Baden.

Gauge: 750 mm　　　　　Locn: ?　　　　　　　　　　　　　Date: 03.2002

| ANGELO | 0-4-0T | Bn2t | Heilbronn II | 274 | 1891 | new | (1) |

(1) 1894 to Aebli, Rossi & Krieger, Schaffhausen.

CEMENTFABRIK HOLDERBANK-WILDEGG AG, Wildegg　　　AG015

formerly:　Aargauische Portland Cementfabrik Holderbank Wildegg (15/2/1912-4/8/1930)
alternative:　Cementwarenfabrik Holderbank

This cement works was brought into operation in 1913, enlarged 1954-1955 and closed in 1979. The branch from the SBB dates from 1901; it also served Kalkfabrik Holderbank.

Gauge: 1435 mm   Locn: 224 $ 6550/2531   Date: 05.1992

	1	0-6-0WT	Cn2t	Krauss-M XXIV a	1150	1882	(a)	(1)
Ed 3/4	51	2-6-0T	1Cn2t	SLM Ed 3/4	1726	1906	(b)	(2)
	1	4wDM	Bdm	Kronenberg DT 180/250	141	1956	new	(3)
	2	6wDM	Cdm	Kronenberg 3 L 600	144	1961	new	(4)
		4wCE	B-el	Vollert Robot	?	1974	(d)	s/s

Gauge: 600 mm

4wDM	Bdm	MBA MD 1	2003	c1950	(e)	s/s
4wDM	Bdm	O&K MV0	25045	1951	(f)	(6)

(a) purchased from Seetalbahn (SeTB) "BEINWYL 3" /1912.
(b) purchased from BSB "51" /1932.
(d) new; into service 1962.
(e) purchased via MBA, Zürich.
(f) purchased new via Wander-Wendel/MBA, Dübendorf.

(1) 1962 preserved on site; 1988 donated to Historische Seethalbahn, Bremgarten (West).
(2) 1970 to Heimatmuseum Schwarzenburg; 1998; to Verein Dampfbahn Bern (DBB).
(3) 1981 to Holderbank AG, Rekingen AG; 1992 returned for a short period.
(4) 1979 to Holderbank AG, Rekingen AG.
(6) 1957 to Portlandcementwerk, Laufen AG, Laufen.

CHEMIE UETIKON AG (CU), Werk Full   AG016
formerly: Chemische Fabrik Uetikon (CFU), Werk Full (1948-1990)

A chemical factory constructed in 1948 and rail connected to the SBB between Felsenau and Leibstadt stations. It was closed in 2002 and demolished in 2003. Locomotive "104" figured in the records kept by CFU; but may never have worked here (see Leim- und Düngerfabrik Märstetten, Märstetten).

"3" the ex Escher-Wyss Krauss 7899 of 1921 (type LXI dm) was positioned for photography on 27 May 1973 – it is now an active preserved locomotive in Germany.

				Locn: 215 $ 6566/2732				Date: 07.1999
Gauge: 1435 mm								
Tm	2		4wDM	Bdm	RACO 45 PS	1237	1935	(a) (1)
	3		0-4-0T	Bn2t	Krauss-S LXI dm	7899	1921	(b) (2)
Tm 2/2	03	FULLINA	4wDH	Bdh	Gmeinder Köf II	4678	1951	(c) (3)
Tm	5		4wDM	Bdm	RACO RA11	1364	1948	(d) 1989 scr
	7		4wDH	Bdh	KHD A8L614 R	56511	1957	(e) (5)
Tm 237 911-3	9	ARGOVIA	0-4-0DH	Bdm	KHD A12L714 R	57069	1960	(f) (3)

(a) new to SBB "Tm 516" as 4wPM Bpm, ex CFU, Uetikon "102" /1989.
(b) purchased from Escher Wyss, Zürich /1948 via Uetikon.
(c) new to DB "Köf 6129", purchased from DB "322 175-1" /1987 via Layritz.
(d) ex CFU, Uetikon "5" /1948.
(e) ex CFU, Uetikon "7" /1961.
(f) ex CFU, Uetikon "9" /1990.

(1) 19xx returned to CFU, Uetikon.
(2) 1958-78 used as steam generator and reserve locomotive; 1990 sold to J. Gaillard, Yverdon; 1997 to Institut für Fördertechnik und Schienenfahrzeuge (IFS), Aachen West [D].
(3) 2002 donated to EUROVAPOR, Sektion Balsthal, Balsthal.
(5) 1992 donated to EUROVAPOR, Sektion Haltingen (Kandertalbahn) [D] "V7".

CHIRESA AG, Turgi AG017

The company operates a chemical waste handling facility. Address: Landstrasse 2, 5300 Turgi.

				Locn: 215 $ 6626/2597			Date: 08.2008
Gauge: 1435 mm							
Tm^{II} 637		4wDM	Bdm	RACO 95 SA3 RS	1512	1958	(a)

(a) ex SBB "Tm^{II} 637" /2008.

CHOCOLAT FREY AG, BUCHS AG/Suhr AG018

formerly: Jowa AG, Buchs AG (1965-1966)

A chocolate factory constructed in 1965 as part of the Migros group.

				Locn: 224 $ 6488/2483			Date: 08.1997
Gauge: 1435 mm							
Tm		4wDH	Bdh	RACO 40 MA4 H	1766	1968	(a) 1997 scr
LIESELI		4wDM	Bdm	RACO 85 LA7	1603	1961	(b)

(a) ex Migros, Schwerzenbach /1985-86.
(b) new to SBB "Tm^I 333"; ex SBB "Tm^I 433" /1996.

COMOLLI AG, Bremgarten AG AG019 C

formerly: Comolli Erben from 1929
            Herman Comolli from 1901
            Baugeschäft und Steinhauer Comolli, founded in 1876
later: Comolli Baustoff AG (by 2007)

This public works company also operates a gravel pit (Locn: 225 $ 6683/2456) and performs stone dressing.

			Locn: 225 $ 6674/2450			Date: 05.2002
Gauge: ? mm						
	4w?M	B?m	O&K(M) S5	?	19xx	(a) s/s

(a) purchased via O&K/MBA, Zürich /19xx.

DACHZIEGELWERK, Frick AG020

A tile works taken over by Keller AG in 1922. In 1934 a long aerial ropeway was built to bring clay to a narrow-gauge rail-served mixing area. The standard gauge was used for despatch and, with the use of a flat wagon, for internal transport. The narrow gauge closed in c1971 and the works in 1995 or 1996. The site is now cleared.

Gauge: 1435 mm Locn: 214 $ 6432/2621 Date: 10.1990

		4wPM	Bbm	Breuer I, II or III	?	192x		
	rebuilt	4wBE	Ba1	Stadler	21	1947		(1)

Gauge: 500 mm Locn: 214 $ 6439/2610 Date: 01.1992

4wDM	Bdm	O&K(M) RL 1	3474	1929	(b)	s/s
4wDM	Bdm	O&K(M) RL 1	3816	1929	(c)	s/s
4wDM	Bdm	O&K(M) RL 1a	4003	1930	(d)	(4)
4wDM	Bdm	O&K(M) RL 1a	4439	1931	(c)	(5)
4wDM	Bdm	O&K(M) RL 1a	5310	1934	(f)	s/s
4wBE	Ba1	Stadler,BBC	26, ?	1948	new	(7)

(b) originally 600 mm; purchased from Jos. Baumer, Ziegelei Rheinfelden.
(c) purchased via O&K/MBA, Zürich.
(d) transferred from Ziegelei Köllikon.
(f) gauge not confirmed; purchased via O&K/MBA, Zürich.
(1) OOU; 1995 or 1996 scrapped.
(4) preserved on site; by 2003 to Bergwerksilo, Herznach.
(5) by 2007 to Bergwerksilo, Herznach; AG.
(7) 1977 returned to Stadler.

DEMUTH & KLEMENSIEWICZ, Lenzburg? AG021 C

A public works concern. The spelling of Klemensiewyz is in the SLM records.

Gauge: 750 mm Locn: ? Date: 04.2002

0-4-0T	Bn2t	SLM	76	1875	new	s/s

DRAISINEN-SAMMLUNG FRICKTAL (DSF), Koblenz AG022 M

This group has a workshop within the Kera Werk Laufenburg site. It acquired the Koblenz depot of the SBB on 15 December 2006. Probably once the depot is repaired; all the stock will migrate there. At the time of writing, the group also owns at least 14 motor draisines and a road/rail Unimog, which are not listed in this Handbook.

Gauge: 1435 mm Locn: 215 $ 6595/2727 Date: 10.2007

Tm		4wDM	Bdm	RACO 45 PS	1232	1935	(a)	
Tm	1 ATLAS	4wDM	Bdm	RACO 45 PS	1247	1937	(b)	(2)
Tm^{II}		4wDM	Bdm	RACO 95 SA3 RS	1453	1955	(c)	
		4wDM	Bdm	Kronenberg DT 60 P	112	1948	(d)	
		4wDM	Bdm	RACO 95 SA3 RS	1481	1956	(e)	
SBB RBe 4/4 1404		BoBoWERC		MFO,SWS,SIG	?	1959		
SBB RBe 4/4 1405		BoBoWERC		MFO,SWS,SIG	?	1959		(6)
MO ABDe 4/4 8		BoBoWERC		SIG,SWS,SAAS,BBC,MFO	?	1965		
WM BDe 4/4 2		BoBoWERC		SIG,SWS,SAAS,BBC,MFO	?	1966		(8)
Te^{III} 130		0-4-0WE	Bw1	SLM,SAAS Te^{III}	3946, ?	1946	(k)	
Tem^{I} 275		0-4-0W+DEBw1		Tuchschmid,MFO Tem^{I}	?	1957	(k)	

(a) new to SBB "Tm 511"; ex Nordostschweizerischen Kraftwerken (NOK), Beznau /2000.
(b) new to SBB "Tm 324" as 4wPM Bpm; ex Maschinenfabrik und Eisengiesserei Ed. Mezger AG (MEKA), Kallnach /2001.
(c) new to SBB "Tm 784"; ex Kiosk AG, Pratteln /2000.
(d) ex Scherer & Bühler AG, Meggen 2/2001.
(e) ex SBB "Tm^{II} 617" /2006.
(j) new to WM "BDe 4/4 2"; SOB "90" /1997; later "556 042"; to DSF /12/2004.

(k)	ex Centralbahn, Freiburg (Breisgau) [D] 27/12/2008.						
(2)	1/2008 scrapped after cannibalisation for spares.						
(6)	5/5/2008 scrapped after cannibalisation for spares.						
(8)	1/2008 scrapped after cannibalisation for spares and repatriating the transformer to Martigny.						

see: www.draisine.ch/index.html

DSM NUTRITIONAL PRODUCTS AG, Sisseln AG   AG023

formerly: Hoffmann-La Roche AG
          Roche AG

The Basel-based chemical products group of Roche started a new chemical product factory here around 1965. There are 2.8 kilometres of track and a locomotive depot.

Gauge: 1435 mm   Locn: 214 $ 6417/2662   Date: 07.1999

Em 2/2 FRIDOLIN	4wDE	Bde	SIG,BBC	? ,7701	1967	new

ELEKTRIZITÄTSWERK LAUFENBURG (EGL), Laufenburg   AG024

formerly: Kraftwerk Laufenburg (1915-55)
          Deutsch-Schweizerische Wasserbau-Gesellschaft (1908-15)

A hydroelectric power station constructed 1908-14 and rail connected from the start.

Gauge: 1435 mm   Locn: 214 $ 6457/2672   Date: 08.1998

1	0-4-0T	Bn2t	O&K 100 PS	3081	1909	new	(1)

(1) 1972 donated to Verein Dampfbahn Bern (DBB); 1997 EUROVAPOR-Wutachtalbahn [D].

ELEKTROCHEMISCHES WERK LAUFENBURG, Laufenburg AG027

The branch to this factory was put into service in 1917.

Gauge: 1435 mm   Locn: ?   Date: 01.1996

	0-4-0TVB	Bn2t	Cockerill II	2104	1898	(a)	s/s

(a) purchased from BBC, Baden /1917.

ELEKTROZINN AG, Oberrüti   AG028

This firm specialises in the recovery of non-ferrous metals and is rail connected just north of the station.

Gauge: 1435 mm   Locn: 235 $ 6732/2244   Date: 08.2005

	4wDH	Bdh	Windhoff Lg II	1036	1948	(a)	(1)
1	4wDH	Bdh	Jung RK 8 B	14127	1971	(b)	

(a) planned as works number "910", new to Trümmerverwertung, Senator für Bauwesen, Bremen [D]; later Ruhr-Lippe-Eisenbahn [D] "50"; purchased from Hermann Sprenger KG, Essen [D] /1969.
(b) purchased from Kabel- und Lackdraht-Fabriken GmbH, Mainz-Gustavsburg [D] /c1980.
(1) 2006 to ? for preservation (to be confirmed).

ERNE AG, Laufenburg   AG029 C

formerly: Josef Erne-Speiser (from 1906)

This building firm (Bauunternehmung) has its main offices in Laufenburg and at least four subsidiaries in Kantons Basel Stadt and Basel Land. There are several sections of the firm in the area around the Kera Werke in Laufenburg. Office Address: Baslerstrasse 5, 5080 Laufenburg.

Gauge: 600 mm   Locn: 214 $ 6462/2674   Date: 02.2003

| | 4wDM | Bdm | O&K(M) RL 1c | 8558 | 1937 | (a) | (1) |

(a) purchased via MBA to construct the waterworks building "Obere Richi" to the east of Rheinfelden.
(1) ca. 1950 used for a circular railway in connection with the Laufenburg Autumn fair; stored; placed on display outside Erne Holzbau AG, Werkstrasse 14, Laufenburg.

ERNE & ZUMSTEG, Leibstadt   AG030 C

A public works concern that was wound up ca. 1977. Erne AG of Laufenburg carried on the business.

Gauge: 600 mm   Locn: 214 $ 6553/2713   Data: 5.2002

| | 4wDM | Bdm | O&K(M) RL 1a | 4900 | 1933 | (a) | (1) |
| | 4wDM | Bdm | O&K(M) RL 1c | 7811 | 1937 | (a) | s/s |

(a) purchased via O&K/MBA, Zürich.
(1) reported to have been placed on display; not verified; perhaps an error for O&K 8558; s/s.

EUROPEAN VINILS CORPORATION (EVC), Sins   AG031

formerly: Lonza AG, Werk Sins
           Imperial Chemical Industries (ICI)

A nitric acid plant constructed in 1941 and closed in 199x.

Gauge: 1435 mm   Locn: 235 $ 6727/2279   Date 01.2002

| | 4wPM | Bbm | Breuer | ? | 19xx | | (1) |
| | 4wDH | Bdh | Henschel DH 240B | 31106 | 1965 | (a) | (2) |

(a) from Henschel via Asper /1970.
(1) 1970-87 s/s.
(2) 2001 to Leon D'Oro, Marmirolo [I].

EW NOSTALGIE, Brittnau-Wikon   AG032 M

An organisation formed to preserve EWI and EWII coaches, but with a locomotive to move them about.

Gauge: 1435 mm   Locn: ?   Date 03.2009

| | 4wDM | Bdm | RACO 95 SA3 | 1562 | 1959 | (a) | |

(a) ex SBB "Tm^{II} 685" ?1/2009 (withdrawn 12/2008).

FELDSCHLÖSSCHEN-GETRÄNKE AG   AG033

Jung 668 of 1903 returns to Brauerei Salmen from the SBB station on 11 November 1966.

Baslerstrasse, Rheinfelden

formerly: Brauerei Feldschlösschen (1991-2000)
Brauerei Cardinal Rheinfelden AG (1971-1991)
Brauerei Salmen (1799-1971)

This brewery was constructed for Brauerei Salmen in 1884 and was rail connected from 1885. A six hundred metre long branch formed the connection to Rheinfelden station. The site lies on the other side of the SBB from the Feldschlösschen establishment into which it was absorbed in 1991. The fireless locomotive ceased operating in 1984 and the brewery closed in 2002.

Gauge: 1435 mm　　　　　　　Locn: 214 $ 626020,266910 + 626060,266810　　　Date: 05.2009

SALMEN	0-4-0F	Bf2	Jung	668	1903	(a)	display
1	0-6-0DH	Cdh	Henschel DH 500Ca	30303	1961	(b)	(2)

Krauss 5666 of 1907 propels empty beer containers up to Brauerei Feldschlösschen on 7 September 1969. The "castle" can be seen in the left background

Theophil Roniger-Strasse, Rheinfelden

formerly:　　Brauerei Feldschlösschen (until 2000)

The brewery commenced operations in 1876; the rail connection down to Rheinfelden station was installed in 1889. Passenger trains are sometimes operated from a separate station adjoining the SBB one. The rail system originally entered most buildings. In the case of the old buildings this was often over turntables or via sharp curves. Within the old buildings the locomotive kept to the main line and its principle branch, road vehicles fitted with railway buffers manoeuvred wagons over the difficult sections of track. With the advent of a new rail-served warehouse (around ?1990) the need for this has finished.

Gauge: 1435 mm　　　　　　Locn: 214 $ 626180,266170 + 625950,266100　　　Date 05.2009

FELDSCHLÖSSCHEN	1907-1965						
	0-4-0WT	Bn2t	Krauss-S XXVIII p	5666	1907	(c)	display
FELDSCHLÖSSCHEN	1907-1965						
8481	0-6-0WT	Cn2t	SLM E 3/3	1877	1907	(d)	
837 801-0	0-6-0DH	Cdh	Henschel DH 500Ca	30303	1961	(e)	OOU
837 952-1 SVENJA	6wDH	Cdh	KrMa ML 700 C	19089	1963	(f)	

(a)　　purchased via Fritz Marti AG, Winterthur; the "SALMEN" nameplates were removed /1971; stored /1984; transferred to Brauerei Feldschlösschen, Rheinfelden; 1displayed on site; displayed on Lokikreisel (Quellenstrasse/Kaiserstrasse, outside Salmen/Cardinal buildings) with new "SALMEN" plates /2009.
(b)　　new to Eisenwerke Gelsenkirchen [D]; purchased from Thyssen Schalker Verein, Gelsenkirchen [D] "51" via NEWAG.
(c)　　display on site from /1995-96.

(d) hired from SBB "E 3/3 8481" /1964; purchased /1965.
(e) ex SIBRA (formerly Brauerei Cardinal, Rheinfelden) /1995.
(f) new to Ewald-Kohle AG, Herten [D] "D 06"; ex SERSA by /2007; also recorded as type M 700 C.
(g) ex Brauerei Feldschlösschen, Baslerstrasse, Rheinfelden for display in Feldschlösschen 'castle' /2002.

(2) 1995 transferred to Brauerei Feldschlösschen, Rheinfelden.

see: Eisenbahn Amateur p.61, Feb 1969

FERRO AG   AG034

A metal re-cycling firm established in 1934.

Baden

This site is located by Baden-Oberstadt station on the line from Baden to Wettingen which no longer has a passenger services.

Gauge: 1435 mm   Locn: 215 $ 6654/2580   Date: 09.2004

Tm	SEPP	4wDM	Bdm	RACO 45 PS	1225	1934	(a)	
	rebuilt	4wDE	Bde1	Stadler,BBC	-	1967		(1)
2		4wDH	Bdh	Gmeinder Köf II	5051	1958	(b)	

Birr

This operation started in 1989 and is located within the Alstom site at Birrfeld.

Gauge: 1435 mm   Locn: 215 $ 6587/2540   Date: 09.2002

Tm	SEPP	4wDM	Bdm	RACO 45 PS	1225	1934	(c)
	rebuilt	4wDE	Bde1	Stadler,BBC	-	1967	

(a) new to SBB "Tm 506", finally "Tm 553", purchased from BBC, Münchenstein /1987.
(b) new to DB "Köf 6339"; later "323 651-0"; purchased from Layritz /1989-91.
(c) ex Ferro AG, Baden-Oberstadt /1990.

(1) 1990 transferred to Ferro AG, Birr.

FERROFLEX AG, Rothrist   AG035

A steel trading concern established in 1968.

Gauge: 1435 mm   Locn: 224 $ 6324/2390   Date: 07.1999

	4wDH	Bdh	Schöma CFL 60DR	3136	1969	(a)

(a) purchased from Bahnhofkühlhaus, Basel Bad Bahnhof "2" /1990; Schöma lists give the type as CFL DBR.

FERROSTAHL, Lupfig   AG036

A new site, first reported in 2009. Full details have yet to be determined.

Gauge: 1435 mm   Locn: ?   Date: 04.2009

Tm 236 313-3	4wDM	Bdm	RACO 95 MA3 RS	1798	1971	(a)

(a) new to EBT "Tm 13"; on hire from Stauffer 3/2009.

E. FLÜCKIGER AG, Rothrist   AG037 M

This scrap merchant has one locomotive stored on site. At least one other locomotive has come here for scrapping. Address: Industrieweg 12, 4852 Rothrist.

Gauge: 750 mm		Locn: 224 $ 6330/2396					Date: 02.1991
	4wDM	Bdm	Gmeinder		4545	1949	stored

ERNST FREY AG, Kaiseraugst   AG038 C

A civil engineering firm.

Gauge: 1435 mm		Locn: 214 $ 6224/2656					Date: 10.1986
206	4wDM	Bdm	Breuer VL		3067	1953	(a) (1)

(a)   probably ex French Military authorities, Rühl [D].
(1)   c1972 sold to Klingenthal Mühle AG, Kaiseraugst.

GEMEINDE AUENSTEIN, Auenstein   AG039 M

The exact location of this locomotive had not been established when this Handbook closed for pubication.

Gauge: 750 mm		Locn: ?					Date: 03.2008
AUENSTEIN	0-4-0T	Bn2t	SLM		3833	1943	(a)

a)   new for building Kraftwerk Rupperswil-Auenstein, Rupperswil; ex Schinznacher Baumschulbahn (SchBB), Schinznach Dorf /2007.

HEINZ GERBER, Unterentfelden   AG040 M

This locomotive is part of a private collection.

Gauge: 700 mm		Locn: ?					Date: 05.2009
	4wDM	Bdm	Schöma CDL 28		2069	1957	(a)

(a)   new to Oving, Rotterdam [NL]; ex ODS Hendrik Ido, Ambracht [NL] /200x.

GIEZENDANNER AG, Rothrist   AG041

This transport firm has many sites throughout Switzerland. Address: 75 Juraweg 26, 4852 Rothrist.

Gauge: 1435 mm		Locn: 224 $ 6332/2398				Date: 06.2008
	4wDM R/R Bdm		Mercedes Unimog	?	19xx	

GIPS UNION, Felsenau   AG042

Two separate systems were in operation at different times.

- Roadside tramway from quarry to station. It was originally animal worked and closed in the 1930s.

- Underground workings in Felsenau started in 1923. The last tracks on site were removed 2003-2004.

Gauge: 600 mm		Locn: 215 $ 6587/2731					Date: 08.2004
	0-4-0T	Bn2t	Hohenzollern	462	1888	(a)	(1)
	4wBE	Ba	Stadler	19	194x	(b)	s/s
	4wBE	Ba	unknown	?	1xxx		s/s

(a)   acquired from F. Marti, Winterthur /1908.
(b)   probably Stadler 19 of 194x, acquired from Ziegelei Wiesenthal, Chur /1951.
(1)   1936 removed from boiler register.

ERNST HAMMANN, Laufenburg   AG043 C

A building contractor (Bauunternehmung).

Gauge: 750 mm   Locn: ?   Date: 04.2002

	0-4-0T	Bn2t	Heilbronn 80 PS	584	1913	new	(1)
BERNA	0-4-0T	B?2t	O&K 70 PS	6572	1913	(b)	(2)
	0-4-0T	B?2t	O&K 70 PS	6573	1913	(b)	(3)

(b)   delivered to Rubigen (Kanton Bern).
(1)   1916 sold to Ilseder Hütte [D] (780 mm).
(2)   before 1919 sold to ?Belart & Cie, ?Olten-Klingnau.
(3)   before 1932 sold to Baumann, Wädenswil.

HEMLING & STRAKS, Lenzburg   AG044 C

A civil engineering concern.

Gauge: 750 mm   Locn: ?   Date: 01.1997

0-4-0T	Bn2t	SLM	83	1875	new	s/s

HERO AG, Lenzburg   AG045

formerly: Hero Schweiz AG
             Hero Conserven Lenzburg (from 1946)
             Konservenfabrik Lenzburg

A factory producing conserves and rail connected from 1898; latterly a supply company.
Address: Niederlenzer Kirchweg 3, 5600 Lenzburg.

Gauge: 1435 mm   Locn: 214 $ 6556/2491   Date: 04.2008

4wPM	Bbm	MStG	?	1910	new	s/s
4wDM R/R	Bdm	Mercedes Unimog	?	19xx		

HISTORISCHE SEETHALBAHN, Bremgarten West   AG046 M

A group with a base in the old BDWM standard gauge shed at Bremgarten (West).

Gauge: 1435 mm				Locn: 225 $ 6676/2442				Date: 05.2005
Tm^I	465	Till	4wDM	Bdm	CMR Tm^I	'Tm^I 365'	1962	(a)
	3	BEINWYL	0-6-0WT	Cn2t	Krauss-M XXIV a	1150	1882	(b)
BDWM Em 2/2 103			4wDE	Bde	Stadler,BBC	162,7640	1984	

(a) new to SBB "Tm^I 365"; ex SBB "Tm^I 465" /2001.
(b) new to Seethal-Bahn "3 BEINWYL"; /?1884; to Cement Holderbank AG, Wildegg /1912; to Historische Seethalbahn /1988.

see: www.historische-seethalbahn.ch

HOCH- & TIEFBAU BAUUNTERNEHMUNG AG, Aarau   AG047 C

This is civil engineering firm with a large presence in Germany. One task that they performed near Full is recorded in more detail. Address: Rohrerstrasse 20, 5000 Aarau.

Gauge: 1000 mm				Locn: 224 $ 6467/2495				Date: 05.2002
		0-6-0T	Cn2t	O&K 50 PS	2110	1906	(a)	(1)
Gauge: 900 mm at Full				Locn: ?				Date: 05.2008
		0-4-0T	Bn2t	Henschel Helfmann	13250	1914	(b)	(2)
		0-4-0T	*Bn2t*	*Borsig*	*6977*	*1909*	*(c)*	*1933 s/s*
		0-4-0T	*Bn2t*	*Borsig*	*7258*	*1910*	*(d)*	*1933 s/s*
Gauge: 600 mm				Locn: 224 $ 6467/2495				Date: 05.2002
RITOM		0-4-0T	Bn2t	Maffei 40 PS	2920	1909	(e)	(1)
		4wDM	Bdm	Deutz PME117 F	9279	1929	(f)	s/s

(a) purchased from Fasoletti & Malfanti Impresa costruzioni, Lugano-Viganello /1931-43.
(b) purchased from B. Wittkop, Berlin [D] /1932.
(c) new to Julius Blenke, Berlin [D]; ex ? /1932.
(d) new to B. Wittkop, Groß Lichterfelde [D]; ex ? /1932.
(e) purchased from Gebr. Caprer Erben, Baugeschäft, Chur /1943.
(f) new via Martin Kallmann, Mannheim [D] /1930.

(1) 1954 removed from boiler register.
(2) 1933 removed from boiler register.

HOLCIM AG   AG048

Rekingen AG

formerly: Holderbank Cement und Beton (HCB), Rekingen (1992-2001)
            Cementfabrik Holderbank, Rekingen-Mellikon (till 1991)

A cement works of the Holderbank group constructed 1973-1975. Production ceased in 1998. The site has not been dismantled; part of it provides facilities for the Hochrhein Swiss Terminal – an inter-modal terminal.

Gauge: 1435 mm   Locn: 215 $ 6659/2698   Date: 12.1998

			6wDE	Cde	Moyse CN60EE500D	3548	1974	(a)	(1)
	1		4wBE	Bo-a	Vollert V 7000	74/049 I	1974	new	(2)
	2		4wBE	Bo-a	Vollert V 7000	74/049 II	1974	new	(2)
Tm	1	OMA	4wDM	Bdm	Kronenberg DT 180/250	141	1956	(d)	(4)
Tm	2	OPA (ex GROSI)							
			6wDM	Cdm	Kronenberg 3 L 600	144	1961	(e)	(5)
			4wBE	Ba	Stadler	132	1970	(f)	(2)
		GUSTI	6wDH	Cdh	MaK G 763 C	700100	1991	new	(7)
			4wDH	Bdh	O&K MB 7 N	26606	1966	(h)	

Würenlingen-Siggenthal

formerly: Portland-Cement-Werk, Würenlingen-Siggenthal AG

This cement works was founded in 1912 and became part of HOLCIM in 2001. It is rail connected to Würenlingen-Siggenthal station.

Gauge: 1435 mm   Locn: 215 $ 6603/2637   Date: 10.2007

		4wBE	Ba	Stadler	132	1970	(i)
EMIL 1	J21-4R1	4wBE	Bo-a	Vollert V 7000	74/049 I	1974	(i)
EMIL 2		4wBE	Bo-a	Vollert V 7000	74/049 II	1974	(i)

Gauge: 600 mm   Locn: ?   Date: 05.2002

4wPM	Bbm	O&K(M) M	2680	1927	new	s/s
4wDM	Bdm	O&K(M) RL 1a	4398	1931	(m)	s/s
4wDM	Bdm	O&K(M) RL 1a	4615	1931	(m)	s/s
4wDM	Bdm	O&K(M) RL 1a	4639	1931	(m)	s/s
4wDM	Bdm	O&K(M) RL 1c	8103	1937	(m)	s/s
4wDM	Bdm	O&K(S) RL 1c	1640	194x	(n)	s/s

(a)   into service /1975.
(d)   ex Cementfabrik Holderbank-Wildegg AG, Wildegg /1981.
(e)   ex Cementfabrik Holderbank-Wildegg AG, Wildegg /1979.
(f)   purchased from Cementfabrik Holderbank, Liesberg /1989.
(h)   ex HOLCIM AG, Roche VD /2002.
(i)   ex HOLCIM AG, Rekingen AG /2001-02.
(m)   purchased via MBA, Zürich.
(n)   ex ?, via MBA, Zürich /1952; the source data does not quote O&K(S) but this seems likely for an RL 1c with the given number being traded in 1952.

(1)   1999 to ARGE Filderen, Birmensdorf ZH; 2001 to ArGe AMA, Amsteg.
(2)   2001-02 transferred to HOLCIM AG, Siggenthal.
(4)   1997 to Schnyder Gotthard, Emmen via LSB.
(5)   1991 to Rhenus AG, Basel Kleinhüningen.
(7)   1998 transferred to HOLCIM AG, Kieswerk Hüntwangen, Hüntwangen.

HOTEL STALDEN, Berikon   AG049 M

The locomotive stands in a 'station' and appears to be pulling a train. The coach is, in fact, one of the hotel's restaurants. Address: Friedlisbergstrasse 9, 8965 Berikon.

Gauge: 1435 mm				Locn: 225 $ 6709/2446				Date: 05.2005	
HEINRICH 2		0-4-0F	Bf2	Jung 80 PS		911	1905	(a) display	
(a)	ex Viscose Suisse, Emmenbrücke /1976, via private owner, Alpnach-Dorf.								

AG HUNZIKER & CIE   AG052 C

formerly: Hunziker & Cie, Brugg (1907-14)
This firm deals in building materials (Brugg AG) and equipment for the cement industry (Olten).

Brugg AG

Gauge: 750 mm		Locn: 215 $ 6573/2584				Date: 05.2002	
	0-4-0WT	Bn2t	O&K 50 PS	2187	1906	(a)	s/s
	0-4-0WT	Bn2t	O&K 50 PS	2448	1908	(a)	s/s

Döttingen

A site operated by AG Hunziker & Cie, Brugg AG.

Gauge: 600 mm		Locn: ?				Date. 05.2002	
	4wDM	Bdm	O&K(S) RL 1c	1554	1947	(c)	(3)

(a)   purchased from Bauabteilung für den Simplontunnel II, Brig /1920-29.
(c)   purchased via O&K/MBA, Zürich.
(3)   sold to Holzverzuckerung AG (HOVAG), Ems.

IZAG, Zofingen   AG053

formerly: Imprägnieranstalt AG (IZ)
           Egg-Steiner Imprägnieranstalt

The company initially made preserving casks and barrels, but latterly produces packing materials. The factory has been rail connected since 1895. By 2007 there was about fifty metres of street-running track connected to a main-line siding via a wagon table.

Gauge: 1435 mm		Locn: 224 $ 6381/2372				Date: 05.1992	
Tm	4wDM	Bdm	RACO 25 PS	1066	1933	new	(1)
FERDI	0-4-0DE	Bde1	SLM,BBC Tm^{III}	4112,5459	1953	(a)	

(a)   ex BBC, Baden /1989.
(1)   by 2005; to Swisstrain, Rikon; later Eisenbahn Amateure Zofingen.

JÄGGI AG, Brugg AG   AG054 C

This is now a builder and window-making company, but has apparently been active in the public works sector. Office address: Feerstrasse 16, 5200 Brugg; locomotive located in Schipinstrasse.

Gauge: 600 mm		Locn: 215 $ 6576/2589				Date: 04.1996	
	0-6-0T	Cn2t	WN	5597	1918	(a)	s/s
	4wDM	Bdm	Brun B20/2HK65	48	19xx	(b)	display

(a)   new to k.u.k. Heeres-Rollbahn [A] "218"; purchased from ? /1926.
(b)   Fabr. No. 58603.

JARDINI, Wohlen AG　　　　　　　　　　　　　　　　　　　　AG055 C

This was probably a local depot of the Basel-based construction engineers that may have been used in the construction of the Aarau-Muri railway line.

Gauge: 750 mm　　　　　　　Locn: ?　　　　　　　　　　　　　　Date: 03.2002

0-4-0T	Bn2t	Krauss-S XXVII	434	1874	new	s/s

JURA-BERGWERKE AG, Herznach　　　　　　　　　　　　　AG056

This ironstone mine operated 1932-67. The surface buildings have been taken over by Hohl AG.

Gauge: 600 mm　　　　　Locn: 215 $ 6451/2583　　　　　　　Date: 05.2002

4wDM	Bdm	O&K(M) RL 1a	4005	1930	(a)	(1)
4wDM	Bdm	O&K(M) RL 1a	4835	1933	(b)	s/s
4wDM	Bdm	O&K(M) RL 1a	5681	1934	(b)	s/s

(a)　purchased from Zement- und Kalkwerk Schinznach Bad.
(b)　purchased from O&K/MBA, Zürich.

(1)　sold to P. Schlatter, Münchwilen AG.

JURA-CEMENT-FABRIKEN (JCF)　　　　　　　　　　　　　　AG057

formerly:　Cementfabrik Zurlinden & Cie, Aarau (from 1883 to 1897)

Aarau

This cement works was operational from 1883-1929. The enlarged works at Wildegg replaced it. Some of the locomotives may have been used at Wildegg during the rebuilding phase. A narrow gauge line connected the quarry at Küttingen with the works. An aerial ropeway replaced it in 1907.

Gauge: 750 mm　　　　　Locn: 224 $ 6460/2499　　　　　　　Date: 03.2002

0-4-0T	Bn2t	Krauss-S IV o	430	1874	(a)	(1)
0-4-0T	Bn2t	Krauss-S IV kl	2960	1893	(b)	(1)
4wPM	Bpm	O&K(M) S 10 a	2572	1927	(c)	s/s

Gauge: 500 mm　　　　　Locn: ?　　　　　　　　　　　　　　Date: 03.2002

4wPM	Bbm	O&K(M) S 10 a	2346	1926	(c)	s/s

Wildegg

Rail connected since 1890, the plant opened in 1891 and was enlarged in the period 1927-1929.

Gauge: 1435 mm　　　　Locn: 224 $ 6546/2517　　　　　　　Date: 04.2007

	4wBE	Ba1	JCF,BBC	?	1xxx	(e)	s/s a1962
	4wBE	Ba2	von Roll	-	19xx	(f)	s/s c1978
	0-4-0DH	Bdh	KHD A6M517 R	55125	1952	(g)	(7)
	0-4-0DE	Bde1	SLM,BBC Tm^{III}	4440,6183	1962	new	
	4wDH	Bdh	Diema DVL200/2	5239	1993	(i)	
237 924-6	ESPERANZA						
	4wDH	Bdh	O&K MB 220 N	26721	1972	(j)	

(a)　purchased from St. Gallische Rhein-Korrektion /1913.
(b)　purchased from Baumann & Stiefenhofer, Wädenswil /1926.
(c)　purchased via O&K, Zürich; may have been used at Wildegg.
(e)　conversion of a wagon frame.
(f)　conversion of a self-propelling crane.
(g)　purchased via Hans Würgler, Zürich-Albisrieden.
(i)　new via Asper.
(j)　ex Gemeinschaftkraftwerk Weser, Porta Westfalica-Veltheim [D] "1", via LSB /2003.

(1) 1930 removed from boiler register; 1932 sold to AG Hunziker & Cie Baustofffabrik, Olten.
(7) 2001-3 scrapped.

KABELWERK BRUGG AG, Brugg AG   AG058

A cable works originally founded in 1908 and reorganised in 1970 when it acquired the premises of the adjoining firm of Zschokke-Wartmann, together with a locomotive.

Gauge: 1435 mm   Locn: 215 $ 6581/2591   Date: 10.1996

31	4wBE	Ba1	unknown,BBC	?	1xxx	(a)	1996 s/s
2	4wDH	Bdh	O&K MV6a	25408	1953	(b)	(1)

(a) ex Zschokke-Wartmann, Brugg AG /1971.
(b) purchased from Verenigde Cooperative Suikerfabriek, Roosendaal [NL] /1996 via Asper.
(1) 2005 or 2006 to Swisstrain, Rikon.

A. KÄPPELI'S SÖHNE AG, Wohlen AG   AG059 C

formerly: A. Käppeli & Co (till 1930)

A yard of this firm is situated just north of Wohlen SBB station. A preserved British-built steamroller can be seen from passing trains. The remaining diesel locomotives were removed 2003-2006.

Gauge: 600 mm   Locn: 225 $ 6623/2448   Date: 08.2006

4wDM	Bdm	O&K(M) MD 1	5645	1934	(a)	s/s
4wDM	Bdm	O&K(M) MD 1	8767	1938	(b)	s/s
4wDM	Bdm	MBA MD 1	2005	c1950	(c)	s/s

(a) purchased from O&K/MBA, Dübendorf; gauge not confirmed.
(b) new to Hermann Schäler, Baugeschäft, Berlin-Schmargendorf [D]; purchased from MBA, Dübendorf.
(c) purchased from MBA, Dübendorf /1951; gauge not confirmed.

KALKFABRIK SPÜHLER AG, Rekingen AG   AG060

This company operated a factory producing chalk products. It started operation in 1888. The railway originally connected the factory with the quarry and the station, latterly only the latter. The factory was bought by Cementfabrik Holderbank in 1980. It closed in 1988 and the railway was removed.

On 23 March 1968 O&K 12030 (type MD 2) stands alongside two trains of two wagons each.

Gauge: 600 mm   Locn: 215 $ 6664/2690   Date: 04.1987

4wDM	Bdm	O&K(M) MD 2	12030	194x	(a)	(1)
4wDM	Bdm	O&K(M) S 10 a	2196	c1926	(b)	(2)

(a) fitted with Deutz A2L514 motor /1957.
(b) new as 4wPM Bpm; later fitted with Deutz A4L514 motor; probably rebuilt by KHD.

(1) 1989-90 sold to Hr. Victor Longo.
(2) 1989-90 sold to Reinhard Kern Strassenbau AG, Bülach (Herr Kern private).

KELLER AG, TONWERKE OBERFELDEN, Kölliken   AG061

formerly: Ziegelei Kölliken AG (from 1923)
Ziegelei Hillfiker (from 1823)

This clay-pit line was about five hundred metres long and closed ca. 1975.

Gauge: 500 mm   Locn: 224 $ 6443/2417   Date: 01.1992

4wBE	Ba	unknown	?	1xxx		(1)
4wDM	Bdm	O&K(M) RL 1a	4003	1930		(2)
4wBE	Ba1	Stadler	25	1948	new	(3)

(1) 19xx transferred to Dachziegelei Frick, Frick.
(2) 1991 transferred to Dachziegelei Frick, Frick; by 1996 to Bergwerksilo, Herznach.
(3) 1977 sold to Stadler.

KERA WERKE AG, Laufenburg   AG062

A ceramic works, operating since 1932 and specialising in sanitary products. In 1971 it became part of the holding company Keramik Laufen. Works closed 1999.

The Platformwagen seen stored (together with the Draisenen Sammlung Fricktal stock) on 24 October 2007.

Gauge: 1435 mm			Locn: 214 $ 6465/2672				Date: 10.2007	
	2	4wBE	Ba2	SWS,MFO		?	1942	new (1)

(1) stored on site for preservation.

KETTNER AG, Muri AG   AG063 C

also: Baugesellschaft Muri/Rotkreuz

This is a construction firm. As built Heilbronn 69 had boiler 296; the records show Heilbronn 296 and a boiler changing hands.

Gauge: 750 mm			Locn: ?				Date: 03.2002	
	0-4-0T	Bn2t	Heilbronn II			69	1878	(a) (1)

(a) purchased from Rommel & Schoch, Stuttgart-Heslach [D] (800 mm).
(1) 1883 sold to St. Gallische Rheinkorrektion (SGRK), Rheineck.

KONSORTIUM BÄRENGRABEN, Würenlingen   AG064 C

This project was for decontaminating the rubbish tip at Würenlingen (Sanierung Deponie Bärengraben).

Gauge: 900 mm			Locn: ?				Date: 05.2009
	4wDH	Bdh	Schöma CFL 100DCL	5242	1991	(a)	s/s

(a) ex Theiler & Kalbermatter, Luzern by /1995.

KFOR (KOSOVO FORCE), Effingen Bahnhof   AG065

Military installations of the KFOR, a subsidiary of the "Waffenplatz" of Brugg AG.

Gauge: 1435 mm			Locn: 214 $ 6500/2584				Date: 04.2007
675		4wDM	Bdm	RACO 95 SA3	1552	1959	(a)

(a) ex SBB Tm" 675" /2001.

KRAFTWERK BRUGG-WILDEGG, Villnachern   AG066 C

An electrified line (six kilometres long) between a quarry at Villnachern brought stone to the construction site for this power station at Au (1949-1953); almost certainly for construction of the water channels.

Gauge: 750 mm 600V DC			Locn: 215 $ 6552/2557?				Date: 01.2000	
1	URSULA	4wWE	Bg2	Stadler		33	c1949	new (1)
2	ERIKA	4wWE	Bg2	Stadler		34	c1949	new (1)
3	THERESE	4wWE	Bg2	Stadler		35	c1949	new (1)

(1) 195x sold to Schafir & Mugglin, Liestal; after 1/2000 scrapped following a failed sale to Sri Lanka.

KRAFTWERK RUPPERSWIL-AUENSTEIN, Rupperswil   AG067 C

A railway was installed in 1945 as part of the work necessary for the infrastructure of this hydroelectric power station. This system employed thirtyfive kilometres of track and thirty steam locomotives as well as three diesel locomotives. The majority of the locomotives were made available for the project from various civil engineering firms and some were returned whence they came on completion of the project.

Gauge: 750 mm   Locn: 224 $ 6509/2513   Date: 01.2000

	Name	Type	Cyl	Builder	No.	Year	Note	Fate
		0-4-0T	Bn2t	Freudenstein	106	1901	(a)	s/s
		0-4-0T	Bn2t	Freudenstein	210	1905	(b)	s/s
	BIRS	0-4-0T	B?2t	Henschel Monta	17674	1920	(c)	(3)
	ST. GALLEN	0-4-0T	Bn2t	Jung 80 PS	1516	1910	(d)	(4)
		0-4-0T	B?2t	Jung Helikon	3024	1919	(e)	(5)
	WEISSENSTEIN	0-4-0T	Bn2t	Krauss-S IV bh	5295	1905	(f)	(6)
		0-4-0T	Bn2t	Maffei	3505	1909	(g)	(7)
		0-6-0T	Cn2t	Maffei	3508	1909	(h)	(8)
	VALLONE	0-4-0T	B?2t	Maffei	3609	1911	(d)	(9)
		0-4-0T	B?2t	Maffei	4124	1920	(g)	(10)
		0-4-0T	Bn2t	O&K 40 PS	550	1900	(k)	(8)
	TOGGENBURG	0-4-0T	Bn2t	O&K 50 PS	1897	1906	(f)	(6)
		0-4-0T	B?2t	O&K 50 PS	4382	1911	(m)	(13)
6	MURI	0-4-0T	B?2t	O&K 70 PS	6020	1914	(c)	(15)
7	ZÜRICH	0-4-0T	B?2t	O&K 70 PS	4375	1911	(d)	(15)
10	AUENSTEIN	0-4-0T	Bn2t	SLM	3833	1943	(p)	(16)
11	WILDEGG	0-4-0T	Bn2t	SLM	3834	1943	(p)	(16)
12	RUPPERSWIL	0-4-0T	Bn2t	SLM	3835	1943	(p)	(16)
14	ALVIER	0-4-0T	Bn2t	Jung 80 PS	1682	1911	(d)	(19)
15	SÄNTIS	0-4-0T	Bn2t	Jung 80 PS	1622	1911	(g)	(20)
16	FALKNIS	0-4-0T	Bn2t	Jung 80 PS	1683	1911	(d)	(19)
17	AARE	0-4-0T	Bn2t	O&K 80 PS	1247	1904	(d)	s/s
19	LIMMAT	0-4-0T	B?2t	O&K 70 PS	6573	1913	(c)	(23)
20	BADEN	0-4-0T	Bn2t	O&K 60 PS	4281	1911	(d)	(24)
31	AARAU	0-4-0T	Bn2t	O&K 70 PS	4374	1911	(m)	s/s

(a) made available by Hans Rüesch, St. Gallen (600 mm) /19xx.
(b) made available by VEBA, Zürich /19xx.
(c) made available by Schafir & Mugglin, Zürich /1944.
(d) see note for this locomotive under Robert Aebi & Cie AG, Regensdorf; made available by Schafir & Mugglin, Liestal /1944.
(e) made available by Robert Aebi & Cie AG, Zürich /19xx.
(f) ex Constantin von Arx, Olten /1939-44.
(g) made available by IRR, Rorschach /19xx.
(h) see note for this locomotive under Robert Aebi & Cie AG, Regensdorf, new as 750 mm; made available by Locher & Cie, Zürich /1943.
(k) see note for this locomotive under Maschinen- und Bahnbedarf (MBA), Dübendorf; made available by Locher & Cie, Zürich /1943.
(m) made available by Losinger & Cie, Bern /1944.
(p) new; into service /1944.

(3) 1944-52 returned to Schafir & Mugglin, Zürich; 1953 removed from boiler register.
(4) 1944-48 returned to Schafir & Mugglin, Liestal.
(5) 19xx probably returned to Robert Aebi & Cie AG, Zürich.
(6) 1944-48 to VEBA GmbH, Wildegg.
(7) very shortly returned to IRR, Rorschach.
(8) 1950 removed from boiler register.
(9) 1944-59 returned to Schafir & Mugglin.
(10) 1944-49 returned to IRR, Rorschach.
(13) 1944-56 returned to Losinger & Cie, Bern.
(15) 1944-53 returned to Schafir & Mugglin, Liestal.
(16) 19xx sold to VEBA GmbH, Wildegg or Zürich.
(19) 1944-53 returned to Schafir & Mugglin, Liestal; after 1959 scrapped.
(20) before 1950 returned to IRR, Rorschach.
(23) 1944-59 returned to Schafir & Mugglin, Liestal.
(24) 1944-53 returned to Schafir & Mugglin, Liestal; after 1953 scrapped.

KURATLE & JAECKER AG, Leibstadt   AG068

This firm operates a timber distribution centre (Holzwerkstoff Zentrum) and was related to Kuratle AG of Laufenburg. It has been rail connected since 1993. The locomotive also works traffic to the nearby Knecht Mühle.

Gauge: 1435 mm		Locn: 214 $ 6459/2717				Date: 07.1999
	0-4-0DE	Bde1	SLM,BBC Tm^{III}	4129,6003	1954	(a)

(a) purchased from ABB, Münchenstein "MAX" /1994 via P. Schlatter, Münchwilen AG.

LOGISTIK- UND GEWERBE-ZENTRUM (LGZ) HOCHRHEIN AG, Rekingen AG   AG069

An inter-modal terminal (logistics centre) established in 2002 within part of the HOLCIM cement works. Trains are moved between the reception sidings and one of at least two SBB stations by locomotives hired as required.

Gauge: 1435 mm   Locn: 215 $ 6659/2698   Date: 05.2006

DR. ING. GOTTLIEB LÜSCHER, Aarau   AG070 C

A construction firm.

Gauge: 750 mm		Locn: ?					Date: 03.2002
ANGELO	0-4-0T	Bn2t	Heilbronn II	274	1891	(a)	(1)
	0-4-0T	Bn2t	O&K 40 PS	717	1900	(b)	(2)

(a) purchased from Minder & Galli, Seftigen, Aebli /1902-20.
(b) purchased from F. & A. Bürgi, Bern /1905-24.

(1) 1936 removed from boiler register; s/s.
(2) 1927 removed from boiler register; s/s.

MEIER SÖHNE AG, Schwaderloch   AG071 M

formerly: Albin Meier AG (renaming took place before 2003)

A firm founded in 1935 and specialising in roads and road surfaces. The locomotive is displayed at a house (Albin Meier?) across the road from the firm, where a Kaelble road roller is on show. Address: Hauptstrasse 85, 5326 Schwaderloch.

Gauge: 600 mm		Locn: 214 $ 6530/2707				Date: 02.2003
	4wDM	Bdm	O&K(M) MD 1	7901	1937	(a) display

(a) purchased via MBA, Zürich for building waterworks in Schwaderloch.

MEIER + JÄGGI AG, Zofingen   AG072 C

A public works company, also working in the railway field. When not required the locomotive was often loaned to the OeBB. Office address: Chorgasse 4, 4800 Zofingen.

Gauge: 1435 mm		Locn: ?				Date: 10.1996
	4wDH	Bdh	Jung RK 20 B	13634	1963	(a) (1)

(a) new to Nienburger Glass Himly Holscher GmbH & Co, Nienburg [D] " UVK 04182"; ex NEWAG /1984.
(1) 22/6/2007 to OeBB.

GEBR. MESSING, Böttstein + Laufenburg + Koblenz   AG073 C

The spelling MESSIN has also been recorded: it is not clear if there were two different similarly named concerns or not. Messing was active in Germany. This was probably also a civil engineering concern. Some of the locomotives were used in the construction of the Beznau power station in 1899.

	Gauge: 900 mm		Locn: ?				Date: 03.2002	
		0-4-0T	Bn2t	Heilbronn III alt	93	1880	(a)	(1)
ELISE		0-4-0T	Bn2t	Heilbronn III neu	256	1890	(b)	(2)
HELVETIA		0-4-0T	Bn2t	Heilbronn III	269	1891	new	(3)

(a) used at Beznau /1899.
(b) purchased new for work at Koblenz; then Beznau /1899.
(1) 1899 sold to Hoch-Tiefbau Gebr. Helfmann, Frankfurt-am-Main [D].
(2) before 1910 sold to Robert Aebi & Co., Zürich.
(3) before 1907 sold to Baumann & Stiefenhofer, Wädenswil (or Ad. Baumann?).

MIGROS-GENOSSENSCHAFT AARGAU/SOLOTHURN, Suhr
AG074

A warehouse and distribution centre.

Gauge: 1435 mm		Locn: 224 $ 6491/2479				Date: 03.1995	
	0-4-0DH	Bdh	Cockerill	4143	1964	(a)	(1)
	4wDH	Bdh	O&K MB 360 N	26751	1972	(b)	(2)

(a) ex ? /1987, via Locorem.
(b) ex Georg Fischer, Mettman [D] "2", via OnRail and LSB /1994.
(1) 1994 to J. Müller AG, Effretikon.
(2) 2003 to Migros Genossenschaft Bern, Schönbühl.

GOTTLIEB MÜLLER & CIE, Zofingen AG077 C

formerly: Gottlieb Müller & Tottoli, Zofingen (1898-1912)
later: Gottlieb Müller AG, Zofingen (from 1967)

A civil engineering firm. Address (from 1967): Mühlethalstrasse 17, 4800 Zofingen.

Gauge: 600 mm		Locn: ?				Date: 05.2002	
	4wDM	Bdm	O&K(M) LD 2	8588	1937	(a)	s/s

(a) new via O&K/MBA, Zürich.

NORDOSTSCHWEIZERISCHE KRAFTWERKE (NOK),
Döttingen-Klingnau AG078

A short branch line was built 1946-1949 to serve the Beznau/Würenlingen power station. It was then extended to serve the nuclear power station of Würenlingen. Near its junction with the SBB it also serves a store and workshops of the NOK.

Gauge: 1435 mm			Locn: 215 $ 6605/2678			Date: 10.2007	
Tm	4wDM	Bdm	RACO 45 PS	1232	1935	(a)	(1)
TM 237 918-6 JANKA	0-4-0DH	Bdh	KHD A12L714 R	57198	1961	(b)	

(a) new as 4wPM Bdm; purchased from SBB "Tm 511" 5/1964.
(b) new to Klöckner Georgsmarienwerke AG, Georgsmarienhütte [D] (as A12L714 R), purchased from Krupp-Klöckner, Osnabrück [D] "3" via OnRail/LSB /1991; also referred to as type KS230 B.
(1) 2000 donated to Draisinen-Sammlung Fricktal (DSF), Laufenburg.

OEHLER & CIE, Aarau AG079

This mechanical workshop and foundry was started in 1881 in Wildegg and transferred to Aarau in 1894 where it was connected to the rail network. From 1970 it formed part of the +Georg Fischer+ group. Products covered rail vehicles of all sorts, including electric cranes and small locomotives. The works closed in 1983. At least one locomotive for their own use was registered with the authorities. The factory had extensive narrow gauge systems with different gauges within the works.

Gauge: 1435 mm   Locn: 215 $ 6471/2493   Date: 11.1986

			unknown		?	1xxx	(a)	s/s

(a)   registered /1918.

PROVIMI KLIBA SA, Kaiseraugst   AG080

formerly: Kliba Mühlen AG (from 1995 till a merger on 23/4/2001)
             Klingenthalmühle AG (till 1995)

These mills also use the trade name of Kliba Futter.

Gauge: 1435 mm   Locn: 214 $ 622450,265675   Date: 12.2002

	4wDM	Bdm	Breuer VL	3067	1953	(a)	(1)
	4wDH	Bdh	KHD A8L614 R	56794	1957	(b)	

(a)   ex Ernst Frey AG, Kaiseraugst "206" by /1972.
(b)   new to Stadtwerke Wesel [D]; purchased from NEWAG /1988; also referred to as type KK130 B.
(1)   1988 sold to Thommen AG, Kaiseraugst and converted to an electrically operated robot.

RAILARENA, Brittnau-Wikon   AG081 M

The Zofinger Eisenbahn-Amateure Klub was founded in 1966 and has its center in Walterswil, SO. The locomotives which it has collected are based in the developing museum here, but may be loaned elsewhere. The direction is under Herr R. Fischer.

Gauge: 1435 mm   Locn: 224 $ 6392/2348   Date: 01.2009

	4wBE	Ba	Olten?,MFO	?	c1932	(a)	
Tm	4wDM	Bdm	RACO 25 PS	1066	1933	(b)	
SBB Tm^{III} 9526	4wDH	Bdh	RACO 225 SV4 H	1931	1986	(c)	(3)

(a)   new to von Roll, Bern; ex Swisstrain /2008.
(b)   ex IZAG, Zofingen by /2007.
(c)   ex SBB "Tm^{III} 9526" /2008.
(3)   2009 donated to Swisstrain.

     see: www.Railarena.ch

REICHHOLD CHEMIE AG, Hausen bei Brugg   AG082

formerly: Zementwerk Hausen bei Brugg (1930-31)

A cement works on this site was opened in 1930 and then closed shortly after - following a take-over. The site was re-used for a chemical factory from 1938, which in turn closed in 1993. It was rail-connected to Birrfeld SBB station, renamed Lupfig since 1994.

Gauge: 1435 mm   Locn: 215 $ 6585/2561   Date: 03.2008

Ee 2/2		0-4-0WE	Bw1	SLM,SAAS Ee 2/2	3415, ?	1930	new	(1)
Tm 2/2 REICHHOLD		4wDE	Bde1	Stadler,BBC	111, ?	1960	(a)	(2)

(a)   purchased from Stadler AG, Bussnang /1990.
(1)   1931 sold to SBB "Te 71".
(2)   1997 sold to Auto- und Metallverwertung Ostschweiz (AVO), Schwarzenbach SG.

ROTHPELZ, LIENHARD + CIE AG (RL) AG083 C

A public works company with locomotives stored between contracts.
Address: Schifflandestrasse 35, 5000 Aarau.

Aarau

Gauge: 750 mm   Locn: 224 $ 6454/2495   Date: 02.2003

	4wDM	Bdm	O&K MD 2b	25132	1951	(a)	s/s

Gauge: 600 mm   Locn: ?   date: 02.2003

	4wDM	Bdm	GIA?	?	19xx		(2)
	4wDM	Bdm	GIA?	?	19xx		(2)
	4wDM	Bdm	O&K(M) RL 1a	4612	1931	(d)	
10.3.8	4wBE	Ba1	SIG ETB 70	578 301	1957		(2)
	4wBE	Ba1	SIG ETB 70	?	19xx		(2)
	4wBE	Ba1	SIG ETB 70	?	19xx	(g)	

Material store areal, Asp

Gauge: 600 mm   Locn: 214 $ 6466/2556   Date: 02.2002

10.3.4	4wBE	Ba1	Oehler	?	c1960	stored
10.3.5	4wDM	Bdm	O&K	?	19xx	stored
10.3.10	4wDM	Bdm	O&K	?	19xx	stored
10.3.12	4wDM	Bdm	O&K	?	19xx	stored

(a) built by Schöma as their number 1240; purchased new via Wander-Wendel/MBA, Dübendorf.
(d) purchased from ? /1952.
(g) SIG reference lists give three ETB 70 locomotives delivered to this company; this one has not been reported.
(2) 8/1994 stored; since s/s.

HANS SALM, Brugg AG AG084

The business of this concern is not known.

Gauge: 1435 mm   Locn: ?   Date: 11.1988

4wDM	Bdm	Windhoff 6 II	211	1931	new	s/s

AD. SCHÄFER & CIE AG, Aarau AG085 C

A public works company. Address: Buchserstrasse 12, 5000 Aarau.

Gauge: 600 mm   Locn: 225 $ 6469/2492   Date: 07.1995

4wDM	Bdm	O&K 3 D	21504	1939	(a)	(1)
4wDM	Bdm	O&K MV1	25140	1952	(a)	(1)

(a) purchased from Schafir & Mugglin, Liestal after closure of that site.
(1) 1994 sold to FWF, Otelfingen (but not identified there).

SCHAFIR & MUGGLIN, Klingnau AG086 C

A civil engineering firm participating in the installation of the hydroelectric plant at Klingnau. A number of locomotives were made available from other Schafir & Mugglin depots and returned either to the original or another Schafir & Mugglin base afterwards.

Gauge: 750 mm   Locn: 215 $ 6592/2718   Date: 03.2002

BIRS	0-4-0T	B?2t	Henschel Monta	17674	1920	(a)	(1)
SÄNTIS	0-4-0T	Bn2t	Jung 80 PS	1622	1911	(b)	(2)

	ALVIER	0-4-0T	Bn2t	Jung 80 PS	1682	1911	(b)	(3)
	FALKNIS	0-4-0T	Bn2t	Jung 80 PS	1683	1911	(b)	(2)
	MUTTENZ	0-4-0T	Bn2t	Jung 20/25 PS	3620	1924	(b)	(1)
	SPERANZA	0-4-0T	Bn2t	Krauss-S XXVII vv	3254	1895	(f)	(1)
	LICHTENSTEIG	0-4-0T	Bn2t	Krauss-S XXVII bg	6173	1909	(f)	(1)
	VALLONE	0-4-0T	B?2t	Maffei	3609	1914	(h)	(8)
		0-4-0T	Bn2t	O&K 40 PS	1106	1903	(i)	(9)
	AARE	0-4-0T	Bn2t	O&K 80 PS	1247	1904	(j)	(10)
		0-4-0T	Bn2t	O&K 40 PS	1588	1906	(i)	s/s
	BASEL	0-4-0T	Bn2t	O&K 80 PS	1622	1905	(j)	(1)
	VEVEY	0-4-0T	Bn2t	O&K 50 PS	2907	1908	(j)	(1)
	KLINGNAU	0-4-0T	Bn2t	O&K 50 PS	3702	1910	(j)	(14)
	MURI	0-4-0T	B?2t	O&K 70 PS	6020	1914	(j)	(1)
	BERNA	0-4-0T	B?2t	O&K 70 PS	6572	1913	(j)	(1)
	LIMMAT	0-4-0T	B?2t	O&K 70 PS	6573	1913	(q)	(1)
	GWATT	0-4-0T	B?2t	O&K 50 PS	7520	1916	(f)	(1)

Gauge: 600 mm   Locn: ?   Date: 03.2002

	BUBI	0-4-0T	Bn2t	Jung 10 PS	1349	1909	(s)	(1)
	SCHWEIZ	0-4-0T	Bn2t	Jung 80 PS	1515	1910	(b)	(20)
		0-4-0T	Bn2t	O&K 40 PS	5225	1912	(u)	(21)

(a) made available by Schafir & Mugglin, Zürich /1931.
(b) made available by Schafir & Mugglin (no given location) /1931.
(f) made available by Schafir & Mugglin, Muri AG /1931.
(h) see note for this locomotive under Robert Aebi & Cie AG, Regensdorf; made available by Schafir & Mugglin, Muri AG /1931.
(i) made available by Arbeitsgemeinschaft Stauwehrbau Klingnau, Klingnau /1932.
(j) made available by Schafir & Mugglin, Muri AG /1931-33.
(q) made available by Schafir & Mugglin, Muri AG /1932.
(s) made available by Schafir & Mugglin, Muri AG /?1931.
(u) new to Ph. Holzmann, Danzig [D]; made available by Arbeitsgemeinschaft Stauwehrbau Klingnau, Klingnau /1932.

(1) 1933-39 to Schafir & Mugglin, Zürich.
(2) 1933-39 to Schafir & Mugglin, Liestal.
(8) 1933-39 to Schafir & Mugglin, ?Zürich.
(9) 1933-39 to Kies AG Bollenberg, Tuggen.
(10) 1937 to Schafir & Mugglin, Liestal.
(14) 1933 to Schafir & Mugglin, Zürich.
(20) 1935 to Schafir & Mugglin, Liestal.
(21) 1933 removed from boiler register.

SCHAFIR & MUGGLIN & GOTTLIEB MÜLLER, Klingnau   AG087 C

Apparently a separate company involved in the Klingnau dam and associated lake though a relation with the previous entry may be deduced.

Gauge: 600 mm   Locn: ?   Date: 02.2008

	4wDM	Bdm	Jung MS 131	4935	1930	(a)	s/s
	4wDM	Bdm	Jung MS 131	4936	1930	(a)	s/s
	4wDM	Bdm	Jung MS 131	4940	1930	(a)	s/s
	4wDM	Bdm	Jung MS 131	4941	1930	(a)	s/s

(a) new via Robert Aebi & Cie AG, Zürich.

SCHINZNACHER BAUMSCHULBAHN (SchBB), Schinznach Dorf
AG088 M

formerly: Zuglauf AG

A non-locomotive worked railway was installed in the nursery of Baumschule Hermann Zuglauf in 1928. Starting ca. 1976 the rail layout was extended and locomotives introduced as an attraction. The railway is no longer regularly used for transporting the nursery's products and operation of the trains is now the responsibility of Verein Schinznacher Baumschulbahn. Ownership of the locomotives is not always SchBB. Address: Degerfeldstrasse 4, 5107 Schinznach Dorf.

Gauge: 600 mm Locn: 214 $ 6534/2559 Date: 09.2006

AZELEA		0-4-0DM	Bdm	KHD OMZ122 F	39663	1940	(a)	
	rebuilt			Ruhrthaler D54Z	3325	1955		
DRAKENSBERG		2-6-2+2-6-2 4c 1C1+1C1h4t		Hanomag	10551	1927	(b)	
PINUS		0-4-0T	Bn2t	Henschel Riesa	23672	1937	(c)	
LISELI		0-4-0WT	Bn2t	Jung 20 PS	1693	1911	(d)	(4)
1		4wDM	Bdm	Jung ZL 105	5810	1934	(e)	(5)
10		4wDM	Bdm	Jung ZL 105	8208	1938	(e)	(6)
SYRINGA		4wDM	Bdm	Jung ZL 105 Brun	9515	1941	(g)	
	rebuilt							
MAGNOLIA		4wDM	Bdm	Jung ZL 105 Brun	9519	1941	(g)	
	rebuilt							s/s
		4wDM	Bdm	JW		1950	(h)	s/s
TAXUS		0-8-0T	Dn2t	Krauss-M LXXXI c	7349	1917	(i)	
EMMA		0-4-0T	Bn2t	Maffei	4144	1925	(j)	
LUKAS		0-4-0T	Bn2t	O&K 50 PS	7479	1918	(k)	
SEQUOIA		0-6-0TT	Cn2	MBA	13585	1944	(l)	
ABELIA		0-4-0DM	Bdm	LKM Ns2f	262005	1958	(m)	
		0-4-0T	Bn2t	SLM	3833	1943	(n)	(14)
MOLLY		0-4-0T	Bn2t	SLM	3834	1943	(o)	
		0-4-0T	Bn2t	SLM	3835	1943	(n)	2007 scr
		0-4-0DM	Bdm	Ruhrthaler G42Z	3278	1955	(q)	(17)
PAEONIA		0-4-0DM	Bdm	Ruhrthaler G42Z	3291	1955	(r)	
KALAHARI		2-8-2	1D1h2	Franco-Belge	2686	1953	(s)	(19)
		4wDH	Bdh	Diema DS90/1	2987	1958	(t)	

(a) purchased from Barbara Rohstoffbetriebe GmbH, Grube Fernie, Grossen Linden [D] "DL 9" /1977 (or 1979?); plates of Klöckner & Co., Duisburg [D] (agent).
(b) purchased from SAR "NGG13 60" (610 mm) /1986.
(c) purchased from Kies- und Schotterwerk Nordmark, Lürschau bei Schleswig [D] "3" /1977.
(d) privately owned; made available by Thomas Brändle, Geroldswil /2000.
(e) acquired from Ziegelei Hochdorf, Hochdorf /1976.
(g) rebuilt by Brun, identity of original Jung diesel not confirmed; ex Ziegelei Hochdorf /1976.
(h) as well as the type JW8 locomotives listed in the miscellaneous section for 1951 this might be 8066 of 1950.
(i) purchased from DR (DDR) Waldeisenbahn Muskau [D] "99 3311" /1977, the locomotive now has the frame and motion of a Jung locomotive (HFB "2373").
(j) new to Robert Aebi & Cie, Zürich; to Holzverzuckerung AG (HOVAG), Ems /1941; acquired from Oswald Steam Samstagern /1992.
(k) ex Kieswerk Kissing, Augsburg [D], via Herr Bauer, Lüsslingen "EVI 1"; SchBB by 4/1996.
(l) purchased from PKP Jarocin Light Railway [PL] "Ty3-194" /1978.
(m) ex Granitwerk Wiesa/Sachsen [D] /1996.
(n) new to Flusskraftwerke, Rupperswil-Auenstein (750 mm); ex Messerli AG, Kaufdorf /2006.
(o) new to Flusskraftwerke, Rupperswil-Auenstein (750 mm); ex Schuljugend Turgi put on display at Turgi SBB Bahnhof 8/1967-/1994.
(q) purchased from Barbara Rohstoffbetriebe GmbH, Grube Fernie, Grossen Linden [D] "100" /1976.
(r) purchased from Barbara Rohstoffbetriebe GmbH, Grube Fernie, Grossen Linden [D] "DL 20" /1976; property of Zuglauf AG.
(s) ex SAR "NG 136" (610 mm) /1998.
(t) ex Zürcher Ziegeleien AG, Schinznach Dorf "TANTE EMMI" /2005.
(4) 2/2005 to Parc d'Attractions du Châtelard VS SA, Château d'Eau.
(5) 1998 to Minièresbunn, Dhoil [L].
(6) c1985 donated to Oliver Weder, Diepoldsau; later Hr. Bernard Könen, Losheim [D]. Carries a plate of Max Giese Stahlbeton GmbH.
(14) 2007 to Gemeinde Auenstein for restoration and display.
(17) used as spare parts for PAEONIA; 1998 remains to Minièresbunn, Dhoil [L].
(19) 2008 returned to South Africa for preservation.

see: www.schbb.ch

SCHLATTER PETER AG, Rheinsulz — AG089 C

formerly: Eisen und Maschinen Schlatter AG (from 18 March 1977 till February 1997)
P. Schlatter, Münchwilen AG (from 1960)

Peter Schlatter (senior) has been in the business of hiring and providing pipes and other material for tunnel construction and structural engineering since 1960. The transfer to Rheinsulz took place in 1991 and the field of operation now includes the provision of railway track items. The list of locomotives is not complete and there may be errors in the entries. Those locomotives remaining are held in a store in Frick. Address: Bahnhofstrasse 12, 5084 Rheinsulz.

Gauge: 1435 mm   Locn: 214 $ 6493/2677   Date: 02.2003

		0-4-0D	Bdm	Demag	2937	*1941*		(1)

Gauge: 900 mm

2		4wDH	Bdh	*ASEA*	?	19xx		store

Gauge: 750 mm

		4wDH	Bdh	Ageve ?type	?	19xx	(aa)	s/s
		4wDH	Bdh	Ageve ?type	707	1972	(ab)	s/s
181017		4wDH	Bdh	Ageve D12	779	1976		store
		4wDH	Bdh	Ageve ?type	780	1976	(aa)	s/s
		4wDH	Bdh	Ageve D12	951	1984		store
		4wDH	Bdh	ASEA?	?	19xx		store
		0-4-0DM	Bdm	KHD A4M517 G	47169	1951	(ag)	(33)
		4wDH	Bdh	KHD A6M517 G	56209	1955	(ah)	(33)
		0-4-0DH	Bdh	GIA DHD35	3591	1993		(35)
		0-6-0DH	Cdh	Gmeinder HF 130 C	4005	1943	(aj)	(36)
		0-6-0DH	Cdh	Gmeinder HF 130 C	4233	1946	(ak)	(37)
		0-6-0DH	Cdh	Gmeinder HF 130 C	4234	1946	(ak)	(38)
		4wDH	Bdh	Levahn	?	19xx		store
1	225	0-4-0DH	Bdh	O&K MV9	25967	1960	(an)	(39)
2	227	0-4-0DH	Bdh	O&K MV9	26135	1961	(an)	(39)
3	228	0-4-0DH	Bdh	O&K MV9	26136	1961	(an)	(39)
4	232	0-4-0DH	Bdh	O&K MV9	26181	1962	(an)	(39)
5	233	0-4-0DH	Bdh	O&K MV9	26182	1963	(an)	(39)
6	235	0-4-0DH	Bdh	O&K MV9	26216	1963	(an)	(39)
7	236	0-4-0DH	Bdh	O&K MV9	26217	1963	(an)	(39)
8	238	0-4-0DH	Bdh	O&K MV10	26285	1964	(an)	(39)
9	239	0-4-0DH	Bdh	O&K MV10	26286	1964	(an)	(39)
10	240	0-4-0DH	Bdh	O&K MV10	26287	1964	(an)	(39)
11	241	0-4-0DH	Bdh	O&K MV10	26288	1964	(an)	(39)
		0-4-0DM	Bdm	O&K(M) MD 3	11713	1943	(ay)	(51)
		4wDM	Bdm	RACO 50 PS	1312	1945	(az)	s/s
		4wBE	Ba	SIG ET 8	701 107	1973	(ba)	store
		4wBE	Ba	SIG ATS 610	705 714	1975	(ba)	store

Gauge: 600 mm

	4wDM	Bdm	Gmeinder	?	19xx		s/s
	4wDM	Bdm	Gmeinder	998	193x		s/s
	4wDM	Bdm	JW DM500.8.10	3.453.31	1967	(cc)	s/s
	4wDM	Bdm	JW DM100.6.10	3.453.36	1970	(cd)	s/s
	0-4-0DMSO	Bdm	O&K(M)	780	1914	(ce)	(83)
	4wDM	Bdm	O&K(M) RL 1a	4005	1930	(cf)	s/s
	4wDM	Bdm	O&K(M) MD 2	10701	1940	(cg)	s/s
	?4wDM	Bdm	Plymouth	7305	19xx		s/s
	?4wDM	Bdm	Plymouth	7571	19xx		s/s
	0-4-0DM	Bdm	Plymouth WL2770	7731	19xx	(cj)	s/s
	4wDM	Bdm	R&H 48DL	249517	1947	(ck)	(89)
	4wDM	Bdm	R&H LBU	386623	1955	(cl)	(90)
	4wDH	Bdh	Schöma CFL 200DCL	4665	1983	(cm)	(91)

CH 47 AG

130	4wDH	Bdh	Schöma CFL 200DCL	4667	1983	(cn)	(92)
131	4wDH	Bdh	Schöma CFL 200DCL	4669	1983	(co)	(93)
129	4wDH	Bdh	Schöma CFL 200DCL	4670	1983	(cp)	(94)
	4wBE	Ba	SIG	?	19xx	(cq)	s/s
2	4wBE	Ba	SIG	?	19xx		s/s
5	4wBE	Ba	SIG	?	19xx		s/s
1	4wBE	Ba	SIG	701 110	1973	(cq)	(98)
3	4wBE	Ba	SIG	707 728	1977	(cq)	(98)
1	4wBE	Ba	SIG	709 446	1977		(98)
	4wBE	Ba	Schalke,SSW 3A16	67040,6301	1967	(cw)	(101)
	0-4-0DM	Bdm	Spoorijzer A2L514	60001	1960	(cx)	(102)

Gauge: 500 mm

	4wDM	Bdm	O&K(M)	7806	1937	(da)	(201)

Gauge: ? mm

	4wDH	Bdh	Ageve D12	1201	1994	new	(301)
	4wDH	Bdh	Ageve D12	1202	1994	new	(301)

(aa) may have been 600 mm.
(ab) may have been 600 mm: ex GIA.
(ac) ex EEG Henriksen.
(ag) new to Eisenbergwerk Gonzen AG, Sargans (probably 600 mm); ex St. Gallische Rhein-Korrektur "KAMOR 114" (750 mm) /197x.
(ah) new to Eisenbergwerk Gonzen, Sargans (probably 600 mm); ex St. Gallische Rhein-Korrektur "GONZEN 115" (750 mm) /197x (?4wDM Bdm).
(aj) new to HF [D] "M13808"; ex Dyckerhoff Zementwerke AG, Wiesbaden [D] (600 mm) /196x.
(ak) new to Glaser & Pflaum, Mannheim [D] for ?.
(an) ex Rheinisch-Westfälische Kalkwerke AG, Werk Hönnetal [D] /1968.
(ay) ex IRR, Widnau (or SGRK, Rorschach) /197x.
(az) ex SGRK, Rorschach /1972.
(ba) marked Orr, probably ex Porr Deutschland GmbH, München [D].
(cc) new to ÖAMAG, Radenthein [A]; ex ÖAMAG [A] "18321".
(cd) new to ÖAMAG, Radenthein [A]; ex ÖAMAG [A] "21146".
(ce) almost certainly 4wPM Bbm when new.
(cf) ex Jurabergwerke AG, Herznach.
(cg) ex Joseph Klug, Regensburg [D].
(cj) also quoted as "CA42".
(ck) new via R. Aebi & Co, Zürich; ex ? /19xx (existence here to be confirmed).
(cl) ex Zschokke AG, Gondo /19xx (to be confirmed).
(cm) new to Tunnel Makkah Taif [Saudi Arabia] (750 mm); ex Conrad Zschokke, Genève /1992 (may have been 750 mm).
(cn) new to Tunnel Makkah Taif [Saudi Arabia] (750 mm); ex Chantier Matmata [Morocco] (may have been 750 mm).
(co) new to Tunnel Makkah Taif [Saudi Arabia] (750 mm); ex Chantier Matmata [Morocco] /1992 (may have been 750 mm).
(cp) new to Tunnel Makkah Taif [Saudi Arabia] (900 mm); ex Chantier Matmata [Morocco] /1992(may have been 750 mm).
(cq) ex Dywidag Bau AG.
(cw) ex BAUMEX Bauunion [A].
(cx) supplied to Zürcher Ziegeleien, Ziegelei Brunau, Zürich-Giesshübel /19xx and returned /19xx.
(da) new to O&K, Amsterdam [NL] (700 mm); ex O&K, Zürich.
(1) by 1973 derelict; s/s.
(33) reported at Locher AG, Stettbach and Gäbris (1000 mm); by 1986 at Museumsbergwerk Gonzen, Sargans (600 mm).
(35) 2003 to RhB Workshops, Landquart for overhaul and then to construction plant in Singapore [Malaysia].
(36) 1968 to St. Gallische Rhein-Korrektur, "CHURFISTEN"; later IRR; 1976 returned to P. Schlatter; 1991 Gemeinde Rittersgrün (for Schmalspurmuseum Oberrittersgrün) [D].
(37) 1968 to St. Gallische Rhein-Korrektur, "FALKNIS"; later IRR; 1976 returned to P. Schlatter; 1991 to Jörg Siedel, Köln [D]; kept on the island of Rügen [D].
(38) 1968 to St. Gallische Rhein-Korrektur, "PIZOL"; later IRR; 1976 returned to P. Schlatter; 1991 to IG Preßnitztalbahn, Jöhstadt [D].
(39) after 1974 to M. K. Makama, Ikeja [Nigeria].
(51) 19xx to Verein zur Förderung von Klein- und Lokalbahnen (VFKL), Groß-Schwechat [A] "5".

- (83) by 2006 to FWF, Otelfingen.
- (89) 19xx to Zementwerk Mannerdorf [A]; 1970 to Montan- und Werksbahnmuseum Graz, Graz [A].
- (90) by 2008 to Montan- und Werksbahnmuseum Graz, Graz [A].
- (91) hire to Nuova ORMEF, Pontida [I] "R 053"; later ArGe Mont Terri Sud, Saint-Ursanne, JU "601" (750 mm).
- (92) hire to Cariboni spA Colico "130"; later Nuova ORMEF SrL Gestione Mezzi Ferroviarie [I].
- (93) hire to Cariboni spA Colico "131"; later Nuova ORMEF SrL Gestione Mezzi Ferroviarie [I] "Machina 84" or "054" (900 mm).
- (94) may have been 750 mm; hire to Cariboni spA Colico "129" (900 mm); later Nuova ORMEF SrL Gestione Mezzi Ferroviarie [I] "Machina 71".
- (98) reported sold to Austria [A].
- (101) reported sold to Italy [I].
- (102) 19xx to FWF, Otelfingen.
- (201) by 2007 to Verein zur Förderung von Klein- und Lokalbahnen (VFKL), Groß-Schwechat [A]
- (301) hired to Dyckerhoff & Widman for Zeulenroda [D] contract.

   see: Eisenbahn Amateur 11/1973 p493
   see: schlatter-peter-ag.ch/firmenhistory.html

SCHWEIZERISCHES BUNDESBAHNEN (SBB)   AG090 M

Bahnhof, Baden

Gauge: 1435 mm   Locn: 215 $ 6654/2586   Date: 04.1996

Eb 3/5	5811	2-6-2T	1C1h2t	SLM Eb 3/5	2212	1911	(a)	(1)
Be 4/6	12332	2-4-4-2WE	1BB1w2	SLM,BBC Be 4/6	2809,1804	1922	(b)	(2)
Ae 3/6	10601	2Co1WE	2Co1w3	SLM,BBC Ae 3/6	2740,1541	1920/21	(c)	(3)

Brugg AG

Gauge: 600 mm   Locn: ?   Date: 02.2008

		0-4-0T	Bn2t	Maffei	4144	1925	(d) (4)

Bahnhof, Turgi

Gauge: 750 mm   Locn: 215 $ 6615/2604   Date: 09.1968

11	MOLLY	0-4-0T	Bn2t	SLM	3834	1943	(e) (5)

- (a) ex SBB "5811" /1965 with plates of SLM 1798 of 1907 (SMB "1").
- (b) ex SBB "12332" by /1974.
- (c) new to SBB "10301" 19/8/1921; ex SBB "10601" /1983.
- (d) hired from Robert Aebi & Cie, Zürich for use on a construction site after /1925.
- (e) ex VEBA, Niederhasli /1967.
- (1) 1974 returned to SBB; stored in Glarus and used as spare parts bank.
- (2) 1981 to Verkehrshaus, Luzern.
- (3) 1994-99 to SBB depot in Winterthur; later Swisstrain (via Classic Rail), Le Locle.
- (4) by 1938 returned to Robert Aebi & Cie, Zürich.
- (5) 1994 to Schinznacher Baumschulbahn, Schinznach Dorf.

SCHWEIZERISCHES MILITÄRMUSEUM (SMM), Full-Reuenthal
AG091 M

This military museum was established in or after 2004. One part of it is situated in the former CFU factory lying between the two constituent parts of the town, and is usually referred as being in Full.

Gauge: 1435 mm   Locn: 215 $ 6566/2732   Date: 10.2007

Tm	1	4wDH	Bdh	Jung ZN 233	13061	1958	(a)
		4wDM	Bdm	RACO 80 SA3	1397	1950	(b)

(a) new to Züricher Freilager AG, Zürich ; Südostbahn "Tm 32" /1975; renumbered "236 010" /2002; to SMM /2004. Note that SOB have numbered another tractor "236 010".
(b) ex Walter Solenthaler, Winkeln AG, Gossau SG /2006.

SOLVAY (SUISSE), Rekingen AG   AG092

formerly: Schweizerische Sodafabrik Zurzach (1917-1989)

Originally opened as a plant to produce soda, the factory then diversified into soda products. Production started in 1916 and the concern was sold to the Belgian firm of Solvay in 1922. A separate Swiss operation was formed in 1989. The narrow gauge line served a chalk quarry and was connected to the works by an aerial ropeway.

Gauge: 1435 mm   Locn: 215 $ 6660/2677   Date: 06.2007

Ed 3/4	1	2-6-0T	1Cn2t	SLM Ed 3/4	1425	1902	(a)	1965 scr
Em 2/2	ANTONIO	4wDE	Bde2	SIG,BBC	? ,6130	1960	new	(2)
Em 837 903-4		6wDE	Cde2	Moyse CN60EE500D	3537	1973	(c)	
		4wDH	Bdh	Vollert Diesel, FFS	77/013	1977	new	2007 str
		0-4-0DH	Bdh	Henschel DH 240B	29705	1958	(e)	(5)
Em 847 901-6		BBDH	BBdh	U23A LDH370	23879	1979	(f)	

Gauge: 600 mm   Locn: 215 $ 665x/269x   Date: 12.1997

	0-4-0DM	Bdm	Deutz MLH228 F	9318	1929	new	s/s
	0-4-0DM	Bdm	SLM	3771	1941	(h)	s/s

(a) new to Seetalbahn (SeTB) "Ed 3/4 13"; purchased /1912.
(c) new via LSB; was initially named GIOVANNI.
(e) hired from LSB /1991.
(f) hired from LSB /2006.
(h) new, also reported as of 1940.
(2) 1973 sold to BT "Tm 2/2 10 ANTONIO"; 1997 to SERSA.
(5) 1992 returned to LSB.

SPIELPLATZ, Wettingen   AG093 M

Gauge: 1000 mm   Locn: 215 $ 6615/2604   Date: 01.2008

HGe 2/2	2	0-4-0WRE	Baed2	SLM HGe 2/2	1140	1898	(a)	1968 scr

(a) originally StEB "2"; wdn /1964; ex BBC, Baden /1966.

STAHLROHR AG, Rothrist   AG094

formerly: Rothrist Rohr

A factory manufacturing parts for the automobile industry (and other industrial purposes) from steel tubes. There is some internal 600 mm track with bogies for carrying tubes.

Gauge: 1435 mm   Locn: 224 $ 6330/2398   Date: 08.1987

Tm		4wPM	Bbm	Breuer IV	?	1930	(a)	1968 scr
Tm		0-4-0PM	Bbm	RACO 45 PS	1212	1933	(b)	(2)
		4wDH	Bdh	Ruhrthaler NO1206V	3530	1957	(c)	c2002 scr
		4wDH	Bdh	KHD KG230 B	57667	1964	(d)	

(a) purchased from SBB "Tm 409" 1/1963.
(b) new to SBB "Tm 505"; purchased from SBB "Tm 873" 12/1967.
(c) new to Rheinstahl Hüttenwerke, Witten-Annen [D]; purchased from Rothrist Rohr GmbH, Bottrop [D] "1" /1982.
(d) ex Pestalozzi, Dietikon /2002.
(2) 19xx to Dampflokfreunde, Langenthal.

STRUMPFFABRIK ARGO AG, Möhlin   AG095

The following locomotive was hired, probably for use as a stationary boiler.

Gauge: 750 mm   Locn: ?   Date: 02.2008

              0-4-0T   Bn2t   Henschel Danzig   17270   1919   (a)   1956 scr

(a)   ex Robert Aebi & Cie AG /1936.

TEE-CLASSICS, Eiken   AG096 M

A group with a collection of rolling stock built during the TEE period. While the clubroom (an ICE bar car) is at Eiken the locomotive has been kept elsewhere and by 2009 was in the Dreispitz-Areal in Basel. Address: TEE-CLASSICS, Postfach, 8050 Zürich.

Gauge: 1435 mm   Locn: ?   Date: 10.2007

Re 4/4^I 10034   BoBoWE   BBw4   SLM,BBC Re 4/4^I 4013, ?   1949   (a)

(a)   new to SBB "Re 4/4 434"; renumbered "Re 4/4^I 10034" 24/10/1962.

THOMMEN AG, Kaiseraugst   AG097

formerly: Thommen & Co

This scrap merchant (Recycling, Eisenhandel und Altmetalle) has two sites alongside the railway to the east of Kaiseraugst SBB station. Locomotives may be present for scrapping from time to time; but there are also resident locomotives.

Jung 14129 of 1973 (type RK 8 B) awaits its next duty on 24 April 2009.

Gauge: 1435 mm   Locn: 214 $ 6217/2654   Date: 12.2002

Tm   4wDM   Bdm   unknown   ?   19xx   (a)   s/s
                4wDH   Bdh   Jung VN 234   8812   1940   (b)   (2)
                4wDH   Bdm   Schöma CFL 120DBR   2399   1961   (c)   2000 scr
                4wDH   Bdm   Jung RK 8 B   14129   1973   (d)

	4wCE	Bw	Breuer VL (reb Robot)	3067	1953	(e)	s/s?	
	4wDM	Bdm	RACO 95 SA3 RS	1451	1955	(f)	s/s?	
	4wDH	Bdh	Kronenberg DL 200	147	1964	(g)	2006 scr	
	4wDH	Bdh	Henschel DHG 160B	31085	1966	(h)		
Tem^{III} 333	0-4-0W+DEBw2		SLM,SAAS Tem^{III}	4199, ?	1956	(i)		

(a) new to SBB, Breuer type; purchased from Debrunner & Co AG, St. Gallen-Haggen, one of "Tm 402, 404 or 406" all Breuer of 1930.
(b) new to Reichsbahn Ausbesserungswerk, Stadt Friedrichshafen [D]; ex Stadtwerke Heidenheim [D]. Plates of R. Dollberg, Berlin [D]. Some sources quote two similar machines.
(c) purchased from Eisenwerke Nürnberg, Nürnberg [D] "1" /1984.
(d) purchased from Lonza, Schweizerhalle /2000, via ETRA.
(e) ex Klingenthalmühle AG, Kaiseraugst /1988; converted to electric robot.
(f) ex CFF "Tm 789" /2000-04.
(g) new to Feldmühle AG, Rorschach "2"; ex SBB "Tm 940" /2005 (withdrawal date 31/12/2004).
(h) ex SATRAM-Huiles SA, Bâle Petite Hunigue by /2009.
(i) new to SBB "Tem^{III} 33"; ex SBB "Tm^{III} 333" by /2008 (withdrawn 6/2006).

(2) 1986 sold to Dietiker Metallhandel AG, Regensdorf.

TONWERK ED. MEIER AG, Wettingen Dorf   AG098

A clay works that closed c1975.

Gauge: ?600 mm   Locn: ?   Date: 01.1988

?4wDM	?Bdm	Diema		?	19xx	(1)

(1) donated to Bergwerkverein Käpfnach, Horgen. The history of the locomotive is in doubt as the locomotive cannot be positively identified at Käpfnach. There is a Diema there with a plate corresponding to the Diema DS12 locomotive (1280 of 1948) from Ziegelei Bacher, Freudenstadt [D]. Two possibilities are that the locomotive from Wettingen Dorf has inherited the Diema plate from Freudenstadt, or that the Freudenstadt locomotive came to Wettingen before going to Horgen (to be verified).

VALLI AG STRASSENBAU, Aarau   AG099 C

A road builder.

Gauge: 600 mm   Locn: ?   Date: 02.2008

4wDM	Bdm	Jung MS 131	?	1929-33	(1)

(1) 1988 to FWF, Otelfingen.

VEREINIGTE BAUUNTERNEHMER AG, Wildegg   AG102 C

alternative: VEBA (= **VE**reinigte **BA**unternehmer)

The main offices of this construction firm are located in Zürich. There was a related operation in Wildegg.

Gauge: 750 mm   Locn: ?   Date: 05.2002

TOGGENBURG	0-4-0T	Bn2t	O&K 50 PS	1897	1906	(a)	(1)
WEISSENSTEIN	0-4-0T	Bn2t	Krauss-S IV bh	5295	1905	(a)	(2)
	0-4-0T	Bn2t	Zobel	569	1909	(c)	(3)

(a) purchased from the building site for the Kraftwerk Rupperswil-Auenstein, Rupperswil /1948.
(c) purchased from Franz Stirnimann, Olten-Hammer /1948 (Moser refers to 589).

(1) 1950 removed from boiler register.
(2) 195x removed from boiler register.
(3) 1953 removed from boiler register.

VEREINIGTE SCHWEIZER RHEINSALINEN, Möhlin   AG103

formerly: Schweizer Saline, Rheinfelden (1874-1909)

The salt extraction process is owned by all the Cantons, with the exception of Vaud (which has its own salt supply). The concern operated salines at Kaiseraugst (1843-1909) and at Schweizerhalle (from 1837 till the present day in BL). The installation at Rheinfelden (6278 2682) was worked 1845-1942; but since then only Riburg (in use since 1848) remains in operation.

Saline Riburg

The works at Riburg have been connected to the SBB by a branch from Möhlin station since 1891. The first locomotive arrived in 1892. The branch now also serves the railway wagon builder Meyer AG.

Gauge: 1435 mm   Locn: 214 $ 6293/2684   Date: 04.2000

	2	0-4-0T	Bn2t	Krauss-S XVIII bb	2605	1891	new	(1)
E 2/2	1	0-4-0T	Bn2t	SLM E 2/2	2532	1915	new	1958 scr
	1 (ex 3) RIBURGERLI	4wDM	Bdm	Kronenberg D 250/300	143	1958	new	
	2	0-6-0DH	Cdh	MaK G 500 C	500042	1966	new	(4)
HERKULES		6wDH	Cdh	JW DH660C54	3.684.080	1982	(e)	
Em 847 901-6		BBDH	BBdh	U23A LDH370	23879	1979	(f)	(6)

(e)   purchased from ÖMV Zentraltanklager, Lobau [A] "HERKULES" /2000 via Gmeinder/LSB.
(f)   hired from LSB /2001.

(1)   1954 OOU; ?1958 scrapped.
(4)   2000 sold to LSB; 2001 to VSFT (as a hire locomotive).
(6)   2001 returned to LSB.

VEREIN MIKADO 1244, Brugg AG   AG104 M

A preservation group with a base in the locomotive shed at Brugg AG, previously at Rapperswil SG. SBB-Historic also base stock here.

Gauge: 1435 mm   Locn: 225 $ 6577/2587   Date: 09.2007

141.R.1244	2-8-2	1D1h2	MLW 141.R	75053	1946	(a)
Ae 4/7 11026	2Do1WE	2D1w4	SLM,BBC Ae 4/7	3547,3723	1932/33	(b)
Tm^{II} 813	4wDM	Bdm	RACO 95 SA3 RS	1744	1965	(c)

(a)   ex SNCF "141.R.1244" /1975.
(b)   new to SBB "Ae 4/7 11026"; wdn 30/11/1996; ex Adtranz /2001.
(c)   ex SBB Tm^{II} 813 /2003.

see: www.mikado1244.ch

VEREIN 241.A.65, Full-Reuenthal   AG105 M

The society operates a SNCF "Mountain" on chartered trains. *The stock is kept on the old CFU site at Full (to be confirmed).* Address: Seetalstrasse 4, 5706 Boniswil.

Gauge: 1435 mm   Locn: 215 $ 6566/2732 (to be confirmed)   Date: 09.2008

241.A.65	4-8-2 4cc	2D1h4v	Fives-Lille	4714	1931	(a)
Roger		4wDM	Bdm	RACO 95 SA3 RS	1620	1961	(b)

(a)   new to C.F. de Etat "241.001"; imported to Switzerland by H. Glaser by /1968; ex Oswald Steam, Samstagern (OSS) about /1992.
(b)   ex SBB "Tm^{II} 717".

WASSERBAUAMT, Frick — AG106

Gauge: 500 mm Locn: ? Date: 02.2008

950 or 951		4wDM	Bdm	Jung EL 105	5799	1934	(a) (1)

(a) new via U. Ammann, Langenthal.
(1) by 1999 to FWF, Otelfingen (600 mm).

WASSER-KRAFTWERK-BAU (WKW), Laufenburg — AG107 C

A locomotive used in connection with building the Laufenburg power station.

Gauge: 600 mm Locn: ? Date: 04.2002

LUCIA		0-4-0T	Bn2t	Heilbronn III	477	1907	(a) (1)

(a) made available by G. Bampi, Bauunternehmung, Rheinfelden [D].
(1) 19xx sold to Kurt Stein, Feld- und Industriebahnen, Essen [D].

X. WIEDERKEHR AG, Waltenschwil — AG108

A metal recycling plant (Shredder- und Scherwerk) opened in 1960 and rail connected since 1994.

Gauge: 1435 mm Locn: 225 $ 6647/2425 Date: 03.2006

Xaver		4wDH	Bdh	Schöma CFL 250DVR	4452	1994	(a)

(a) new as CFL 200DVR; ex Michelin Reifenwerke, Bad Kreuznach [D] /1994 via LSB.

ROBERT WILD AG, Muri AG — AG109

This was a mechanical workshop. The location given may not be entirely accurate.

Gauge: 1435 mm Locn: 225 $ 6682/2355 Date: 12.1988

		4wPM	Bbm	Breuer III	?	1929	(a) a1980 s/s

(a) purchased from SBB "Tm 452" 1/1962.

ZEMENT- & KALKWERK SCHINZNACH BAD, Schinznach Bad — AG110

A chalk works, bankrupt in 1936.

Gauge: 600 mm Locn: 215 $ 6559/2654 Date: 05.2002

		4wDM	Bdm	O&K(M) RL 1a	4005	1930	new (1)

(1) sold to Jurabergwerke AG, Herznach.

ZIEGELEI OTTO BENZ, Birmenstorf AG — AG111

A brickworks opened in 1938 and closed in 1967. The connection from the works to the clay pit (location: 215 $ 6608 2575) was made with a cable worked incline and an adhesion line of some hundred metres.

Gauge: ? mm Locn: 215 $ 6650/2574 Date: 05.2002

| | 4wDM | Bdm | O&K(S) RL 1c | 1505 | 194x | (a) | s/s |

(a) purchased via O&K/MBA, Zürich.

ZIEGELEI RHEINFELDEN (JOSEF BAUMER), Rheinfelden — AG112

A brickworks.

Gauge: 600 mm Locn: ? Date: 05.2002

| | 4wDM | Bdm | O&K(M) RL 1 | 3474 | 1929 | new | (1) |

(1) 19xx sold to Stahlton AG, Frick.

ZUCKERMÜHLE RUPPERSWIL AG, Rupperswil — AG113

A factory producing lump sugar.

Gauge: 1435 mm Locn: 224 $ 6517/2504 Date: 06.1995

| Tm | 4wDM | Bdm | RACO 60 LA7 | 1500 | 1958 | (a) |

(a) new to SBB "Tm 305"; ex SBB "Tm 894" /1994.

ZÜRCHER ZIEGELEIEN AG, Schinznach Dorf — AG114

A short line linked the quarry with the staithes located at the SBB Schinznach Dorf station. The line closed ca. 1999. A private operation has since been started using the railway, but not the staithes.

Gauge: 600 mm Locn: 214 $ 6522/2564 Date: 04.1996

| TANTE EMMI | 4wDH | Bdm | Diema DS90/1 | 2987 | 1968 | (a) | (1) |

(a) new via Asper, name "TANTE EMMI" added between 1981 and 1996. There are reports that in /1974-75 the locomotive was replaced by a similar one. If so the plates were transferred from the first to the second. Some modifications to the frames were made - but after 1981.

(1) 2005 to Schinznacher Baumschulbahn (SchBB), Schinznach Dorf.

ZSCHOKKE-WARTMANN, Brugg AG — AG115

formerly: Zschokke-Wartmann AG (till 1970)
Wartmann & Valette (from 1899)
Wartmann & Cie (1896-99)

A building firm founded in 1896, the buildings, site and remaining locomotive were incorporated in 1971 into the Kabelwerke Brugg AG.

Gauge: 1435 mm Locn: 215 $ 6581/2591 Date: 10.1996

| 31 | 4wBE | Ba1 | unknown,BBC | ? | 19xx | (1) |

(1) 1971 to Kabelwerke Brugg AG, Brugg AG.

Appenzell (Innerrhoden + Ausserrhoden)   AI+AR

Appenzell - Innerrhoden   AI

DAMPF-LOKI-CLUB (DLC), Appenzell   AI1 M

A preservation group that operates tourist trains over public railways in the region. See also the entry under Herisau in Kanton Appenzell - Ausserrhoden.

Gauge: 1000 mm   Locn: ?   Date: 09.1969

VEREIN HISTORISCHE APPENZELLER BAHNEN, Wasserauen
    AI2 M

This concern operates a heritage train as required.

Gauge: 1000 mm   Locn: ?   Date: 09.1969

CFe3/3 2 (ex SGA CFeh 2/3 17)					
	6wWERRC	SIG,SLM,EGA	?	1911	(a)

(a) bearing "ALTSTAETTEN-GAIS" identification.

Appenzell - Ausserrhoden   AR

APPENZELLER BAHNEN (AB), Gais   AR1 M

formerly: Appenzeller Bahn (AB)

The heritage stock includes the following motor coach.

Gauge: 1000 mm   Locn: ?   Date: 10.2007

AB CFe4/4 5	BoBoWERRC BbBbeg2			
		SLM,SIG,BBC BCFeh 4/4		
			3463,3366	1930

DAMPF-LOKI-CLUB (DLC)   AR2 M

A preservation group that operates tourist trains over public railways in the region. See also the entry under Appenzell Inner Rhoden. There are three separate depots. Metre gauge stock is kept in the old AB depot at Herisau, while the standard gauge stock is kept in the old SOB one, where the SOB also keeps some of its own heritage stock. Metre gauge stock has also been based in Appenzell.

Gauge: 1435 mm   Locn: 217 $ 739220,250670   Date: 09.1968

BT Eb3/5 9	2-6-2T	1C1h2t	Maffei	3129	1910	(a)
BT Be3/4 43 556 013-1	1A'BoWERC		SIG,SAAS	?	1938	(b)

Gauge: 1000 mm   Locn: 217 $ 739137,250600   Date: 04.2008

G 3/4 14	2-6-0T	1Cn2t	SLM G 3/4	1479	1902	(c)
AB BCFm 2/4 56	Bo2DERC		SIG,MFO	?	1929	(d)
AB BCe 4/4 30	BoBoWERC		SIG,MFO	?	1933	(e)

(a)   new to BT "9"; later SBB "5889"; on loan or purchased from BT "9" /1985.
(b)   new to BT "Be3/4 43", on loan or purchased from SOB /200x as "Be 556 013-1".

(c) ex RhB "14".
(d) ex AB "BCFm 2/4 26".
(e) renumbered AB "BCe 4/4 43" /195x - /200x.

EUROPÄISCHE VEREINGUNG VON EISENBAHNFREUNDEN FÜR DEN ERHALT VON DAMPLOK (EUROVAPOR), Sektion Sulgen/Rorschach, Heiden   AR3 M

see under Appenzeller Bahnen (Rorschach-Heiden-Bergbahn), Rorschach

KUMMLER + MATTER AG, Heiden   AR4 C

see: Appenzeller Bahnen (Rorschach-Heiden-Bergbahn), Rorschach

PREISIG AG, Teufen AR   AR5 C

formerly: Paul Preisig, Stein auf der Taufe (till 1956)
later: Paul Preisig AG (from 1/1/2001)

A public works company. It started in 1945 as a family concern with head offices in Stein auf der Taufe and a subsidiary in Teufen. It was reorganised in 1956 to become a company with headquarters in Teufen and a subsidiary in Stein AR. Other subsidiaries have been opened and closed as required. The location given is that of the offices in 2007. Address: Hauptstrasse 39, 9053 Teufen + Preisig AG Schachen 62, 9063 Stein.

Gauge: 600 mm   Locn: 217 $ 7265/2506   Date: 07.1999

4wDM	Bdm	O&K(M) MD 2	?	19xx		(1)
4wDM	Bdm	O&K(M) LD 2	8400	1937	(b)	(2)
4wDM	Bdm	O&K(M) RL 1c	8422	1937	(c)	(1)

(b) existence at Teufen AR not confirmed, the locomotive may have been at another branch of Preisig AG; ex Bauamt Leichtenstein [FL] /19xx. This locomotive is probably the one illustrated in the referenced Web-site.
(c) ex Prader & Co, Zürich (as RL 1c).
(1) 198x to Galli Egli, Besazio.
(2) 1956 resold to Bauamt Leichtenstein [FL].

see: www.preisigbau.ch/firma/firmengeschichte/p1_3_3.php

SCHWEIZERISCHE SÜDOSTBAHN AG (SOB), Herisau   AR6 M

Heritage stock of this Samstagern based company may be found in the old depot at Herisau. This, shared by the Dampf-Loki Club (DLC), is adjacent to the main depot.

Gauge: 1435 mm   Locn: 217 $ 739220,250670   Date: 10.2008

Be 4/4 BT-11	BoBoWE	BBw4	SLM,SAAS Be 4/4	3509,2080-1	1931

Bern   BE

In 1979 the Canton of Jura was created from the western part of this Canton.

AMEBA AG, Konolfingen   BE001 C

This track maintenance firm has its head offices in Bauerstrasse 126, Zürich, but the workshops are in the old EBT workshops in Konolfingen. Address: Bernstrasse 8, 3510 Konolfingen.

Gauge: 1435 mm   Locn: 233 $ 6135/1920   Date: 05.2000

	4wDM	Bdm	O&K(M) H 2	2385	1927	(a)	(1)

(a)   ex Berner Alpenmilchgesellschaft, Konolfingen /1990-97.
(1)   2001 to LSB; 2003 to Bahn Museum Kerzers (BMK), Kerzers.

ALBERT BAUER, Lüsslingen   BE002 M

Hr. Bauer started a collection and to construct a little of his projected railway. However the plans ran foul of nature conservation policies and the scheme was abandoned.

Gauge: 600 mm   Locn: 233 $ 6045/2269   Date: 05.2002

0-4-0WT	Bn2t	O&K 50 PS	7479	1918	(a)	(1)
4wDM	Bdm	JW JW8	225	1954	(b)	(2)

(a)   ex Kieswerk A. Bauer, Kissing [D] /1965.
(b)   ex Ciments de Saint-Ursanne /19xx.
(1)   by 2002 donated to Schinznacher Baumschulbahn (SchBB), Schinznach Dorf.
(2)   1996 to FWF, Otelfingen.

ARGE UNTERTAGEBAU FRUTIGER AG, Kandersteg   BE003 C

This was a specialised contract involving the firm of Frutiger AG of Uetendorf. Full details of the consortium are not known.

Gauge: 900 mm   Locn: ?   Date: 05.2009

4wDH	Bdh	Schöma CFL 100DCL	5197	1991	new	(1)
4wDH	Bdh	Schöma CFL 100DCL	5198	1991	new	(2)

(1)   probably returned to Schöma or s/s
(2)   probably returned to Schöma; 2006 to MCG Malmö City Tunnel Group HB, Malmö [S].

ARGE HONDRICH-TUNNEL-NORD, Hondrich   BE004 C

The consortium Rothpelz-Lienhard-Zschokke performed work connected with replacing a single tunnel of the BLS with a double track section. Diema records indicate that the locomotives were converted to 800 mm gauge, but the gauge of 750 mm is usually quoted.

Gauge: 750 mm   Locn: ?   Date: 06.2008

4wDH	Bdh	Diema DFL 90/0.2	2360	1960	(a)	(1)
4wDH	*Bdh*	*Diema DFL 90/0.2*	*2715*	*1964*	(b)	s/s
4wDH	Bdh	Diema DFL 90/0.2	2729	1964	(c)	s/s

(a)   new to Munitions-Depot Aurich-Tannenhausen [D] "2" as DS90 600 mm; rebuilt by Diema to DFL 90 800 mm /1981; ex Arge Bietschtal "1" /1984.
(b)   new to Munitions-Depot Aurich-Tannenhausen [D] "8" as DS90 600 mm; rebuilt by Diema to DFL 90/0.2 /1981; not all sources give this locomotive as coming here.

(c) new to Munitions-Depot Aurich-Tannenhausen [D] "9" as DS90 600 mm; rebuilt by Diema to DFL 90/0.2 800 mm /1981; ex Baustelle Baltschieder /1984.
(1) 198x to Conrad Zschokke AG, Genève.

ARGE PIERRE PERTUIS, Sonceboz BE005 C

Murer AG, Erstfeld made at least one locomotive available to this consortium.

Gauge: 750 mm Locn: ? Date: 12.2007

1675	4wDH	Bdh	Schöma CFL 180DCL	4999	1989	new	(1)

(1) to Associazione Lavori Piora, Faido.

BAU GRIMSEL STAUMAUER AG, Meiringen BE006 C

The firm carried out the Grimsel Barrage construction 1927-31.

Gauge: 900 mm Locn: ? Date: 03.2002

	0-4-0T	B?2t	Henschel Brauns	17581	1920	(a)	(1)
	0-4-0T	B?2t	Henschel Helfmann	17603	1921	(b)	(2)
	0-4-0T	B?2t	Krauss-S VXI de	7759	1920	(c)	(3)
	0-4-0T	B?2t	Krauss-S VXI de	7760	1920	(d)	(3)
	0-4-0T	B?2t	Krauss-S VXI dl	7884	1921	(e)	(5)
	0-4-0T	B?2t	Krauss-S VXI dl	7889	1921	(e)	(3)
VALLONE	0-4-0T	B?2t	Maffei	3609	1911	(g)	(7)
	0-4-0T	Bn2t	O&K 80 PS	1247	1904	(h)	(8)

(a) ex H. Elias & Co., Dortmund [D] /1928.
(b) ex Glaser & Pflaum, Berlin [D] /1928.
(c) ex Edwards & Hummel-A. Kunz, München [D] /1927.
(d) ex Edwards & Hummel-A. Kunz, München [D] /1928.
(e) ex Futter, Hirsch & Co., Berlin [D] /1928.
(g) see note for this locomotive under Robert Aebi & Cie AG, Regensdorf; ex Impresa Ing. Antognini & Noli, Bellinzona (750 mm) /1927.
(h) ex H. & F. Frutiger & Lanzrein, Bern (750 mm) /1927.

(1) 1929 returned to Germany.
(2) 1931 returned to Germany or s/s (reports differ).
(3) 1931 returned to Germany.
(5) 1931 removed from boiler register.
(7) 1932 to Schafir & Mugglin, Klingnau.
(8) 1932 or 1933 to Schafir & Mugglin, Klingnau (750 mm).

BAUKONSORTIUM, Zweilütschinen BE007 C

Probably a public works concern.

Gauge: 600 mm Locn: ? Date: 05.2002

4wDM	Bdm	O&K(M) LD 2	8341	1937	(a)	(1)

(a) purchased via O&K/MBA, Dübendorf.
(1) to Hatt-Haller AG, Zürich.

BERNISCHE BRAUNKOHLENGESELLSCHAFT AG, Gondiswil
BE008

There was a short-lived lignite mining operation at the end of World War I. It appears that plans were made to restart it in 1945 at the end of World War II, but there is no indication that that was done although a locomotive was earmarked for it.

Gauge: 750 mm			Locn: ?			Date: 05.2002		
SAN PAOLO		0-4-0T	Bn2t	O&K 50 PS	4347	1911	(a)	(1)
(a)	ex Fritz Marti, Bern /1918.							
(1)	1919 to Antognini & Noli, Bellinzona.							

BERNISCHE KRAFTWERKE AG, Bern — BE009 C

The railway was used ca. 1923 for the building of the Mühleberg power station, 14 kilometres west of Bern.

Gauge: 750 mm			Locn: ?			Date: 03.2002		
1	0-4-0T	Bn2t	O&K 20 PS	690	1901	(a)	(1)	
2	0-4-0T	B?2t	Maffei	3513	1911	(b)	(2)	
3	0-4-0T	Bn2t	O&K 50 PS	1136	1903	(c)	(3)	
7	0-4-0T	Bn2t	Borsig	7842	1911	(d)	(4)	

(a) ex Fritz Marti, Bern /1922.
(b) see note for this locomotive under Robert Aebi & Cie AG, Regensdorf; owned by Fritz Marti, Bern from /1919; ex Dauster (details unknown) /1922-23.
(c) ex Jardini, Basel /1923-27.
(d) ex K. Schmidt, Hannover [D] /1920.
(1) 1924 to Losinger & Cie, Bern.
(2) 1923-32 to E. Bellorini + Oyex, Chessex & Cie, Lausanne.
(3) 1923-27 to Robert Aebi & Co, Zürich.
(4) 1923-28 to Robert Aebi & Co, Zürich.

BERN-LÖTSCHBERG-SIMPLON-BAHN (BLS), Spiez — BE010 M

Heritage stock includes the following locomotives.

Gauge: 1435 mm			Locn: ?			Date: 02.2008		
BLS Ae 6/8 205	1CoCo1WE							
	1CC1ew12							
			SLM,SAAS Ae 6/8	3678, ?	1939			
3	0-6-0WT	Cn2t	SLM E 3/3	1332	1901	(b)	(B)	
BLS Tm 64	4wDH	Bdh	BLS	-	1962	(c)		
Ae 4/4 251	BoBoWE	BBw4	SLM,BBC Ae 4/4	3883,4507	1944			
Ae 8/8 273	BoBo+BoBoWE							
			BBBBw8 SLM,BBC Ae 8/8	4443,6407	1963			

(b) new to Gürbetalbahn "3"; ex Zellulosefabrik, Attisholz "26" /1977.
(c) ex BLS "Tm 64"; with SLM motor and hydrostatic transmission.
(B) on loan to Verein Dampfbahn Bern (DBB), Burgdorf.

BERNMOBIL, Burgernziel — BE011 M

formerly: Städtische Verkehrsbetriebe Bern (SVB)
Strassenbahn Bern
Berner Tramway Gesellschaft

The heritage stock is kept in the Burgenziel tram depot and includes one locomotive in addition to several trams. Address: Thunstrasse, Bern.

Gauge: 1000 mm				Locn: ?			Date: 02.2008	
G 3/3	12	0-6-0Tm	Cn2t	SLM G 3/3	863	1894	(a)	

(a) 1908 to Renfer & Co. AG, Biel-Bözingen; returned from Technorama, Winterthur /1998.

M.H. BEZZOLA AG, Biel — BE012 C

A public works company.

Gauge: 600 mm Locn: ? Date: 05.2005

| | 4wDM | Bdm | O&K MV0 | 25102 | 1951 | (a) | (1) |

(a) ex UTAS (or UTALS), Brig /1953.
(1) by 2005 to Schumacher AG, Ziegelei Körbligen, Gisikon.

BRISSARD & CRÉPEL, Biel — BE013 C

This contractor participated in the construction of the Biel-La Neuville section of railway.

Gauge: 1435 mm Locn: 242 $ 5758/1956 Date: 05.2008

| L'ARDENNAISE | 0-4+4 | B2n2 | Esslingen Engerth | 481 | 1859 | (a) | (1) |

(a) from work on the Ardennenbahn [B] /1860.
(1) 1861 to SCB "56 LINTH".

BÜHLMANN RECYCLING AG, Münchenwiler — BE014

formerly: Bühlmann Alteisen AG (1985-2006)

A scrap merchant.

Gauge: 1435 mm Locn: 242 $ 5758/1956 Date: 12.2004

| Em 2/2 | 4wDH | Bdh | Moyse BN24HA150GM | 1262 | 1973 | (a) |

(a) ex Karton Deisswil AG, Stettlen /1994.

F. & A. BÜRGI, Bern — BE015 C

The firm may have been engaged in construction.

Gauge: 750 mm Locn: ? Date: 03.2002

| | 0-4-0T | Bn2t | O&K 40 PS | 717 | 1900 | (a) | (1) |

(a) ex Ing. Galli & Co (which branch?) /1900-05.
(1) 1905-24 to Löscher & Gottlieb, Aarau.

BUNDESTANKANLAGEN (BTA) — BE016

Huttwil

Fuel store administered by Carbura, Zürich, closed c2000 and later reused by EUROVAPOR.

Gauge: 1435 mm Locn: 234 $ 6294/2176 Date: 08.1995

| 3 | 4wDH | Bdh | RACO 80 ST4 | 1408 | 1951 | (a) | (1) |
| | 0-4-0DH | Bdh | Henschel DH 240B | 29968 | 1959 | (b) | (2) |

Zollikofen

This was a fuel oil store operated by Carbura, Zürich. Built 1940 and dismantled 2003.

Gauge: 1435 mm Locn: 243 $ 6021/2065 Date: 08.2003

	6	4wDM	Bdm	Kronenberg DT 90/120	123	1952	(c)	(3)
Tm	6	4wDM	Bdm	Kronenberg DT 90/120	127	1952	new	(4)
		4wDM	Bdm	RACO 95 SA3 RS	1259	1940	new	(5)

(a) ex BTA, Zollikofen /19xx.
(b) new to Kraftwerk Kassel, Kassel, D "3"; via OnRail and LSB /1991.
(c) transferred from OKK, Interlaken West /1982.
(1) 1991 to LSB; Classic Rail, Le Locle.
(2) 2000 to RM "Tm 236 340-6".
(3) 200x disposal unknown.
(4) 2000 privately owned (Herr M. Eichenbergerr, Hasle-Rüegsau).
(5) 1955 to OKK, Laupen.

CIMENTS VIGIER SA, Péry   BE017

Part of the same group as the Fabrique de Chaux et Ciments Pecks (opened 1860, purchased by Vigier in 1902 and closed c1930) at Rondchâtel.
Located at Reuchenette-Péry Gare this factory started production in 1890 and was rail connected in 1895. Traffic to Rondchâtel SA used the same branch line. The narrow gauge system was closed by c1970.

Gauge: 1435 mm   Locn: 233 $ 5877/2261   Date: 04.2005

1	4wBE	Ba2	SIG,MFO	?	1942	new	1973 scr
2	0-4-0T	Bh2t	SLM Ed 2/2	2590	1917	(b)	1961 scr
2	0-6-0WT	Cn2t	SLM E 3/3	1974	1909	(c)	(3)
3	0-6-0WT	Cn2t	SLM E 3/3	2341	1913	(d)	(4)
"4"	4wDH	Bdh	SLM Tm^{IV}	4951	1973	new	
"5"	4wDH	Bdh	SLM Tm^{IV}	4982	1973	new	

Gauge: 750 mm

4wDM	Bdm	O&K MD 2b	25125	1951	(e)	s/s

(b) purchased from Uerikon-Bauma-Bahn (SBB) "23" /1950.
(c) purchased from SBB "E 3/3 8494" /1961.
(d) purchased from SBB "E 3/3 8518" /1965.
(e) locomotive built by Schöma as their 1230; new via Wander-Wendel/MBA, Dübendorf.
(3) 1973 to Alusuisse SA, Chippis "36".
(4) 1974 to DVZO.

CLUB SALON BLEU, Spiez   BE018 M

The club maintains a collection of mainline locomotives and railcars from the BLS and EBT and their predecessors. Not all locomotives are at Spiez; some are located at Huttwil or Interlaken West.

Gauge: 1435 mm   Locn: 253 $ 6189/1701   Date: 03.2008

Te 2/3 31	2-4-0WE 1Bw1	SLM,MFO,BLS	2998, ? 1925(56)	(a)		
RM Be 4/4 102	BoBoWE BBw4	SLM,SAAS Be 4/4	3552, ?	1932	(b)	
BLS Ce 4/6 307	2-4-4-2WE 1BB1w2	SLM,MFO Ce 4/6	2689, ?	1920	(c)	
BLS Ce 4/4 313	0-4-4-0WE BBw2	SLM,BBC	2695,1321 1919/20	(d)	(4)	
RM BDe 2/4 240	Bo2WERC B2w2	SIG,SWS,BBC	3851	1932	(e)	
RM De 4/4 257	BoBoWERC					
	BBw4	SIG,SWS,BBC	3846 1932(81)	(f)		
RM De 4/4 258	BoBoWERC					
	BBw4	SIG,SWS,BBC	3844 1932(81)	(g)	(4)	
VHB De 4/4 259	BoBoWERC					
	BBw4	SIG,SWS,BBC	3850 1932(81)	(h)		

(a) new to BSB "CFe 2/5 785"; rebuilt BLS "Te 2/3 32"; renumbered /1995; ex BLS "Te 2/3 31" /2002.
(b) ex RM "Be 4/4 102" /2006.
(c) new to "SEZ Ce 4/6 307"; indefinite loan from BLS /2003.
(d) new to "GBS Ce 4/6 313"; ex BLS after /2000.
(e) new to EB "CFe 2/4 124"; EBT "BDe 2/4 224" /1963; EBT "232" /1981; EBT "240" /1985; ex RM /2000.

(f) new to Bergdorf-Thun Bahn (BTB) "CFe 2/4 130"; EBT "230" /1963; EBT "De 4/4 236" /1981; RM "De 4/4 257" /1997; ex RM /2005.
(g) new to Bergdorf-Thun Bahn (BTB) "CFe 2/4 128"; EBT "223" /1963; VHB "De 4/4 266" /1981; RM "De 4/4 258" /1997, ex RM /2005.
(h) new to EB "CFe 2/4 123"; EBT "228" /1963; VHB "De 4/4 267" /1981; RM "De 4/4 259" /1997; ex RM /2005.
(4) 22/09/2008 for scrap.

see: www.salonbleu.ch

COOP, Bern-Riedbach   BE019

A distribution centre near Niederbottigen and rail connected to Bern-Bümpliz Nord.
Address: Riedbachstrasse 165, 3027 Bern.

Gauge: 1435 mm   Locn: 243 $ 5943/1991   Date: 10.2007

Tm^1 496   4wDM   Bdm   RACO 85 LA7   1690   1964   (a)

(a) ex SBB "Tm^1 496" /2004 (withdrawn 29/2/2004).

DAMPFLOK-FREUNDE LANGENTHAL, Huttwil   BE020 M

A group of three families has been responsible since 1966 for the maintenance of a steam locomotive and its associated train: the operating base is Huttwil depot.

Gauge: 1435 mm   Locn: 234 $ 6304/2182   Date: 09.1967

2   2-6-0T   1Cn2t   SLM Ed 3/4   1799   1907   (a)
         0-4-0PM   Bbm   RACO 45 PS   1212   1933   (b)   (2)
         0-4-0PM   Bbm   O&K(M) L308   1520   1922   (c)   (3)

(a) new to SMB "Ed 3/4 2" ;1932 to Gaswerk Zürich, Schlieren; /1946 to Holzverzuckerungs AG (HOVAG), Ems; withdrawn /1971; 1973 to this group.
(b) new to SBB "Tm 505"; ex Stahlrohr, Rothrist /1986.
(c) ex Mühlen AG, Interlaken /1994.
(2) experimentally converted to diesel, then scrapped
(3) 2005 to Bahn Museum Kerzers (BMK), Kerzers.

see: www.dfl-langenthal.ch
see: Eisenbahn Amateur 12/73 p.543.

DHL SOLUTION, Ostermundigen   BE021

formerly: SWISSCOM, Ostermundigen
            Schweizerische Post-, Telephon- und Telegraphenbetriebe (PTT) (1985 till 31/12/1997)

Telecommunication service stores.

Gauge: 1435 mm   Locn: 243 $ 6026/2006   Date: 07.2005

12 ALEX   4wDH   Bdh   Schöma CFL 80DR   2151   1958   (a)   (1)

(a) ex Hamburger Gaswerke, Kokerei Kattwyk [D] /1983, via Asper.
(1) 2005-6 Stauffer Schienen- und Spezialfahrzeuge, Schlieren; 2007 to HOLCIM, Schwarzenbach SG (Jonschwil).

Die Schweizerische Post, Bern   BE022

formerly: (Schweizerische) Post-, Telephon- und Telegraphenbetriebe (1985 till 31/12/1997)

Rail served postal centre attached to the Hauptbahnhof and sometimes referred to as Bern-Schanze. The facility closed on 7.3.2009.

	Gauge: 1435 mm			Locn: 243 $ 5998/1996			Date: 12.2007	
	5	0-4-0WE	Bw2	SLM,MFO Te[III]	4582, ?	1965	new	(1)
	9	6wWE	Cw3	SLM,BBC Ee 3/3	5287, ?	1985	(b)	(2)
	10	6wWE	Cw3	SLM,BBC Ee 3/3	5288, ?	1985	(c)	(3)
	11	6wWE	Cw3	SLM,BBC Ee 3/3	5289, ?	1985	new	(2
	14	6wWE	Cw3	SLM,BBC Ee 3/3	5467, ?	1992	(b)	(2)

(b) ex Die Schweizerische Post, Zürich-Mülligen, /2007.
(c) ex Die Schweizerische Post, Däniken SO /2000.
(1) 1986 to PTT, Lausanne "5".
(2) 2009 to RM workshops, Oberburg.
(3) 2008 to RM workshops, Oberburg.

EIDGENÖSSISCHE PULVERFABRIK (P+F), Eifeld, Wimmis   BE023

Powder factory, rail connected in 1917. The branch has been worked by the SEZ since 1989.

	Gauge: 1435 mm			Locn: 253 $ 6157/1706			Date: 08.1998	
		1ABE	1Aa1	BBC Akku-Fahrz.	1214	1918	new	s/s
	1	4wDM	Bdm	Kronenberg DT 60P	103	1940	new	1971 scr
	2	4wDE	Bde2	Stadler	103	1953	new	(3)

(3) 1989 to SEZ "Tm 75"; 2000 to Verein Dampfbahn Bern (DBB).

ENTREPRISE GÉNÉRALE DU LÖTSCHBERG (EGL), Frutigen + Brig   BE024

A consortium of civil engineering firms formed for the building of the Frutigen-Brig line including the Lötschberg tunnel. Formed in 1906 it was dissolved in 1913. The northern base was located in Kandergrund and the southern one at Naters. For more details see Canton Valais.

Gauge: 750 mm			Locn: ?			Date: 10.2001	
1	0-8-0T	Dn2t	O&K 150 PS	2413	1907	new	s/s
2	0-8-0T	Dn2t	O&K 150 PS	2414	1907	new	s/s
3	0-8-0T	Dn2t	O&K 150 PS	2415	1907	new	(3)
4	0-8-0T	Dn2t	O&K 150 PS	2416	1907	new	(3)
5	0-6-0T	Cn2t	Borsig	6989	1908	new	(5)
6	0-6-0T	Cn2t	Borsig	6990	1908	new	(6)
7	0-6-0T	Cn2t	Borsig	6991	1908	new	(7)
8	0-6-0T	Cn2t	Borsig	6992	1908	new	(6)
11	0-4-0WT	Bn2t	O&K 50 PS	2187	1906	new	(9)
12?	0-4-0WT	Bn2t	O&K 50 PS	2197	1908	new	s/s
13?	0-4-0WT	Bn2t	O&K 50 PS	2449	1908	new	(11)
14	0-4-0WT	Bn2t	O&K 50 PS	2448	1908	new	(12)
15?	0-4-0WT	Bn2t	Borsig	6485	1908	new	(6)
16	0-4-0WT	Bn2t	Borsig	6487	1908	new	s/s
17	0-4-0WT	Bn2t	Borsig	6486	1908	new	s/s
18?	0-4-0WT	Bn2t	Borsig	6488	1908	new	s/s
21	0-4-0CA	Bp2	Thébault	?	1908	(q)	s/s
22	?	?	Thébault	?	1908	(r)	s/s
23	?	?	Thébault	?	*1908*	(r)	s/s
24	?	?	Thébault	?	*1908*	(r)	s/s
25	*0-6-0CA*	Cp2	Thébault	?	*1908*	(q)	s/s
26	*0-6-0CA*	Cp2	Thébault	?	*1908*	(q)	s/s
27	*0-6-0CA*	Cp2	Thébault	?	*1908*	(q)	s/s
28	*0-6-0CA*	Cp2	Thébault	?	*1908*	(q)	s/s
31	0-8-0CCA	Dp2v	O&K 200 PS	2676	1907	new	s/s
32	0-8-0CCA	Dp2v	O&K 200 PS	2677	1907	new	s/s
34	*2-4-0CA*	2Bp2	Thébault	?	*1908*	new	s/s
36	0-8-0T	Dn2t	Maffei	*3725*	*1911*	(ab)	s/s
41	0-8-0T	Dn2t	O&K 200 PS	4053	1910	new	s/s

| | | 42 | 0-8-0T | Dn2t | O&K 200 PS | 4054 | 1910 | new | s/s |
| | | | 4wDM | Bdm | Oberursel | ? | 19xx | | s/s |

(q) new; unknown if with coupling rods or not.
(r) variously quoted as Bp2 or Cp2; two or three axles.
(ab) purchased new by Léon Chagnaud, Paris [F] for L'Estaque, Marseille [F] (760 mm) in 1907. Identity confusion here; Moser gives this as Maffei 3555 of 1910; according to Merte this was a 0-4-0T Bn2t; the 0-8-0T Dn2t supplied to Léon Chagnaud, Paris [F] being 3724, 3725, 3734 of 1911 and 3779 of 1912. The given numbers are postulates.

(3) to Léon Chagnaud & Fils [Algeria].
(5) to ? (possibly RUBAG?); 1939 to Sawmill Svalava [Carpathian Ruthenia] "9"; later MAV 390.201.
(6) 1914 to RUBAG.
(7) by 1921 to IRR.
(9) 1913 to département de construction du tunnel du Simplon II "2".
(11) probably 1913 to département de construction du tunnel du Simplon II; 1921 to Jacquet, Vallorbe.
(12) 1913 to département de construction du tunnel du Simplon II "4".

EUROPÄISCHE VEREINGUNG VON EISENBAHNFREUNDEN FÜR DEN ERHALT VON DAMPLOK (EUROVAPOR), Worblaufen

BE027 M

A now terminated operation on the Solothurn-Zollikofen-Bern-Bahn (SZB) (from 1984 RBS).

Gauge: 1000 mm   Locn: ?   Date: 04.2009

| | | 0-4-0T | Bn2t | KrMa KDL 10 | 17627 | 1949 | (a) | (1) |

(a) ex Mittelbadische Eisenbahn [D]"101" /1971.
(1) 2002 to Interessengemeinschaft historischer Schienenverkehr IHS, Selfkantbahn, Schierwaldenrath [D] "101".

FAVETTO, BOSSHARD, STEINER & CIE, Brienz BE

BE028 C

A public works company, probably builders of the Brienz-Interlaken line.

Gauge: 750 mm   Locn: ?   Date: 04.2002

| ENGADIN | 0-4-0T | Bn2t | Krauss-S IV zs | 4589 | 1901 | (a) | (1) |

(a) new to construction of RhB; ex Johannes Rüesch, St. Gallen /1912 or /1913.
(1) 1913-29 to Hunziker & Cie, Olten.

FELDSCHLÖSSCHEN-GETRÄNKE AG, Biel-Mett

BE029

A distribution centre with a shunting device for positioning wagons in a siding.

Gauge: 1435 mm   Locn: ?   Date: 11.2007

| | 4wPH | Bbh | Zagro Maxi Rangierer | 80063 | 9.2007 | new |

FISCHER & CIE AG, Langnau-im-Emmental

BE030

formerly: Rittmann et Cie

This firm trades in metals, construction material and fuels.

Gauge: 1435 mm   Locn: 244 $ 6269/1985   Date: 12.1988

| | 4wPM | Bbm | Breuer IV | ? | 1931 | (a) | (1) |

(a) new to SBB "Tm 416" later "Tm 892", ex SBB 1/1965.
(1) by 1968 present on the premises of Röthlisberger & Sohn, Langnau-im-Emmental; 1974 scrapped.

FRUTIGER & LANZREIN, Bern   BE031 C

also Frutiger, Lüthi & Lanzrein
A public works concern.

Gauge: 750 mm Locn: ? Date: 04.2002

0-4-0T	Bn2t	Märkische 60 PS	22	1893	(a)	(1)
0-4-0T	Bn2t	O&K 30 PS	375	1899	(b)	s/s
0-4-0T	Bn2t	O&K 80 PS	1247	1904	(c)	(3)
0-4-0T	Bn2t	O&K 50 PS	4347	1911	(d)	(4)
0-4-0T	Bn2t	O&K 70 PS	4375	1911	(e)	(5)
0-4-0T	B?2t	O&K 70 PS	6020	1914	(f)	(5)
0-4-0T	B?2t	O&K 50 PS	7520	1916	(g)	(7)

(a) ex RUBAG /1919-23.
(b) new to Krause & Co, Berlin [D]; ex Minder & Galli (which branch?) by /1925.
(c) new to J. + F. Heinke, Lege [D]; ex ? /1922-27.
(d) ex Dupont & Schaffner, Eggerberg to Frutiger, Lüthi & Lanzrein /1914.
(e) ex RUBAG /1922-26.
(f) new to Gebrüder Klose, Posen [D], ex ? /1914-26.
(g) new to Ostdeutsche Eisenbahn Gesellschaft, Königsberg [D]; ex ? /1916-22.

(1) 1923-29 to Losinger & Cie, Bern.
(3) 1922-27 to Grimsel Barrage construction, Meiringen.
(4) 1918 to Fritz Marti, Bern.
(5) 1926 to Schafir & Mugglin, Muri bei Bern.
(7) 1922-31 to Schafir & Mugglin, Muri bei Bern.

FRUTIGER SÖHNE AG, Uetendorf   BE032 C

alternatively: Frutiger AG, Thun

This civil engineering firm is located near Thun; locomotives have been used on contracts as required. See also ARGE Untertagebau Frutiger AG. The SIG list of references gives one ET 35L locomotive delivered here.

Gauge: 1435 mm Locn: 225 $ 6821 2492 Date: 04.2005

MO TM 511	4wDM	Bdm	RACO 45 PS	1248	1937	(a)	(1)

Gauge: 750 mm Locn: ? Date: 12.2001

10 AUENSTEIN	0-4-0T	Bn2t	SLM	3833	1943	(aa)	(27)
11 WILDEGG	0-4-0T	Bn2t	SLM	3834	1943	(aa)	(28)
12 RUPPERSWIL	0-4-0T	Bn2t	SLM	3835	1943	(aa)	(27)
	4wDM	Bdm	Brun FL 12	36	1943	new	(30)
	4wDM	Bdm	O&K MD 2b	25204	1951	(ae)	s/s
281.24.001	4wDM	Bdm	O&K MD 2b	25205	1951	(af)	(30)
	4wBE	Ba	Schalke,SSW EL 9	57610,6067	1960	(ag)	(33)

(a) hired from Asper /1994 (the unusual inscription was inherited from the locomotive's MO days.
(aa) new for Kraftwerk Rupperswil-Auenstein contracts.
(ae) new via Wander-Wendel and MBA, Dübendorf; locomotive built by Schöma as 1273.
(af) new via Wander-Wendel and MBA, Dübendorf; locomotive built by Schöma as 1274.
(ag) new via MBA.

(1) 1994 returned to Asper.
(27) to Messerli AG, Kaufdorf via Debrunner & Co AG.
(28) to Schuljungend, Turgi and displayed at SBB Bahnhof Turgi.
(30) 1983 to Ziegelei Schumacher, Körbligen, Gisikon (500 mm, later 600 mm).
(33) by 2003 to FWF, Otelfingen.

FURRER & FREY, Gwatt (Thun)   BE033 C

formerly: Furrer & Frey, Bern-Bümpliz Süd (till 1990)
Overhead construction and maintenance concern founded in 1923 with officrs in Bern. Address: Eisenbahnstrasse 62, 3645 Gwatt (Thun)

Gauge: 1435 mm   Locn: 243 $ 5963/1981   Date: 12.2002

1	A1DH	A1dh	Windhoff	2253	1973	(a)	
2	A1DH	A1dh	Windhoff	?	1974	(b)	
132	4wDH	Bdh	KHD A6M617 R	57309	1960	(c)	(3)
133	4wDH	Bdh	KHD A6M617 R	57338	1960	(d)	(4)
135	4wDH	Bdh	Gmeinder Köf III	5355	1965	(e)	(5)
	4wDH	Bdh	RACO 225 SV4 H	1870	1981	(f)	
	4wDH	Bdh	RACO 225 SV4 H	1881	1982	(g)	
	4wDH	Bdh	RACO 225 SV4 H	1928	1986	(h)	

Gauge: 1000 mm   Locn: RhB   Date: 03.2000

50	4wDH	Bdh	Gmeinder V12-16	5446	1985	(i)	
75	4wDH	Bdh	Gmeinder V12-16	5445	1985	(j)	
100	4wDH	Bdh	Schöma CFL 150 DCL	4807	1985	(k)	(11)

(a)   ex DB via NEWAG /1989.
(b)   ex DB via NEWAG /1995.
(c)   new to DB "Köf 6451"; ex DB "323 207-1" /1990.
(d)   new to DB "Köf 6480"; ex DB "323 235-2" /1991.
(e)   new to DB "Köf 11215"; ex DB "333 215-3" /2000.
(f)   ex SBB "Tm 9452" /2008.
(g)   ex SBB "Tm^{III} 9502" /2008.
(h)   ex SBB "Tm^{III} 9530" /2008.
(i)   ex Stadtwerke München, München [D] "8902" (1435 mm) via LSB /1997.
(j)   ex Stadtwerke München, München [D] "8901" (1435 mm) via LSB /1998.
(k)   new to Ferrocarril de Tajuña [E]; ex LSB 07/2000.
(3)   2003 to WBB/PACTON.
(4)   2003 to Stauffer; 29/10/2007 to Unirail; seen 10/2007 at Schweizerisches Militärmuseum (SMM), Full.
(5)   2006 to EUROVAPOR, Sektion Haltingen (Kandertalbahn) [D].
(11)   2004 to Chemin de fer de La Mure, Saint Georges de Commiers [F].

GASWERKE DER STADT BERN, Wabern bei Bern   BE034

Gasworks built in 1875, enlarged and rail connected between 1905 and 1906; final closure was in 1969. The branch from the mainline at Wabern bei Bern was about 2.5 kilometres in length.

Gauge: 1435 mm   Locn: 243 $ 6003/1986   Date: 06.1988

1		0-6-0WT	Cn2t	SLM		1901	1908	new (1)
	MUTZ	4wDE	Bde2	SIG,BBC		? ,6149	1960	new (2)

(1)   1969 donated to Verein Dampfbahn Bern (DBB).
(2)   1969 to Sihltalbahn (SiTB) "MUTZ 7".

GEMEINDE OSTERMUNDIGEN, Schule Ostermundigen   BE035 M

This locomotive is on display outside the school, adjacent to the Zollgasse bus stop. Address: Bernstrasse 60, Ostermundigen.

Gauge: 1435 mm + Riggenbach Rack   Locn: 243 $ 6038/2007   Date: 10.2007

HG 2/2 2 ELFE	0-4-0RT	Baz2t	Aarau	10	1876	(a)

(a)   new to Steinbruchgesellschaft Ostermundigen, Ostermundigen; sold to von Roll, Gerlafingen "6" /1907; returned for display /1981.

GRIBI & HASSLER, Burgdorf BE036 C

formerly: Gribi (ca. 1863)
Gribi & Wütrich (1874 - ca. 1905)

A public works concern.

Gauge: 750 mm (mainly)　　　Locn: ?　　　Date: 04.2002

1 WEHRDICH	0-4-0T	Bn2t	Olten	7	1863	(a)	s/s
2 KEHRDICH	0-4-0T	Bn2t	Olten	8	1863	(a)	s/s
	0-4-0T	Bn2t	SLM	35	1874	new	s/s
	0-4-0T	Bn2t	SLM	42	1874	new	s/s
	0-4-0T	Bn2t	SLM	49	1874	new	s/s
SPERANZA	0-4-0T	Bn2t	Krauss-S XXVII vv	3254	1895	(f)	(6)

(a) new, gauge not confirmed.
(f) ex Fischer, Schmutziger & Co, Zürich /1895-1905.
(6) 1905-08 to Buchser & Broggi, Degersheim.

HALTER ROHSTOFFE, Biel-Mett BE037

formerly: Halter Eisen + Metalle AG (till before 1999)

A scrap dealer.

Gauge: 1435 mm　　　Locn: 223 $ 5887/2230　　　Date: 03.2007

1	4wDM	Bdm	Kronenberg DT 60	118	1951	(a)	(1)
	4wDH	Bdh	Gmeinder Köf III	5391	1965	(b)	

(a) ex Feldmühle AG, Rorschach "Tm 1" /1978.
(b) new to DB "Köf 11 225"; ex DB "335 225-2" /1998.
(1) 1997-99 scrapped.

HEIMATMUSEUM SCHWARZENBURG BE038 M

Near the terminal station of the Gürbetal-Bern-Schwarzenburg section of the BLS.

Gauge: 1435 mm　　　Locn: ?　　　Date: 10.2007

Ed 3/4 51	2-6-0T	1Cn2t	SLM Ed 3/4	1726	1906	(a)	(1)

(a) new to Bern-Schwarzenburg-Bahn "51"; 1930 sold to Cementfabrik Holderbank-Wildegg AG; to Schwarzenburg for display /1970.
(1) 1998 to Verein Dampfbahn Bern (DBB).

HELLER BAU, Bern BE039 C

This may have been a civil engineering contract for a building of this name. Heller is also a common surname in Switzerland. In Bern in the 1950s there was a construction company with the name Heller Bauunternehmer, Bern. This might also have been the company involved.

Gauge: 600 mm　　　Locn: ?　　　Date: 10.2001

	4wDM	Bdm	Ruhrthaler G22Z	3257	1954	new s/s
	4wDM	Bdm	Ruhrthaler G22Z	3286	1955	new s/s

WALTER J. HELLER AG, Bern BE040 C

A public works company. One contract was the Vinglez Tunnel, near Biel (232 $ 5823/2193), completed in 1970. The SIG list of references gives one ET 20L5, one ETS 35 and two ETE 70 locomotives delivered here.

Gauge: 750 mm	Locn: ?	Date: 05.2002

4wDM	Bdm	O&K(M) MD 2	12009	194x	(a)	s/s
4wBE	Ba1	Stadler	24	1947	(b)	s/s
4wBE	Ba1	Stadler	36	8/1948	new	(3)

(a) new, via MBA, Dübendorf.
(b) new, delivered to Räterichboden (name of a work site).
(3) 1951 or 1952 to J. Seeberger, Frutigen BE.

HOCH & TIEFBAU AG   BE041 C
A public works company.

Biel/Bienne

Gauge: 600 mm	Locn: ?	Date: 05.2002

4wDM	Bdm	O&K(M) RL 1c	7576	1937	(a)	(1)

Interlaken

Gauge: 600 mm	Locn: ?	Date: 05.2002

4wDM	Bdm	O&K(M) LD 2	8341	1937	(b)	s/s

(a) purchased from MBA, Dübendorf /1951.
(b) ex Heinrich Hatt-Haller AG, other sources quote MD 2. The firm's head office is in Zürich; where the locomotive came from is not known.
(1) to Otto Schachtler, Ziegel- und Backsteinfabrik, Burgdorf.

E. HOFMANN & Co, Bern   BE042
also: Ziegelei Rehag AG
This is a tileworks near Bern. This concern imported from Altstadt Baugesellschaft, Walltrop [D] a 500 mm gauge Diema HK 5/1, number 3030 of 1968. (HK = Hydraulik-Kipper mit Eigenhydraulik). It is not known what sort of railway was operated.

HOLZWERKE RIEDER AG, St. Stephan   BE043
This sawmill was rail connected by a short spur to the MOB. Traffic was mainly standard gauge wagons on transporter wagons.

Gauge: 1000 mm	Locn: 263 $ 5972/1505	Date: 08.1988

4wDM	Bdm	O&K RL 2a	20200	1931	(a)	(1)

(a) new to La Grande Dixence SA (as 4wPM Bbm); ex MOB /1950.
(1) 1979 returned to MOB "Tm 1"; 1995 withdrawn; 2006 to BC.

GUSTAV HUNZIKER AG, Müntschemier   BE044
The full name of the firm and location is not confirmed. The O&K/MBA records refer to Hunziker, Münchemier. The above company (Baubedarf und Zementwaren Gustav Hunziker AG), with several locations in Switzerland, operated a sand and gravel pit in the area.

Gauge: ? mm	Locn: ?	Date: 05.2002

4wDM	Bdm	MBA MD 1	2010	c1950	(a)	s/s

(a) second-hand from MBA, Dübendorf /1954.

RUDOLF JENZER AG, Frutigen   BE045 C

A building contractor. Address: Tellenfeldstrasse 5, 3714 **Frutigen**.

Gauge: 600 mm   Locn: 253 $ 6164/1588   Date: 09.2008

		4wDM	Bdm	O&K (M) RL 1a	4210	1930	(a)	(1)
(a)	ex MBA, Dübendorf /1996.								
(1)	2006 donated to FWF, Otelfingen.								

JURAGEWÄSSER KORREKTION (JGK), Aarberg   BE046 C

The project was for work connected with modifying water channels in the Jura area, including work on the Aare-Hagneck canal. The locomotives were purchased by Gustave Bridel, the engineer for the JGK who was also responsible for the completion of the Gotthard line. The wagons were provided by the Olten workshops.

Gauge: 1435 mm   Locn: ?   Date: 12.1993

1		0-4-0T	Bn2t	Köchlin 68bis	1289	1870	(a)	(1)
2		0-4-0T	Bn2t	Köchlin 68bis	1290	1870	(a)	(1)
(a)	see text.								
(1)	1890 one locomotive offered for sale in Aarberg; s/s.								

KÄSTLI & SPYCHER, Bern   BE047 C

A public works concern. Probably related to Kästli AG Bauunternehmung, Bern und Rubigen.

Gauge: 500 or 600 mm   Locn: ?   Date: 03.2002

		4wBE	Ba1	Stadler	30	1951	(a)	(1)
(a)	ex F. Ramseier & Cie, Bern Wyler.								
(1)	sold second-hand to Ostermundigen or Bollingen.								

KARL KAUFMANN AG, Thörishaus   BE048

This scrap metal dealer has a yard alongside the mainline. Locomotives are present from time to time to be scrapped. A few, listed below, have been used on site.

Gauge: 1435 mm   Locn: 243 $ 5936/1945   Date: 03.2002

4wDM	Bdm	Kronenberg DT 120	117	1950	(a)	1975 scr
4wDM	Bdm	O&K MV4a	25701	1956	(b)	1992 scr
4wDH	Bdh	O&K MB 5 N	26583	1966	(c)	(3)
4wDH	Bdh	KHD KG230 B	57668	1964	(d)	2003 scr

(a) ex AG für Petroleum-Industrie AG (IPSA), Rotkreuz /1961.
(b) ex Aktien-Zuckerfabrik Schöppenstadt [D] /1974 via MBA; there may have been another locomotive, but it is thought to have been this one reported twice.
(c) new to Dynamit Nobel AG 253, Rheinfelden [D] "253""; ex Hüls AG, Rheinfelden [D] /1992, via LSB/WBB; also recorded as type MB 5 N ex.
(d) new to Société de traction et d´électricite, Bruxelles [B]; ex Sociétés réunies d'énergie du bassin de l'Escaut, Langerbrugge (EBES) [B] /1993 via OnRail/LSB.
(3) 1993 returned to LSB.

KARTON DEISSWIL AG (KD), Stettlen BE049

formerly: Karton- und Papierfabrik Deisswil AG, Deisswil

An internal rail system was installed in 1913, and was electrified at 800V DC till 1975. Originally traffic was brought in on transporter trucks over the Worbental line of the metre gauge VBW using locomotives constructed from electric motor coaches. The running numbers "1" and "2" were each carried by two different locomotives at the same time. Since 1974 the section of track to the SBB has been equipped as dual gauge (with a third running rail for this traffic). The locality name has also been altered as shown.

Gauge: 1435 mm   Locn: 243 $ 6057/2008   Date: 10.2003

	4	4wWETm	Bg2	unknown	?	1913	(a)	(1)	
	3	4wWETm	Bg2	unknown	?	19xx	(b)	(1)	
	2	4wWETm	Bg2	SZB	-	1955	(c)	1975	scr
	1	4wWETm	Bg2	SZB	-	1958	(d)	(4)	
Em 2/2		4wDH	Bdh	Moyse BN24HA150GM	1262	1973	new	(5)	
		4wDH	Bdh	O&K MB 9 N	26683	1970	(f)		

(a) wooden bodied; new; numbered "4" /197x.
(b) wooden bodied; originally numbered "3"; renumbered "2" /197x.
(c) steel bodied; constructed using parts of SZB "19"; parts of SZB "1" were incorporated before /1959; originally numbered "1"; renumbered "2" /1959.
(d) steel bodied; ex SZB "xx".
(f) ex Maschinenfabrik Deutz-Fahr, Gottmadingen [D] "292" /1994, via LSB.

(1) 1970-75 s/s.
(4) 1975 withdrawn; converted into social rooms on site.
(5) 1994 to Bühlmann Alteisen AG, Münchenwiler.

KENTAUR AG, Lützelflüh-Goldbach BE052

formerly: Hafermühle Lützelflüh AG
            Fritz Bichsel & Cie

These flourmills and cereal factory has been rail connected since 1914.

Gauge: 1435 mm   Locn: 233 $ 6186/2061   Date: 12.2002

	1ABE	1Aa1	BBC Akku-Fahrz.	879	1914	new	(1)
236 341-4	4wDM	Bdm	RACO 85 LA7	1610	1961	(b)	

(b) new to SBB "Tm¹ 336", later "Tm¹ 436"; ex RM "236 341" (wdn /2003).
(1) 2001 to Swisstrain, Rikon; by 10/2007 to Bahn Museum Kerzers (BMK), Kerzers.

KIESWERK STEINIGAND AG (KIESTAG), Wimmis BE053

The company operates quarries and a stocking yard alongside the R. Kander between Spiezwiler and Wimmis. These have been rail connected to the BLS since 1990 by a kilometre long branch from Eifeld, using the former formation of the Spiez-Thun line, which was replaced by a deviation in 1987. Address: Hauptstrasse, 3752 Wimmis.

Gauge: 1435 mm   Locn: 253 $ 6167/1704   Date: 02.2004

260 355-3	0-6-0DH	Cdh	Jung V 60	12485	1957	(a)

(a) new to DB "260 355-3"; hired and then purchased from ETRA, Zürich /1990.

KRAFTWERKE OBERHASLI AG (KWO) BE054

Guttanen

Within the mountain is an underground railway linking two funiculars. It is 4915 metres long with a maximum gradient of 8.8%.

Gauge: 500 mm   Locn: ?   Date: 10.1994

1		4wBE	Ba2	von Roll,SAAS	-, ?	1928	new	
2	rebuilt	4wBE	Ba2	SIG,SAAS	?	1943		(1)
3		4wBE	Ba2	SIG,SAAS	?	1948	new	(2)
1		A1BE	A1a	Stadler,SIG	241, ?	1994	new	
2		A1BE	A1a	Stadler,SIG	242, ?	1994	new	
3		A1BE	A1a	Stadler,SIG	243, ?	1994	new	

Innertkirchen

This concern started to build a series of dams and power stations in the upper valley of the R. Aare and adjacent valleys in 1925. Currently there are eight lakes and six power stations. A series of funiculars and téléphériques remain for service purposes. A five kilometre long line was built from Meiringen to Innertkirchen along the R. Aare gorge to connect the area with the SBB. This is now a public railway, the Meiringen-Innertkirchen-Bahn (MIB). This part of the story is excluded from this Handbook. The SIG list of references gives one ET 15KL locomotive delivered here.

Gauge: 1000 mm   Locn: ?   Date: 05.2008

23	0-4-4-2T4cc BB1n4vt	SLM		958	1896	(f) 1940 s/s
24	0-4-4-2T4cc BB1n4vt	SLM		959	1896	(g) 1937 s/s
	4wBE	Ba	EFAG	?	1931	s/s

(f) purchased from RhB "G 2/2 + 2/3 23 MALOJA" /1926.
(g) purchased from RhB "G 2/2 + 2/3 24 CHIAVENNA" /1926.

(1) to VHS, Luzern /19xx.
(2) on display Innertkirchen.

see: Eisenbahn Amateur 12/89 p964
see: Voie Étroite 160 p25-6

LAGERHAUS AG, Buchmatt   BE055

This is a subsidiary of Hermann Dür AG, Burgdorf. This warehouse was built in 1939. Address: Kirchbergstrasse 179, 3401 Burgdorf.

Gauge: 1435 mm   Locn: 233 $ 6128/2126   Date: 03.2007

4wDM	Bdm	RACO 95 SA3 RS	1521	1958	(a)	(1)
4wDM	Bdm	RACO 95 SA3 RS	1790	1968	(b)	

(a) ex SBB "Tm^{II}" 644" /1995.
(b) hired from SBB "Tm^{II}" 847" /2002.

(1) 2002 OOU.

LAUTERBURG & THOMMEN, Bern   BE056 C

A public works concern.

Gauge: 750 mm   Locn: ?   Date: 04.2002

0-4-0T	Bn2t	SLM		61	1875	new s/s

Biel/Bienne

This is the head office. The SIG list of references gives one ET 20L3 and one ETB 50 locomotive delivered here. Address Länggasse 9, 2504 Biel/Bienne.

Gauge: 600 mm (mainly)   Locn: ?   Date: 05.2002

	Type		Works	Number	Year	Note	Status
	4wDM	Bdm	O&K(S) MD 2s	1475	194x	(a)	s/s
	4wDM	Bdm	O&K(S) RL 1c	1641	194x	(b)	s/s
	4wPM	Bbm	O&K(M) S5	1751	192x	(b)	s/s
	4wDM	Bdm	O&K(M) RL 1a	4376	1931	(d)	s/s
	4wDM	Bdm	O&K(M) LD 2	6141	1935	(d)	s/s
	4wDM	Bdm	O&K(M) LD 2	7465	1937	(f)	s/s
	4wDM	Bdm	O&K MV0	25075	1951	(g)	s/s
	4wBE	Ba	SIG ES 50	1165	19xx		(8)

Pieterlen

A subsidiary since 1996, it declared bankruptcy in 2005. Address: Freidorfweg 1, 2545 Pieterlen

Gauge: 600 mm (probably)   Locn: ?   Date: 2.1996

1	4wDM	Bdm	RACO 12/16PS	1301	1945		(9)
	4wBE	Ba	SIG	?	19xx	(j)	s/s?

(a) gauge not confirmed, purchased from MBA, Dübendorf /1951.
(b) gauge not confirmed, purchased via O&K/MBA, Dübendorf.
(d) purchased via O&K/MBA, Dübendorf.
(f) ex Locher & Co, Zürich.
(g) new via Wander-Wendel and MBA, Dübendorf.
(j) gauge not confirmed.
(8) 1985 to Fabrique de Pâte de Bois, (Rondchâtel SA), Frinvillier.
(9) to FWF, Otelfingen.

RENFER & CO. AG, Biel-Bözingen   BE081

alternative: Bienne-Bojean

This wood yard (Holzwerk) shared a 1.5 kilometre long metre gauge line to Biel-Mett station with Vereinigte Drahtwerke AG. Traffic was in standard gauge wagons on transporter trucks. The works and line closed in 1994.

Mixed gauge track and the locomotive shed seen on 31 July 1966.

Gauge: 1000 mm				Locn: 233 $ 5872/2220				Date: 08.1998	
	12	0-6-0Tm	Cn2t	SLM G 3/3	863	1894	(a)	(1)	
	6	0-6-0T	Cn2t	SLM G 3/3	1511	1903	(b)	1943 scr	
	6	0-6-0T	Ch2t	SLM G 3/3	1341	1901	(c)	(3)	
	8	0-6-0T	Cn2t	SLM G 3/3	2095	1910	(d)	(4)	
		4wDE	Bde2	Moyse BN34E168D	1157	1967	(e)	(5)	

(a)	purchased from Städtischen Strassenbahnen Bern "12" /1908.
(b)	purchased from LEB "6" /1924.
(c)	purchased from BAM "6" /1943.
(d)	purchased from LEB "8" /1946.
(e)	new, ex Moyse /19xx; plates read BN34E180D.
(1)	1943 set aside for VHS, Luzern; 19xx to Technorama, Winterthur; 1971-83 loan to Blonay-Chamby (BC); 1993 to Bernmobil, Bern.
(3)	1967 donated to Blonay-Chamby (BC).
(4)	1977 returned to LEB.
(5)	1994 to CJ "Tm 501".

see: l'Escarbille p.19, Dec. 1970

RONDCHÂTEL SA, Frinvillier   BE082
also   Fabrique de Pâte de Bois

A subsidiary of Papierfabrik Biberist, rail connected in 1895. The factory lies between Reuchenette-Péry and Frinvillier stations. The products were brought to the finishing factory by funicular. Thence a short narrow-gauge line took the products to a warehouse. This had a standard gauge connection via a relatively long branch line, latterly serving only the oil store of Ciments Vigier SA. A fire in the finishing factory in 2002 caused a closure of the whole plant and the eventual insolvency of the company.

Gauge: 1435 mm				Locn: 233 $ 5855/2248				Date: 04.2005	
		0-6-0WT	Cn2t	SLM E 3/3	1194	1899	new	(1)	
Gauge: 640 mm				Locn: 233 $ 5855/2248				Date: 04.2005	
		4wDM	Bdm	R&H 48DL	296044	1952	(b)	(2)	
		4wBE	Ba	SIG ES 50	1165	19xx	(c)	(2)	

(b)	constructed by Robert Aebi & Co, Zürich from R&H frame 296044 of 1950 delivered without motor and separate motor 257653 of 1952; ex ? via Robert Aebi & Co, Zürich /19xx.
(c)	ex Ing Reifer & Guggisberger, Biel "7001" /1985.
(1)	1940 to Papierfabrik Biberist.
(2)	2005 to Parc d'Attractions du Châtelard VS SA, Le Châtelard VS.

ROSSI, PERUSSET & CIE, Ostermundigen BE083 C

Probably a public works company.

Gauge: 750 mm   Locn: ?   Date: 05.2002
    VEVEY   0-4-0T   Bn2t   O&K 50 PS   2907   1908   (a)   (1)
(a)   ex Lindenmeyer, Boulenaz & Cie, Vevey /1910-17.
(1)   1910-17 to Schafir & Mugglin, Muri bei Bern.

RUWA DRAHTSCHWEISSWERK AG, Sumiswald BE084

formerly: Rudolph Ruch AG (RUWA), Wasen im Emmental (till 1962)

A wire and metal factory was founded by Hans Ruch in 1881. From this, in 1962, a new firm was created to weld the forms and latticework for ferro-concrete structures. The site lies to the west of the line to Wasen im Emmental at the former halt of Burghof. Address: Burghof, 3454 Sumiswald.

Gauge: 1435 mm   Locn: 234 $ 624400,209000   Date: 01.2009
                    0-6-0WE   Cw1   SLM,BBC Ee 3/3   3899,4529   1945   (a)
(a)   ex SBB "Ee 3/3 16403" /2007.

OTTO SCHACHTLER, ZIEGEL- UND BACKSTEINFABRIKEN, Burgdorf BE085

Johannes Schachtler purchased and modernised in 1891-3 an existing brick and tile works (built in two parts from 1786-8 and 1835) lying alongside the route d'Heimiswil (6150/2107). A horse-worked line was installed to connect these works to the claypit (6154/2105). This is stated to have been three kilometres long - perhaps total track? In 1949 a new works was built just south of the claypit and a new pit started south of both. The railway was adapted to suit the new requirement. The older works closed in 1966 and the new works were extended in 1979. The use of rail had ceased before the brickworks closed in 1998.

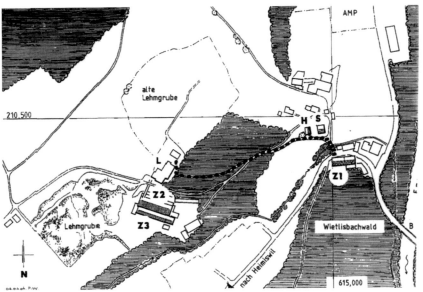

Z1	original location of the brickworks, eastern part from 1786-8, western part from 1835.
Z2	brickworks from 1949
Z3	extension from 1979
L	clay preparation area
....	railway

Gauge: 600 mm   Locn: 233 $ 6154/2107   Date: 05.2008

	4wDM	Bdm	MBA MD 1	2006	c1950	(a)	a1976 s/s
	4wDM	Bdm	O&K(M) RL 1c	7576	1937	(b)	(2)
	4wDM	Bdm	Brun	?	19xx		(3)
(a)	new to Zbinden, Erlach.
(b)	ex Hoch- & Tiefbau, Biel.
(2)	19xx donated to Schumacher, Körbligen, Gisikon (to be confirmed).
(3)	initially preserved on site; later s/s.

SCHAFIR & MUGGLIN, Muri bei Bern   BE086 C

This civil engineering firm had a depot near Bern.

Gauge: 750 mm   Locn: ?   Date: 03.2002

BUBI	0-4-0T	Bn2t	Jung 10 PS	1349	1909	(a)	(1)
LICHTENSTEIG	0-4-0T	Bn2t	Krauss-S XXVII bg	6173	1909	(b)	(1)
BASEL	0-4-0T	Bn2t	O&K 80 PS	1622	1905	(c)	(3)
VEVEY	0-4-0T	Bn2t	O&K 50 PS	2907	1908	(d)	(1)
ZÜRICH	0-4-0T	Bn2t	O&K 70 PS	4375	1911	(e)	(5)
MURI	0-4-0T	B?2t	O&K 70 PS	6020	1914	(e)	(1)
BERNA	0-4-0T	B?2t	O&K 70 PS	6572	1913	(c)	(1)

Gauge: 600 mm   Locn: ?   Date: 05.2002

| | 4wDM | Bdm | O&K(M) RL 2 | 2698 | 1927 | (h) | (8) |
| | 4wDM | Bdm | O&K RL 2a | 20196 | 1931 | (i) | s/s |

Gauge: ? mm   Locn: ?   Date: 05.2002

| | 4wDM | Bdm | O&K | ? | 19xx | (j) | s/s |
| | 4wDM | Bdm | O&K | ? | 19xx | (j) | s/s |

	4wPM	Bbm	O&K(M) S5		1491	192x	(l)	s/s

- (a) ex H. Piot & Ch. Piguet, Bavois /1926.
- (b) ex ? /1931.
- (c) ex Belart & Cie, ?Olten-Klingnau /1919.
- (d) ex Rossi, Perusset & Cie, Ostermundigen /1917.
- (e) ex Frutiger & Lanzrein, Bern /1926.
- (h) ex Mangold & Co, Zürich.
- (i) ex A. Hauser, Oberwinterthur.
- (j) ex ? /1951.
- (l) purchased via O&K/MBA, Dübendorf.
- (1) from 1931 used on the hydroelectric contract at Klingnau; /1933-39 transferred to Schafir & Mugglin, Zürich.
- (3) from 1931 used on the hydroelectric contract at Klingnau; /1933-37 transferred to Schafir & Mugglin, Zürich.
- (5) 1926-39 transferred to Schafir & Mugglin, Zürich.
- (8) before 1947 transferred to Schafir & Mugglin, Liestal "2.21.02".

SCHLACHTHOF, Biel — BE087

alternative: Abbatoirs, Bienne
The city slaughter-house.

Gauge: 1435 mm   Locn: ?   Date: 01.1992

	4wPM	Bbm	Breuer		?	c192x	(a)	s/s

- (a) mentioned in the Breuer list of references of about 1930.

SCHMALZ H.R. AG, Bern — BE087a C

The SIG list of references gives one ETB 70 and four ETS 100 locomotives delivered here.

WALTER SCHMUTZ AG, Heimberg, Steffisburg — BE088

A dealer in steel: trading ceased ca. 1997.

Gauge: 1435 mm   Locn: 253 $ 6132/1807   Date: 09.1994

	0-4-0DM	Bdm	KHD A4L514 R	55182	1952	(a)	(1)

- (a) new to Eisenlager GmbH, Essen [D]; ex Hubert Schulte, Bochum-Dahlhausen [D] 1988 via WBB/LSB. Also referred to as type KS55 B.
- (1) 1999 to Hafenverwaltung Kehl, Kehl [D].

HERR A. SCHWARZ, Sumiswald — BE089 M

This private individual has created a 400 mm system. Most, if not all, the locomotives have been converted from larger gauges.

Gauge: 400 mm   Locn: 234 $ 623x/208x   Date: 12.1999

4wDM	Bdm	JW JW8	051	1950	(a)	(1)
4wDM	Bdm	JW DM20/1	2450	1965	(b)	
4wDM	Bdm	RACO	?	19xx	(c)	(3)
0-4-0DM	Bdm	LKM Ns2f	262057	1959	(d)	
4wDM	Bdm	O&K(M) MD 1	8606	1938	(e)	

- (a) ex Ziegelei Oberdiessbach /1982 (500 mm).
- (b) ex von Roll, Gerlafingen /1995 (500 mm).
- (c) ex Bergverein, Käpfnach, Horgen /1995 (600 mm).
- (d) ex FWF, Otelfingen /2005 (600 mm).

(e) new to Armeeflugzeugplatz Dübendorf; ZH (600 mm); ex ? /199x (gauge?).
(1) 2008 to FWF, Otelfingen (600 mm).
(3) 1996 to FWF, Otelfingen (600 mm).

SCHWEIZERISCHE MUNITIONSFABRIK, Lerchenfeld, Thun BE090

formerly: Eidgenössische Militärwerkstätten und Sektionen der K.T.A. Munitionsfabrik
alternative: Eidgenössische Militärbetriebe und Waffenplatz

A munitions stocking site. An extensive internal system was visible from public streets notably around an under-bridge near the station. The system closed at the end of 1997.

Gauge: 750 mm Locn: 253 $ 6129/1795 Date: 09.1999

	4wBE	Ba	SWS,MFO	?	1917	new	(1)
	4wBE	Ba	SWS,MFO	?	1917	new	(1)
32.023	4wDM	Bdm	RACO RA7	104	1933	new	(3)
	4wDH	Bdh	Schöma CFL 60DZ	3686	1973	(d)	(4)

(d) new, via Asper.
(1) 1970 seen in a scrap yard.
(3) 1999 to Hr. Jürg Meili, Arch; 2005 to Bahn Museum Kerzers (BMK), Kerzers.
(4) 2005 to Bahn Museum Kerzers (BMK), Kerzers.

SCHWEIZERISCHE STRASSENBAU- UND TIEFBAU-UNTERNEHMUNG AG (STUAG), Bern BE091

later: Batigroup AG (from 1999)

This is a public works company with subsidiaries throughout Switzerland.

Gauge: 600 mm (or unknown) Locn: ? Date: 05.2002

4wDM	Bdm	O&K(M) LD 2	7127	1936	(a)	s/s
4wDM	Bdm	O&K(M) MD 2	12039	194x	(b)	s/s
4wDM	Bdm	O&K(M) MD 2	12058	194x	(b)	s/s
4wBE	Ba1	Stadler	31	1948	(d)	(4)
4wBE	Ba1	Stadler	32	1948	(d)	(5)
4wWE	Bg	Oehler	582	1944	(f)	s/s
4wWE	Bg	Oehler	?	1950	(f)	s/s

Gauge: 500 mm Locn: ? Date: 05.2002

4wDM	Bdm	O&K(M) RL 1a	4440	1931	(a)	s/s

(a) new via O&K/MBA Zürich.
(b) new via O&K/MBA Zürich, gauge unknown (500 mm or 600 mm).
(d) new for building work at the electricity plant at Lavey.
(f) new; gauge unknown.
(4) to Hew & Co, Chur "7".
(5) to Hew & Co, Chur "8".

SEEBERGER & JORDI AG, Frutigen BE092 C

formerly: J. Seeberger

A public works company.

Gauge: 750 mm Locn: ? Date: 04.2002

1	0-4-0WT	B?2t	Maffei	4203	1921	(a)	(1)
2	0-4-0WT	B?2t	Maffei	4204	1921	(b)	(1)
3	0-4-0WT	B?2t	Maffei	4206	1921	(c)	(3)
4	0-4-0WT	B?2t	Maffei	4208	1922	(d)	(3)
ROSINA	0-4-0WT	Bn2t	O&K 50 PS	2391	1907	(e)	(3)

TÄUFFELEN	0-4-0WT	Bn2t	Märkische 50 PS	302	1898	(f)	(6)
	4wBE	Ba2	Stadler	36	8/1948	(g)	(7)
	4wDM	Bdm	Gmeinder	?	19xx		s/s
	0-4-0DM	Bdm	Deutz OME117	?	1937		s/s

- (a) ex Martin, Baratelli & Cie, Lausanne (Barberine contract) /1929.
- (b) ex Martin, Baratelli & Cie, Lausanne (Barberine contract) /1927.
- (c) ex Martin, Baratelli & Cie, Lausanne (Barberine contract) /1928.
- (d) ex Martin, Baratelli & Cie, Lausanne (Barberine contract) /1922.
- (e) ex Robert Aebi & Co, Zürich /1928.
- (f) ex Schafir & Müller, Generalunternehmer, probably in Täuffelen /1928.
- (g) ex Walter J. Heller /1951-52.
- (1) 1952 removed from boiler register.
- (3) 1952 removed from boiler register; c1970 to von Roll, Gerlafingen for scrap, but by c1979 present in various playgrounds in Gerlafingen till after 1982 when all three were finally scrapped.
- (6) 1939-57 to VEBA AG, Zürich (for a contract in Rheinau).
- (7) 1995 donated to FWF, Otelfingen.

SIGRIST-MERZ & GRÜEBLER, Wasen im Emmental — BE093 C

A construction site with the name of "Reuss" operated by the firm of Sigrist-Merz & Grüebler [Gräbler ?], St. Gallen.

Gauge: 600 mm Locn: ? Date: 06.2002

	4wDM	Bdm	O&K MV0	25043	1951	(a)	s/s

- (a) new via MBA, for this site.

P. SIMON, Bern — BE094 C

A public works concern.

Gauge: ? mm Locn: ? Date: 03.2002

	0-4-0T	Bn2t	O&K 20 PS	689	1900	new	s/s

SORREL & FASOLA, Frutigen — BE095 C

One of a number of firms participating in the construction of the BLS.

Gauge: 750 mm Locn: ? Date: 04.2002

	0-4-0T	B?2t	Maffei	3513	1911	(a)	(1)

- (a) see note for this locomotive under Robert Aebi & Cie AG, Regensdorf; hired from Robert Aebi & Co, Zürich /1911 and used on BLS construction work.
- (1) 19xx returned to Robert Aebi & Co, Zürich.

SONDIERSTOLLEN FRUTIGEN-KANDERSTEG (ARGE SFK), Frutigen — BE096 C

Gauge: 750 mm Locn: ? Date: 01.1996

1 MARION	4wDH	Bdh	Schöma CFL 180DCL	5192	1991	(a)	s/s
2 NIKOLINA	4wDH	Bdh	Schöma CFL 180DCL	5193	1991	(a)	s/s
3 DANIELLA	4wDH	Bdh	Schöma CFL 180DCL	5196	1991	(a)	s/s
4 BRIGITTA	4wDH	Bdh	Schöma CFL 180DCL	5194	1991	(a)	s/s
DL2	4wDH	Bdh	Schöma CFL 180DCL	5176	1990	(e)	s/s
	4wDH	Bdh	Bedia D105/17B - Umbau	313	1995	(f)	s/s

- (a) hired from ArGe Untertagebau Schweiz, Schachen LU.
- (e) ex Trans-Manche Link (TML), Sandgatte [F] "TU 20".

(f)   also quoted as type D105/15 of 1985; hired from Schlatter Peter AG, Rheinsulz.

STAUMAUER OBERAAR, Guttanen   BE097 C
This was a construction project for a barrage completed in 1953.
Gauge: 750 mm   Locn: ?   Date: 08.2002

	4wDM	Bdm	O&K MV0	25093	1951	(a)	s/s?
	4wDM	Bdm	O&K MV0	25094	1951	(a)	s/s?
	4wDM	Bdm	O&K MV0	25095	1951	(c)	(3)

(a)   new via Wander-Wendel and MBA, Dübendorf.
(c)   new via Wander-Wendel and MBA, Dübendorf; not confirmed as being here.
(3)   unconfirmed report of it being at Ziegelei Vöhrum, Sankt Georg [A].

FERDINAND STECK, MASCHINENFABRIK AG, Bowil   BE098 M
This engineering firm has built at least one diesel locomotive and maintained some steam locomotives. The following locomotives are on display. Address: Bahnhofstr. 3, 3533 Bowil.

The one locomotive built by the company and numbered Steck 001 passes Otelfingen station on 27 September 2004. At that time it was owned by Benkler AG, Villmergen.

Gauge: 1435 mm   Locn: 243 $ 6198/1944   Date: 09.2007
        01 180   4-6-2   2C1h2   Henschel BR 01   22923   1936   (a)   display
Gauge: 800 mm + rack   Locn: 243 $ 6198/1944   Date: 01.2008
        HG 2/3 3   0-4-2RT   3bn2t   SLM H   719   1892   (b)   display
(a)   ex DB (wdn 24/8/1973) /1975.
(b)   ex BRB /2005.

STEINBRÜCHE HERBRIG AG, Zementfabrik, Därligen   BE099
Quarrying for clay was started in 1900: a change to chalk extraction followed. This operation closed in 1990. The name of the last operator is given, but this may not apply to the period when a railway was in use. The grid reference applies to the quarry. An aerial ropeway, a kilometre long, connected this with a works located alongside the railway and the Thunersee.

Gauge: 600 mm		Locn: 254 $ 6280 1674					Date: 04.2009
	4wPM	Bbm	O&K(M) M	2743	1928	(a)	s/s

(a) purchased via O&K, Dübendorf.

see: www.ksebern.ch/d/portraet/grubenplan/oberacher/findex.html

STEINBRUCHGESELLSCHAFT OSTERMUNDIGEN, Ostermundigen
BE102

The 1350 metre long access line to the quarry contained a 480 metre long rack section. The quarry was closed in 1902 and the society liquidated in 1907.

Gauge: 1435 mm + Riggenbach rack			Locn: 243 $ 6045/2011				Date: 03.2008
1	GNOM	2-2-0RT	1Azn2t	Olten	20	1870	new (1)
2	ELFE	0-4-0RT	Baz2t	Aarau	10	1876	new (2)

(1) 1902 withdrawn; 1904 at Nestlé AG, Neuenegg; 1907 to von Roll, Rondez, Delémont (later JU) "7".
(2) 1907 to von Roll, Gerlafingen "6".

STRAFANSTALTEN, Witzwil
BE103

The prison was opened in 1895 and was rail served, from Gampelen station, between 1910 and 1998. The 2.8 kilometre long line was used 1914-54 for conveying refuse from the city of Bern. The section Chützstrasse-Birkenhof closed in 1980, Witzwil-Rive du lac in 1985 and the final section of Gampelen-Witzwil in 1998. A track length of 3.3 kilometres is quoted by Schweers+Wall; almost certainly a total track length. The data from various sources for the Breuer is not consistent. Clarification is required.

Gauge: 1435 mm			Locn: 232 $ 5715/2059				Date: 12.2002
	4wPM	Bbm	Breuer II or III	?	1927	(a)	1947 scr
	4wDM	Bdm	RACO 80 SA 3 (80 PS)	1343	1947	new	1980 scr
2	4wDM	Bdm	RACO RA11	3021	1949	(c)	(3)

(a) ex SBB "Tm 897" / 1935; another states that it was ex SBB "Tm 895" /1935.
(c) ex Bundesamt für Militärflugplätze (BAMF), Dübendorf /1980 via Stadler.
(3) c1998 OOU; 2005 to Bahn Museum Kerzers (BMK), Kerzers.

SUCHARD TOBLER AG, Bern
BE104

formerly: Chocolat Tobler AG (till 1970)

The company manufactures chocolate products.

Gauge: 1435 mm			Locn: 243 $ 5984/1996			Date: 08.1989
	4wBE	Ba	SWS,SAAS	?	1921	b1985 s/s

TELA AG, Niederbipp
BE105

formerly: Papierfabrik Tela AG

Stocking yard and manufacturing plant for "Balsthal" paper. The factory is rail connected to Oensingen station in Canton Solothurn. Address: Rothboden 1, 4705 Niederbipp.

Gauge: 1435 mm			Locn: 223 + 224 $ 6200/2362			Date: 10.2001
4wDH	Bdh	Moyse BL18HS80D	137	1966	new	(1)
4wDH	Bdh	KHD KG230 B	57938	1965	(b)	
4wCEM	B-em	Windhoff RW 40 EM	260087	1992	new	
4wDH	Bdh	Henschel DHG 200B	30876	1964	(d)	(4)

(b) ex Oberpostdirektion, Köln-Deutz [D] "3" /1978, via WBB, Hattingen [D].

(c)	hired from Asper /1986.
(1)	1980 to PESA, Chavornay.
(4)	1986 returned to Asper.

THOMMEN-FURLER AG, Rüti bei Büren BE106

formerly: Thommen & Cie (till 2003)

This company makes and distributes chemical products, heating oils and fuels.

Gauge: 1435 mm   Locn: 233 $ 5976/2229   Date: 05.2008

	4wDM	Bdm	RACO 85 LA7	1660	1963	(a)	2002 scr
	4wDM	Bdm	RACO 85 LA7	1657	1963	(b)	s/s
	4wDE	Bde	Stadler,BBC	143, ?	1974	(c)	
	4wDM	Bdm	RACO 85 LA7	1695	1964	(d)	

(a)	new to SBB "Tm^I 361"; ex SBB "Tm^I 461" /1994.
(b)	new to SBB "Tm^I 351"; ex SBB "Tm^I 451" /2002.
(c)	ex Maggi (Nestlé AG), Kemptthal /2006.
(d)	ex SBB "Tm^I 498" (wdn 30/9/2005) /2005-8.

TIGER KÄSE AG, Langnau-im-Emmental BE107

formerly: Röthlisberger & Sohn AG (till after 1963)

ex SBB Breuer Tm 411 performs some shunting on 29 October 1982. By this time it had acquired the asymmetric cab extension that eliminated the possibility of entering the cab on that side.

This firm has a factory for producing (processed) cheese in boxes. Latterly other Breuer locomotives came here for scrapping and recovery of spare parts. See for instance, Fischer & Cie AG, Langnau-im-Emmental and Zent AG, Ostermundigen.

Gauge: 1435 mm   Locn: 244 $ 6268/1985   Date: 12.2002

	4wPM	Bbm	Breuer IV	?	1931	(a)	(1)
3	4wDM	Bdm	RACO 95 SA3 RS	1708	1964	(b)	

(a)	ex SBB "Tm 411" 1/1963; from at least 1982 the locomotive had an extension on one side that eliminated the cab door on that side – did it have a replacemnt diesel engine?
(b)	ex SBB "Tm^{II} 735" /1993.
(1)	1993 scrapped, following an accident.

UFA AG, Herzogenbuchsee — BE108

formerly: Orador (till 1998)
Verband Landwirtschaftlicher Genossenschaften (VLG), Niederlassung Herzogenbuchsee

The concern prepares and provides feeding material for livestock. It is part of the VLG/UFA union of agrarian confederations. Initially it used silos adjacent to Herzogenbuchsee station (6198/2264); then those at "Hoffmatt" situated on the old line to Solothurn (6196/2271). Finally a third plant was added in 2002-3 at "Biblis" (6195/2270). Since then the locomotive has been stationed at "Biblis" but also shunts at "Hoffmatt", where a road/rail device may also be in use.

Gauge: 1435 mm   Locn: 233 $ 6196/2271 + 6195/2270   Date: 03.2008

	4wD	Bd	O&K	?	?	19xx	(1)
	4wDM	Bdm	O&K MV3	26596	1966	new	(2)
2	4wDH	Bdh	O&K MB 200 N	26804	1975	new	
	4wDM R/R	Bdm	unknown R/R	?	xxxx		

(1) 19xx sold for scrap to Kaufmann AG, Thörishaus.
(2) 1976 to Migros Genossenschaft Basel AG, Münchenstein.

UNTERNEHMUNG FÜR AAREKORREKTION, Bern — BE109 C

The one known locomotive for this river grading work is recorded as 750 mm gauge. Jung records show it as built as 600 mm and it is now (2008) at 600 mm.

Gauge: ?750 mm   Locn: ?   Date: 03.2002

0-4-0WT	Bn2t	Jung 20 PS	1693	1911	(a)	(1)

(a) completed 12/03/1912 new via Fritz Marti, Bern /1912; also stated also reported as new to A.H. Bürgi, Bern via Fritz Märti, Bern for use here.
(1) 1922 to Robert Aebi & Co, Zürich.

VANNI, BASSO & CIE, Frutigen — BE110 C

One of a number of firms participating in the construction of the BLS.

Gauge: 600 mm   Locn: ?   Date: 04.2002

0-4-0T	Bn2t	Jung 50 PS	1684	1911	(a)	(1)

(a) ex Fritz Marti, Bern, used on BLS construction at Mitholz and then the new station at Biel/Bienne.
(1) 1916 to Fritz Marti, Bern for construction of the Simplon II tunnel; 1996 to Technorama, Winterthur via AG Heinrich Hatt-Haller.

VEREIN 241.A.65, Burgdorf — BE111 M

The society operates a SNCF "Mountain" on chartered trains and shared the old EBT depot with the Verein Dampfbahn Bern (DBB). During 2008 the stock was moved to Full-Reuenthal and the locomotive details are given there (Kanton Aargau). Address: Seetalstrasse 4, 5706 Boniswil.

Gauge: 1435 mm   Locn: 233 $ 6136/2122   Date: 09.2008

VEREIN DAMPFBAHN BERN (DBB), Bern — BE112 M

This society has centres in two locations, namely: Burgdorf (old EBT depot) and Spiez. It has operated trains from both. Until 12/2008 it also had a base at Laupen, BE. At least one additional locomotive is (2008) on loan from each of the Bern-Lötschberg-Simplon-Bahn (BLS) and Swisstrain. Address: Postfach 5841, 3001 Bern.

Gauge: 1435 mm   Locn: ?   Date: 04.2008

10	'853'	0-6-0WT	Cn2t	SLM E 3/3	629	1890	(a)

11	'855'	0-6-0WT	Cn2t	SLM E 3/3	631	1890	(b)	(2)	
	51	2-6-0T	1Cn2t	SLM Ed 3/4	1726	1906	(c)		
	LISE	0-6-0WT	Cn2t	SLM	1901	1908	(d)		
		0-4-0T	Bn2t	O&K 100 PS	3081	1908	(e)	(5)	
	11	2-8-0T 2cc	1Dh2vt	SLM Ec 4/5	2160	1911	(f)		
	5810	2-6-2T	1C1h2t	SLM Eb 3/5	2211	1911	(g)		
	8	2-8-0T	1Dn2t	SLM Ed 4/5	2427	1914	(h)		
EBT Te 155		0-4-0WE	Bw1	SLM,MFO Te^l	3928, ?	1945	(i)		
EBT Te 157		0-4-0WE	Bw1	SLM,MFO Te^l	3930, ?	1945	(j)		
		4wDM	Bdm	Breuer V	3039	1951	(k)		
Tm	75	4wDE	Bde	Stadler	103	1953	(l)		

(a) new to JS "853"; ex von Roll AG, Gerlafingen "10" /1973.
(b) new to JS "855"; ex von Roll AG, Gerlafingen "11" /1987.
(c) ex display at Bahnhof Schwarzenburg /1998.
(d) ex Gaswerk Bern, Wabern /1969.
(e) ex construction of Kraftwerk Laufenburg /1972.
(f) ex display at Oberdorf /1986.
(g) new to SBB "Eb 3/5 5810"; ex MThB "Eb 3/5 5810" /1974.
(h) ex EBT "Ed 4/5 8" /1972.
(i) ex RM "216 325" /2007.
(j) ex RM "216 327" /2007.
(k) ex von Roll AG, Olten /1999.
(l) ex BLS "Tm 235 075" /2000.
(2) 1988 on loan to VVT, St. Sulpice NE; 3/2008 to Bahn Museum Kerzers (BMK), Kerzers.
(5) 1997 to EUROVAPOR, Wutachtalbahn [D].

see: www.dbb.ch

VEREIN HISTORISCHE EISENBAHN EMMENTAL (VHE), Huttwil
BE113 M

formerly: Vereinigte Dampf-Bahnen (VDB) (till 3/2005)
EUROVAPOR, Sektion Emmental (till 1997)

The group operates secondary mainline locomotives over the lines of the EBT/SMB/VHB. Address: Postfach 1574, 3401 Burgdorf.

Gauge: 1435 mm   Locn: 234 $ 6316/2182   Date: 01.2009

Ed 3/4	11	2-6-0WT	1Ch2t	SLM Ed 3/4	1904	1908	(a)		
	64 518	2-6-2T	1C1h2t	Jung BR 64	9268	1941	(b)		
	262	2-8-2T	1D1h2t	Henschel	25263	1954	(c)	(3)	
Tm^{III}	9527	4wDH	Bdh	RACO 225 SV4 H	1932	1986	(d)		
Te 216 321		4wWE	Bw1	SLM,MFO Te^l	3846, ?	1944	(e)		
Te 216 324		4wWE	Bw1	SLM,MFO Te^l	3927, ?	1945	(f)	7/2008 scr	

(a) new to LHB "11" as a saturated steam locomotive (1Cn2t); ex von Roll, Gerlafingen "17" /1973 via EUROVAPOR.
(b) ex DB "64 518"(fitted with boiler Meiningen 1513) via EUROVAPOR.
(c) ex Frankfurt-Königstein (FK) [D] via DME/Oswald Steam/EUROVAPOR.
(d) ex SBB "Tm^{III} 9527"; privately owned.
(e) new to EBT "Te 151"; ex RM.
(f) new to EBT "Te 154"; ex RM.
(3) 1996 to private ownership.

see: www.historische-eisenbahn-emmental.ch

VEREINIGTE DRAHTWERKE AG (VDW) BE114
alternative: Tréfileries Réunies SA
formerly: Blösch, Schwab & Cie (till 1914)

Bielwerk
alternative: Usine de Bienne

One plant was situated in the Schwanengasse, as part of the Güterbahnhof; this closed in 1992 and was demolished in 1999.

Gauge: 1435 mm   Locn: 233 $ 5858/2213   Date: 08.1997

	0-4-2D	B1dm	SAMHUL	?	19xx	(a)	a1948 s/s
	4wDM	Bdm	RACO RA11	1365	1948	new	(2)

Bözingenwerk
alternative: Usine de Boujean

The Usine de Boujean was connected to Biel-Mett station by a metre gauge line operated jointly with Renfer AG. This latter company took over all workings in 1989. The factory and line closed in 1993.

Gauge: 1000 mm   Locn: 233 $ 5877/2220   Date: 08.1988

	4wBE	Ba2	SWS,MFO	?,19	1909	new	(3)

(a) constructed by SAMHUL /1924-30 using parts of Baldwin 600 mm locomotives.
(2) 1992 stored; c1999 scrapped.
(3) 1989 withdrawn following collision; 18/6/1993 donated to La Traction, Pré-Petitjean.

Bibliography: L'Escarbille p.19, Dec. 1970

VEREIN PACIFIC 01 202, Lyss BE115 M

The society operates a DB Pacific on chartered trains. Address: Wehrstrasse 14, 3203 Mühleberg.

Gauge: 1435 mm   Locn: 233 $ 5897/2143   Date: 10.2007

01 202	4-6-2	2'C1'h2	Henschel BR 01	23254	1937	(a)

(a) new to DRG "01 202"; ex DB "01 202-1"; EUROVAPOR /1975; privately owned Münsingen BE /1980; to VVT, St. Sulpice NE for disposal /1988; Verein Pacific 01 202 2/6/1990.

VON ROLL AG, Werk Bern BE116
formerly: L. von Roll'sche Eisenwerke, Giesserei Bern
           Maschinenfabrik Bern AG (from 1870 till c1890)

This foundry was started by A. Marquard in 1870 and was rail connected from 1873. It absorbed the Fabrik für Eisenbahnmaterial (1872-1877) before being purchased by von Roll in 1894. The concern has constructed mountain railway equipment of all sorts, and also rail-cranes, lifts etc. The non-railway production was transferred to Thun and production of railway equipment to Ambri-Piotta (as Tensol Rail) in 1997.

Gauge: 1435 mm			Locn: 243 $ 5988/2003				Date: 08.1989
	4wBE	Ba	Olten?,MFO		?	c1932	(1)
(1)	1997 donated to Swisstrain, Liesberg; then Bodio.						

J. WAMPLER SA, Bienne   BE117 C

A public works company.

Gauge: 750 mm			Locn: ?				Date: 04.2002
	0-4-0T	Bn2t	Krauss-S IV fg		2811	1893	(a) (1)
(a)	ex Aebli, Rossi & Krieger, Schaffhausen.						
(1)	in or before /1923 to A. Baumann, Bauunternehmer, Wädenswil.						

DR. A. WANDER AG, Neuenegg   BE118

Originally a factory of Nestlé AG, opened and rail connected in 1904 and closed in 1921. It was re-opened in 1927 by the present concern. Rail operations have been provided by the Sensetalbahn (STB) since 1989.

Gauge: 1435 mm			Locn: 243 $ 5894/1937				Date: 03.2008
	4wPM	Bbm	Breuer II		631	*1925*	(a) a1937 s/s
2	1ABE	1Aa	?				new
rebuilt			Stadler		-	1951	c1960 s/s
	4wDE	Bde1	Stadler,BBC		111, ?	1960	new (3)
(a)	ex SBB "Tm 471" or "Tm 472" after /1933.						
(3)	1989 to Stadler AG, Bussnang.						

WORBLENTALBAHN, Bern   BE119 C

One locomotive from the construction period has been identified.

Gauge: 1000 mm			Locn: ?				Date: 05.2008
	0-6-0Tm	Cn2tk	SLM G 3/3		456	1887	(a) (1)
(a)	provided by F. Marti, Bern /1908.						
(1)	19xx returned to F. Marti, Bern.						

GEBR. WÜTRICH AG, Langnau-im-Emmental BE120

A firm making timber products (Holzwarenfabrik) with uses manually worked railway which includes wagon tables. Address: Hohgantweg 10, 3550 Langnau-im-Emmental.

Gauge: 600 mm   Locn: 244 $ 626245,198700   Date: 10.1982

FRITZ WYSS AG, Leuzigen BE121

Operating: Kieswerk Leuzigen

A gravel extraction plant that closed ca. 1970.

Gauge: 600 mm   Locn: 233 $ 6012/2254   Date: 01.1994

0-4-0DM	Bdm	Deutz ML128 F	6454	1923	(a)	(1)
0-4-0DM	Bdm	Deutz MLH322 F	10003	1931	(b)	(2)
0-4-0DM	Bdm	KHD A2L514 F	55827	1954	(c)	(3)

(a) via Leipziger & Co., Köln [D] and Fritz Marti, Bern.
(b) new via Robert Aebi & Co, Zürich.
(c) new, via Hans Würgler, Zürich.
(1) to Mainische Feldbahnen, Hattingen [D], possibly via FWM, Oekoven [D].
(2) 1986 to Bernhard van Engelen, Worpswede [D] via Mainische Feldbahnen, Hattingen [D]; later in the Netherlands.
(3) 19xx on display locally; 1993 donated to FWF, Otelfingen.

ZBINDEN, Erlach BE122 C

A public works concern. The firm also has sites elsewhere in Switzerland.

Gauge: ? mm   Locn: ?   Date: 05.2002

4wPM	Bpm	O&K(M) S5	1545	192x	(a)	(1)
4wDM	Bdm	MBA MD 1	2006	c1950	(b)	(2)

(a) purchased via O&K/MBA, Dübendorf.
(b) purchased via MBA, Dübendorf.
(1) to MBA; fitted with a MD 2S motor; 17/03/1953 sold to Hew & Co, Chur.
(2) to Otto Schachtler, Ziegel- und Backsteinfabriken, Burgdorf.

ZENT AG, Ostermundigen BE123

formerly: Zentralheizungsfabrik AG

A central heating construction factory that closed in 197x.

Gauge: 1435 mm   Locn: 243 $ 6027/2006   Date: 12.1988

4wPM	Bbm	Breuer IV		?	1931	(a) (1)

(a) ex SBB "Tm 413" 1/1963.
(1) 1975 demolished for spare parts on the land of Röthlisberger & Sohn AG, Langnau-im-Emmental.

ZIEGEL- & BACKSTEINFABRIK AG, Langenthal BE124

This company owned a single brickworks that closed in the 1930s.

Gauge: ? mm   Locn: ?   Date: 04.2002

	0-4-0T	Bn2t	Krauss-S VXXII g	3839	1898	(a)	1935 s/s

(a) ex IRR " SEELACHE" (750 mm) /1917.

ZIEGELEI GEBRÜDER FINK, Riedtwil BE127

This concern operated a single brickworks. Horse traction was used till about 1950. The works was demolished in 1958.

Gauge: 500 mm   Locn: ?   Date: 12.1999

	4wDM	Bdm	JW JW8	051	1950	new?	(1)

(1) 1958 to Ziegelei Fink, Oberdiessbach.

ZIEGELEI OBERDIESSBACH AG, Oberdiessbach BE128

A brickworks; the rail system was abandoned in the late 1970s or early 1980s.

Gauge: 500 mm   Locn: 243 $ 6134/1866   Date: 12.1999

	4wBE	Ba	Oehler ELG1042	701	1946		(1)
	4wDM	Bdm	JW JW8	051	1950	(b)	(2)

(b) ex Ziegelei Gebrüder Fink, Riedtwil /1958.
(1) 198x to Karl Heinz Rohrwild, Nürnberg [D].
(2) 1982 to A. Schwarz, Sumiswald (400 mm).

ZIEGELEI TIEFENAU AG, Worblaufen BE129

This firm operated a brickworks.

Gauge: ? mm   Locn: ?   Date: 05.2002

	4wDM	Bdm	O&K(S) RL 1c	1503	194x	(a)	s/s?

(a) ex ?, via MBA.

ZIEGELWERKE ROGGWIL AG, Roggwil BE130

This brickworks was owned by Ziegel- und Backsteinfabrik, Langenthal.

Gauge: 600 mm   Locn: 224 $?6283/2333?   Date: 05.2002

	4wDM	Bdm	O&K(M) RL 1a	4405	1931	(a)	s/s
	4wDM	Bdm	O&K(M) LD 2	4911	1933	(a)	s/s
	4wDM	Bdm	O&K MV0	25103	1951	(c)	a1985 s/s
	4wDM	Bdm	O&K MV1	?	19xx		(3)

(a) purchased via O&K/MBA, Dübendorf.
(c) ex Hew & Co, Chur /1957.
(3) still in existence 1983-5; most likely the MV0 reported again.

Basel - Landschaft (Basel Land)   BL

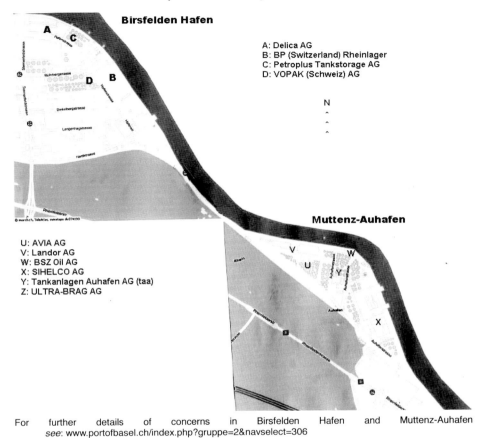

For further details of concerns in Birsfelden Hafen and Muttenz-Auhafen
see: www.portofbasel.ch/index.php?gruppe=2&navselect=306

AKTIEN-ZIEGELEI ALLSCHWIL, Allschwil   BL01

alternative: Tuilerie par Actions Allschwil
later   AZA Immobilien AG (from 1997)

This brickworks was founded before 1931. Address: Binningerstrasse 74.

Gauge: 600 mm   Locn: 213 $ 6081/2662   Date: 02.2007
                4wDM   Bdm   O&K(M) RL 1a   4614   1931   new s/s

ARGE ADLERTUNNEL, Muttenz   BL02 C

A consortium formed to construct the rail tunnel between Muttenz and Liestal. Benkler and Vanoli were part of this consortium.

Gauge: 1435 mm		Locn: ?					Date: 01.2008	
Vreneli		BBDH	BBdh	KHD V 100	57361	1962	(a)	(1)

(a) provided by Vanoli /199x.
(b) ?2000 returned to Vanoli on completion of tunnel (opened 2000).

ARGE KW BIRSFELDEN, Birsfelden   BL03 C

KW = Kraftwerk

The railway served works in connection with the modification of the watercourses for the Birsfelden hydroelectric power station. The German firm of Christian Krutwig carried out the work; the railway ceased operating in 1953.

Gauge: 900 mm				Locn: 213 $ 6138/2668 approx			Date: 04.2002	
16		0-4-0T	Bn2t	Henschel Deutschland	12779	1914	(a)	(1)
17		0-4-0T	Bn2t	Henschel	?	*1914*	(b)	(1)
21		0-4-0T	Bn2t	Henschel	?	192x	(b)	(1)
22		0-4-0T	Bn2t	Henschel Klettwitz	17725	1920	(b)	(1)
46	FRITZ LEOPOLD	0-4-0T	Bn2t	Henschel Klettwitz	24087	1938	(b)	(1)
47	RUDOLF FÖRSTER	0-4-0T	Bn2t	Henschel Klettwitz	24129	1939	(b)	(1)

(a) ex Christian Krutwig, Köln, [D] /1953.
(b) ex Christian Krutwig, Köln, [D] /195x.
(1) returned to Christian Krutwig, Köln [D].

ASEA BROWN BOVERI (ABB), Münchenstein   BL04

formerly: Brown Boveri & Cie (BBC) (1913-1988)
           Elektricitäts-Gesellschaft Alioth (EGA) (1895-1913)
later: EBM Elektizitätsmuseum, Münchenstein

The factory produced electrical equipment. A branch line was installed in 1894 and was electrified in 1899. The works closed progressively 1988-94.

Gauge: 1435 mm 500V AC				Locn: 213 $ 6132/2621			Date: 12.1994	
1		4wWE	Bg2	SIG,EGA	?	1899	new	s/s
1		1ABE	1Aa1	BBC Akku-Fahrz.	1210	1917	new	s/s
3		4wBE	Ba	Jäger,BBC Akku-Fahrz.	? ,1010	c1914	(c)	(3)
4	MAX	0-4-0DE	Bde	SLM,BBC Tm^{III}	4129,6003	1954	new	(4)
13	SEPP	4wDM	Bdm	RACO 45 PS	1225	1934	(e)	
	rebuilt	4wDE	Bde1	Stadler,BBC	-	1967		(5)

(c) ex BBC, Baden /1966; the exact building date is in doubt, BBC quote 1915, Jäger 1916.
(e) new to SBB as "Tm 506", purchased from SBB as "Tm 553" 8/1965.
(3) 1966 sold to WM "31".
(4) 1994 sold to Kuralit AG, Leibstadt.
(5) 1987 sold to Ferro AG, Baden; by 1990 to ABB- & Alstom, Werk Birrfeld.

AVIA AG, Muttenz-Auhafen   BL05

formerly: Schweizerische Reederei AG (SRAG) (1939-1969)

A fuel depot. There are also two cable operated mules made by Windhoff.

Gauge: 1435 mm				Locn: 213 $ 6162/2660			Date: 09.2008	
		0-4-0DH	Bdh	KHD T4M625 R	56464	1956	(a)	(1)
Tm 2/2	TONY	4wDH	Bdh	Jung RC 24 B	14025	1969	new	(2)
Tm 237 867-7	RONNY	4wDH	Bdh	Gmeinder D 35 B	5468	1970	(c)	

(a) purchased from Säurefabrik Schweizerhalle (SFS), Schweizerhalle /1961.
(c) ex Mobil AG, Raffinerie Wörth GmbH, Wörth/Karlsruhe [D] "2" by /2007.

(1) 1977-82 transferred to Schweizerische Reederei & Neptun AG (SRB), Basel Kleinhüningen.
(2) to Imbach Logistik AG, Schachen by /2004 via LSB.

JULIUS BERGER TIEFBAU AG, Tecknau   BL06 C

This firm held contracts for the first Hauenstein Tunnel (Basel to Olten). The northern portal (Läufelfingen) lies in Basel Land, the other (and Hauenstein itself) in Canton Solothurn. In this Handbook there are entries under both Cantons. Both Tecknau and Sissach are variously quoted as the location of the base at the northern end.

Gauge: 900 mm   Locn: ?   Date: 05.2008

	0-4-0T	Bn2t	Borsig	7263	1910	(a)	s/s

Gauge: 750 mm   Locn: ?   Date: 01.2008

0-4-0T	Bn2t	Borsig	6849	1908	(b)	

(a) ex Julius Berger, Berlin [D] /1913.
(b) new to Julius Berger, Bromberg [D]; to Sissach for this project /1912.

BP (SWITZERLAND) RHEINLAGER, Birsfelden Hafen   BL07

formerly: Geldner Rheinlager (GRL) (1969-74 to about 1996)
            Kohlen Union Geldner (KUG) (1941 to 1969-74)

These warehouses and port bordering the Rhein were closed and demolished in 1998.

Gauge: 1435 mm   Locn: 213 $ 6152/2672   Date: 08.2008

		4wDM	Bdm	Gebus Lokomotor	564	1957	(a)	(1)
		4wDH	Bdh	Henschel DHG 200B	30876	1964	new	(2)
	Lotti	4wDH	Bdh	SLM Tm^{IV}	4983	1973	new	(3)

(a) Breuer licence; new to Lonza Werke, Waldshut [D] via Ing. Struppe, Wien [A]; ex RACO, Zürich /19xx (to be confirmed).
(1) before 1962 sold; by 1967 at BRAG, Basel Kleinhüningen.
(2) 1973 sold to Schlachthof, Zürich.
(3) 1997 sold to Tankanlage BP-ARAL (taa), Muttenz-Auhafen.

BSZ OIL AG, Tankanlage West, Muttenz-Auhafen   BL08

formerly: BP (Schweiz) (2000-2001)
            Mobil-Exxon (Switzerland) (19xx-1999)
            Mobil Oil (Switzerland) (1969-19xx)
            Hanniel (1956-1969)
owned by: Tankanlagen Auhafen AG (taa), Muttenz

This is an oil transhipment and storage facility. The locomotive is now operated by taa, but is kept on BSZ premises.

Gauge: 1435 mm   Locn: 213 $ 6166/2660   Date: 01.2007

Tm 237 956-8 PEGASUS

	4wDH	Bdh	KHD A6M617 R	55754	1955	(a)	(1)

(a) new to DB "Köf 6207", ex DB "323 085-1" /1990.
(1) 2002 to LSB and later FEBEX, Bex, via Tafag.

     see: zefix.admin.ch/shabpdf/current/2005/2004/2003/2002/2001/079-25042001-1.pdf

BÜROCCA AG, Liesberg   BL09 M

A firm occupying several of the buildings of the Portland Cementfabrik Laufen AG, Werk Laufen and operating the Little Nashville restaurant. The locomotives are intended to become part of the "Wild West" decor.

Gauge: 1435 mm   Locn: 223 $ 6004/2501   Date: 11.2007

	0-4-0DM	Bdm	SLM Tm	3935	1946	(a)	store
	4wDH	Bdh	KHD KK140 B	57900	1965	(a)	store

(a)   ex Papierfabrik Zwingen, Zwingen /2007.

CEMENTFABRIK HOLDERBANK   BL10

Werk Liesberg

formerly: Portland Cementfabrik Laufen AG, Werk Liesberg (1929-1985)
              Cement- & Kalkwerk Liesberg (1872-1929)
alternative: Portlandcementwerk Laufen
              Portlandcementwerk Liesberg

The cement works was operational 1872-1986. After closure and from about 1995 Betriebsgesellschaft Historischer Schienenfahrzeuge (BHS) used the site as a store till the group changed its title to Vereinigung Swisstrain Schienenfahrzeuge and moved the locomotives to Bodio in 1998. Since then Burocca SA has brought other locomotives onto the site.

Gauge: 1435 mm   Locn: 223 $ 5998/2494   Date: 08.1990

	4wBE	Ba	Rastatt,MFO	?,21	1910	(a)	(1)
	4wPM	Bbm	Breuer III	1096	c192x	(b)	(2)
	4wBE	Ba2	Stadler	132	1970	new	(3)

Gauge: 600 mm   Locn: ?   Date: 05.2002

	4wDM	Bdm	O&K(M) RL 1a	4615	1931	(d)	s/s

Werk Laufen

Gauge: 600 mm   Locn: ?   Date: 08.2002

	4wDM	Bdm	O&K MV0	25045	1951	(e)	s/s

(a)   ex Cellulose Attisholz AG, Attisholz after /1952.
(b)   mentioned in the Breuer list of references of about 1930.
(d)   purchased via O&K/MBA, Dübendorf.
(e)   ex Cementfabrik Holderbank-Wildegg AG, Wildegg /1957.
(1)   after 1970 scrapped.
(2)   1980-86 scrapped.
(3)   1989 sold to Cementfabrik Holderbank-Wildegg AG, Rekingen AG.

CHEMISCHE FABRIK SCHWEIZERHALL (CFS), Schweizerhalle
                                                                 BL11

later   Schweizerhall Chemie AG (in 2001)
           Brenntag (from 2007)

This fertiliser plant was created in 1845 and connected (between 1889 and 1890) to the Pratteln-Schweizerhalle line (built 1872 to serve the Saline). Since 1935 the locomotives of Saurerfabrik Schweizerhalle have performed the necessary shunting and from 1976 the branch line has been connected to Muttenz II yard instead of Pratteln. For locomotive "1" see CFS, Basel St. Johann.

Gauge: 1435 mm				Locn: 213 $ 6179/2644			Date: 08.1989	
2		0-4-0T	Bn2t	SLM	1439	1902	(a)	1935 wdn

(a) ex Dreispitz-Verwaltung, Basel "1" /1926.

CHIESA ALTMETAL AG, Pratteln — BL12

A scrap merchant. Address: Dammweg 97, 4133 Pratteln.

Gauge: 1435 mm			Locn: 213 $ 6182/2643			Date: 01.2000	
	4wDM	Bdm	RACO 95 SA3 RS	1474	1956	(a)	

(a) ex SBB "Tm^{II} 624" /2004.

DELICA AG, Birsfelden Hafen — BL13

formerly: Migros Betriebe Birsfelden AG (MBB) (1/1/1987-31/12/2006)
Migros Lagerhaus Genossenschaft Birsfelden (MLG) (1954-1987)
alternative: Migros Zentralpackerei AG (seen in 1977)

Warehouse and distribution centre for the alliance of the Migros co-operatives. Locomotives work both the sidings in front of the building and those on the quays alongside the Rhein.

On the quayside on 17 October 2007 237 824-8 (Henschel 30868 of 1964) is passed by SBB Bm 4/4 18435.

Gauge: 1435 mm			Locn: 213 $ 6147/2675 + 6146/2675			Date: 10.2007		
BL 372A209	4wDM R/R	Bdm	Unilok B6000S	?	19xx	(a)	s/s?	
2	4wDM	Bdm	Gebus Lokomotor	565	1957	(b)	(2)	
1	4wDH	Bdh	Jung RK 8 B	13691	1964	(c)	(3)	
Tm 2/2 1 237 822-1	4wDH	Bdh	O&K MB 220 N	26799	1975	(d)		
Tm 2/2 2 237 824-8	4wDH	Bdh	Henschel DHG 240B	30868	1964	(e)		
4	4wDH	Bdh	Henschel DHG 160B	31084	1966	(f)	(6)	

(a) seen in 1977.

(b) Breuer licence, new via Robert Aebi & Co, Zürich.
(c) new, no running number before /1984.
(d) purchased from Schwäbische Zellstoffwerke AG, Dettingen [D] "807" /1996.
(e) purchased from Shell, Ludwigshafen [D] via NEWAG /1993.
(f) short term loan/hire only; ex Rheinstahl Hanomag AG, Hannover-Linden [D] "4" by /1998.
(2) 1984 stored; 1991 donated to Historische Eisenbahn Gesellschaft (HEG), Grenchen; 1995 to Swisstrain; 1997 to Veluwsche Stoomtrein Maatschappij (VSM), Beekbergen [NL].
(3) 1995 sold to Martig AG, Basel.
(6) after 1998 to Eisenwerk Sulzau-Werfen R. & E. Weinberger AG, Tenneck [A].

DEBRUNNER ACIFER AG, Frenkendorf   BL14

formerly: Debrunner & Cie. AG (till 1999)

A steel handling company, rail connected to Frenkendorf-Füllinsdorf SBB station. Railway usage was discontinued in 2002. Address: Bächliackerweg 4402 Frenkendorf.

Gauge: 1435 mm   Locn: 214 $ 6211/2609   Date: 04.2000

4wPM	Bbm	Breuer	?	19xx	(a)	(1)
4wPM	Bbm	RACO 45 PS	1224	1934	(b)	?1970 s/s
0-4-0DM	Bdm	Gmeinder WR 200 B 14	4272	1947	(c)	(3)

(a) may have come from Debrunner & Co AG, St. Gallen-Haggen and may have been 4wDM, Bdm.
(b) new to SBB "Tm 507"; purchased from SBB "Tm 879" /1967.
(c) new to Bayerisches Hafenamt Regensburg [D]; ex J. Wallner, Deggendorf Hafen [D] via Thommen AG, Kaiseraugst /1968.
(1) 1968 scrapped by Thommen AG, Kaiseraugst.
(3) 2002 sold to Süddeutsches Eisenbahnmuseum Heilbronn eV (SEH), Heilbronn-Böckingen [D].

EBM ELEKTIZITÄTSMUSEUM, Münchenstein   BL15 M

A building in the old Alioth (EGA) factory has been adapted as a museum.
Address: Weidenstrasse, 4142 Münchenstein.

Gauge: 800 mm + rack   Locn: 213 $ 6136/2627   Date: 11.2007

WAB HGe 2/2 55	4wWRE	2/bg2	SLM,EGA He 2/2	2086, ?	1910	(a)	display

(a) new to WAB "55", SPB "15" /1964; ex Jungfraubahnen AG /2007.

EUROPÄISCHE VEREINGUNG VON EISENBAHNFREUNDEN FÜR DEN ERHALT VON DAMPLOK (EUROVAPOR), Waldenburg
BL16 M

This society operates trains on the Waldenburgerbahn (WB).

Gauge: 750 mm   Locn: 224 $ 6234/2485   Date: 08.1983

298 14		0-6-2T	C1n2t	Krauss-L AV m	3816	1898	(a) (1)
G 3/3	5 GEDEON THOMMEN						
		0-6-0T	Cn2t	SLM G 3/3	1440	1902	(b)

(a) new to kkStB [A] "U14"; ex ÖBB "298 14" (760 mm).
(b) ex WB "5".
(1) 1984 to Öchsle Schmalspurbahn e.V. [D] "99 7843".

FIRESTONE (SCHWEIZ) AG, Pratteln   BL17

A tyre factory created in 1935: production ceased in 1978.

Gauge: 1435 mm			Locn: 213 $ 6180/2635				Date: 04.1989
		4wD	Bd	unknown		?	19xx	s/s
Tm	1	4wDM	Bdm	Kronenberg DT 90	145	1962	new	(2)

(2) 1980 sold to Borner AG, Trimbach via Giezendanner, Rothrist and Meier & Jäggi, Reiden.

FLORETTSPINNEREI RINGWALD AG, Füllinsdorf — BL18

formerly: Glasfabrik Liestal, Louis Morin, Basel (1917-18) when the location was called Niederschöntal-Frenkendorf

The original glass factory became a weaving mill and operated as such in the period 1918-1956. The site was later re-used by the engineering firm of Schafir & Mugglin. It has not been possible to determine when the locomotive finally vanished.

Gauge: 1435 mm			Locn: 214 $ 6213/2606			Date: 04.1989
	1ABE	1Aa1	BBC Akku-Fahrz.	1209	1917	new s/s

FLORIN OEL- UND FETTWERK AG, Muttenz — BL19

A firm producing and marketing vegetable oil. It was constructed in 1993-94 and rail connected from 1994.

Gauge: 1435 mm			Locn: 213 $ 6152/2651			Date: 01.2003	
	4wDH	Bdh	Schöma CFL 150DBR	2292	1960	(a)	(1)
837 802-8	6wDH	Cdh	KrMa M 350 C ex	19399	1969	(b)	

(a) ex Kalkwerke Oker, Adolph Willikens AG [D] "2" /1990 via HFB/Asper.
(b) ex Lonza AG, Waldshut, [D] "289 later 443" via LSB /1993.
(1) 1994 sold to LSB; later MF. Hügler AG, Dübendorf "237 900-6".

GALVASUISSE AG, Prattelen — BL20

formerly: Verzinkerei Prattelen AG

A short branch connects the firm to Prattelen station. A rail-crane has been used for shunting. Address: Kunimattweg 10.

Gauge: 1435 mm		Locn: 213 $ 6184/2638			Date: 09.1969
	4wDcrane		Bell & Cie, Kriens	?	19xx

GIPS UNION AG, Werk Läufelfingen — BL21

A depot of this Zürich-based firm (founded 1895) lay beside the station of the Hauenstein line (Sissach-Olten). Typical of many installations of Gips Union it was rail-served, but without locomotives. The track layout was unusual being almost symmetrical about a line perpendicular to the main line. An aerial ropeway, 4.5 kilometres long, brought ore from the quarries at Zeglingen via the works lying about a kilometre from the railway until they closed in 1983. The narrow gauge system is surmised to have been in the quarries; the continued existence of the narrow gauge locomotives has not been verified.

Gauge: 1435 mm			Locn: 224 $ 6316/2490			Date: 11.1970	
Gauge: 600 mm			Locn: 224 $ 6361/2511 ?			Date: 05.2002	
	4wDM	Bdm	MBA MD 1	2009	c1950	(a)	
	4wDM	Bdm	O&K(M) RL 1a	4210	1930	(b)	(2)
	4wDM	Bdm	O&K(M) RL 1a	4640	1931	(c)	

(a) ex MBA, Dübendorf as second hand locomotive.
(b) ex MBA, Dübendorf /1949.

(c) via O&K/MBA, Dübendorf.
(2) 19xx returned to MBA, Dübendorf; 1996 on display there.
     see: www.baselland.ch/docs/archive/chronik/chro1983/jun1983.htm

HAUENSTEIN, Sissach   BL22 C

Some locomotives were made available for the construction of the Hauenstein line.

Gauge: ?1000 mm   Locn: ?   Date: 05.2008
            0-4-0T   Bn2t   Heilbronn III   343   1898   (a)   (1)
Gauge: 600 mm   Locn: ?   Date: 03.2002
1   0-4-0T   Bn2t   Freudenstein   106   1901   (b)   (2)
3   0-4-0T   Bn2t   Maffei   2833   1908   (b)   (2)
4   0-4-0T   Bn2t   Jung Helikon   1427   1910   (b)   (2)

(a) from Sissach-Gelterkinden (SG) (1000 mm) on closure /1916.
(b) made available by W. &. J. Rapp AG, Basel.
(1) 19xx to construction of Chur-Arosa (ChA).
(2) probably returned to W. &. J. Rapp AG, Basel.

HENKEL & CIE AG, Pratteln   BL23

The Swiss branch of a detergent making concern with head office at Düsseldorf [D] established in 1913. Production stopped in 1997 and the factory was partially demolished in 1998.

Gauge: 1435 mm   Locn: 213 $ 6184/2638   Date: 12.2001
            4wBE   Ba   SSW   ?   1930   new   (1)
            4wDE   Bde   Stadler   ?   1963   (b)   (2)
3   4wDE   Bde   Stadler   131   1969   (c)   (3)

(b) constructed on the chassis of a battery locomotive from Metallwerke Dornach AG.
(c) ex Schlachthof Basel via Stadler /1976.
(1) ca. 1963 scrapped.
(2) 1976 returned to Stadler and scrapped.
(3) 1998 sold to la Fonte électrique, Bex (Fébex).

JURASSISCHE STEINBRUCH CUENI AG, Laufen   BL24

formerly: Steinbruch Friedrich (from 1875)

A stone quarry at Lochfeld. A rail connection was installed in 1877 and equipped with a 160 metre long rack section; later the rack section was extended. The quarry closed in 1930.

Gauge: 1435 mm + Riggenbach rack   Locn: 223 $ 6055/2526   Date: 09.1993
     2   0-4-0VBRT Bzn2t   Aarau   13   1878   new   (1)
(1) 1918 withdrawn; c1932 scrapped.

KERAMIK LAUFEN, Laufen   BL25

formerly: Tonwarenfabrik Laufen

The quarries were situated above the village to the west and were connected to the storage and drying sheds by a kilometre long tramway that closed in 1969.
From the works area an aerial ropeway ran down to the station: this closed before the railway.

Gauge: 600 mm   Locn: 223 $ 6050/2512   Date: 05.1988
            4wDM   Bdm   O&K(S) RL 1c   1552   194x   (a)   s/s

4wDM	Bdm	O&K(M) RL 1	3356	1929	(a)	s/s	
4wDM	Bdm	O&K(M) RL 1	3466	1929	(a)	s/s	
4wDM	Bdm	O&K(M) RL 1a	4602	1931	(a)	s/s	
4wDM	Bdm	O&K MV0	25044	1951	(e)	(5)	
4wDM	Bdm	O&K MV0	25046	1951	(f)	(6)	
4wDM	Bdm	O&K MV0	25074	1951	(g)	(7)	
4wDM	Bdm	O&K MV1	25136	1951	(h)	(8)	
4wDM	Bdm	O&K MV0	25155	1951	(i)	1974 s/s	
4wDM	Bdm	O&K MV0	25165	1951	(j)	s/s	
4wDM	Bdm	O&K MV1a	25542	1953	(k)	1974 s/s	
4wDM	Bdm	O&K MV1a	25738	1957	(l)	1992 scr	

- (a) purchased via MBA, Dübendorf.
- (e) purchased via Wander-Wendel.
- (f) new to Wander-Wendel.
- (g) purchased via Wander-Wendel and MBA, Dübendorf.
- (h) purchased from MBA, Dübendorf /1954.
- (i) new to Pletz & Bender, Meppen/Ems [D]; ex ? /19xx.
- (j) new via Wander-Wendel and MBA, Dübendorf.
- (k) new to Dipl. Ing. Hans Oehm, Meppen/Ems [D]; ex ? /19xx.
- (l) new to Max Hohl, Baumschule (tree nursery), Stuttgart-Degerloch [D]; ex Tonwerk Kandern GmbH, Kandern [D] /198x.

- (5) ?1974 sold to Kaolins de Bauvoir, Echassières, Bellenaves, Allier [F].
- (6) 1969 sold to Mechanische Ziegelei AG, Oberwil; returned; 1988 to FWF, Otelfingen.
- (7) 19xx sold to Ziegelei Landquart (750 mm).
- (8) 1974 sold to Tonwerke Kandern GmbH, Kandern [D].

KIOSK AG, Muttenz   BL26

A distribution centre for a chain of kiosks constructed in 1993 and rail connected in 1994. It is understood that rail traffic has ceased though the siding is still in place (to be confirmed).

Gauge: 1435 mm   Locn: 213 $ 6154/2650   Date: 09.1999

4wDM	Bdm	RACO 95 SA3 RS	1453	1955	(a)	(1)

- (a) ex SBB "Tm^{II} 784" /1995.
- (1) 2000 donated to Draisinen-Sammlung Fricktal (DSF), Laufenburg.

KOHLER BAUBEDARF AG, Frenkendorf   BL27

A distribution centre for building materials.

Gauge: 1435 mm   Locn: 214 $ 6212/2621   Date: 09.1999

4wDM	Bdm	RACO 85 LA7	1577	1960	(a)	(1)

- (a) ex SBB "Tm^{I} 414" by 4/1997 (possibly only on hire?).
- (1) before 1998 returned to SBB.

LANDOR AG, Muttenz-Auhafen   BL28

formerly: Union Umschlag und Spedition AG (19xx-1986)
           Brikett- Umschlags- und Transporte AG

These harbour installations date from 1941.

Gauge: 1435 mm   Locn: 213 $ 6161/2661   Date: 06.2001

Tm 2/2 Joggeli	4wDM	Bdm	Kronenberg DT 50/90	122	1951	(a)
Tm 2/2 Marco	4wDH	Bdh	Jung RK 8 B	13272	1960	(b)

- (a) new; unnamed till 1991.
- (b) new to A.S. Det Danske Stalvalsvaerk [DK]; ex Container-Depot, Niederhasli via LSB /1990; unnamed till /1995; then "Hitsch"; renamed "Marco" by /2007.

LONZA AG, Werk Schweizerhalle — BL29

A plant for producing and storing chemical products. It operated 1967-98.

Gauge: 1435 mm Locn: 213 $ 6179/2645 Date: 05.1996

Tm	0-4-0PM	Bbm	SLM Tm	3581	1934	(a)	(1)
	4wDH	Bdh	Jung RK 8 B	14129	1973	(b)	(2)
	4wDM R/R	Bdm	unknown R/R	?	19xx		s/s?

(a) new to SBB "Tm 210", ex SBB "Tm 557" /1968; SLM used this works number twice.
(b) ex Oskar Kiehm by /1977.

(1) by 1974 in a playground at Frenkendorf; later scrapped.
(2) 1998 sold to ETRA, Zürich; 2000 to Thommen AG, Kaiseraugst.

MECHANISCHE ZIEGELEI AG, Oberwil — BL30

A brickworks with a line of three hundred metres in length, apparently joining two parts of the works and running roughly parallel to Hohestrasse. Address: Hohestrasse 131 (offices).

Gauge: 500 mm Locn: 213 $ 6086/2632 Date: 04.2002

4wDM	Bdm	O&K(M) RL 1	3660	1929	(a)	s/s
4wDM	Bdm	O&K MV0	25046	1951	(b)	(2)
4wDM	Bdm	Diema DS20	2733	1964	new	(3)

(a) purchased via MBA, Dübendorf (originally 600 mm).
(b) ex Keramik Laufen, Laufen (600 mm) /1969.

(2) before 1988 to Keramik Laufen, Laufen and thence FWF, Otelfingen (600 mm).
(3) 199x sold to M. Strebel, Mägenwil.

MIGROS GENOSSENSCHAFT BASEL, Münchenstein — BL31

Distribution centre constructed in 1977-78 and connected from the start to the Dreispitz industrial zone.

Gauge: 1435 mm Locn: 213 $ 6127/2640 Date: 08.1998

M	4wDM	Bdm	O&K MV3	26596	1966	(a)	(1)

(a) O&K delivery list indicates type MV3: ex VLG, Herzogenbuchsee /1976.

(1) 1998 donated to Transcontinental Museum Club and stored at Henkel, Pratteln; 2003 to Altola AG, Olten via Stauffer.

PAPIERFABRIK ZWINGEN, Zwingen — BL32

formerly: Holzstoff und Papierfabrik AG

A kilometre long line across the valley connected the station to the factory. This closed in 2005. The line crosses some tracks of a long abandoned 600 mm system on the way.

Gauge: 1435 mm Locn: 223 $ 6063/2539 Date: 08.2001

214	4wPM	Bbm	Breuer	?	1930	new	(1)
	0-4-0DM	Bdm	SLM Tm	3935	1946	new	(2)
	4wDH	Bdh	KHD KK140 B	57900	1965	(c)	(2)
402	4wDH	Bdh	Henschel DH 160B	26924	1960	(d)	(4)

(c) ex Ziegler Papier AG, Grellingen /1991.
(d) hired from Asper /1991.

(1) 1946 to WM.
(2) 2007 to Bürocca AG, Liesberg.
(4) 1991 returned to Asper.

PASSAVANT-ISELIN AG, Allschwil   BL33
formerly: Ziegelei Passavant-Iselin & Co AG
A brickworks founded in 1878. Address: Binningerstrasse 112.

Gauge: 600 mm   Locn: 213 $ 6084/2662   Date: 02.2007

4wDM	Bdm	O&K(M) MD 1	10502	1940	new	s/s
4wDM	Bdm	O&K(M) MD 1	11415	1939	new	s/s
4wDM	Bdm	O&K MV0	25164	1952	new	s/s

PETROPLUS TANKSTORAGE AG, Birsfelden Hafen   BL34
formerly: Shell (Switzerland) (1952-2000)
This fuel storage depot has rail connections to both the railway in Hafenstrasse and the railway on the quay side. Address: Hafenstrasse 92, 4127 Birsfelden.

Gauge: 1435 mm   Locn: 213 $ 6149/2674 + 6149/2675   Date: 06.2008

	4wDM	Bdm	Breuer	?	c1943		(1)
	4wDM	Bdm	Breuer VL	3088	1955	(b)	(2)
	4wDM	Bdm	RACO VL82PS	1490	1957	(c)	(3)
	4wDH	Bdh	Henschel DHG 240B	30867	1964	(d)	
Tm 2/2 1	4wDH	Bdh	Henschel DHG 240B	31089	1965	(e)	
Bärli Em 847 856-2	BoBoDH	BBdh	TransLok TL-DH440	180	2008	new	

(b)   new via Robert Aebi & Co, Zürich.
(c)   new; Breuer style.
(d)   new, the delivery list indicates DHG 240ex.
(e)   hired and later purchased from Raffinerie de Cressier, Cornaux NE (RCSA) "1" /1981, the delivery
      list indicates DHG 240ex.

(1)   1963 sold to Raffinerie de Cressier, Cornaux NE.
(2)   by 1972 sold to Shell AG, Niederhasli.
(3)   1979 sold to Haute- und Fettwerke, Zürich via Asper.

SCHAFIR & MUGGLIN AG (S&M), Liestal   BL35 C
A public works concern that had a workshops and storage area adjacent to Frenkendorf-Füllinsdorf station. This site was established in 1921 and gradually closed 1987-92.

600 mm gauge O&K 2698 of 1927 (type RL 2) was modified for use on the standard gauge. It is seen here outside the workshops area on 17 July 1979.

Gauge: 1435 mm				Locn: 214 $ 6212/2608				Date: 04.1989	
	2.21.02	4wDM	Bdm	O&K(M) RL 2		2698	1927	(a)	(1)
Gauge: 800 mm				Locn: ?				Date: 04. 2002	
1.03.1	2.20.01	4wBE	Ba1	Stadler,BBC		20, ?	1946	(b)	s/s?
Gauge: 750 mm				Locn: ?				Date: 04. 2002	
	BUBI		0-4-0T	Bn2t	Jung 10 PS	1349	1909	(c)	(3)
	SCHWEIZ	0-4-0T	Bn2t	Jung 80 PS		1515	1910	(d)	(3)
15	SÄNTIS	0-4-0T	Bn2t	Jung 80 PS		1622	1911	(e)	s/s?
14	ALVIER	0-4-0T	Bn2t	Jung 80 PS		1682	1911	(d)	(3)
16	FALKNIS	0-4-0T	Bn2t	Jung 80 PS		1683	1911	(d)	a1959 scr
	MUTTENZ	0-4-0T	Bn2t	Jung 20/25 PS		3620	1924	(c)	(3)
	SPERANZA	0-4-0T	Bn2t	Krauss-S XXVII vv		3254	1895	(c)	(7)
	LICHTENSTEIG	0-4-0T	Bn2t	Krauss-S XXVII bg		6173	1909	(c)	(7)
	VALLONE	0-4-0T	B?2t	Maffei		3609	1911	(d)	a1959 scr
17	AARE	0-4-0T	Bn2t	O&K 80 PS		1247	1904	(l)	s/s?
	BASEL	0-4-0T	Bn2t	O&K 80 PS		1622	1905	(m)	(13)
	VEVEY	0-4-0T	Bn2t	O&K 50 PS		2907	1908	(c)	s/s?
20	BADEN	0-4-0T	Bn2t	O&K 60 PS		4281	1911	(o)	(3)
31	AARAU	0-4-0T	Bn2t	O&K 70 PS		4374	1911	(o)	a1959 scr
7	ZÜRICH	0-4-0T	B?2t	O&K 70 PS		4375	1911	(q)	a1959 scr
6	MURI	0-4-0T	B?2t	O&K 70 PS		6020	1914	(q)	(13)
19	LIMMAT	0-4-0T	B?2t	O&K 70 PS		6573	1913	(q)	(13)
	GWATT	0-4-0T	B?2t	O&K 50 PS		7520	1916	(t)	(13)
1.03.2	2.20.02 URSULA								
		4wWE	Bg2	Stadler		33	c1949	(u)	(21)
1.03.3	2.20.03 ERIKA								
		4wWE	Bg2	Stadler		34	c1949	(u)	(21)
1.03.4	2.20.04 THERESE								
		4wWE	Bg2	Stadler		35	c1949	(u)	(21)
1.02.6		4wDM	Bdm	O&K(M) RL 2		2698	1927	(x)	(24)

CH 108 BL

Gauge: 600-750 mm				Locn: ?			Date: 04. 2002	
1.03.5	2.20.05	4wBE	Ba1	SIG ET 20L	?	19xx	new?	s/s?
1.02.4	2.21.01	4wDM	Bdm	O&K(M) RL 2a	20196	1931	(y)	s/s?

Gauge: 600 mm				Locn: ?			Date: 04. 2002	
	2.21.04	4wDM	Bdm	O&K 3 D	21504	1939		(27)
		4wDM	Bdm	O&K MV1	25140	1952	(ab)	(27)

(a) rebuild of 750 mm gauge locomotive "1.02.6" /1947; first used on the Etzelwerk contract near Rupperswil.
(b) new for work at Julien/Tiefencastel then stored at Frenkendorf.
(c) ex Schafir & Mugglin, Zürich /1939-53.
(d) see note for this locomotive under Robert Aebi & Cie AG, Regensdorf; ex Schafir & Mugglin, Muri AG /1933-35, after of the reconstruction of the watercourses for the hydroelectric power station at Klingnau.
(e) ex Schafir & Mugglin, Muri AG /1933-35, after of the reconstruction of the watercourses for the hydroelectric power station at Klingnau; used at Zollbrück /1948.
(l) ex Schafir & Mugglin, Muri AG /1933-37, after of the reconstruction of the watercourses for the hydroelectric power station at Klingnau.
(m) ex Schafir & Mugglin, Zürich /1939-59.
(o) ex Robert Aebi & Co, Zürich /1942 and used on the construction of the watercourses for the hydroelectric power station at Rupperswil-Auenstein.
(q) ex Schafir & Mugglin, Zürich /1939-59 and used on the construction of the watercourses for the hydroelectric power station at Rupperswil-Auenstein.
(t) ex Schafir & Mugglin, Zürich /1939-59.
(u) new for construction of the electric power station of Wildegg then stored at Frenkendorf.
(y) new to A. Hauser, Oberwinterthur.
(x) ex Mangold &Co, Bauunternehmer, Zürich.
(ab) ex Hew & Co, Chur after /1970.
(1) 1988 stored off the track.
(7) 1953 removed from boiler register; after 1959 scrapped.
(13) 1960 removed from boiler register.
(21) stored for ten years; intended for sale to Ceylon; 1975 scrapped at Stadler.
(24) 1947 rebuilt to standard gauge locomotive "2.21.02".
(27) 198x sold to Ad. Schäfer & Cie AG, Aarau.

SF CHEM AG, Pratteln BL36

formerly: Säurefabrik Schweizerhalle (SFS)

A factory constructed in 1917 to produce sulphuric acid from salt obtained from the adjacent Saline.

Gauge: 1435 mm				Locn: 213 $ 617900,264230			Date: 07.1998		
	8552	0-6-0WT	Cn2t	SLM E 3/3	898	1894	(a)		1940 scr
		0-6-0WT	Cn2t	SLM E 3/3	679	1891	(b)		1942 scr
		0-6-0WT	Cn2t	SLM E 3/3	1009	1896	(c)		1957 scr
		0-4-0DH	Bdh	KHD T4M625 R	56464	1956	new	(4)	
Em 3/3 1	837 813-5	0-6-0DE	Cdh	Jung R 30 C	12998	1960	(e)		
Tm 237 859-4		4wDH	Bdh	KrMa ML 225 B	18701	1960	(f)		
	8481	0-6-0WT	Cn2t	SLM E 3/3	1877	1907	(g)	(7)	
		0-4-0DE	Bde1	SLM,BBC Tm^{III}	4564,7516	1965	(h)		
237 866-9		0-4-0DE	Bde1	SLM,BBC Tm^{III}	4586,7539	1965,66	(i)		

(a) new to NOB "254", later "454"; purchased from SBB "E 3/3 8552" /1935.
(b) purchased from SOB "5" /1940.
(c) new to SCB "74"; purchased from SBB "E 3/3 8419" /1942.
(e) into service /1961.
(f) new to Westfälische Union AG, Hamm [D]; later Thyssen Draht, Hamm [D] "2"; then OnRail; ex LSB /1997.
(g) on loan from Feldschlösschen Brauerei, Rheinfelden /1970.
(h) new to SBB "Tm^{III} 907"; hire from Stauffer 10/2007.
(i) new to Dreispitz-Verwaltung, Basel "12"; hire from Stauffer /2006-7.

(4) 1961 to Schweizerische Reederei AG (SRAG), Muttenz-Auhafen.
(7) returned to Feldschlösschen Brauerei, Rheinfelden.

SI GROUP - SWITZERLAND GmbH, Pratteln BL37

SI = Schenectady International
formerly: Schenectady Pratteln AG (1994-96)
  Alphen Pratteln AG (1991-93)
  Schweizerische Teerindustrie AG (STIA) (1937-91)
  Schweizerische Industriegesellschaft für Prodorite (SIP) (1926-37)

The plant initially made products from the tar resulting from conversion of (pit-) coal to gas. Tar came from all over Switzerland for processing and the site took a leading role in providing such diverse products as road surfaces and cobblers' wax. At the end of the 1960s chemical products (acrylics) slowly replaced the output and the name was changed to suit (Alkylphenolen). In 1994 Schenectady International Inc took control and the parent company, in turn, made a name change in 2006.

The second locomotive at this location (Stadler 137 of 1971) is seen at work on 9 September 1996.

Gauge: 1435 mm   Locn: 213 $ 6180/2636   Date: 09.2007

	4wBE	Ba2	SIG,MFO	? ,22	1912	(a) c1972 scr
Tm 2/2	4wDE	Bde	Stadler	137	1971	new

(a) originally A1BE A1a; also reported as MFO 32; fitted with second motor /1927. ; ex Maggi AG, Kemptthal /19xx.

see: www.industrieweg.ch/cms/front_content.php?idcat=103

SIHELCO AG, Muttenz-Auhafen BL38

SIHELCO = SIlice HElvétique COmpagnie

This company performs transhipment and storage of industrial silica in various forms.

The Italian version of a KHD locomotive, Greco 2658 of 1966 was photographed in the loading bay on 13 November 2007.

Gauge: 1435 mm			Locn: 213 $ 6167/2656					Date: 06.2001
Tm 2/2	1	4wDH	Bdh	Greco		2658	1966	(a)
(a)	built under Deutz licence; ex SIBELCO Italia, Sessa Aurunca, Napoli [I] /1968.							

SISSACH-GELTERKINDEN-BAHN, Sissach — BL41 C

This line operated from 1891-1916 when it was superseded by a standard gauge line. During construction, the following locomotive has been recorded.

Gauge: 1000 mm		Locn: ?					Date. 05.2008
	2-4-2Tm	1B1n2tk	SLM		256	1882	(a) s/s
(a)	made available by Pümpin & Herzog, Bern.						

SPIELPLATZ, Frenkendorf — BL42 M

Gauge: 1435 mm		Locn: ?					Date. 01.2008
	0-4-0PM	Bbm	SLM Tm		3581	1934	(a) 19xx scr
(a)	new to SBB "Tm 210"; ex Lonza AG, Schweizerhalle by /1974; SLM used this works number twice.						

TANKANLAGEN AUHAFEN AG (taa), Muttenz-Auhafen BL43

formerly: Tankanlage BP-Aral (1997-1999)
Tankanlage Aral-Shell (1984-1997)
Gemeinschaftstankanlage Aral-Gulf (1976-1984)
Gulf (Schweiz) AG, Basel (c1965-1976)
Robert Jecker Mineralöl - Benzin AG, Zürich (RIMBA AG) (1952-c1965)

This fuel storage depot operates locomotives in two separate areas. There are also two cable worked mules.

Gauge: 1435 mm Locn: 213 $ 616400,265900 + 616460,266050 Date: 12.2008

James	4wDM	Bdm	Breuer VL	3093	1955	(a)	1978 scr
Fridolin	4wDH	Bdh	Krupp 225 PS	3340	1956	(b)	2003 scr
Tm^{IV} 237 814-9 Lotti	4wDH	Bdh	SLM Tm^{IV}	4983	1973	(c)	
Tm 237 868-5	4wDH	Bdh	Henschel DHG 275B	31098	1967	(d)	

(a) new via Robert Aebi & Co, Zürich.
(b) ex B.V. Aral AG, Bochum - Tanklager Gelsenkirchen [D] "1" /1977.
(c) ex BP, Birsfelden Hafen /1997.
(d) ex Merck KGaA, Darmstadt [D] (delivered 29/06/1968) /200x via LSB.

TOTTOLI & MÜLLER, Pratteln BL44 C

The work performed was the construction of the goods station at Pratteln. The name of the firm is not confirmed.

Gauge: 600 mm Locn: ? Date: 02.2008

	0-4-0T	B?2t	Maffei	3564	1911	(a) s/s

(a) ex construction of the station at Sursee /19xx.

TRANSCONTINENTAL MUSEUM CLUB, Pratteln BL45 M

A preservation group concerned with both steam and electric locomotives from European countries. At least one locomotive belonging to the club was located on the premises of the former Henkel SA in the company of locomotives from other groups.

Gauge: 1435 mm Locn: 213 $ 6184/2638 Date: 12.2001

M	4wDM	Bdm	O&K MV3	26596	1966	(a)	(1)

(a) ex Migros Genossenschaft Basel, Münchenstein /1998.
(1) 2003 to Altola AG, Olten via Stauffer.

ULTRA-BRAG AG, Muttenz-Auhafen BL46

formerly: Basler Rheinschiffahrt AG (BRAG) (1925-1974)
Umschlags- Lagerung- und Transporte (ULTRA) AG (1953-1974)

This company was formed in 1974 by the merger of two neighbouring concerns. The port installations and storage facilities were installed in 1953. ULTRA and BRAG tanks followed in 1956; though the latter company had been established much earlier.

Gauge: 1435 mm Locn: 213 $ 6167/2655 + 6169/2654 Locn: 06.2001

Tm 2/2 1	4wDH	Bdh	Jung RK 15 B	13397	1961	(a)	1999 scr
Tm 2/2 2	4wDH	Bdh	Jung RC 24 B	14154	1973	(b)	
1 Tm 237 857-8	4wDH	Bdh	O&K MHB 101	26971	1981	(c)	
Tm 237 864-4	4wDM	Bdm	Jung RK 15 B	13399	1961	(d)	(4)

(a) ex Chemische Fabrik Schweizerhall (CFS), Basel-St. Johann "3" /1974.
(b) into service 1974.
(c) new to Maschinenfabrik Fahr AG, Gottmadingen [D] "390"; ex Kverneland Gottmadingen GmbH & Co. KG (formerly Greenland) /1997 via METRAG.
(d) ex Ultra-Brag AG, Basel Kleinhüningen "3" /1999.

(4) by 2007 to Ultra-Brag AG, Hafen St. Johann.

VICARINO & CURTY, Pratteln BL47

This concern (Baugesellschaft) was involved in the construction of the Bötzbergbahn, from Pratteln to Rheinfelden. The exact location of the site is not known.

Gauge: 1435 mm Locn: ? Date: 05.2008

        0-4-0T   Bn2t   Heilbronn II alt   28   1873   (a)   (1)

(a) ex Wolf & Grafs, Zürich /18xx.
(1) 18xx to Baugesellschaft Flüelen-Göschenen, Flüelen.

VIGIER CEMENT AG, Lausen BL48

formerly: Wilhelm Brodtbeck AG (1893-31/12/1999).

A Portland cement factory originally located in Liestal and transferred to Lausen (three kilometres away) in 1922. It lay out of use 1932-48. The ovens were stopped again in 1967 and the factory abandoned ca. 2000.

Gauge: 1435 mm Locn: 214 $ 6251/2576 Date: 05.2000

        4wDE   Bde   Moyse BN28E168D   1155   1967   (a)   OOU

(a) new via Asper; the type has also been quoted as "BN24E 180D".

VOPAK (SCHWEIZ) AG, Birsfelden Hafen BL49

formerly: Van Ommeren-Packoed
          Van Ommeren (Schweiz) AG (1997-?31/12/1999)
          Van Ommeren-Bragtank AG (1992-97)
          Bragtank AG (1972-1992)
          Basler Tankanlager AG (BRAG) (1956-72)

This is a fuel storage depot. Address: Hafenstrasse 87-89, 4127 Birsfelden

Gauge: 1435 mm Locn: 213 $ 6151/2671 Date: 08.2008

4wDM	Bdm	Gebus Lokomotor	564	1957	(a)	1977 scr
4wDH	Bdh	Henschel DHG 240Bex	31091	1965	(b)	(2)
0-4-0DH	Bdh	Henschel DH 240B	29705	1958	(c)	(3)
4wDH	Bdh	Henschel DHG 300B	31555	1973	(d)	

(a) Breuer licence; ex BRAG, Basel Kleinhüningen /1968.
(b) or DHG 240B ex; purchased new via Asper.
(c) hire from LSB /1990.
(d) ex Thyssen Sonnenberg GmbH, Duisburg-Ruhrort [D] "5", via OnRail/LSB /1992.
(2) 1992 sold to LSB; 1995 to OnRail; 1998 to KM Europa Metall AG, Osnabrück, [D].
(3) 1990 returned to LSB.

ZIEGLER PAPIER AG, Grellingen BL52

formerly: Papierfabrik Albert Ziegler AG

This paper factory was founded in 1861. Location: Bahnhofstrasse 21.

Gauge: 1435 mm Locn: 213 $ 6117/2546 Date: 12.2002

1	4wDM	Bdm	Kronenberg DT 90	115	1949	(a)	(1)
2	4wDH	Bdh	KHD KK 140B	57900	1965	(b)	(2)
1	4wDH	Bdh	KHD A8L614 R	57216	1961	(c)	

(a) purchased from Acciaierie e Laminatori Monteforno, Bodio "1" /1960.
(b) new to Müller & Sohn AG, Hamburg-Waltersdorf [D]; ex Silo P. Kruse, Eversween,

(c)	Hohe Schaar, Hamburg [D] "2" via WBB /1983. new to Glas- & Spiegel-Manufaktur AG, Gelsenkirchen [D] "2"; ex E. Zimmermann-Mollet, Reinach BL /1990 via Asper.
(1)	1983 OOU; 1987 scrapped in Laufen.
(2)	1991 sold to Papierfabrik Zwingen, Zwingen.

E. ZIMMERMANN-MOLLET, Reinach BL BL53 C

An enterprise with a locomotive intended to provide shunting capability. The locomotive was used at Basel Wolf and then stored at Basel St. Johann and from 1988 at Pratteln.

Gauge: 1435 mm Locn: ? Date: 05.1989

2	MIGGELI	4wDH	Bdh	KHD A8L614 R	57216	1961	(a)	(1)

(a) ex Glas- und Spiegelmanufaktur, Gelsenkirchen-Schalke [D] "2" /1984 via WLH and Asper.
(1) 1990 to Ziegler Papier AG, Grellingen via Asper.

Basel Stadt   BS

In Basel the DB Badischer Bahnhof is located inside the city boundaries. Consequently locomotives on industrial sidings in the area are registered with the Baden area of the DB, but listed here. The locomotives usually carry the corresponding three-digit number allocated by Baden-Württemberg as well as any other insignia.

All three main-line terminal stations within Basel have moved location during the period under review; this has affected the traffic flows.

BAHNHOFKÜHLHAUS AG   BS01

Basel Badischer Bahnhof

Cold storage warehouses. Ice for refrigerated wagons passing through Basel was made here 1935-8x.

Gauge: 1435 mm   Locn: 213 $ 6127/2649   Date: 05.1989

514	243	4wPM	Bbm	RACO 45 PS	1235	1935	(a)	(1)
1	230	4wDH	Bdh	Schöma CFL 40DR	2801	1964	(b)	(2)
2	286	4wDH	Bdh	Schöma CFL 60DR	3136	1969	(c)	(3)

Basel Wolf

Gauge: 1435 mm   Locn: 213 $ 6126/2664   Date: 04.1989

		4wPH	Bbh	Stemag, von Roll	?	19xx	(d)	(4)
		4wPM	Bbm	Breuer III	?	1928	(e)	(5)
		4wPM	Bbm	Breuer III	?	1931	(f)	(5)
		4wPM	Bbm	Breuer III	?	1929	(g)	(5)
		4wPM	Bbm	Breuer III	?	1928	(h)	(5)
	514	4wPM	Bbm	RACO 45 PS	1235	1935	(i)	(9)

(a) originally SBB "Tm 514"; ex Bahnhofkühlhaus, Basel-Wolf /1966.
(b) new, via Asper.
(c) new, via Asper; Schöma lists give the type as CFL DBR.
(d) constructed from a four-wheeled hopper-wagon on which a small petrol engine (VW Beetle?) was mounted. This drove a hydraulic pump and thence both axles.
(e) ex SBB "Tm 446" /1961.
(f) ex SBB "Tm 463" /1961.
(g) new to SBB "Tm 455"; ex SBB "Tm 894" /1964.
(h) ex SBB "Tm 442" /1964.
(i) ex SBB "Tm 514" /1964.

(1) 1969 to SATRAM Huiles SA, Basel Kleinhüningen.
(2) 1990 to Sika Norm, Düdingen.
(3) 1990 to Ferroflex, Rothrist.
(4) before 1960-66 to Basaltstein AG, Buchs SG.
(5) 19xx s/s; but may have been transferred to Bahnhofkühlhaus, Basel Badischer Bahnhof first.
(9) 1966 to Bahnhofkühlhaus, Basel Badischer Bahnhof.

BASEL GAS- UND WASSERWERK, GASKOKEREI,
Basel Kleinhüningen   BS02

formerly: Gaswerk + Wasserwerk Basel (G+W) (from 1875)
            Gaswerk Basel

Gasworks initially at Basel St. Johann (1860-1931) and rail connected from 1882 till closure. A new factory was constructed in Basel Kleinhüningen in 1931; this was demolished in 197x.

Gauge: 1435 mm			Locn: 213 $ 6118/2703					Date: 06.1888
1 (later GAS)		0-4-0T	Bn2t	SLM	1438	1902	new	(1)
2		0-4-0T	Bn2t	SLM	1837	1907	new	(2)
1		0-6-0WT	Cn2t	SLM E 3/3	727	1892	(c)	1948 scr
1		0-6-0WT	Ch2t	SLM E 3/3	2239	1912	(d)	(4)
2		0-6-0WT	Cn2t	SLM E 3/3	795	1893	(e)	(5)
73		0-6-0WT	Cn2t	SLM E 3/3	38	1874	(f)	(6)
		4wPM	Bbm	Breuer III	?	c1930	new	(7)
		0-4-0DH	Bdh	KHD A12L614 R	56282	1956	new	(8)

(c) purchased from Sihltalbahn (SiTB) "1" /1924.
(d) new to Sihltalbahn (SiTB) "6"; purchased from UeBB "6" /1948.
(e) purchased from Sihltalbahn (SiTB) "2" /1924.
(f) new to ARB "1"; later "11"; purchased from BLS "73" /1931.
(1) 19xx sold to Schweizerische Reederei AG, Strasbourg, Bas-Rhin [F].
(2) 1926 sold to Schweizerische Reederei AG, Basel "2".
(4) 1956 withdrawn; one unconfirmed report suggests it survived till at least 1969.
(5) 1979 to Rhôna SA, Le Bouveret for Rive Bleue Express "HANSLI"; 2005 Zürcher Museumsbahn, Sihlwald.
(6) 1934 to von Roll, Gerlafingen "73".
(7) 1933 to SBB "Tm 2/2 465".
(8) 1971 to Schweizerische Reederei AG (SRAG), Basel Kleinhüningen.

BRENNTAG SCHWEIZERHALL, Basel St. Johann   BS03

formerly: Chemische Fabrik Schweizerhall (CFS)

Just over the border from France an industrial line goes to St. Johann Docks. A spur from it serves this site. The factory built for producing fertiliser was rail connected in 1900. It has been used as a store and distribution centre since the 1920s. By 2008 the locomotive shed had been disconnected and adapted for road/rail vehicles. Address: Elsässerstrasse 231, Basel.

Gauge: 1435 mm				Locn: 213 $ 6103/2692				Date: 10.2007
	1	0-4-0F	Bf2	SLM	2096	1910	new	1951 scr
No	2	4wDM	Bdm	Kronenberg DT 60/90P	121	1951	(b)	(2)
	3	4wDH	Bdh	Jung RK 15 B	13397	1961	new	(3)
	4	4wDH	Bdh	Jung RK 24 B	14137	1972	new	(4)
		4wDH	Bdh	Henschel DHG 200	30874	1964	(e)	(5)
		4wDM R/R	Bdm	Mercedes Unimog 1600	?	200x		
		4wDM R/R	Bdm	Mercedes MB Trac 700	?	200x		

(b) new, also recorded as DT50/90P.
(e) new to E. Merck, Darmstadt [D] "10" as DHG 160B; from NEWAG after reconditioning /1993; also recorded as DHG 240.
(2) 1995 donated to Betriebsgesellschaft Historischer Schienenfahrzeuge (BHS), Bodio.
(3) 1972 to Jung; 1974 to Umschlags- Lagerung- und Transporte (ULTRA) AG, Muttenz Auhafen "1".
(4) 1993 to NEWAG; 1994 Norddeutsche Affinerie, Hamburg [D] "EM 245 510".
(5) 11/2003 to LSB.

ALBERT BUSS & CIE, Basel   BS04 C

This public works concern had recorded contracts in connection with the hydroelectric power stations at Augst-Wyhlen, Oberdorf and St. Gallen/St. Fiden, as also for railway construction work. Additionally the company acquired a locomotive for building the electric power station at Rheinfelden [D], details of that are omitted here.

Gauge: 1000 mm			Locn: ?				Date: 03.2008
	0-6-0Tm	Cn2tk	SLM G 3/3	473	1887	(a)	s/s
	0-6-0Tm	Cn2tk	SLM G 3/3	548	1888	(b)	(2)

Gauge: 900 mm			Locn: ?				Date: 03.2002
	0-4-0T	Bn2t	Heilbronn IV	215	1885	(c)	s/s
	0-4-0T	Bn2t	Heilbronn III	340	1898	new	(4)
	0-4-0T	Bn2t	Heilbronn III	341	1899	new	s/s
BASILEA?	0-4-0T	Bn2t	Henschel	5408	1900	new	s/s
BASILEA?	0-4-0T	Bn2t	Henschel	5409	1900	new	s/s

Gauge: 750 mm			Locn: ?				Date: 03.2002
WEISSENSTEIN	0-4-0T	Bn2t	Krauss-S IV bh	5295	1905	(h)	(8)
OBERDORF	0-4-0T	Bn2t	Krauss-S LXV n	5492	1905	(h)	s/s

Gauge: 725 mm			Locn: ?				Date: 03.2002
	0-4-0T	Bn2t	Heilbronn III	47	1875	(j)	(10)
	0-4-0T	Bn2t	Heilbronn III	74	1878	(k)	s/s

Gauge: 600 mm			Locn: ?				Date: 03.2002
	0-4-0T	Bn2t	O&K 30 PS	517	1900	new	s/s
	0-4-0T	Bn2t	O&K 40 PS	551	1900	(m)	(13)

(a) ex Birsigtalbahn (BTB) "G 3/3 1 BASEL" /1909 for Bernina Bahn construction.
(b) ex Birsigtalbahn (BTB) "G 3/3 3 BIRSIG" /1907 for Bernina Bahn construction.
(c) ex Philipp Holzmann, Frankfurt am Main [D] "MANNHEIM 31" by /1899; used on construction the goods station of Basel Badische Bahnhof c/1904.
(h) new; used during the construction of the railway lines Solothurn-Moutier /1905 and St. Fiden-Wittenbach /1909.
(j) ex P. Dinndorf & Franke, Föhren [D] (725 mm) /1889-1913; the locomotive received a new boiler from the SCB workshops, Olten /1889.
(k) ex P. Dinndorf & Franke, Föhren [D] (730 mm) /1900; used on the construction of the BT?
(m) new, used during the construction of the railway line St. Fiden-Wittenbach /1909.

(2) 1915 sold to Robert Aebi & Cie, Zürich.
(4) 1924 sold to Societa pro l'Esovcizio della Minere dre Valdarno Vastellnarona, Avozzo [I].
(8) 1914 sold to Robert Aebi & Cie, Zürich.
(10) 1908 scrapped by the builders at Heilbronn [D].
(13) either s/s or 1922 to O&K, Zürich where it was referred to as O&K 550 thereafter.

CHANVIE-ZALIE & CIE, Basel — BS05 C

Probably a public works concern.

Gauge: 600 mm			Locn: ?				Date: 05.2002
	0-4-0T	Bn2t	O&K 20 PS	2459	1908	new	(1)

(1) 1910 to O&K, Zürich.

DREISPITZ-VERWALTUNG, Basel Dreispitz — BS06

An industrial and commercial area under the aegis of Basel-Stadt, rail connected in 1900 and put into service in 1901. In 1989 there were 18.5 kilometres of track including 77 sidings. Lately spare space has been used to store stock for other operators (for example Classic Rail, Schorno Locomotive Management, SBB, and TEE-Classics).

Gauge: 1435 mm			Locn: 213 $ 6126/2652				Date: 10.2007
1 DREISPITZ	0-4-0T	Bn2t	SLM	1439	1902	(a)	(1)
2 RUCHFELD	0-4-0T	Bn2t	SLM E 2/2	2592	1917	new	(2)
3 CHRISTOPH MERIAN	0-6-0WT	Cn2t	SLM E 3/3	6	1873	(c)	1948 scr
4 DREISPITZ	0-6-0WT	Cn2t	SLM E 3/3	11	1873	(d)	1960 scr
5 ALLAINE	0-6-0T	Cn2t	SACM-M	3963	1886	(e)	1948 scr
6 RUCHFELD	2-6-0T	1Cn2t	SLM Ed 3/4	1798	1907	(f)	(6)
7 PAUL SPEISER	0-6-0WT	Cn2t	SLM E 3/3	1293	1900	(g)	1966 scr
8 RUCHFELD	0-6-0WT	Cn2t	SLM E 3/3	1292	1900	(h)	1966 scr

9 CHRISTOPH MERIAN								
	0-6-0WT	Cn2t	SLM E 3/3		901	1894	(i)	1962 scr
10 RUCHFELD	0-4-0DE	Bde1	SLM,BBC Tm^{III}	4438,6179		1962	(j)	(10)
11 CHRISTOPH MERIAN								
	0-4-0DE	Bde1	SLM,BBC Tm^{III}	4439,6180		1962	(k)	(11)
237 866-9 12 PAUL SPEISER								
	0-4-0DE	Bde1	SLM,BBC Tm^{III}	4586,7539	1965,66	new	(12)	
237 865-1 13 LUKAS BURCKHARDT								
	0-4-0DE	Bde1	SLM,BBC Tm^{III}	4587,7540	1965,66	(m)	(13)	
Tm 237 860-2	4wDE	Bde1	Stadler		325	1998	new	
Tm 237 869-3	4wDE	Bde1	Stadler		838	2003	new	

(a) ordered by Basel Gas- und Wasser AG, Basel "2".
(c) new to SCB "82"; purchased from SBB Werkstätte Olten "E 3/3 8581" /1926.
(d) new to SCB "87"; purchased from SBB "E 3/3 8586" /1926.
(e) new to PV later JS "751"; then RPB "1 ALLAINE", purchased from RPB "751" /1929.
(f) purchased from SMB "1" /1934.
(g) new to GTB "1", purchased from BLS "76" /1943.
(h) new to GTB "3", purchased from BLS "75" /1945.
(i) new to NOB "257", purchased from SBB Werkstätte Chur "E 3/3 8555" /1948.
(j) new, named "11 DREISPITZ" till /1975.
(m) new, named "13 RUCHFELD" till /1975.
(1) 1926 sold to Chemische Fabrik Schweizerhalle (CFS) "2".
(2) 1930 sold to Papierfabrik Biberist.
(6) 1945 sold to Lonza AG, Visp; 1972 to DVZO.
(10) 1997 withdrawn; 1998 to CJ "237 480-9".
(11) 1999 to Stadler "11".
(12) 2006 to Stauffer, Schlieren, overhauled and later on hire to SF Chem, Schweizerhalle.
(13) 2006 to Meyerhans, Weinfelden via Stauffer, Schlieren.

ALB. GEISSBERGER, Basel   BS07 C

A public works concern.

Gauge: 600 mm   Locn: ?   Date: 05.2002

	4wDM	Bdm	O&K(M) RL 1a	4170	1930	(a)	(1)
	4wDM	Bdm	O&K(M) RL 1a	4835	1933	(b)	(2)
	4wDM	*Bdm*	*O&K(M) RL 1c*	*7859*	*1937*	*(b)*	*(1)*
	4wDM	Bdm	O&K(M) RL 1c	7879	1937	(b)	(4)
	4wDM	Bdm	Brun		1965		(1)

(a) imported via Lagerhausgesellschaft, Basel; purchased via O&K/MBA, Dübendorf.
(b) purchased via O&K/MBA, Dübendorf.
(1) 1970 seen at Thommen AG, Kaiseraugst for scrap.
(2) to Jurabergwerke AG, Herznach.
(4) 19xx to Thommen AG, Kaiseraugst - for scrap? Could there a duplication between O&K(M) 7859 and 7879?

INTERFRIGO (IF), Basel Badischer Bahnhof   BS08

Repair shops for the Interfrigo fleet of wagons.

Gauge: 1435 mm   Locn: 213 $ 6129/2683   Date: 04.1989

270	4wDM	Bdm	O&K MV2a	25728	1957	(a)	(1)
330	4wBE	Ba	AEG Ks	4561	1930	(b)	(2)
906	4wDH	Bdh	Gmeinder Köf II	4976	1957	(c)	(3)

(a) ex Società Anomina Cementi Armati Centifugati (SACAC), Bodio /?1967.
(b) new to DB "Ka 4015", ex DB "381 101-5" /1983 (or /?1977).
(c) new to DB "Köf 6276", ex DB "323 593-4" /1991.
(1) 1984 to O&K-Reparaturwerkstatt, Dortmund-Derne [D].
(2) 1992 donated to Bayrisches Eisenbahnmuseum Nördlingen [D].

(3)   1997 to EUROVAPOR, Sektion Haltingen (Kandertalbahn) [D] '906'.

JARDINI & CIE, Basel   BS09 C

Probably a public works company that is also recorded as Jardin & Co.

Gauge: 750 mm			Locn: ?				Date: 03.2002
BIRS	0-4-0T	Bn2t	Henschel Monta	17674	1920	(a)	(1)
	0-4-0T	Bn2t	O&K 50 PS	1136	1903	new	(2)

Gauge: 720 mm			Locn: ?				Date: 05.2008
	0-4-0T	Bn2t	Krauss-S XXVII b	559	1876	(c)	s/s

Gauge: 700 mm			Locn: ?				Date: 05.2008
	0-4-0T	Bn2t	Heilbronn II ur	12	1870	(d)	s/s

(a)   ex F. Marti, Bern /1920-30.
(c)   new; presence at this location or even wih this concern not confirmed.
(d)   ex ? Grubitz & Ziegler, Solothurn /1876 (presence at this depot of Jardini & Cie not confirmed).

(1)   1930 to Schafir & Mugglin, Zürich "10".
(2)   1920-23 sold for use in the construction of the electric power station at Mühleberg.

LANDI REBA AG, Basel Dreispitz   BS11

An agricultural supplier located within the industrial zone of Dreispitz.

Gauge: 1435 mm			Locn: 213 $ 6127/2649			Date: 11.2007
	1wPH	Aph	Zagro MR 200 B	90135	7.2005	new

P. LEIMGRUBER & CO AG, Basel Dreispitz   BS12

A transport concern located within the industrial zone of Dreispitz that transfers containers.

Gauge: 1435 mm			Locn: 213 $ 6127/2649			Date: 04.2000
	4wDH	Bdh	Jung Köf II	13156	1960	(a)
	4wDH	Bdh	Jung Köf II	13172	1960	(b)

(a)   new to DB "Köf 6718"; ex DB "323 788-0" /1986; wdn /1998 and later used for spares.
(b)   new to DB "Köf 6734"; ex DB "323 804-5" /1998.

LIPS-PORTMANN, Basel Kleinhüningen   BS13

A firm trading in gravel and similar material.

Gauge: 600 mm			Locn: ?				Date: 05.2002
	4wDM	Bdm	O&K(M) RL 1	3826	1929	(a)	s/s

(a)   new via O&K/MBA, Dübendorf.

NEPTUN AG, Basel Kleinhüningen   BS14

Founded in 1920, it became part of Schweizerische Reederei & Neptun AG (SRN) in 1975 (which see for further details).

Gauge: 1435 mm			Locn: 213 $ 6113/2701			Date: 06.2001	
	4wDM	Bdm	Kronenberg DT 90	110	1947	new	(1)
	4wDH	Bdh	KHD A8L614 R	56837	1958	new	(2)

(1)   1960 to Migros, Taverne.

(2) 1975 to Schweizerische Reederei & Neptun AG (SRN), Basel Kleinhüningen.

NOVARTIS SERVICES AG, Basel Kleinhüningen   BS15

formerly: CIBA Spezialitätenchemie AG (1996-1998)
            CIBA-GEIGY AG (1970-1996)
            Chemische Industrie Basel (CIBA) (1946-1970)
            Gesellschaft für Chemische Jndustrie Basel (GCJB) (1898-1946)

A complex for making chemicals divided into two areas, one is the chemical plant at Rosental connected to Basel Badische Bahnhof in 1882 and the other at Klybek connected to the station at Basel Kleinhüningen.

Gauge: 1435 mm   Locn: 213 $ 6113/2693   Date: 12.1999

1			0-4-0DM	Bdm	SLM Em 2/2	3464	1930	new (1)
2			4wBE	Ba2	SLM,MFO Ta	2622, ?	1917	new
3		rebuilt	4wDE	Bde2			1953	1971 scr
3	MAX		0-4-0DM	Bdm	SLM Tm	3690	1938	(c) (3)
282			4wDE	Bde1	Moyse BN38E220DGM	1141	1968	(d)
329			BBDH	BBdh	U23A LDH240	21236	1971	(e)
			4wDH	Bdh	Henschel DHG 200B	30876	1964	(f) (6)

(c)   new as "2"; renumbered "3" ca. /1971.
(d)   new as "1"; renumbered "282" /198x.
(e)   new; into service /1972; unnumbered till /198x.
(f)   hired from Asper /1988.

(1)   19xx to CIBA, Monthey.
(3)   1999 donated to Swisstrain, Bodio.
(6)   1988 returned to Asper.

RAPP AG, Basel   BS16 C

formerly: W. & J. Rapp AG (1896-1992)

This construction firm may have been related to a firm in Germany with the name W. Rapp AG. It was established in the Basel area by 1860. The SIG list of references gives one ET 15KL locomotive delivered here and one ETS 50 to Muttenz. Address: Hochstrasse 100, 4053 Basel.

Gauge: 600 mm   Locn: ?   Date: 03.2002

0-4-0T	Bn2t	Freudenstein	106	1901	(a)	(1)
0-4-0T	Bn2t	Maffei 40 PS	2833	1908	(b)	(2)
0-4-0T	B?2t	Maffei	4264	1925	(c)	(2)
0-4-0T	Bn2t	Jung Helikon	1427	1910	(d)	s/s
0-4-0T	Bn2t	O&K 50 PS	4382	1911	(d)	(5)
0-4-0T	Bn2t	O&K 30 PS	995	1902	(f)	s/s

(a)   purchased /1902; used during the construction of the Birseckbahn /1902; and the Hauenstein line /1914 "1".
(b)   purchased new via Leipziger, Köln [D]; used during the construction of the second track between Nebikon and Sursee /1908 and the Hauenstein line /1913-19.
(c)   new; *also stated to be 4224 of 1924.*
(d)   purchased new via Leipziger, Köln [D]; used during the construction of the Hauenstein line /1913-19.
(f)   hired from O&K, Zürich during the construction of the second track between Dagmarsellen and Wauwil /1907.

(1)   1916 sold to RUBAG, Zürich.
(2)   1956 removed from boiler register.
(5)   1916-39 sold to Losinger & Cie, Bern.

RHENUS PORT LOGISTICS, Basel Kleinhüningen BS17

formerly: RHENUS Alpina AG (2000-2006)
RHENUS AG (1983-2000)
Compagnie Suisse de Navigation et Neptune (SRN-Alpina) (1992-2000)
Swiss Shipping and Neptune Company Ltd
SATRAM-Stahl AG (19xx till 1982)
Schweizerische Reederei & Neptun AG (SRN) (1976-1992)
Neptun AG (1920-1975)
Schweizerische Reederei AG (SRAG) (1938-1975)
Schweizerische Schleppschiffahrts-Genossenschaft (SSG) (1919-1938)

Several concerns have amalgamated to reach the present situation. The present title of this company dates from 2007. This is but one group within a company that handles both river and high sea shipping and associated docks and warehousing.

In 1975 two companies (SRAG and Neptun) with adjacent sites alongside the R. Rhein amalgamated to form SRN. In 1983 SRN and SATRAM-Stahl AG became part of Rhenus AG and this section deals only with locomotives after that time. In 2000 a merger with Alpina took place.

Work is concentrated around "Bassin 1" built in 1919-26 and "Bassin II" from 1936-1942. The rail connection to Basel Badischer Bahnhof was inaugurated on 1 March 1924.

Gauge: 1435 mm   Locn: 213 $ 6114/2706 + 6118/2708   Date: 10.2007

		4wDH	Bdh	SATRAM	-	1972	(a)	(1)
2		6wDM	Cdm	Kronenberg 3 L 600	144	1961	(b)	(1)
1		4wDH	Bdh	O&K MB 200 N	26873	1976	(c)	
Tm 283 000-8 ex 4		4wDE	Bde	Gmeinder DE-SF3	5490	1972	(d)	
Tm 2/2 T5203		0-4-0DH	Bdh	KHD A12L614 R	56282	1956	(e)	2007 str
Tm 2/2 Rh.B 20 T1B204								
		4wDH	Bdh	Jung Köf II	13174	1960	(f)	
		4wDH	Bdh	KHD A8L614 R	56837	1958	(g)	1985 scr
		0-4-0DH	Bdh	KHD T4M625 R	56464	1956	(h)	
1		6wDH	Cdh	KHD MG600 C	58164	1967	(f)	

(a) rebuild of Moyse BL14HS52G 154 1970 /1972.
(b) ex Cementfabrik Holderbank, Rekingen AG /1991.
(c) ex Migros, Schönbühl /1998.
(d) new to Les Fils d'Auguste Scheuchzer SA, Bussigny-près-Lausanne "4"; purchased from SBB "283 000-8" 9/1997.
(e) new to Basel Gas- und Wasserwerk, Basel Kleinhüningen; ex Schweizerische Reederei & Neptun AG (SRN), Basel Kleinhüningen /1992.
(f) ex Schweizerische Reederei & Neptun AG (SRN), Basel Kleinhüningen /1992.
(g) ex Neptun AG, Basel Kleinhüningen /1975.
(h) ex Schweizerische Reederei & Neptun AG (SRN), Basel Kleinhüningen by /2000.
(1) 1998 to Thommen AG, Kaiseraugst for scrap.

SATRAM-HUILES SA, Bâle Petite Hunigue (Basel Kleinhüningen) BS18

formerly: SATRAM
SATRAM = S.A. de TRAnsbordement et Manutention

The original SATRAM split into two separate companies, namely SATRAM-Huiles SA and SATRAM-Stahl AG. The latter went on to become part of the Rhenus handling firm by 1983, taking some of the locomotives with it. The original installation in "Bassin 1" is still in use by SATRAM-Huiles SA, but they have also acquired part of the former gasworks in "Bassin 2" and one locomotive is kept there. Address: Hafenstrasse 25 and Sudquaistrasse 25.

Gauge: 1435 mm   Locn: 213 $ 6115/2706 + 6118/2703   Date: 10.2007

		4wPM	Bbm	RACO 45 PS	1235	1935	(a)	a1973 s/s
Tm 2/2	Nr. 1	4wDH	Bdh	Jung Köf II	13171	1960	(b)	
Tm 2/2		4wDH	Bdh	Henschel DHG 160B	31085	1966	(c)	(3)

		4wDH	Bdh	Vollert DR 10 000	76/006	1976	new	
(a)	ex Bahnhofkühlhaus AG, Basel Badischer Bahnhof /1969.
(b)	originally DB "Köf 6733", ex DB "323 803-7" /1993.
(c)	ex ESSO (Schweiz), Basel Kleinhüningen /2002.
(3)	2007 stored at Basel-SBB depot; by 2009 to Thommen AG, Kaiseraugst.

SATRAM-STAHL AG, Basel Kleinhüningen   BS19

Split from SATRAM SA in 19xx and sold to Rhenus AG in 1983.

Gauge: 1435 mm   Locn: 213 $ 6118/2708 + 6114/2706   Date: 06.2001

	4wDH	Bdh	Moyse BL14HS52G	154	1970	(a)	
rebuilt	4wDH	Bdh	SATRAM	-	1972	(1)	
	4wPM	Bbm	Breuer	?	19xx	(b)	
rebuilt	4wDM	Bdm	SATRAM	-	c1955		1996 scr

(a) new via Asper.
(b) ex ? [NL] ca. /1950.

SCHIFFERKINDERHEIM, Basel Kleinhüningen   BS20 M

Address: Weilerweg, 4019 Basel Kleinhüningen.

Gauge: 1435 mm   Locn: 213 $ 6113/2703   Date: 11.2007

	0-6-0WT	Cn2t	SLM E 3/3	897	1894	(a)	playgrd

(a) ex Schweizerische Reederei AG, Basel Kleinhüningen "8551" /1963.

SCHLACHTHOF BASEL, Basel St. Johann   BS21

The slaughterhouses, which dated from 1860, were rail connected from 1902 (with 1.1 kilometres of track) and closed in 1970 when a new establishment (without rail access) replaced them. The locomotive lists for BBC include a SIG,BBC locomotive with number 7780 and probable building date of 1970. If it was ever built it was not delivered here.
Address: Schlachthofstrasse 55.

Gauge: 1435 mm   Locn: 213 $ 6097/2694   Date: 08.1988

	4wDE	Bde	Stadler	131	1969	new	(1)

(1) 1976 to Henkel AG, Pratteln via Stadler (practically unused).

W. SCHMIDLIN, Basel   BS22 C

Probably a public works concern.

Gauge: 870 mm   Locn: ?   Date: 04.2002

	0-4-0T	Bn2t	Krauss-S IV m	330	1874	new	s/s
	0-4-0T	Bn2t	Krauss-S IV n	400	1874	new	s/s
	0-4-0T	Bn2t	Krauss-S IV n	401	1874	new	s/s

SCHWEIZERISCHE REEDEREI AG, Basel Kleinhüningen   BS23

This company became part of Schweizerische Reederei & Neptun AG (SRN) in 1976 (which see).

Gauge: 1435 mm   Locn: 213 $ 6113/2703 + 6117/2706   Date: 06.2001

2	0-4-0T	Bn2t	SLM	1837	1907	(a)	1956 scr
8551	0-6-0WT	Cn2t	SLM E 3/3	897	1894	(b)	(2)

		0-6-0WT	Cn2t	Krauss-S XLV ac	5331	1905	(c)	(3)
	8474	0-6-0WT	Cn2t	SLM E 3/3	1805	1907	(d)	(4)

(a) purchased from Gaswerk Basel "2" /1926.
(b) new to NOB "253", later "453"; purchased from SBB "E 3/3 8551" /1935.
(c) new to Gaswerk Zürich, Werk Schlieren "2"; purchased from Kieswerk Hardwald, Dietikon via AG für Metallverwertung, Zürich /1947.
(d) purchased from SBB "E 3/3 8474" /1961.
(2) 1963 to Schifferkinderheim, Basel Kleinhüningen.
(3) 1971 to private ownership; by 2004 stored at Zürcher Freilager AG (ZF), Zürich-Albisrieden.
(4) 1978 to private ownership.

SCHWEIZERISCHE REEDEREI & NEPTUN AG (SRN), Basel Kleinhüningen
BS24

Founded in 1976 from SRAG and Neptun AG and, in turn, formed part of SRN-Alpina in 1992.

Gauge: 1435 mm   Locn: 213 $ 6113/2703 + 6117/2706   Date: 06.2001

FLIEGENPILZLEIN		0-4-0DH	Bdh	KHD A12L614 R	56282	1956	(a)	(1)
		0-4-0DH	Bdh	KHD T4M625 R	56464	1956	(b)	(2)
Rh.B	20	4wDH	Bdh	Jung Köf II	13174	1960	(c)	(1)
	1	6wDH	Cdh	KHD MG600 C	58164	1967	(d)	(1)

(a) ex SRAG, Basel Kleinhüningen /1976.
(b) ex AVIA, Muttenz-Auhafen /1982.
(c) new to DB "Köf 6736"; ex DB "323 806-0" /1985.
(d) new to Unterharzer Berg- und Hüttenwerke, Zinkhütte, Harlingerode [D]; ex Preussag Metall, Oker [D] /1986 via Asper.
(1) 1992 to SRN-Alpina, Basel Kleinhüningen; 2000 RHENUS Alpina AG, Basel-Kleinhüningen.
(2) 1984 OOU and into store; 2000 RHENUS Alpina AG, Basel-Kleinhüningen.

STAMM BAUGESELLSCHAFT, Basel
BS27 C

A construction concern.

Gauge: 600 mm   Locn: ?   Date: 05.2002

4wDM	Bdm	O&K(M) MD 1	7902	1937	(a)	s/s

(a) purchased via O&K/MBA, Dübendorf.

TANKLAGER MIGROL, Basel Kleinhüningen   BS28

formerly: ESSO (Schweiz) (19xx-2001)
            Lumina SA (1926-19xx)

One of the pair of identical electrically operated (with cable connection) Vollert Kabeltrommel Robots in use here. This one is 01/023-2 and was seen on 15 October 2007.

The harbour facilities and fuel depot date from 1926, they were sold to Migrol in 2001 (without the remaining locomotive).

Gauge: 1435 mm   Locn: 213 $ 6111/2700   Date:10.2007

4wDM	Bdm	RACO 80 SA3	1388	1949	new	(1)
4wDH	Bdh	Henschel DHG 160B	31085	1966	(b)	(2)
4wCE	B-el	Vollert Kabeltrommel	01/023-1	2001	new	
4wCE	B-el	Vollert Kabeltrommel	01/023-2	2001	new	

(b)   new to Henschel hire fleet ; via Asper /1969.
(1)   1969 to Migros, Schönbühl.
(2)   2002 to SATRAM-Huiles SA, Basel Kleinhüningen.

ULTRA BRAG AG   BS29

formerly: Umschlags- Lagerungs- und Transporte AG (ULTRA) (till 1974)
            Basler Tankschiffahrt AG (BRAG) (1947-74)
            Basler Rheinschiffahrt AG (BRAG) (1925-47)

The two firms ULTRA and BRAG merged in 1974 to form the current concern. There are sites in Basel Kleinhüningen and Basel St. Johann as well as Muttenz-Auhafen.

"4" a two-axled Moyse (type BN34E260D number 1414 of 1977) waits on the quayside in the Kleinhüningen dock area to allow an SBB operated train to overtake on the principle track on 15 October 2007.

Basel Kleinhüningen

The main site is located in Hafenbecken 2. Address: Südquaistrasse 55, 4057 Basel.

Gauge: 1435 mm   Locn: 213 $ 6119/2705   Date: 08.2008

	1	4wDM	Bdm	Kronenberg DT 90	113	1948	new	(1)
		4wDM	Bdm	Gebus Lokomotor	564	1957	(b)	(2)
158	3	4wDE	Bde	Henschel,SSW DEL150				
					24929,3674	1941	(c)	(3)
	4	4wDE	Bde1	Moyse BN34E260D	1414	1977	(d)	
	3	4wDM	Bdm	Jung RK 15 B	13399	1961	(e)	(5)
3	Em 837 815-0	0-6-0DH	Cdh	Henschel DH 500Ca	30515	1963	(f)	

Basel St. Johann

It is intended that the firm should work the sidings serving the silos in Hafen St. Johann till 2009.

Gauge: 1435 mm   Locn: 213 $ 6108/2690   Date: 10.2007

5	Tm 237 864-4	4wDM	Bdm	Jung RK 15 B	13399	1961	(g)

(b)   ex Kohlen Union Geldner (KUG), Birsfelden Hafen /196x.
(c)   new to Himmelsbach, Freiburg im Breisgau [D]; ex Shell, Ludwigshafen [D] "158" via Thommen AG, Kaiseraugst /1968.
(d)   new via Asper, type may be BN34E260B.

(e)	ex Brohtal-Deumag AG, Urmitz [D] /1988 via Jung.
(f)	new to Rheinisch-Westfälische Elektrizitätswerke AG (RWE), Kraftwerk Frimmersdorf [D] "4"; ex LSB /1998.
(g)	ex Ultra Brag AG, Muttenz-Auhafen by /2007.
(1)	1966 scrapped following accident.
(2)	1968 to BRAG, Birsfelden Hafen.
(3)	1988 withdrawn; 1989 sold for scrap.
(5)	1999 to BRAG, Muttenz-Auhafen "5".

WERFT, Basel   BS28

Which particular ship yard is not known; the locomotive was not acquired for rail haulage purposes.

Gauge: 750 mm   Locn: ?   Date: 01.2008

G 3/3		0-6-0T	Cn2t	Krauss-S XVIII m	1049	1882	(a) s/s
(a)	ex WB /1940 "3 DÜBS".						

Fürstentum Liechtenstein FL

This principality is not a Swiss canton though it has monetary and customs union with Switzerland; in consequence it had several industrial locomotives provided via agents in Switzerland. The Austrian authorities provide main-line rail services.

FÜRSTLICHES BAUAMT LIECHTENSTEIN, Vaduz FL1 C

The public works department of the principality.

Gauge: 600 mm Locn: ? Date: 05.2002

4wDM	Bdm	O&K(M) RL 1a	4377	1931	(a)	s/s
4wDM	Bdm	O&K(M) RL 1a	5363	1934	(a)	(2)
4wDM	Bdm	O&K(M) RL 1c	7810	1937	(a)	s/s
4wDM	Bdm	O&K(M) RL 1c	7859	1937	(d)	(4)
4wDM	Bdm	O&K(M) LD 2	8400	1937	(e)	(5)

(a) purchased via O&K/MBA, Dübendorf.
(d) purchased via O&K/MBA, Dübendorf, recorded as 6859.
(e) sold to Preisig (Stein AR or Teufen AR); repurchased /1956.
(2) 1951 sold to Hr. Wander (MBA, Dübendorf?).
(4) *19xx to Alb. Geissberger, Basel.*
(5) 1956 to Preisig für Fürstliches Bauamt Liechtenstein, Vaduz.

FÜRSTLICH LIECHTENSTEINISCHE EISENBAHN ROMANTIK STIFTUNG FL2 M

Gauge: 1435 mm Locn: ? Date: 10.2002

77.250		4-6-2T	2C1h2t	Krauss(L) BBÖ 629	1430	1927	(a) (1)

(a) new to BBÖ "629.65"; later DR "77 250" (wdn 22/5/1973); stored at Buchs SG 12/2008.

SCHAAN-VADUZ BAHNHOF, Schaan FL3 M

Gauge: 1435 mm Locn: 227 $ 7569/2264 and 7571/2265 Date: 10.2002

77.250		4-6-2T	2C1h2t	Krauss(L) BBÖ 629	1430	1927	(a) (1)

(a) new to BBÖ "629.65"; later DR "77 250" (wdn 22/5/1973); on display at Schaan-Vaduz Bahnhof from /1974; after withdrawal was renumbered "77 244" and later "77 249".
(1) 1999 displayed at Werkhof, Schaan (Feuerwehrhaus) - the second location above; 21/8/2005 to ŽOS České Velenice [CZ] for repair; 2008 return to Fürstlich Liechtensteinische Eisenbahn Romantik Stiftung.

see:
http://www.eisenbahnclub.li/Joomla/index.php?option=com_content&task=view&id=4&Itemid=25

BAPTIST WILLE, Balzers FL4 C

A public works contractor?

Gauge: 600 mm Locn. ? Date: 05.2002

4wDM	Bdm	O&K(M) RL 1a	5643	1934	(a)	s/s?

(a) purchased via O&K/MBA, Dübendorf.

Fribourg

FR

BAHN MUSEUM KERZERS (BMK), Kerzers — FR01 M

As a separate operation from the preserved signal box here was this developing private collection. It was open to the public on limited occasions, but closed in 2008 as it is intended to move the collection. As well as the locomotives and railcars listed there are also railcars, trams, a steam-roller and a draisine; some stock was, in 2006, stored in the nearby wood-yard. Some material is stored at Kallnach.

Gauge: 1435 mm   Locn: 242 $ 5816/2032   Date: 08.2006

	4wDM	Bdm	Breuer III	?	1931(69)	(a)
	4wDM	Bdm	Breuer VL	3088	1944	(b)
	1ABE	1Aa1	BBC Akku-Fahrz.	879	1914	(c)
3	4wDE	Bde	Olten,BBC	? ,1945	1923	(d)
	0-4-0PM	Bbm	O&K(M) L308	1520	1922	(e)
	4wDM	Bdm	O&K(M) H 2	2385	1927	(f)
	4wDM	Bdm	RACO RA11	3021	1949	(g)
11	0-6-0WT	Cn2t	SLM E 3/3	631	1890	(h)
ABDe 4/8 244	2 car set		SIG,SAAS	?	1945	(i)

Gauge: 1000 mm   Locn: 242 $ 5816/2032   Date: 08.2006

He 2/2 6	0-4-4RWE	2deb2	SLM,BBC He	1533, ?	1903	(aa)
HGe 3/3 26	0-6-0RWE	Czzw2	SLM,MFO HGe 3/3	2370, ?	1913	(ab)
Ge 6/6 406	0-6-6-0WE	CCw2	SLM,BBC Ge 6/6	2758,1550	1920,21	(ac)
BCFe 2/4 155	Bo2WERC	Bo2w2	SIG,BBC	?	1931	(ad)

Gauge: 750 mm   Locn: 242 $ 5816/2032   Date: 08.2006

	4wDM	Bdm	RACO RA7	104	1933	(ba)
	4wDH	Bdh	Schöma CFL 60DZ	3686	1973	(bb)

Gauge: 600 mm   Locn: 242 $ 5816/2032   Date: 08.2006

	4wDM	Bdm	Jung EL 105	10966	1948	(ca)

(a) new to EBT "Tm 10" as 4wPM Bbm; ex Coop, Olten by 10/2007 via Swisstrain, Bodio and a period at Winpro, Winterthur.
(b) ex Shell AG Niederhasli 20/10/2003.
(c) ex Kentaur AG, Lützelflüh-Goldbach /2005.
(d) new to CFF "Ta 44" (4wBE Ba); ex la Fonte Électrique, Bex (FEBEX) /1999.
(e) ex Mühlen AG, Interlaken West by /2005.
(f) ex LSB /2003.
(g) ex Strafanstalten, Witzwil "2" /2005.
(h) new to JS "855"; later SBB "8575"; ex Von Roll, Gerlafingen "11" 3/2008 via VHS/DBB/VVT.
(i) new to BLS "BCFe 4/8 742", renumbered "744" /1990, ex OeBB "ABDe 4/8 244" /2008 (withdrawn /2005).
(aa) ex JB "He 2/2 6" /2007.
(ab) ex BOB "He 3/3 26" by /2005.
(ac) ex RhB "Ge 6/6 406" by /2005.
(ad) new to Fribourg-Morat-Anet (FMA, third rail electric); donated by Club du Tramway de Fribourg (CTF), Düdingen /2009.
(ba) ex Schweizerische Munitionsfabrik, Lerchenfeld, Thun /2005 via Hr. Jürg Meili, Arch.
(bb) ex Schweizerische Munitionsfabrik, Lerchenfeld, Thun by /2005.
(ca) ex Chemie Uetikon (CU), Uetikon am See by /2002.

see: www.bahnmuseum-kerzers.ch

CLUB DU TRAMWAY DE FRIBOURG (CTF), Düdingen — FR02 M

The club owns TF trams "9" (repatriated from France), "10" and other historic vehicles and keeps them in a hangar at Düdingen CFF station. It has owned a rail-car, but this has been disposed of.

Gauge: 1435 mm			Locn: 242 $ 5809/1887					Date: 06.2006	
BCFe 2/4 155		Bo2WERC Bo2w2	SIG,BBC		?	1931	(a)	(1)	

(a) new to Fribourg-Morat-Anet (FMA, third rail electric); ex GFM /2002.

CONSERVES ESTAVAYER SA (CESA), Estavayer-le-Lac   FR03

A conserve factory, part of the Migros Empire, built in 1956. Shunting on the site was performed by the CFF until 1984 and again after 2002.

Gauge: 1435 mm			Locn: 242 $ 5551/1884				Date: 04.2002
1	0-6-0DH	Cdh	KrMa ML 500 C	18356	1957	(a)	(1)
	0-8-0DH	Ddh	LEW V60	12246	1968	(b)	(2)

(a) new to Wilhelmsburger Industriebahn [D] "30"; DB "V50 001" from 1962 till 1963; ex Verden-Walsroder-Eisenbahn [D] "3"/1984 via NEWAG; also recorded as type ML 500.
(b) probably the locomotive new to VEB Kalikombinat Südharz, Werk Zielitzer [D] "2" as a V60D; ex Zielitzer Kali AG, Zielitz [D] "2"; rebuilt by NEWAG under their reference "0031/94" /1994.
(1) 1994 to NEWAG; later to Adolf Scheufelen GmbH & Co KG, Lenningen-Oberlenningen [D].
(2) 2002 to Perret SA, Chavornay.

DESPOND SA, Bulle   FR04

formerly: L. Despond Fils
            Lucien Despond

A timber yard started in 1896 by Lucien Despond.

Gauge: 600 mm			Locn: 252 $ 5703/1627				Date: 10.1967
	4wDM	Bdm	O&K(M) RL 1a	4361	1931	(a)	s/s

(a) purchased via O&K/MBA, Dübendorf, ZH.
     see: www.despond.ch/Allemand/main/index.html (and then Geschichte)

ENTREPRISES ÉLECTRIQUES FRIBOURGEOISES (EEF), Rossens FR   FR05 C

Construction of the hydroelectric power station scheme at Rossens FR. Preliminary work started in 1944, but the main construction took place 1945-48. A four kilometre long line connected the gravel pit at Momont near Pont-la-Ville with the dam construction site. At least one steam locomotive was used on the project.

Gauge: 750 mm			Locn: 252 $ 5750/1742				Date: 05.2008
	0-4-0T	Bn2t	unknown	?	19xx		s/s
	4wDM	Bdm	RACO 50PS	1313	1945	new	(2)
	4wDM	Bdm	RACO 50PS	1314	1945	new	s/s
	4wDM	Bdm	RACO 50PS	1315	1945	new	s/s
	4wDM	Bdm	RACO 50PS	1316	1945	new	s/s
	4wDM	Bdm	RACO 50PS	1317	1945	new	s/s
	4wDM	Bdm	RACO 50PS	1321	1945	new	s/s

(2) 19xx to Conrad Zschokke, Näfels; later IRR and SGRK.

ETABLISSEMENTS DE BELLECHASE, Sugiez   FR06

A long siding serves this prison, which farms a large surrounding area. A road vehicle fitted with equipment to operate the brakes works rail traffic on the line.

Gauge: 1435 mm   Locn: 242 $ 5763/2026   Date: 09.1972

INSTITUT LES BUISSONNETS, Fribourg   FR07 M
TF tram "Ce 2/2 10" was displayed in the school grounds from 1965-93.

PFADI BUNDESLAGER, Romanens, Sâles (Gruyère)   FR08
A railway operated here in the summer of 1980 and was then dismantled.

Gauge: 600 mm   Locn: ?   Date: 12.1980

4wDM	Bdm	unknown		?	19xx

PRODO SA, Domdidier   FR09
formerly: Agro-Chemie AG (till 1958)

The factory originally made chemical products for agricultural use and then changed to bituminous products for roads. The siding (187 metres long) was brought into service in 1947. Address: rue de la Distillerie.

Gauge: 1435 mm   Locn: 242 $ 5672/1907   Date: 01.2008

4wPM	Bbm	Breuer III	?	c1930	(a)	1970 scr
0-4-0PM	Bbm	SLM Tm	3599	1935	(b)	(2)
4wDM	Bdm	RACO 85 LA7	1601	1961	(c)	

(a)   new to Gaswerk Basel; ex CFF "Tm 465" /1961.
(b)   new to CFF "Tm 710 later Tm 870"; ex A. Käppeli & Co, Sargans /1970 possibly via P. Schlatter, Münchwilen AG.
(c)   new to CFF "Tm^I 332" ex CFF "Tm^I 432" /1998.
(2)   1997 OOU; 2003 scrapped.

RÉCUPÉRATION RG SA, Sévaz   FR10
formerly: René Goutte SA, Payerne (till 1990)

A scrap merchant's yard with a rail connection between Estavayer and Cugy.

Gauge: 1435 mm   Locn: 214 $ 5567/1874   Date: 06.1999

1	4wDH	Bdh	O&K MB 7 N	26609	1969	(a)	

(a)   ex Régie Fédérale des Alcools, Daillens /1997.

SIKA NORM AG, Düdingen   FR11
The factory of this company produces synthetic material for buildings and additives for cement. It was built in 1967 or 1968 and connected to the CFF from 1989.

Gauge: 1435 mm   Locn: 242 $ 5814/1893   Date: 12.2007

4wDH	Bdh	Schöma CFL 40DR	2801	1964	(a)	2000 scr
4wDE	Bde2	Stadler	118	1965	(b)	
0-4-0W+DEBw2		Tuchschmid,MFO Tem^{II}	?	1967	(c)	

(a)   ex Bahnhofkühlhaus, Basel Bad Bahnhof /1990.
(b)   ex Schweizerische Unternehmung für Flugzeuge, Emmen /2000.
(c)   ex CFF "Tem^{II} 276" by /2007 (30/9/2006 wdn).

SUDAN SA, Broc　　　　　　　　　　　　　　　　　　　　　　　　　　FR12

A line runs from a gravel pit near Grandvillard along the bank of the R. Sarine to the cement works at Pont d' Estavannens, near Enney and located not far from the GFM station of that name. Address: Pont d'Estavannens, Enney.

Gauge: 750 mm　　　　　Locn: 262 $ 5733/1578　　　　　　　　　Date: 04.2005

| | 4wDM | Bdm | Jung ZL 105 | 11024 | 1952 | ?new | s/s |
| | 4wDH | Bdh | O&K MB 125 S | 26680 | 1970 | new | |

TOURBIÈRES DE LA ROUGÈVE, Semsales　　　　　　　　　　　　FR13

Peat was extracted for fuel during World War II. The extraction took place in both Canton Fribourg and Canton Vaud and closed in the 1950s. Although the R&H locomotive could have been operated on 750 mm track this is thought not to have been the case here.

Gauge: 600 mm　　　　　Locn: 262 $ 5585/1575 (approx)　　　　Date: 04.2005

| | 4wDM | Bdm | R&H 30DL | 256190 | 1948 | (a) | (1) |
| | 4wDM | Bdm | Ammann | - | 19xx | | (1) |

(a)　new via RACO.
(1)　in use till the 1950s; ca. /1971 rebuilt with steam outline casings and used on a "Far West" railway - where ?; by /1996 with Momect SA, Collombey; by 6/1999 to FWF, Otelfingen.

TRANSPORTS EN COMMUN (TF), Villars-sur-Glâne　　　　　FR14 M

Initially two trams "Ce 2/2 6 and 9" were displayed from 1977 in front of the new TF Workshops located at Chandolan. "6" is now scrapped and "9" is in France.

TUILERIE DU MOURET, Le Mouret　　　　　　　　　　　　　　　FR15

alternative:　Tuilerie de la Porte des Etangs de Fribourg
alternative:　Tuilerie de Miséricordes

The first proposal for this brickworks dates back to 1626, but the first buildings only appeared in 1763. After a number of vicissitudes it was rebuilt in 1898. It was used for peat harvesting during World War II. Rebuilt again in 1956, it closed in 1963. During the operational period, it lay within the commune of Ferpicloz. Address: Impasse de la Tuilerie.

Gauge: 600 mm　　　　　Locn: 252 $ 5795/1770　　　　　　　　　Date: 04.1999

| | 4wDM | Bdm | O&K MV0 | 25101 | 1951 | (a) | s/s |

(a)　new via Wander-Wendel.
　　　see: www.sodemo.ch/Communes/Ferpicloz/tuilerie.htm

VOIE INDUSTRIELLE DE PÉROLLES, Fribourg　　　　　　　　　FR16

operated by: CFF (from 1998)
　　　　　　Chemins de fer Fribourgeois (GFM)
　　　　　　Entreprises Électriques Fribourgeoises (EEF)
　　　　　　Service des Eaux et Forêts

The industrial network in the south of the town was put into service in 1871 and electrified 1906-98. The successive operators are listed above. The system gradually ran down and closed after 1998. The running numbers are in the GFM fleet series.

Gauge: 1435 mm 600V DC　　　Locn: 214 $ 5783/1825　　　　　　Date: 10.1998

| 51 | 4wWE | Bg2 | CFF (Yverdon),MFO | ? | 1906 | new | 1974 scr |
| 52 | 4w2WE | Bg1 | SLM,SAAS Te | 3979, ? | 1948 | new | 1998 scr |

| | 82 | 4wDH | Bdh | RACO 300 PS HT | 1634 | 1964 | (c) | (3) |

(c) ex von Roll, Gerlafingen "22" /1973.
(3) 1998 to GFM, Bulle.

Genève

GE

AEROPORT DE GENÈVE, Genève — GE01 C
During World War II a railway was used to extend Genève airport. Steam locomotives were used. Details required.

ASSOCIATION DU TRAM 70, Genève — GE02 M
An association running one specific tram over the lines of tramway system of the city.
Gauge: 1000 mm   Locn: ?   Date: 01.2008

ASSOCIATION GENÈVOISE DU MUSÉE DES TRAMWAYS (AGMT), Genève — GE03 M
An association running trams over the lines of tramway system of the city; at least one is on loan from the Blonay-Chamby (BC).
Gauge: 1000 mm   Locn: ?   Date: 01.2008

BUNDESTANKANLAGEN (BTA), La Plaine — GE04
Fuel storage depot built and rail connected in 1953 (1050 metres of track). Carbura SA operated it. Closed and dismantled in 1999.
Gauge: 1435 mm   Locn: 270 $ 4891/1152   Date: 08.1994

8	4wDH	Bdh	RACO 80 ST4	1422	1953	new (1)

(1) 1999 donated to Swisstrain, Bodio.

l'ORGANISATION EUROPÉENNE POUR LA RECHERCHE NUCLÉAIRE (CERN), Meyrin — GE05 C
formerly: Conseil Européen pour la Recherche Nucléaire

With the construction of the Large Electron-Positron Collider (LEP) the site was extended into France. A locomotive worked railway (details required) was used in the construction of this device. Some of the railway stock from the construction period was seen stored alongside the site of EUROAGREGATS at St. Genis, Ain [F] in 1988. This included at least the one locomotive listed. Initially personnel transport and maintenance was served by an underground monorail using battery operated locomotives. By 2000 the system had been changed to use guided road trains.

Gauge: 600? mm   Locn: ?   Date: 12.1988

5	4wBE	Ba	SIG	?	19xx	s/s

CONSORTIUM TUCOL, Genève-La Jonction — GE06 C
A drainage tunnel was built in the La Jonction area in 1964-5. A 600 mm system with at least three battery locomotives was employed. A noteworthy feature was a temporary bridge crossing both the R. Arve and the R. Rhône at their confluence. The locomotive shed was situated on the

south side of the rivers. SIG supplied 2*ETM 50 and one ETB 50 to this consortium (Gini JJ, Genève, Murer SA, Andermatt, Hochtief SA, Essen [D], and Décaillet SA, Martigny). The only known photograph shows an ETM 50. Batteries were provided by Plus AG and Electrona SA, Boudry SA, and these makers and the battery numbers are sometimes used to identify the locomotives.

Gauge: 600 mm   Locn: 270 $ 4985/1174   Date: 04.1965

	4wBE	Ba2	SIG ETB 50	?	19xx	(a)	(1)
	4wBE	Ba	SIG ETM 50	?	19xx	(a)	(1)
	4wBE	Ba	SIG ETM 50	?	19xx	(c)	(1)
	4wBE	Ba	SIG			(d)	(1)
	4wBE	Ba	SIG			(d)	(1)

(a) one of these had battery box Plus AG 2939.
(c) battery box Plus AG "2 2608".
(d) battery boxes Electrona "3834" and "3835".

(1) 19xx to Tuileries et Briqueteries de Bardonnex SA, Bardonnex (to be confirmed).

see: Bulletin technique de la Suisse romande (of 1965)

COOP, Vernier   GE07

This retail centre operated a distribution centre which closed around 2006. Address: Rte de Châtelaine 76, 1214 Vernier.

Gauge: 1435 mm   Locn: 270 $ 4946/1195   Date: 10.2008

	4wDM	Bdm	RACO 85 LA7	1653	1963	(a)	s/s

(a) new to CFF "Tm^I 347"; ex CFF "Tm^I 447" /2002.

ED. CUÉNOD SA, Genève   GE08 C

A civil engineering firm, in liquidation in 1997. Address: Plantaporrêts 8.

Gauge: 750 mm   Locn: 270 $ 4982/1177 (offices)   Date: 05.2002

KRIESSERN	0-4-0T	Bn2t	Krauss-S IV w	1292	1883	(a)	(1)
FLÜELA	0-4-0T	B?2t	Maffei	3624	1910	(b)	(2)

Gauge: 600 mm   Locn: 270 $ 4982/1177 (offices)   Date: 05.2002

	4wDM	Bdm	O&K(M) MD 1	11563	1940	(c)	s/s

(a) new to Ing. Wey, Buchs SG for the St. Gallische Rhein-Korrektion; in service with Cuénod /1922.
(b) new for the construction of the RhB line in the Lower Engadine; purchased /1920.
(c) via O&K, Lager Zürich.
(1) by 1941 at Mineral AG, Brig.
(2) 1939 to Entreprise Erdigt, Russin.

DUMENEST & ECKERT, Genève   GE09 C

Probably a public works concern.

Gauge: 600 mm   Locn: ?   Date: 05.2002

	4wDM	Bdh	O&K(M) RL 1a	4551	1931	(a)	s/s

(a) purchased via O&K/MBA, Dübendorf.

ENTREPRISE ERDIGT, Russin   GE10 C

A firm engaged in the construction of a hydroelectric dam at Verbois.

Gauge: 750 mm			Locn: ?					Date: 08.1994
	0-4-0T	B?2t	Maffei		3624	1910	(a)	(1)

(a) ex Ed. Cuénod, Genève /1939.
(1) 1942 to Braunkohlenwerk Zell AG, Zell LU.

GRANDS TRAVAUX DE MARSEILLE, Meyrin — GE11 C

In the late 1960s this firm were engaged in building an underground accelerator at CERN and used a temporary railway - details required. Shortly afterwards work started on the larger LEP project - see elsewhere in this section.

HENNEBERG, Genève — GE12 C

The business of this concern is not known.

Gauge: 750 mm			Locn: ?				Date: 08.1994
2	0-4-0T	Bn2t	SLM	369	1884	new	s/s

ROBERT HUFSCHMID, Genève — GE13 C

The business of this concern is not known.

Gauge: 1000 mm			Locn: ?					Date: 08.1994
	0-6-0T	Cn2t	SLM		58	1874	(a)	(1)

(a) probably built in /1874 and delivered /1879.
(1) 1894 to Bastin, Sallanches [F] and resold.

IMPLENIA SA, Satigny — GE14 M

formerly: Société Anonyme Conrad Zschokke

This public works concern has many sites throughout Switzerland. It is not clear if the locomotives were at this base in Canton Genève or not.

Gauge: 600 mm			Locn: 270 $ 4934/1192				Date: 06.1997
2	4wDM	Bdm	Diema DS40	2391	1960	(a)	(1)
	4wDM	Bdm	Diema DS30	2590	1963	(a)	(2)
3	4wDM	Bdm	Diema DS30	1883	1956	(c)	(3)
	4wDM	Bdm	O&K(M) MD 1	11366	1940	(d)	s/s
	4wDM	Bdm	O&K(M) MD 1	11367	1940	(d)	s/s
	4wDH	Bdh	Diema DFL 90/0.2	2360	1960	(g)	(6)

(a) from Diema via Asper /1984; used on the Assainissement Nant d'Avanchet (ANA) contract for the Galerie du Nant d'Avanchet.
(c) new to Dampfziegelei Ahrensbök, Ahrensbök/Bezirk Kiel [D]; ex Asper and then stored.
(d) via MBA, Dübendorf.
(f) new Marine-Munitionsdepot, Tannhausen [D] "2" (as type DS90); ex Diema /1983; used on the Pont du Bietschtal contract /1983-85; it is not sure that this locomotive passed through Genève.
(1) to Murer AG, Luzern.
(2) by 2002 to Sauvin Schmidt SA, Ports Franc, La Praille, Genève (a transport firm); by 2006 privately preserved at the Gare CEN, Fillinges, Haute Savoie [F].
(3) 2000 put on display near the depot and offices in the rte du Bois-de-Bay, Satigny.
(6) by 1995 Association Tunnel Mont Terni Sud, Saint-Ursanne "603".

PORTS FRANCS ET ENTREPÔTS DE GENÈVE SA — GE15

formerly Entrepôts de l'État de Genève (from 1888)
and   Port Franc de Genève (from 1964)

Warehouses and bonded storage. Transfer from Cornavin to La Praille took place over a period from 1964. By 2007 rail traffic had ceased and the connection to the CFF removed.

Genève-Cornavin

Gauge: 1435 mm   Locn: 270 $ 5002/1192   Date: 07.1993

1ABE	1Aa1	BBC Akku-Fahrz.	1232	1919	new	s/s
1ABE	1Aa1	BBC Akku-Fahrz.	1339	1921	new	s/s
4wDE	Bde1	Moyse 20TDE	336	1953	new	(3)

Genève-La Praille

Gauge: 1435 mm   Locn: 270 $ 4986/1176   Date: 07.2007

4wDE	Bde1	Moyse 20TDE	336	1953	(d)	1989 scr
4wDE	Bde2	Moyse BN36EE250M	1030	1962	new	
4wDH	Bdh	Schöma CFL 150DBR	2630	1963	(f)	

(d) ex Genève-Cornavin ca. /1964.
(f) new to Feldmühle AG, Düsseldorf [D] for Norddeutsche Papierwerke, Hagen-Kabel [D] "1"; ex Schöma /1989.
(3) ca. 1964 to La Praille.

ROUTORAIL SA, Zimeysa — GE16

A centre for the distribution of materials extracted from gravel pits, in particular ballast. The site lies on the most easterly pair of lines of the ZIMEYSA group, but the locomotive may work throughout the industrial estate. Rail access is from Vernier-Meyrin CFF station. Address: Rue de Turretin 11, 1242 Satigny.

Gauge: 1435 mm   Locn: 270 $ 494000,119450   Date: 10.2008

4wDM	Bdm	R&H 88DS	252840	1949	(a)	display
4wDH	Bdh	Jung RK 20 B	13372	1961	(b)	

(b) new to BTA, Saint-Triphon "2"; ex Carrières du Lessus HB SA, Saint-Triphon by 10/2007.
(c) new to Sihlpapierfabriken, Landquart; ex Carrières du Lessus HB SA, Saint--Triphon by 10/2007.

see: SER 3/2008 page 113

SÉCHERON SA — GE17

formerly: ABB-Sécheron (ABB) (1988-92)
           BBC-Sécheron (BBC) (1970-88)
           SA des Ateliers de Sécheron (SAAS) (1918-70)
           Compagnie de l'Industrie Électrique et Mécanique (CIEM) (1907-18)
           Compagnie de l'Industrie Électrique (CIE) (18xx-1907)

The company makes heavy electrical engineering products, including railway components.

Genève-Sécheron

The site is adjacent to Cornavin station and has been rail connected since 1896. In 1992 some of the work was transferred to a new industrial estate (ZIMEYSA) rail connected to Vernier-Meyrin CFF station.

Gauge: 1435 mm   Locn: 270 $ 5003/1197   Date: 07.1990

'Berthe'	4wBE	Ba2	SAAS		?	1xxx	new	c1963 s/s
	4wDE	Bde1	Moyse 20TDE	368	1960	(b)	(2)

Zimeysa

The depot where the locomotives work is located in Rue des Sablières, Satigny. It is served by a pair of lines roughly in the centre of four groups of lines connected to freight lines south of the CFF mainline. In turn, this line is connected to Vernier-Meyrin CFF station. Office address: Rue du Pré-Bouvier 25, 1217 Meyrin.

Gauge: 1435 mm　　　　　Locn: 270 $ 493590,119380　　　　　　　　Date: 09.2008

'Berthe'	4wDE	Bde1	Moyse 20TDE	368	1960	(d)	OOU
	4wDH	Bdh	ABL VI/C	6171	1980	(d)	

(b)　　new to Sucrerie de Montcornet, Aisne [F] (1000 mm); rebuilt by Moyse before transfer to SAAS /1963. The Moyse plates now show the year as 1963.
(d)　　ex Sécheron SA, Genève /1992.
(2)　　1992 to ZIMEYSA, Vernier-Meyrin.

SCHMIDLIN, Genève　　　　　　　　　　　　　　　　　　　　GE18 C

Probably a civil engineering firm.

Gauge: 600 mm　　　　　Locn: ?　　　　　　　　　　　　　　　Date: 08.1994

	0-6-0T	Cn2t	Krauss-S Zwilling	2936B	1894	(a)	(1)

(a)　　half of a twin, originally Deutsche Feldbahn probably "62"; ex Paul Juillard, Saxon /1943.
(1)　　1944 to Rusterholz, Näfels.

SERVICES INDUSTRIELS DE GENÈVE (SIG), Usine à Gaz, Vernier
GE19

Locomotive "1" SLM 2298 of 1912 was built new as a "Tigerli" for the gasworks. Here it shunts oil tank wagons around the Vernier-Meyrin and Cointrin area on 24 April 1965.

As one of the city services a new gasworks was built 1911-15. It was rail connected from 1913 by a 2.5 kilometre long line to Vernier-Meyrin CFF station. Production ceased in 1973; shortly afterwards the CFF took over the services on the branch, till it closed in 1986. For later developments see Shell AG, Vernier.

Gauge: 1435 mm Locn: 270 $ 4972/1182 Date: 03.1988

	1	0-6-0WT	C2nt	SLM E 3/3	2298	1912	new	(1)
	2	0-6-0WT	C2nt	SLM E 3/3	2570	1916	new	a1960 scr
		0-6-0WT	C2nt	SLM E 3/3	1362	1901	(c)	1950 scr
	3	0-6-0WT	C2nt	SLM E 3/3	2134	1911	(d)	(4)
	1	0-6-0DH	Cdh	Henschel DH 500Ca	31082	1965	new	(5)
	2	0-6-0DH	Cdh	Henschel DH 500Ca	31083	1965	new	(6)

(c) new to SCB "44"; purchased from CFF "8413" /1944.
(d) purchased from SBB "8511" /1960.
(1) 1965 sold to Alusuisse, Chippis "31".
(4) 1967 stored; 1987 loan to VVT, St. Sulpice NE.
(5) 1973 to Société des Chaux et Ciments de la Suisse Romande (SCC), Usine d'Eclépens; Eclépens.
(6) 1973 to Société des Chaux et Ciments de la Suisse Romande (SCC), Usine de Roche, Roche VD.

SHELL AG, Genève-La Renfile   GE20

A fuel depot connected to the branch to the gasworks from Vernier-Meyrin station. The use of private locomotives ceased about 1985, but a large rail-served tank farm for several concerns remains.

Gauge: 1435 mm Locn: 270 $ 4969/1192 Date: 08.1994

	1	4wDH	Bdh	Gmeinder 190 PS	5246	1960	(a)	(1)
	2	4wPM	Bpm	RACO RA11	1260	1940	(b)	(2)

(a) ex Süddeutsche Zucker AG, Plattling [D] "1" /1985.
(b) ex BTA, Gland "2" /1973.
(1) 1986 stored; 1988-9 loaned to and 1992-93 transferred to Shell, Tanklager Rothenburg.
(2) 1985 to Richi AG, Weiningen ZH.

SOCIÉTÉ COOPÉRATIVE MIGROS - GENÈVE, Genève-La Praille
GE21

Gauge: 1435 mm Locn: 270 $ 4991/1154 (approx) Date: 04.1999

GE 33677	4wDM R/R	Bdm	Mercedes Unimog 1400	?	19xx

SOCIÉTÉ ROMANDE DES CIMENTS PORTLAND, Vernier   GE22

A cement works, owned by M. Dionisotti, was rail connected in 1933. A second connection was planned but never completed and the locomotive was never used. The cement works was sold in 1938 to EG Portland (Société des Ciments de la Suisse Romande).

Gauge: 1435 mm Locn: 270 $ 4969/1192 Date: 08.1995

	0-6-0T	Cn2t	SLM E 3/3	1064	1897	(a) (1)

(a) new to Seetalbahn (SeTB) "WILDEGG 8"; then KLB "2"; ex CFF "8652" /1933.
(1) from 1938 stored at EGT, Bussigny-près-Lausanne; 1953 or later to Société de Ciments Portland de Saint-Maurice SA, Saint-Maurice.

SPENO INTERNATIONAL SA, Genève GE23 C

This international concern provides many types of railway and rolling stock maintenance facilities. One facility is the provision of trains and locomotives for use in track reprofiling. These locomotives are not listed in this Handbook. Address: Parc Château-Banquet 26, Genève 1202.

TUILERIES ET BRIQUETERIES DE BARDONNEX SA, Bardonnex GE24

owned by: Morandi Frères SA

A brickworks built in the 1940s. Morandi Frères SA acquired a share in the business in 1953. At this time the logo became MBB (Morandi, Barraud, Bardonnex). The railway brought clay to the works from at least four electrically operated excavators that ran on 1400 mm gauge track. Two sides of the clay-pit adjoin France. After a modernisation of the works in 1986 (including the installation of an inclined conveyor belt) the railway lost its principal task, and bar a section in and around the unloading building was dismantled. By 9/1991, the rails were no longer used. Address: 1257 La Croix-de-Rozon.

Gauge: 600 mm   Locn: 270 $ 496550,111440   Date: 07.2007

	0-4-0DM	Bdm	Deutz OME117 F	10811	1932	(a)	s/s
	4wDM	Bdm	Hatlapa Junior II	4372	1950	new	b1968 s/s
	4wDM	Bdm	Hatlapa Junior II	4374	1950	new	a1968 s/s
9	4wDM	Bdm	Hatlapa Junior II	7915	1954	(d)	a1983 s/s
11	4wDM	Bdm	O&K(M) RL 1c	11049	1939	(e)	s/s
29	4wDM	Bdm	O&K MV0	25047	1951	(f)	OOU
28	4wDM	Bdm	O&K MV0	25100	1951	(g)	OOU
	4wDM	Bdm	Diema DL8	2527	1962	(g)	OOU
	4wDM	Bdm	Diema DL8	2626	1963	(g)	OOU
	4wBE	Ba	SIG ETM 50	?	19xx	(j)	s/s
3	4wBE	Ba	SIG	?	19xx	(k)	s/s
5	4wBE	Ba	SIG	?	19xx	(l)	s/s
	4wBE	Ba	unknown	?	19xx	(l)	s/s
	4wBE	Ba	unknown	?	19xx	(l)	s/s
	4wBE	Ba	unknown	?	19xx	(l)	s/s

(a) new via Robert Aebi & Cie. AG, Zürich.
(d) new to Mefag AG, Luzern.
(e) purchased via MBA, Dübendorf, it is not sure that the locomotive carried "11".
(f) ex Morandi Frères SA, Corcelles-près-Payerne (400 mm) by 23/12/1982. The locomotive may have carried "11" or "29".
(g) ex Morandi Frères SA, Corcelles-près-Payerne (400 mm) by 23/12/1982.
(j) ex Consortium TUCOL, Genève-La Jonction /19xx; battery box carries "2 2608".
(k) ex Consortium TUCOL, Genève-La Jonction /19xx, battery box Plus AG "2939 of 1958".
(l) some, or all, may have come from Consortium TUCOL, Genève-La Jonction.

see: www.morandi.ch/images/pdf/famille.pdf

Glarus GL

ETERNIT AG, Niederurnen   GL01

A narrow gauge railway was used to convey loads to the main line loading sheds. A somewhat complex network evolved, lying beside the road past the station. The system had closed by 1988.

Gauge: 600 mm   Locn: 236 $ 7231/2192   Date: 04.1987

	4wDM	Bdm	unknown?	?	19xx	(a)	19xx scr
	4wDM	Bdm	RACO 12/16 PS	1333	194x		(2)
	4wDM	Bdm	Diema DS28	2580	1962	(c)	(2)

(a)   claimed to be built by RUBAG, but more likely supplied via RUBAG.
(c)   ex Zürcher Ziegeleien AG, Giesshübel via Asper.
(2)   1988 to Schlitter AG, Niederurnen; later FWF, Otelfingen.
(3)   1988 to Schlitter AG, Niederurnen.

AG FÜR HANDEL & INDUSTRIE, Glarus   GL02 C

formerly: Handels- & Industriewerk, Glarus
A retail dealer or agent.

Gauge: 600 mm   Locn: ?   Date: 05.2002

	4wDM	Bdm	O&K(M) RL 1a	5994	1935	new	s/s
	4wDM	Bdm	O&K(M) MD 2	7452	1937	new	(2)

(2)   exported to Bulgaria; this note might apply to O&K 5994 as well.

KALKFABRIK NETSTAL AG, Netstal   GL03

The chalk works were equipped with a small internal 400 mm gauge system and connected to the main line by a standard gauge siding. However, there was additionally a locomotive-worked 600 mm gauge system that used to bring the raw materials to the works. The latter had closed by 1970.

Gauge: 600 mm   Locn: 236 $ 7233/2129   Date: 02.1994

	4wBE	Ba	Oehler		?	19xx	(1)

(1)   by 1970 OOU; since s/s?

KAMM & CO   GL04

The sub-title Schotterwerke indicates that they operated ballast quarries.

Mollis

This ballast quarry lies opposite Netstal in the valley of the R. Linth.

Gauge: 600 mm   Locn: ?   Date: 05.2002

	4wPM	Bbm	O&K(M) H 1	2308	1926	(a)	s/s
	4wDM	Bdm	O&K(M) RL 1a	4826	1932	(a)	s/s

(a)   purchased via O&K/MBA, Dübendorf.

Obstalden

This entry may refer to the works alongside the Walensee, but that is far from sure. There are other concerns with the same name in the area.

Gauge: 600 mm Locn: ? Date: 05.2002

	4wDM	Bdm	O&K(M) RL 1a	5642	1934	(a) s/s?

(a) purchased via O&K/MBA, Dübendorf.

BAUUNTERNEHMUNG ROBERT RÜESCH, SCHWANDEN GL   GL05 C

A public works concern.

Gauge: 600 mm Locn: ? Date: 02.2002

	4wBE	Ba	SSW EL 9	5301	1950	(a) s/s?

(a) new via FKG, Zürich. FKG is the usual intermediary between SSW and MBA, Dübendorf, but that is not so recorded for a batch of three locomotives that includes this one.

RUSTERHOLZ, Näfels   GL06 C

Probably a civil engineering concern.

Gauge: 600 mm Locn: ? Date: 02.2002

	0-6-0T	Cn2t	Krauss-S Zwilling	2936B	1894	(a) s/s

(a) half of a twin, originally Deutsche Feldbahn probably "62"; ex Schmidlin, Genève /1944.

W. SCHLITTER AG, Niederurnen   GL07 C

A construction company.

Gauge: 600 mm Locn: 236 $ 7231/2192 Date: 04.1987

4wDM	Bdm	RACO	?	19xx	(a) s/s?
4wDM	Bdm	RACO 12/16 PS	1333	194x	(b) (2)
4wDM	Bdm	Diema DS28	2580	1962	(c) (3)

(a) Austro-Daimler motor, ex ? /19xx.
(b) ex Eternit AG, Niederurnen /1988.
(2) 19xx to FWF, Otelfingen.
(3) before 2000 to P. Büchel, Gossau SG.

SBB-HISTORIC, Glarus   GL08 M

The locomotive is retained as a spare parts bank.

Gauge: 1435 mm Locn: ? Date: 03.2008

Eb 3/5 5811	2-6-2T	1C1h2t	SLM Eb 3/5	2212	1911

TONEATTI AG, Bilten   GL09 C

formerly: Toneatti & Co AG, Bilten

A civil engineering firm. Address: Tschachenstrasse 9.

Gauge: 600 mm (unless indicated) Locn: 236 $ 7205/2236 Date: 08.1999

281.21.101 No 4	4wDM	Bdm	O&K(M) S5	1543	192x	(a) (1)
281 21 102 1060	4wDM	Bdm	O&K(S) MD 2b	1518	194x	(b) (2)
281 21 103	4wDM	Bdm	O&K(M) RL 1c	8456	1937	(c) (2)

281.21.105 No 5	4wDM	Bdm	O&K(M) LD 2	8587	1937	(d)	(1)	
281 21 106 00117	4wDM	Bdm	O&K(M) ?LD 2	?	19xx		(5)	
	4wDM	Bdm	O&K(M) LD 16	4609	1932	(f)	(6)	
	4wDM	Bdm	O&K(M) LD 2	8191	1937	(f)	(5)	
	4wDM	Bdm	O&K(M) RL 1c	11427	1939	(i)	(2)	
	4wDM	Bdm	O&K(M) S5	?	19xx	(d)	(5)	
	4wDM	Bdm	O&K(M) RL 1a	?	193x	(k)	(5)	
	4wDM	Bdm	O&K(M) RL 1a	?	19xx	(k)	(5)	
	4wDM	Bdm	O&K(M) MD 2	?	19xx	(k)	(5)	
	4wDM	Bdm	O&K(M) MD 2	?	19xx	(k)	(5)	
	4wDM	Bdm	Jung MS 131	4704	1929	(l)	(2)	

(a) new as 4wPM Bpm.
(b) ex Bauunternehmung Zervreila, Chur /19xx.
(c) purchased via O&K/MBA, Dübendorf; O&K lists state type RL 1c, Toneatti records state RL 1a.
(d) purchased via O&K/MBA, Dübendorf.
(f) ex Rathgeber, Zürich /1952.
(i) purchased via O&K, Lager Zürich.
(k) gauge not confirmed.
(l) new to Robert Aebi & Cie AG, Zürich.
(1) 1995 donated to FWF, Otelfingen.
(2) 2007 to FWF, Otelfingen.
(3) after 1985 to a private collector in Kanton Graubünden (GR).
(5) before 1985 s/s.
(6) 19xx to Oliver Weder

AG CONRAD ZSCHOKKE, Näfels   GL10 C

This civil engineering firm has branches throughout Switzerland. The SIG list of references gives nine ETE 70 locomotives delivered here.

Gauge: 750 mm (unconfirmed)   Locn: ?   Date: 05.2008

4wDM	Bdm	RACO 50 PS	1312	1945		(1)	
4wDM	Bdm	RACO 50 PS	1313	1945		(1)	

Gauge: 600 mm (unless indicated)   Locn: ?   Date: 06.2002

4wDM	Bdm	Diema DS30	2054	1957	(c)	(3)	
4wDM	Bdm	Diema DS30	?	19xx		s/s	
4wDM	Bdm	Diema DS30	?	19xx		s/s	
4wBE	Ba	SSW EL 9	5277	1950	(f)	s/s	
4wBE	Ba	SSW EL 9	5278	1950	(f)	s/s	
4wBE	Ba	unknown CD33	1177	195x	(h)	s/s	
4wBE	Ba	unknown CD33	1177	195x	(h)	s/s	

Gauge: 500 mm   Locn: ?   Date: 06.2002

4wBE	Ba	Stadler	?	194x	

(c) new to Flemming'sche Ziegelwerke, Hannover [D] (500 mm); ex Diema /1983 via Asper .
(f) new via Wander-Wendel and MBA, Dübendorf.
(h) gauge not confirmed; new via MBA, Dübendorf.
(1) 19xx to IRR.
(3) 19xx to ?; 2006 to Heinz Gerber, Unterentfelden; 200x to Bözenegg-Eriwis Bahn, Schinznach Dorf via Tafag.

Graubünden (Grisons) GR

ARGE EKW, BAULOS 2, Ramosch — GR01 C

Murer and Zschokke AG provided the locomotives for this building contract.

Gauge: 750 mm   Locn: ?   Date: 12.2007

	4wDH	Bdh	Schöma CFL 180DCL	5189	1991	new	(1)
	4wDH	Bdh	Schöma CFL 180DCL	5190	1991	new	(1)
281.34.104.05	4wWE	Ba	SIG ETB 70	705 409	1976	(c)	s/s?

(c) property of Zschokke AG.
(1) by 1995 to Murer AG, Erstfeld for Associazione Lavori Piora, Faido.

ARGE TRANSCO, Sedrun — GR02 C

A key feature of the construction of the fifty-seven kilometre long Gotthardbasis rail tunnel being built for AlpTransit Gotthard AG is the ability to bore tunnels from intermediate locations. One such is the Zwischenangriff Sedrun. This facility is accessed from the surface at Tujetsch below Sedrun. Four railways, a funicular and two lifts are involved. A branch leaves the FO at Tscheppa (km 89.88) to the east of Bugnei and descends with rack assistance over a distance of 2.20 kilometres to Las Rueras. Although the trains are operated by the MGB (FO) a metre gauge locomotive provides shunting power on the reception sidings to which a funicular also descends from the offices located at Mira. From this location on the valley floor a tunnel has been driven into and finally up and out of the mountain. A 900 mm gauge railway has been laid in this for about a kilometre to where it reaches the two shafts that provide access to what will be the Multifunktionstelle Sedrun and currently gives access to the tunnelling and the construction railways down there. A scheduled workman's transport is provided on this line in addition to carrying all the materials to and from the works. The shafts were built by mining experts from South Africa and amongst the equipment they brought with them (and took back) were track sections to the indigenous three foot six inch gauge.

The much travelled Schöma type D60 railcar and trailer (5022+5023 of 1989) at the head of the shaft to the Multifunktionstelle Sedrun on 11 June 2007

Gauge: 1000 mm				Locn: 256 $ 7022/1704			Date: 06.2007	
LK 7		4wDH?	Bdh?	EMAM,Belloli T250D 030-3025/ ?		1996	(a)	

Gauge: 900 mm				Locn: 256 $ 7022/1704			Date: 06.2007	
TGV Tujetsch		4wDH	Bdh	Schöma D60	5022	1989	(b)	
LK 1		4wDH	Bdh	Schöma CFL 180DCL	5170	1991	(c)	
LK 2		4wDH	Bdh	Schöma CFL 180DCL	4418	1981	(d)	(4)
LK 3		4wDH	Bdh	Schöma CFL 180DCL	5172	1991	(e)	
LK 4		4wDH	Bdh	Schöma CFL 180DCL	4419	1981	(d)	(4)
LK 5		4wDH	Bdh	Schöma CFL 180DCL	5064	1989	(g)	(4)
LK 6		4wDH	Bdh	Schöma CFL 180DCL	5062	1989	(g)	(4)
LK 8		4wDH	Bdh	Schöma CFL 180DCL	5134	1990	(i)	(4)
LK 9		4wDH	Bdh	Schöma CFL 180DCL	5159	1989	(j)	(4)
LK 10		4wDH	Bdh	Schöma CFL 200DCL	5624	1999	(k)	(4)
LK 11		4wDH	Bdh	Schöma CFL 180DCL	5840	2003	new	
LK 12		4wDH	Bdh	Schöma CFL 180DCL	5841	2003	new	
LK 13		4wDH	Bdh	Schöma CFL 180DCL	5842	2003	new	
LK 14		4wDH	Bdh	Schöma CFL 180DCL	5843	2003	new	
LK 15		4wDH	Bdh	Schöma CFL 200DCL	5837	2003	new	(4)
LK 16		4wDH	Bdh	Schöma CFL 200DCL	5838	2003	new	(4)
		4wDH	*Bdh*	*Schöma CFL 200DCL*	*5839*	*2003*	*new*	*(18)*
LK 17		4wDH	Bdh	Schöma CFL 180DCL	5844	2003	new	
LK 18		4wDH	Bdh	Schöma CFL 180DCL	5845	2004	new	
LK 19		4wDH	Bdh	Schöma CFL 180DCL	5846	2004	new	(21)
LK 20		4wDH	Bdh	Schöma CFL 180DCL	5853	2004	new	
LK 21		4wDH	Bdh	Schöma CFL 180DCL	5854	2004	new	
LK 22		4wDH	Bdh	Schöma CFL 180DCL	5855	2004	new	
LK 23		4wDH	Bdh	Schöma CFL 180DCL	5856	2004	new	
LK 24		4wDH	Bdh	Schöma CFL 180DCL	5873	2004	new	
LK 25		4wDH	Bdh	Schöma CFL 180DCL	5872	2004	new	
LK 26		4wDH	Bdh	Schöma CFL 180DCL	5874	2004	new	
LK 27		4wDH	Bdh	Schöma CFL 180DCL	5871	2004	new	
LK 28		4wDH	Bdh	Schöma CFL 180DCL	5925	2004	new	
LK 29		4wDH	Bdh	Schöma CFL 180DCL	5926	2004	new	
LK 30		4wDH	Bdh	Schöma CFL 180DCL	5927	2004	new	
		4wDH	Bdh	Schöma CFL 180DCL	6015	2005	new	
		4wDH	Bdh	Schöma CFL 180DCL	6016	2005	new	

(a) rebuild of EMAM locomotive.
(b) usually coupled to Schöma trailer (D60-Tr) 5023 of 1989, new to Trans-Manche Link (TML), Cheriton [GB] "RU006 and RU006A"; ex ArGe Vereinatunnel Nord (ARGVN), Klosters/Selfranga ca. /2005 after a period in store at Beton+Kies Unterrealta/Casis.
(c) new to CSC Bauunternehmung AG, Pradella "1"; ex ArGe Vereinatunnel Nord, Selfranga "10" 8/2004 after a period in store at Beton+Kies Unterrealta/Casis.
(d) new to Bechlingen [A]; then Trans-Manche Link, Cheriton [GB]; MTG Körsor [DK]; Vereina-Tunnel Nord, Selfranga "11"; SDK, Singapore [Malaysia]; ex Schöma, on loan /2002.
(e) new to CSC Bauunternehmung AG, Pradella "3"; ex ArGe Vereina-Tunnel Nord, Selfranga "10" 8/2004 after a period in store at Beton+Kies Unterrealta/Casis.
(g) new to Trans-Manche Link, Cheriton [GB]; then U-Bahn, München [D]; SDK, Singapore [Malaysia]; ex Schöma, on loan /2002.
(i) new to Trans-Manche Link, Cheriton [GB]; then U-Bahn, München [D]; SDK, Singapore [Malaysia]; ex Schöma, on loan /2003.
(j) new to Trans-Manche Link, Cheriton [GB]; then Costain Taylor [GB]; SDK, Singapore [Malaysia]; ex Schöma on loan /2003.
(k) new to Tubecon [NL], ex Schöma on loan /2003.
(4) 2004 returned to Schöma.
(18) ordered; but never delivered.
(21) 2004 to NESCO, Metro Madrid [E].

ARGE VEREINATUNNEL NORD (ARGVN), Klosters/Selfranga
GR03 C

A construction consortium for the Prättigau end of the RhB Vereina tunnel. As work progressed a number of locomotives changed gauges between the 1000 mm of the RhB and the 900 mm of the tunnel construction. The depot was established at Zugwald between the Zugwald tunnel from Klosters and the main Vereina tunnel. In the latter tracks of both gauges existed simultaneously over part of the distance.

"6" Schöma 5420 of 1995 (type CFL 200DCL) unloads stone on 30 March 1996.

Gauge: 900 mm (or 900/1000 mm)   Locn: 248 $ 7866/1924   Date: 01.1996

No.	Type		Make/Model	Works No.	Year		
1	4wDH	Bdh	Schöma CFL 200DCL	5368	1993	(a)	(1)
2	4wDH	Bdh	Schöma CFL 200DCL	5370	1993	(b)	(2)
3	4wDH	Bdh	Schöma CFL 200DCL	5369	1993	(c)	(1)
4	4wDH	Bdh	Schöma CFL 200DCL	5417	1995	(d)	s/s
5	4wDH	Bdh	Schöma CFL 200DCL	5418	1995	(e)	(5)
6	4wDH	Bdh	Schöma CFL 200DCL	5420	1995	new	s/s
7	4wDH	Bdh	Schöma CFL 200DCL	5419	1995	new	s/s
8	4wDH	Bdh	Schöma CFL 180DCL	5171	1995	(h)	s/s
9	4wDH	Bdh	Schöma CFL 180DCL	5172	1991	(i)	(9)
10	4wDH	Bdh	Schöma CFL 180DCL	5170	1991	(j)	(10)
11	4wDH	Bdh	Schöma CFL 180DCL	4419	1981	(k)	(11)
12	4wDH	Bdh	Schöma CFL 180DCL	4418	1981	(l)	(12)
	4wDH	Bdh	Schöma D60	5022	1989	(m)	(13)

(a) ex ArGe Zugwald-Tunnel, Klosters; converted from 1000 to 900 mm.
(b) ex ArGe Zugwald-Tunnel, Klosters, (gauge 1000 mm).
(c) ex ArGe Zugwald-Tunnel, Klosters, converted from 1000 to 900 mm.
(d) converted from 1000 to 900 mm /1995-96.
(e) converted from 1000 to 900 mm /1995.
(h) ex CSC Bauunternehmung AG, Zürich, Druckstollen San Niclà, Strada en Engadina "2".
(i) ex CSC Bauunternehmung AG, Zürich, Druckstollen San Niclà, Strada en Engadina "3".
(j) ex CSC Bauunternehmung AG, Zürich, Druckstollen San Niclà, Strada en Engadina "1".

(k) new to ArGe Walgaustollen [A]; 1988 to Schöma (converted to CFL 180DCL); later Trans-Manche Link (TML), Cheriton [GB] "RS005"; 1993 to Schöma, ex MT Group IS, Halskov [DK] "45-33" via Schöma (on hire) /1995.
(l) new to ArGe Walgaustollen [A]; 1988 to Schöma (converted to CFL 180DCL); later Trans-Manche Link (TML), Cheriton [GB] "RS002"; 1993 to Schöma; ex MT Group IS, Halskov "45-32" [DK] via Schöma (on hire) /1995.
(m) usually coupled to Schöma trailer (D60-Tr) 5023 of 1989, new to Trans-Manche Link (TML), Cheriton [GB] "RU006 and RU006A"; ex MT Group IS, Halskov [DK] "35-06" via Schöma /1995.
(1) 3/2001 to Schöma.
(2) by 9/1995 to KMW [NL] via Schöma.
(5) 3/2006 to Scandinavia.
(9) 9/2001 to store Beton+Kies Unterrealta/Casis; by 9/2004 to ArGe Transco, Las Rueras, Sedrun "LK 3".
(10) 9/2001 to store Beton+Kies Unterrealta/Casis, by 8/2004 to ArGe Transco, Las Rueras, Sedrun "LK 1".
(11) by 1/1999 to SDK JV Mrt Nel C710, Singapore [Malaysia]; 2000 to KMW [NL] via Schöma.
(12) by 1/1999 to SDK JV Mrt Nel C710, Singapore [Malaysia]; 2006 to NECSO, Metro Madrid [E] via Schöma; 2006 to Acciona [E].
(13) 9/2001 to store Beton+Kies Unterrealta/Casis; to ArGe Transco, Las Rueras, Sedrun "TGV, Tujetsch".

ARGE VEREINATUNNEL SÜD, Susch   GR04 C

alternative: Süs

A consortium responsible for the construction of the Engadine end of the RhB Vereina tunnel.

Gauge: 900 mm   Locn: ?   Date: 04.1994

L1	4wDH	Bdh	Schöma CFL 180DCL	5374	1993	new	
L2	4wDH	Bdh	Schöma CFL 180DCL	5375	1993	new	
L3	4wDH	Bdh	Schöma CFL 180DCL	5376	1993	new	
L4	4wDH	Bdh	Schöma CFL 150DCL	5201	1991	(d)	
L5	4wDH	Bdh	Schöma CFL 150DCL	5147	1990	(e)	
L6	4wDH	Bdh	Schöma CFL 200DCL	5420	1995	new	
L7	4wDH	Bdh	Schöma CFL 200DCL	5419	1995	new	
	4wDH	Bdh	Schöma CFL 180DCL	5424	1995	new	(8)
	4wDH	Bdh	Schöma CFL 180DCL	4495	1981	(i)	

(d) ex Trans-Manche Link (TML), Cheriton [GB] "RS048 INKE" via Schöma /1993.
(e) ex Trans-Manche Link (TML), Cheriton [GB] "RS036 JOANNE" via Schöma /1994.
(i) new to ArGe Walgaustollen, Baulos Beschling [A]; Trans-Manche Link (TML), Cheriton [GB] "RS004 KENNA" via Schöma /1988; MT Group IS, Haldkov [DK] "45-43" via Schöma /1996; via Schöma.
(8) 2003 to Nishimatsu-Skanska-Cementation Joint Venture [GB] for Singapore [Malaysia]; 2006 Sacyr, Abdalajis West [E].

ARGE ZUGWALD-TUNNEL, Klosters   GR05 C

A construction consortium for the first step in building the RhB Vereina tunnel. This included the provision of a new layout and bridges at Klosters station and a curved tunnel up to Selfranga.

Gauge: 1000 mm   Locn: ?   Date: 08.2007

1	4wDH	Bdh	Schöma CFL 200DCL	5368	1993	new	(1)
2	4wDH	Bdh	Schöma CFL 200DCL	5370	1993	new	(2)
3	4wDH	Bdh	Schöma CFL 200DCL	5369	1993	new	(3)

(1) to ArGe Vereinatunnel Nörd, Klosters/Selfranga "1".
(2) to ArGe Vereinatunnel Nörd, Klosters/Selfranga "2".
(3) to ArGe Vereinatunnel Nörd, Klosters/Selfranga "3".

AT COPOUND, San Vittore GR06

formerly: Monteforno Valmoesa SA

This steel works of the Monteforno/von Roll group was situated alongside the 1000 mm Mesocco line of the RhB – (Bellinzona-)Castione-Arbedo-Mesoco. Standard gauge wagons were transferred as required by transporter wagons. The works closed in 1994 and with that the commercial use of the railway ceased.

Gauge: 1435 mm   Locn: 277 $ 7276/1217   Date: 04.1998

2	0-4-0WT	Bn2t	SLM E 2/2	917	1895	(a)	(1)
N1	4wDM	Bdm	O&K MV2a	25729	1957	(b)	s/s
N2	4wDM	Bdm	Kronenberg DT 180/250	130	1953	(a)	197x s/s
	4wDH	Bdh	Diema DVL90/1.1	3516	1974	(d)	(4)

(a) purchased from Monteforno SA, Bodio /1970.
(b) new via MBA, Dübendorf.
(d) new via Asper.
(1) 1971 to Verein Dampfbahn Bern (DBB); 1987 to VVT, St. Sulpice NE.
(4) 1997 to Belloli SA, Grono.

BAUSTELLE VAL SAMPOIR, Samnaun GR07 C

alternative: Val Sampuoir

SIG battery locomotives of type ETS 100 were used on this contract. Other details are not available.

BAUUNTERNEHMUNG STETTER, Chur GR08 C

A public works concern.

Gauge: 600 mm   Locn: ?   Date: 06.1999

	4wDM	Bdm	O&K(M) RL 1c	11434	1940	(a)	s/s
	4wDM	Bdm	O&K(M) RL 1c	11435	1940	(a)	s/s

(a) purchased via O&K, Lager Zürich.

BAUUNTERNEHMUNG ZERVREILA, Chur GR09 C

Either a public works concern or a consortium for the construction of the Zervreila hydroelectric scheme dating from ca. 1958.

Gauge: ? mm   Locn: ?   Date: 06.1999

	4wDM	Bdm	O&K(S) MD 2b	1518	194x	(a)	(1)
	4wDM	Bdm	O&K(M) MD 2r	12234	1952	(a)	s/s?

(a) ex MBA, Dübendorf /1952.
(1) 19xx to Toneatti AG, Bilten.

BELLOLI SA, Grono GR10 C

alternative: Belloli & Cie
alternative: Ferriere Belloli

A firm trading in and repairing equipment (including locomotives) for building sites. Rail traffic was ensured by standard gauge wagons carried on RhB transporter wagons and unloaded onto a standard gauge siding on site. The firm has carried out major rebuilds of some locomotives and has affixed their own plates. Other locomotives have been stored on site: those recorded are listed here.

Gauge: 1435 mm			Locn: 277 $ 7313/1226			Date: 04.2005	
	4wDH	Bdh	Diema DVL90/1.1	3516	1974	(a)	
Gauge: 900 mm			Locn: 277 $ 7313/1226			Date: 04.2005	
	4wDH	Bdh	Schöma CHL 60G	5301	1993	(b)	
Gauge: 750 mm			Locn: 277 $ 7313/1226			Date: 04.2005	
	4wDH	Bdh	Ruhrthaler G100HVG	3943	1970	(c)	
Gauge: 600 mm			Locn: 277 $ 7313/1226			Date: 04.2005	
	4wDH	Bdh	JW DM50F-10	77.08	1950	(d)	
	4wBE	Ba	SIG ETB 70	?	19xx	(5)	
	4wBE	Ba	SIG ETR 70	708 739	1978	OOU	
	4wBE	Ba	SIG ETR 70	?	19xx	(5)	
	4wBE	Ba	SIG ETR 70	?	19xx	(5)	
	4wBE	Ba	SIG ETR 70	?	19xx	(9)	
	4wD	4wD	Diema	?	19xx	(10)	

(a) ex AT Copound, San Vittore /1997; stored since /2001.
(b) new to Campenon Bernard for Paris Metro [F] (?1000 mm); ex MT Group IS, Halskov [DK] "45-50" via Schöma /1997.
(c) new to SA Cementazioni per Opere Pubbliche (S.A.C.O.P.). Roma [I]; by 8/1997 at Grono.
(d) new to Zellulose Lenzing [A] /1951; by 8/1997 at Grono.
(5) seen 8/1997; s/s by 4/2005.
(9) seen 8/1997; scrapped by 4/2005.
(10) seen 12/5/1986; s/s by 8/1997.

BELLORINI, Grono   GR11 C

As an illustration of the difficulties encountered when interpreting the data in the RACO archives is an entry for this firm. It is not otherwise recorded at this location, but E. Bellorini of Lausanne could easily have had a contract in the area during relevant period ca. 1947. The archive entry indicates that RACO 1377 of 1947 was delivered here and that it was also the last 50 PS locomotive constructed. However RACO 1377 of 1949 was a RA11 delivered to the SBB as "Tm 301" and was latterly with Egli-Mühlen AG, Nebikon.

BÜNDNER KRAFTWERKE, Küblis   GR12

later: Rätia Energie

From 1920-99 a branch (about a kilometre in length) connected the hydroelectric station, built 1919-21, to the RhB.

Gauge: 1000 mm			Locn: 248 $ 7785/1986			Date: 05.2006	
	0-6-0T	Cn2t	SLM HG 3/3	614	1890	(a)	19xx s/s
Montania	4wPM	Bbm	O&K(M)	1735	1923	new	(2)

(a) purchased from ChA construction /1920.
(2) 2005 to Mainstation 1901, Chur.

GEBRÜDER CAPREZ ERBEN, Chur   GR13 C

A construction concern.

Gauge: 600 mm			Locn: ?			Date: 06.1999	
RITOM	0-4-0T	Bn2t	Maffei 40 PS	2920	1909	(a)	(1)
	4wDM	Bdm	O&K(M) RL 1	3908	1929	(b)	s/s
	4wDM	Bdm	O&K(M) RL 1a	4020	1930	(b)	s/s

(a) ex Taddei Domenico, Impresa Costr. Castagnola /1940.
(b) supplied via O&K/MBA, Dübendorf.

(1)   1943 sold to Hoch- & Tiefbau AG, Aarau.

CAZIS/REALTA, Rothenbrunnen GR14

A locomotive is made available by the local authorities for working traffic on the roadside branch to the south of the RhB station.

Gauge: 1000 mm   Locn: 257 $ 7512/1812   Date: 06.2007
                     4wD R/R   Zephir Lok 6.130   1440   1997   (a)
(a)   new via Belloli.

CLUB 1889, Samedan GR15 M

Gauge: 1000 mm   Locn: ?   Date: 11.2007
Ge 2/4 205   2-4-2WE   1B1w1   SLM,BBC Ge 2/4   2308,727   1913   (a)
(a)   ex RhB after display at Technikum, Winterthur from /1974-2007.

CONSORZIO PRESA CALANCASA, Buseno GR16 C

Consortium of the public works concerns: Somaini, Guidicetti and Pieracci.

Gauge: 600 mm   Locn.: ?   Date: 05.2002
                     4wDM   Bdm   O&K(M) MD 1   11414   1939   (a) (1)
(a)   purchased via O&K/MBA, Dübendorf.
(1)   1955 sold to Payot & Rochat, Vers l'Eglise.

CSC BAUUNTERNEHMUNG AG, Pradella GR17 C

A Zürich-based building contractor responsible for the boring of the pressure tunnel of the San Niclà hydroelectric station located at Strada en Engiadina.

Gauge: 900 mm   Locn: 249 $ 8281/1941   Date: 01.1996
      1   4wDH   Bdh   Schöma CFL 180DCL   5170   1991   new (1)
      2   4wDH   Bdh   Schöma CFL 180DCL   5171   1991   new (2)
      3   4wDH   Bdh   Schöma CFL 180DCL   5172   1991   new (3)
(1)   1994-95 to ArGe Vereina-Tunnel Nord, Klosters/Selfranga "10".
(2)   1994-95 to ArGe Vereina-Tunnel Nord, Klosters/Selfranga "8".
(3)   1994-95 to ArGe Vereina-Tunnel Nord, Klosters/Selfranga "9".

EMS-CHEMIE AG, Domat/Ems GR18

formerly: Holzverzuckerung AG (HOVAG) (1936-1962)
             Domat (rätoromanisch), Ems (Deutsch)

A factory was built to the west of Ems in 1942 together with the Ems Werk station of the RhB. Originally it produced fuel from timber and was rebuilt 1954-55 to produce chemicals for use in making synthetic fibres. By 2007 one road of the short two-road locomotive-shed had been disconnected. Standard gauge traffic is worked over the northern line of the double track RhB metre gauge line from Chur by the SBB at a reduced voltage. Address: Kugelgasse 22.

Ex SBB E 3/3 8501 (a Tigerli, SLM 2076 of 1910) stands under cover awaiting duties on 26 September 1967. It has since gone into preservation with the Club del San Gottardo.

Gauge: 1435 mm			Locn: 247 $ 7516/1883			Date: 09.2001		
6410	0-6-0T	Cn2t	SLM Ec 3/3	1378	1901	(a)	(1)
1	0-6-0WT	Cn2t	SLM E 3/3	2076	1910	(b)	(2)
2	2-6-0T	1Cn2t	SLM Ed 3/4	1799	1907	(c)	(3)
3	0-6-0WT	Cn2t	SLM E 3/3	1881	1907	(d)	(4)
1	4wDE	Bde1	Moyse BN32E210D	1238	1972	new	s/s?
2	4wDE	Bde1	Moyse BN32E210D	1239	1972	new	s/s?
S01	4wDE	Bde	ČKD T 239.1	?	1994	new	
Tm237 952-7 ERWIN	0-4-0DH	Bdh	Henschel DH 240B	30336	1962	(h)	(8)

Gauge: 1000 mm			Locn: ?			Date: 03.1987		
6 WEISSHORN	0-4-2T	B1n2t	SLM HG 2/3	1410	1902	(i)	(9)

Gauge: 600 mm			Locn: ?			Date: 06.1999		
	0-4-0T	Bn2t	Maffei	4144	1925	(j)	(10)
	4wDM	Bdm	O&K(S) RL 1c	1554	1949	(k)	(11)

(a) new to GB "310"; purchased from SBB "Ec 3/3 6410" /1941.
(b) hired from SBB "E 3/3 8501" /1958; purchased /1963.
(c) purchased from Gaswerk der Stadt Zürich, Schlieren /1946.
(d) purchased from SBB "E 3/3 8485" /1963.
(h) hired from SERSA /1998.
(i) originally VZ "HG 2/3 6 WEISSHORN"; ex BVZ (rack gear removed) /1941.
(j) purchased via RACO /1941.
(k) ex Hunziker AG, Döttingen.

(1) 1958 stored; 1965 scrapped.
(2) 1963 sold; 1976 to Oswald Steam, Samstagern; 1993 to Club del San Gottardo, Mendrisio.
(3) 1973 to Dampflok-Freunde, Langenthal.
(4) 1972 to private ownership; 1977 to EUROVAPOR, Wutachtalbahn [D]; 1981 store Koblenz; 2002 to Historische Eisenbahn Gesellschaft (HEG).
(8) 1998 returned to SERSA.

CH 150 GR

(9)	1956 withdrawn; 1965 preserved at Schule Herold, Chur; 1988 donated to Dampfbahn Furka Bergstrecke (DFB), Realp.							
(10)	1944 OOU, 1979 to Oswald Steam, Samstagern "EMMA", 1992 to Schinznacher Baumschulbahn (SchBB), Schinznach Dorf.							
(11)	1979 to Oswald Steam, Samstagern.							

GALLI & CIE, Bonaduz — GR19 C

A construction company. It may have been a part of the building contractors Ing. Galli & Co/Minder & Galli.

Gauge: 750 mm   Locn: ?   Date: 03.2002

	0-4-0T	Bn2t	O&K 40 PS	717	1900	(a)	(1)

(a) new via O&K Strasbourg [F].
(1) 1900-05 to F. & A. Bürgi, Bern.
(1) 1943 to Hoch- & Tiefbau AG, Aarau.

ANDREA GRUBER, Landquart — GR20 C

This building contractor has sites in Landquart and Grütsch. In store with the locomotive are also four complete tipping skips, one without the body and a flat wagon. Address: Wuhrstrasse.

Gauge: 600 mm   Locn: 248 $ 7623/2042   Date: 09.2006

	4wPM	Bbm	Austro-Daimler PS6	102	19xx	(a)	OOU

a) has carried a plate that appears to read "RACO PH6-102 (or 108?)", so was almost certainly supplied via RACO. Used ca. 1930 during the construction of the Salginatobel bridge near Schiers, and then later for work on the "Wuhrbau" in the Prättigau.

HEINEKEN SWITZERLAND AG, Paleu Sura, Felsberg — GR21

formerly: Calanda Haldengut Breweries (03.12.1999 - 01.05.2001, dissolved 21.5.2002)
Calanda Haldengut Beverages Ltd (31.12.1995 - 03.12.1999)

A distribution centre for drinks stemming from the "Calanda-Haldengut" brewery. In 2007 the locomotive retained its Calanda livery. So that it can shunt both standard gauge and 1000 mm wagons on the mixed gauge sidings it has a RhB coupling as well as the standard gauge one.

Gauge: 1435 mm   Locn: 247 $ 7559/1898   Date: 06.2007

CALANDA	4wDH	Bdh	O&K MB 5 N	26600	1966	(a)	

(a) new to Dortmunder Union-Brauerei AG, Dortmund [D] "1"; ex LSB /1991.

see: www.hr-monitor.ch/f/Heineken_Switzerland_Ltd_CH-350.3.000.684-5_12961305.html (where the title is also given in other languages)

HEW & CO, Chur — GR22 C

A public works concern.

Gauge: 600 mm and possibly others   Locn: ?   Date: 08.2002

	4wDM	Bdm	O&K(S) MD 2s	1545	194x	(a)	s/s	
	4wDM	Bdm	O&K MV0	25100	1951	(b)	(2)	
	4wDM	Bdm	O&K MV0	25103	1951		(3)	
	4wDM	Bdm	O&K MV1	25138	1952		(4)	
	4wDM	Bdm	O&K MV1	25139	1952		s/s	
	4wDM	Bdm	O&K MV1	25140	1952		(6)	
	4wDM	Bdm	O&K MV1	25227	1952	(g)	(2)	
7	4wBE	Ba1	Stadler	31	1948	(h)		

		8	4wBE	Ba1	Stadler		32	1948	(i)	19xx sold

(a) new as type S5; ex MBA with MD 2S motor 17/3/1953.
(b) new via MBA, Dübendorf.
(g) supplied via MBA, Dübendorf to Hew & Co, Ilanz /1953.
(h) purchased via STUAG, Bern, overhauled by Stadler /1963 (600 mm).
(i) purchased via STUAG, Bern.
(2) 1956 to Morandi Tileworks, Corcelles-près-Payerne (600 mm).
(3) 1957 to Ziegelei Roggwil AG, Roggwil /1957.
(4) 1957 to Lazzarini, Samedan.
(6) to Schafir & Mugglin, Trimmis /19xx (600 mm).

HOLCIM (Schweiz) AG, Untervaz   GR23

formerly: HOLCIM Zement AG, Untervaz (8/5/2001-31/12/2003)
              HCB Cement AG, Untervaz (3/12/1999-7/5/2001)
              HCB Untervaz AG, Untervaz (1/10/1999-2/12/1999)
              Bündner Cementwerke AG (BCU) (till 30/9/1999)

This cement works lies on the opposite bank of the Rhein to the SBB and RhB. A mixed gauge branch line connects it with the two. All the permanent locomotives and some of the robots have been fitted with draw-gear for both gauges.

Gauge: 1435 mm   Locn: 248 $ 7612/1985   Date: 06.2007

'1' Inv 31681	0-6-0DH	Cdh	Jung R 40 C	12837	1957	new	(1)	
'2'	0-6-0DH	Cdh	Jung R 30 C	13620	1962	new	(2)	
Em 837 911-7	6wDE	Cde3	MaK,BBC DE501 C	700041, ?	1980	(c)		
Em 847 151	BoBoDE	BBde	Fauvet-Girel	?	1968	(d)	(4)	
1	4wDH	Bdh	O&K MB 9 N	26656	1968	(e)	(5)	
631-4R1	4wCE?	Bel?	Windhoff RW 50 E	130426	1973	new		
601-4R6	4wD	Bd	Vollert	?	19xx			
	4wD	Bd	Vollert	?	19xx			

(c) new, "Inv 31682".
(d) originally WM (BDWM) "Em 4/4 151"; ex Stauffer on loan /2006.
(e) ex Kieswerk Hüntwangen AG, Hüntwangen /2002.
(1) 1982 loan to von Moos, Emmenbrücke for a few months; 1986 to Vorarlberger Zementwerk AG, Loruns [A].
(2) 1983 to Verden-Walsroder-Eisenbahn [D] "1", via NEWAG.
(4) 2006 returned to Stauffer.
(5) by 2/2005 hired to Zementwerk Lorüns/Voralberg [A]; 17/7/2006 to Shunter "304 MYRTHE" via Unirail.

see: SER 5/2006 p219

HOMOPLAX SPANPLATTENWERK, Fideris   GR24

A factory producing pressed boards. The locomotive carried MBA plates, but it is not clear if it started life as an O&K or a Schöma product.

Gauge: 1000 mm   Locn: 248 $ 7748/1994   Date: 03.1987

	4wDM	Bdm	unknown	?	19xx	(a)	a1967 s/s

(a) adapted by RhB for 1000 mm at Landquart /1963, MBA plates.

HOWEG TRANSGOURMET AG, Chur   GR27

A food distribution firm with head office in Dietikon that works closely with Prodega AG.
Address: Grossbruggerweg 2, 7000 Chur.

Gauge: 1435 mm   Locn: 247 $ 7582/1915   Date: 06.2007

4wDM R/R	Bdm	Mercedes Unimog,Zweiweg ZW925	

25 10 006478,1110   1980   (a)

(a) supplied by Robert Aebi AG.

IMPRESA RODARI CASPARE & CIE, Bever   GR28 C
A public works concern.
Gauge: 750 mm   Locn: ?   Date: 01.1979
               0-4-0T   Bn2t   Borsig   7605   1910   new s/s

A. KÄPPELI'S SÖHNE AG, Chur   GR28a C
The SIG list of references gives one ET 15KL and one ETS 25 locomotives delivered here.

KUMMLER + MATTER AG, Chur   GR29 C
This Zürich-based company provides overhead maintenance services (Fahrleitungstechnik) to the RhB and usually keeps its metre gauge locomotives somewhere on the system.
Gauge: 1000 mm   Locn: various   Date: 09.2006
Tm 2/2 20   4wDM   Bdm   ČKD T 211.0   4686   1959   (a)
Tm 2/2 21   0-4-0DH   Bdm   O&K MV9   25833 1958   (b)
(a)   new to CSD 1435 mm "211 0059, later "701 101-8" (1435 mm); ex Tyrex, Praha [CZ] /1994.
(b)   new to Braunkohlen-& Briketwerke Roddergrube AG, Brühl [D] "39" (900 mm); ex Tafag /2000; other sources quote this locomotive as CGS "22" (not "21") and hence O&K 25892 1959, and RBW "159".

KÜNZLI & MAI, Davos   GR30 C
A public works concern.
Gauge: 600 mm   Locn: ?   Date: 06.1999
               4wDM   Bdm   O&K(M) RL 1a   4834   1933   s/s

KURVEREIN DAVOS   GR31

Kehrichtversammlung, Davos-Islen
From about 1990 town rubbish has been compressed into rail-borne containers and taken away by the RhB from a siding worked by this locomotive.
Gauge: 1000 mm   Locn: 248 $ 7814/1939   Date: 07.2006
               4wDM R/R Bdm   Mercedes Benz Unimog   ?   19xx

Kehrichtverbrennungsanlage, Davos-Laret
formerly: Gaswerk Davos AG (1906-55?)
The gasworks commenced operations in 1914. By 1955 the site had passed to the Kurverein Davos and they added a rubbish incineration plant to the remaining installation. The connecting line was probably installed ca. 1906 and was removed 1975-76.
Gauge: 1000 mm   Locn: 248 $ 7857/1912   Date: 03.1987
               4wDM   Bdm   Schöma   1177   1950   (a)   a1980 s/s
(a)   new as 750 mm; obtained new MBA.

G. LAZZARINI & CO AG, Samedan (and Chur)   GR32 M

A public works concern. The company advertises its locations as "Chur and Samedan". The remaining locomotive could not be located in 2007. The SIG list of references gives one ETB 50 locomotive delivered to Chur. Address: Via Nouva 18, Samedan; Grossbruggerweg 1, Chur.

Gauge: 600 mm   Locn: 268 $ 7869/1567   Date: 07.1999

4wDM	Bdm	O&K(M) MD 2	12231	1950	(a)	(1)
4wDM	Bdm	O&K MV1	25138	1952	(b)	s/s?
4wDM	Bdm	unknown	?	19xx		display

(a)   ex Prader & Co, Chur /1957.
(b)   ex Hew & Co, Chur /1957.
(1)   to MBA, Dübendorf for display.

LONZA AG, Thusis   GR33

formerly: Gesellschaft für elektrochemische Industrie (1898-1903)

A carbide factory (Karbidfabrik Thusis) and a power station were built 1898-99. Both were taken over by Lonza AG in 1903 and rail connected from 1911. The power station passed to the Rhätische Werke für Elektrizität, but the carbide factory continued under Lonza AG until closed in 1920.

Gauge: 1000 mm   Locn: 257 $ 7532/1743   Date: 06.1998

0-6-0T	Cn2t	SLM HG 3/3	616	1890	(a)	19xx s/s

(a)   new to BOB "4" (ex Can4t with rack gear removed); purchased from ChA construction /1918, although it started work here /1917.

MAINSTATION 1901, Chur   GR34 M

A cultural centre including a bar and exhibition hall. Address: Spundisstrasse.

Gauge: 1435 mm   Locn: 247 $ 7573/1904   Date: 06.2007

2	0-6-0WT	Cn2t	SLM E 3/3		2507	1914	(a) display

Gauge: 1000 mm   Locn: 247 $ 7573/1904   Date: 06.2007

Montania	4wPM	Bbm	O&K(M)		1735	1923	(b) display

(a)   new to SBB "8527"; ex Messerli AG, Kaufdorf ca. /2005.
(b)   ex Rätia Energie, Küblis /2005.

OBERSAXON, Tavansa   GR34a C

The SIG list of references gives one ETB 70 locomotive delivered here.

PIGNIA, Andeer   GR34b C

The SIG list of references gives four ETB 70 locomotives delivered here.

ANDREA PITSCH AG, Thusis   GR35 C

A public works concern, specialising particularly in tunnel building. One locomotive was obtained for museum purposes and was initially placed on display here. The SIG list of references gives two ETB 70 and two ATS 100 locomotives delivered to this company in St. Moritz and one ETB 50 locomotive to Thusis. Address: Rozaweg 2, 7430 Thusis.

Gauge: 750 mm   Locn: 257 $ 7528/1748   Date: 12.1995

281.03	4wBE	Ba2	SIG ET 70	?	19xx	

281.0x	4wDH	Bdh	Diema DFL90/0.2	2729	1964	(b)	
281.0x	4wDH	Bdh	Diema DFL90/0.2	?	1981	(c)	
281.09	4wBE	Ba2	SIG ATS 100	?	19xx		
281.10	4wBE	Ba2	SIG ATS 100	?	19xx		
ALVIER	4wDH	Bdh	KHD A4M517 G	55442	1953	(f)	display

(b) either "281.06" or "281.08"; ex ArGe Hondrich-Tunnel-Nord /19xx.
(c) either "281.06" or "281.08"; purchased via Asper; could it in fact be Diema 2360 1960 from Hondrich-Tunnel-Nord?
(f) new to Eisenbergwerk Gonzen AG, Sargans (600 mm); ex St. Gallische Rheinregulierung " 117 ALVIER" (750 mm) /1976 (4wDM Bdm?).

PRADER & CIE AG, Chur   GR36 C

This is a subsidiary of the public works company of Prader & Cie AG, Zürich. The firm specialises in tunnel construction. The SIG list of references gives four ET 20L5 and five ETS 50 locomotives delivered here.

Gauge: 600 mm and possibly others   Locn: ?   Date: 06.1999

4wDM	Bdm	O&K(S) MD 2s	1548	194x	(a)	
4wDM	Bdm	O&K(M) RL 1a	4465	1931	(b)	s/s
4wDM	Bdm	O&K(M) MD 2	12231	1950		(3)
4wDM	Bdm	O&K(M) MD 1	11438	1940	(d)	s/s
4wBE	Ba1	Stadler	23	1947	(e)	
4wDH	Bdh	Ageve	811	1978		(6)
4wDH	Bdh	Ageve	?	1978		(6)

(a) ex Bretscher & Sohn, Wallisellen /1953 (unknown gauge).
(b) purchased via O&K/MBA, Dübendorf (600 mm).
(d) purchased via O&K, Lager Zürich; into service /1951 (600 mm).
(e) delivered for the works at Julier/Mons (600 mm).

(3) 1957 to Lazzarini, Samedan.
(6) seen at Salgesch 5/2000.

J. RIETER & CIE AG, Mesocco   GR37 C

alternative   AG vorm J.J Rieter

This was a general business company with offices in Winterthur. The locomotives appear to have been purchased for use on the construction of the Bellinzona-Mesocco railway line.

Gauge: 1000 mm   Locn: ?   Date: 05.2009

0-4-0T	Bn2t	Jung 50 PS	105	1891	(a)	(1)
0-6-0T	Ct	O&K 80 PS	1909	1906	new	(2)

(a) new via F. Marti, Winterthur.

(1) 1923 sold to Impresa costruzioni Muttoni & Cattaneo, Biasca.
(2) 19xx to 19xx Charles Monce.

RHÄTISCHE BAHN   GR38 M

The company operates several heritage locomotives and these are based where most appropriate: Landquart, Samedan and Pontresina being the most common locations.

Gauge: 1000 mm   Locn: Various   Date: 11.2008

G 3/4	1	RHÄTIA	2-6-0T	1Cn2t	SLM G 3/4	577	1899
ABe 4/4	30		BoBoWERC		SIG,SAAS?		1911(53)
ABe 4/4	34		BoBoWERC		SIG,SAAS?		1908(46,47)
G 3/4	11	'HEIDI'	2-6-0T	1Cn2t	SLM G 3/4	1476	1902
G 4/5	107	'ALBULA'	2-8-0 2cc	1Dn2v	SLM G 4/5	1709	1906

G 4/5	108 'ENGIADINA'	2-8-0 2cc	1Dn2v	SLM G 4/5		1710	1906	
Xrotd 9213		0-6-6-0 4c	CCn4	SLM Xrotd		2149	1910	
Ge 4/6 353		2-8-2WE	1D1w2	SLM Ge 4/6		2433	1914	
Ge 6/6 412		0-6-6-0WE	CCw2	SLM,BBC Ge 6/6		3045,2242	1925	(23)
Ge 6/6 414		0-6-6-0WE	CCw2	SLM,BBC Ge 6/6		3297,2967	1929	
Ge 6/6 415		0-6-6-0WE	CCw2	SLM,BBC Ge 6/6		3298,2968	1929	
ABe 4/4	501	BoBoWERC		SWS,MFO,BBC		?	1939	

Unknown use and location.

Gauge: 600 mm Locn: ? Date: 06.1999

	4wDM	Bdm	O&K(M) RL 1c	8457	1937	(m)	s/s

Various public works concerns participated in constructing the component parts of what became the RhB. Those locomotives identified as being concerned are listed here: there are conflicts in the source information, clarification required.

Gauge: 1000 or 750 mm Locn: various Date: 06.1999

		0-4-0T	Bn2t	Freudenstein	210	1905	(aa)	(27)
		0-4-0T	Bn2t	Heilbronn II	46	1876	(ab)	(28)
		0-4-0T	Bn2t	Heilbronn III	48	1875	(ac)	s/s
SANTERNO		0-4-0T	Bn2t	Heilbronn II	227	1886	(ad,U)	(30)
RENO		0-4-0T	Bn2t	Heilbronn II	228	1886	(ad,U)	(31)
HELVETIA		0-4-0T	Bn2t	Heilbronn III	269	1891	(af,T)	(32)
		0-4-0T	Bn2t	Heilbronn III	343	1898	(ag,W)	(33)
		0-4-0T	Bn2t	Jung 40 PS	98	1891	(ah,U)	(30)
		0-4-0T	Bn2t	Karlsruhe	1369	1894	(ai, T)	(35)
HELVETIA		0-4-0T	Bn2t	Krauss-S IV kl	2960	1893	(aj,T)	(36)
ENGADIN		0-4-0T	Bn2t	Krauss-S IV zs	4589	1901	(ak,U)	(30)
ALBULA		0-4-0T	Bn2t	Krauss-S IV zu	4642	1902	(ak,V)	1904 s/s
		0-4-0T	B?2t	Maffei	3624	1910	(am,V)	(39)
	1	0-4-0T	Bn2t	SLM G 2/2	370	1884	(an,W)	(40)
1	BASEL	0-6-0Tm	Cn2tk	SLM G 3/3	473	1887	(ao,X)	s,s
		0-4-0T	Bn2t	SLM	526	1888	(ap,Y)	(42)
3	BIRSIG	0-6-0Tm	Cn2tk	SLM G 3/3	548	1888	(aq,X)	(43)
		0-6-0T	Cn2t	SLM HG 3/3	614	1890	(ar,W)	(44)
		0-6-0T	Cn2t	SLM HG 3/3	616	1890	(as,W)	(45)

(m) purchased via O&K/MBA, Dübendorf.
(aa) ex O&K, Zürich /1908, used for (S); purchased from Ad. Baumann, Wädenswil /1910 and used for (T).
(ab) new to Strauss & Bleibler, Backnang [D] (700 mm); used by Castelli & Rüttiman for RhB construction before /1900 (?800 mm).
(ac) new to Grubitz & Ziegler, Solothurn "SZIGET" (800 mm); used by Castelli & Rütimann for RhB construction before /1900.
(ad) new to G. Norsa + G. Basani, Lavezzola bei Argenta, Bologna [I] (700 mm) /1901.
(af) new to Messing, Laufenburg (900 mm); provided by Baumann & Stiefenhofer, Wädenswil /1907.
(ag) new to Sissach-Gelterkinden (1000 mm); later construction of the Hauenstein line.
(ah) new to Wolf & Weiss, Zürich; /1900.
(ai) ex Lahrer Strassenbahn, Lahr [D].
(aj) ex Fritz Marti, Wallisellen; /190x.
(ak) new to Bauleitung der Rhätischen Eisenbahn.
(am) new via Robert Aebi & Cie, Zürich.
(an) ex TT "1" /1914 (1000 mm).
(ao) provided by Albert Buss & Cie, Basel /1909.
(ap) new to Jakob Mast, Zürich (as 1000 mm).
(aq) provided by Albert Buss & Cie, Basel /1907.
(ar) ex BOB "HG 3/3 2" (ex Can4t with rack gear removed) /1914.
(as) ex BOB "HG 3/3 4" (ex Can4t with rack gear removed) /1914.

(S) used by Baugesellschaft Davos-Filisur during the construction of the Davos-Filisur line.
(T) used by Baumann & Stiefenhofer, Wädenswil at Somix for construction of the Truns-Disentis line.
(U) used for construction of the part of the Albula line in the Val Bever.
(V) used for construction of the line in the Lower Engadine.
(W) used for construction of the Chur-Arosa line.
(X) used for construction of the Bernina line.

(Y)	used for construction of the Landquart-Davos line.
(23)	withdrawn 18/11/2008.
(27)	1946 to VEBA AG, Zürich during the construction of the Rupperswil-Auenstein hydroelectric power plant.
(28)	1900 sold to St. Rossi, St Gallen.
(30)	1904 or later to RTU Rickentunnel-Unternehmung AG, Kaltbrunn.
(31)	1904 or later to RTU Rickentunnel-Unternehmung AG, Kaltbrunn; 1910 to IRR "WIDNAU".
(32)	1937 sold to Ed. Züblin & Cie AG, Zürich.
(33)	to NStCM construction.
(35)	1914 to F. Marti, Winterthur for NStCM construction.
(36)	1926 sold to Jura-Cement-Fabrik, Aarau.
(39)	1920 sold to Ed. Cuenod, Genève.
(40)	1915 returned to TT "1".
(42)	1906-11 sold to Entreprise P. Rossi-Zweifel, St. Gallen (750 mm).
(43)	1915 sold to Robert Aebi & Cie for NStCM construction.
(44)	1920 sold to Bündner Kraftwerke, Küblis.
(45)	1917 to Lonza AG, Thusis; 1918 sold to Lonza AG, Thusis.

ROPI, Scuol GR39a C

formerly: Schuls

The SIG list of references gives one ETS 100 locomotive delivered here.

SCHAFIR & MUGGLIN GR40 C

This public works concern of Liestal was responsible for some projects in this Canton.

Julier

Construction of the hydroelectric scheme on the Julier pass.

Gauge: ? mm Locn: ? Date: 06.1999

1.03.1 (later 2.20.01)   4wBE   Ba1   Stadler,BBC   20, ?   1946   a1965 s/s

Trimmis

A construction site.

Gauge: 600 mm Locn: 248 $ 7610/1975 Date: 05.1967

  4wDM   Bdm   O&K MV1   25140   1952   (a)   (1)

(a) ex Hew & Co, Chur stored after completion of a project.
(1) to Ad. Schäfer, Aarau.

SCHULE HEROLD, Chur GR41 M

Gauge: 1000 mm Locn: 247 $ 7590/1918 Date: 06.1988

6 WEISSHORN   0-4-2T   B1n2t   SLM HG 2/3   1410   1902   (a)   (1)

(a) ex HOVAG, Ems /1965 (Abt rack gear removed) for display.
(1) 03/06/1988 to Dampfbahn Furka-Bergstrecke (DFB), Realp "6 WEISSHORN" (rack gear restored).

SIHLPAPIERFABRIKEN, Landquart GR42

formerly: Papierfabrik, Landquart

Part of the Sihlpapier group. The factory is reached via a 2417m long line installed in 1882.

Gauge: 1435 mm							Date: 05.1988		
	1	0-4-0T	Bn2t	SLM	388	1884	(a)	c1946	scr
	2	0-4-0T	Bh2t	SLM E 2/2	3523	1931	(b)	1962	scr
Tm	2	4wDH	Bdh	Jung RK 20 B	13372	1961	new	(3)	

(a) purchased from SLM /1911.
(b) purchased from VHB "2" /1947.
(3) 1993 OOU; 1999 to SERSA; 2001 to Carrières du Lessus HB SA, Saint-Triphon via METRAG?

F.LLI SOMAINI SA, Grono   GR43 C
F.lli = Fratelli

A civil engineering company with its main base here. It is not known if locomotives were present.
Address: Via cantonale, 6537 Grono.

Gauge: 600 mm			Locn: 277 $ 7334/1255			Date: 02.2008	
	4wDM	Bdm	R&H 30DLU	285352	1950	(a)	(1)

(a) new via RACO to ?
(1) 1993 to FWF, Otelfingen.

TOSCANO SA, Thusis   GR43a C

Among the company's specialities is the erection of steel constructions and earth moving. The SIG list of references gives five ETB 70 locomotives delivered here. Address: Bahnhofstrasse, 7430 Thusis.

VEREIN BAHNHISTORISCHES MUSEUM ALBULA, Bergün   GR44 M

It is intended that seven locomotives will go to this museum.

Gauge: 1000 mm			Locn: 258 $ 7767/1670			Date: 07.2006
Ge 6/6 407	0-6-6-0WE	Ccw2	SLM,BBC Ge 6/6	2839,1886	1922	(a)

(a) ex RhB by /2006.
see: www.bahnmuseum-albula.ch

ZIEGELFABRIK WIESENTHAL AG, Chur   GR45
A brickworks.

Gauge: 600 mm			Locn: ?			Date: 06.1999	
	4wBE	Ba1	Stadler	19	194x	new	(1)
	4wDM	Bdm	O&K MV0	25076	1951	(b)	s/s

(b) purchased via MBA, Dübendorf.
(1) 1951 to Gips-Union, Felsenau /1951.

ZIEGELEI LANDQUART, Landquart   GR46

Until the beginning of the 1950s, a railway line ran eastwards alongside a road from the brickworks to the claypits. More recently, a route was laid across the fields between the same locations until 1985 when the railway from the pits was abandoned. The remaining short section retained in the works to feed the mixing plant closed in 2005 when new equipment was installed. The two locations given correspond to the factory and pits respectively. All locomotives obtained were converted to be able to run on the 750 mm system.

Gauge: 750 mm   Locn: 248 $ 7617/2041, 7624/2022   Date: 06.2008

	4wBE	Ba	unknown	?	19xx	(a)	s/s
4	4wDM	Bdm	O&K MV0	25074	1950	(b)	(2)
2	4wDM	Bdm	O&K MV1a	25348	1955	(c)	OOU
3	4wDM	Bdm	O&K MV1a	25586	1955	(d)	OOU
	4wDM	Bdm	R&H 30DL	256190	1948	(e)	(5)
	4wDH	Bdh	Diema DFL30/1.7	4315	1979	(f)	OOU

(a) a locomotive with central cab and a bonnet at each end existing ca. 1950.
(b) ex Tonwarenfabrik, Laufen (600 mm) by /1966.
(c) purchased from or via O&K Amsterdam (700 mm).
(d) purchased from or via MBA, Dübendorf (600 mm) (in /1959?).
(e) ex Gonzenwerk, Sargans after /1966 (600 mm).
(f) new via RUBAG.
(2) 1986 to M. Weder, Diepoldsau by /1996; believed to one of the unidentified locomotives at FWF, Otelfingen (to be verified).
(5) 1986 donated to Museums-Eisenbergwerk Gonzen, Sargans (600 mm).

ZSCHOKKE LOCHER AG, San Bernardino   GR47 C

This road tunnel project ran from November 2002 to November 2006. SIG battery locomotives of types ETS 100 and ETB 50 (details required) were also used; these have not been identified.

Gauge: 900 mm   Locn: ?   Date: 05.2009

4wDH	Bdh	Schöma CHL 60G	5910	2004	new	s/s

Gauge: 750/900 mm   Locn: ?   Date: 05.2009

4wDH	Bdh	Schöma CFL 180 DCL	5386	1993	(a)	s/s
4wDH	Bdh	Schöma CFL 180 DCL	5387	1993	(a)	s/s

Gauge: 750 mm   Locn: ?   Date: 05.2009

4wDH	Bdh	Schöma CHL 60G	5813	2003	new	s/s

(a) new to Devez Sewer [F]; ex GTM Service Material Sewer Tunnel, Chantier Devez-Bordeaux [F]./xxxx

Jura   JU

This canton was formed in 1979 from the western part of Kanton Bern (Canton Berne).

ALCOSUISSE SA, Delémont   JU01
formerly: Régie Fédérale des Alcools (RFA)
alternative: Régie des Alcools, Entrepôt de Delémont

A storage facility for alcohol was opened in 1889 and connected to the station of Delémont from 1882. It moved to the western edge of the town in 1957. Latterly this has been transferred to the sales organisation of the RFA. Address: Rue de la Communance 58, 2800 Delémont.

Both Orenstein & Koppel diesel locomotives are positioned for photography on 30 May 2008. They are - left to right - 26617 of 1967 (type MB 5 N) and 26709 of 1972 (type MB 170 N).

Gauge: 1435 mm   Locn: 223 $ 5923/2447   Date: 08.1993

4	4wPM	Bbm	Boilot-Pétolat	10783	19xx	(a)	s/s
2	4wPM	Bbm	Kronenberg DT 60P	112	1948	new	(2)
No 2	4wDH	Bdh	O&K MB 170 N	26709	1972	new	
Nr. 3	4wDM	Bdh	O&K MB 5 N	26617	1967	(d)	

(a)   built under STEMAG licence, alcohol (methanol?) powered motor.
(d)   ex Eidg. Alkoholverwaltung, Lagerhaus Romanshorn /2000.
(2)   1973 to Scherer & Bühler, Meggen.

ASSOCIATION TUNNEL MONT TERRI SUD (ATS), Saint-Ursanne
JU02 C

A consortium formed to bore the southern section of the Mont Teri tunnel on the N16 motorway, near to Saint-Ursanne. Members were Bosquet, Delémont / Aseda, Vendlincourt / Rothpelz Lienhard, Aarau / Theiler & Kalbermatter, Luzern / Conrad Zschokke, Genève. The work started in 1990 and lasted a few years. The locomotives have been described as "initially four small locomotives and later four large and one small". This is difficult to reconcile with the known data.

Gauge: 750 mm   Locn: 222 $ 5805/2464   Date: 08.1994

601	4wDH	Bdh	Schöma CFL 200DCL	4665	1983	(a)	s/s?
602	4wDH	Bdh	Schöma CFL 200DCL	4666	1983	(a)	s/s?
603	4wDH	Bdh	Diema DFL 90/0.2	2360	1956	(c)	(3)
604	4wDH	Bdh	Schöma CFL 200DCL	4668	1983	(d)	s/s?
605	4wDH	Bdh	Schöma CFL 200DCL	4762	1984	(e)	(5)

(a) new to Alfred Kunz, München, [D] for Tunnel Makkah Taif [Saudi Arabia]: provided by Conrad Zschokke, Genève ex Arge Bietschtal VS.
(c) new to Munitions-Depot Aurich-Tannhausen, [D] "2" as DS90; rebuilt Diema /1981; ex Rothpelz, Aarau /198x (recorded as DSL90-0.2).
(d) new to Alfred Kunz, München, [D] for Tunnel Makkah Taif [Saudi Arabia]: ex Cogefer, Milano [I] on Chantier Matmata [Morocco].

(3) 1996 to WBB via SERSA.
(5) 12/1996 to Schöma.

CIMENTS DE SAINT-URSANNE, Saint-Ursanne   JU03

North of the town the Delémont-Delle railway climbs out the valley of the R. Doubs up to its summit tunnel. A cement works was established alongside the railway while the quarry lay above. The works closed 30/6/1993.

Gauge: 600 mm   Locn: 222 $ 5797/2464   Date: 05.1996

4wDM	Bdm	JW JW8	225	19xx	(a)	(1)
4wDM	Bdm	JW JW8	?	19xx		s/s

(a) fitted with motor 3780, but often recorded as 4980.

(1) 19xx to Herr Bauer, Lüsslingen; 1996 to FWF, Otelfingen.

HISTORISCHE EISENBAHN GESELLSCHAFT (HEG), Delémont   JU04 M

The society is resident in Biel/Bienne. Its locomotives are kept with those of SBB-Historic at the "la Rotonde" in Delémont.

Gauge: 1435 mm   Locn: 223 $ 5940/2458   Date: 06.2008

8485	0-6-0WT	Cn2t	SLM E 3/3	1881	1907	(a)	
Tm 72	0-4-0DM	Bdm	SLM Em 2/2	3464	1930	(b)	
Tm^{II} 620	4wDM	Bdm	RACO 95 SA3 RS	1476	1956	(c)	
Tm^{II} 633	4wDM	Bdm	RACO 95 SA3 RS	1505	1958	(d)	
4	0-4-0WT	Bn2t	SLM	1267	1900	(e)	
5	0-4-0WT	Bn2t	SLM	1670	1905	(e)	
3 ZEPHIR	0-4-0WT	Bn2t	Krauss-S XIII d	290	1874		(V)

(a) new to SBB "8485"; from EUROVAPOR, Wutachtalbahn [D] /2002 after a period in store at Koblenz.
(b) new to Gesellschaft für Chemische Industrie, Monthey VS; from Silo Olten AG (SOAG), Olten /1997.
(c) ex CFF "Tm^{II} 620" /2004.
(d) ex CFF "Tm^{II} 633" by /2005.
(e) ex La Traction, Pré-Petitjean 6/12/2002.

(V) owned by VHS.

see: www.volldampf.ch/heg

SOCIÉTÉ JURASSIENNE DE MATÉRIAUX DE CONSTRUCTION SA, Delémont   JU05 C

Probably a public works concern. It may have some connection with the "La Ballastière" area.

Gauge: 600 mm   Locn: ?   Date: 05.2002

		4wDM	Bdm	O&K RL 1a	4021	1930	(a)	(1)

(a) via O&K/MBA, Dübendorf.
(1) by 9/2006 at FWF, Otelfingen (if correctly identified).

LA TRACTION, Pré-Petitjean   JU06 M

This preservation group runs trains on the metre gauge CJ, using mainly ex-public railway equipment. The society has established a base for its metre gauge stock in an old wire works connected to the narrow gauge CJ line. As well as the listed locomotives, the club has one metre gauge draisine - Dm7. The standard gauge steam locomotives were stored at Glovelier before being moved to Delémont. Address: La Traction, P.O. Box 611, 2800 Delémont.

Gauge: 1435 mm   Locn: 223 $ 5940/2458   Date: 11.2007

4	0-4-0WT	Bn2t	SLM	1267	1900	(a)	(1)
5	0-4-0WT	Bn2t	SLM	1670	1905	(a)	(1)
9	0-4-0WT	Bn2t	SLM E 2/2	2168	1911	(c)	(3)
12	0-6-0WT	Cn2t	SLM E 3/3	1088	1898	(d)	(4)

Gauge: 1000 mm   Locn: 222 $ 5712/2363   Date: 01.2008

E 206		2-4-6-0T 4cc				
			1B'Cn4vt Henschel	12281	1914	(f)
E 164		0-4-4-0T 4cc				
			BBn4vt Henschel	7022	1905	(g)
Gem 4/4	122	BoBoW+DERC	SWS,MFO	? 1955(73)		(h)
Ge 4/4	4004	BoBoWE BBg4	HStP,BBC	1576,2785	1927/28	(i)
Xeh 4/4	91	0-4-4-0RWE	SLM,SIG,BBC BCFeh 4/4			
				3459, ? 1931(89)		(j)
Ta		4wBE Ba2	SWS,MFO	? ,19	1909	(k)

(a) ex von Roll, Usine de Rondez 6/2000.
(c) ex von Roll, Usine de Choindez 9/1998.
(d) ex von Roll, Usine de Choindez 4/1998.
(f) ex CP "E206" 12/7/1993.
(g) ex CP "E194" 14/12/1992; with low-pressure section from "E169".
(h) new to ESB; ex RBS /1996; a building date of 1916 is also quoted.
(i) new to FV Ferrocarriles Vascongados [E] "4"; ex Euskotrenbideak (ET) /1993.
(j) new to SGA "ABDeh 4/4 1"; ex SGA /2000.
(k) ex Tréfileries Réunies SA, Bienne 18/6/1993.
(1) 6/12/2002 to Historische Eisenbahn Gesellschaft (HEG), Biel (kept at Delémont).
(3) 2008 to custody of Vapeur Val-de-Travers (VVT), St. Sulpice, NE; but still owned by La Traction.
(4) 2008 to Vapeur Val-de-Travers (VVT), St. Sulpice, NE.

see: www.la-traction.ch

GEBRÜDER MESSING, Saint-Brais   JU07 C

A branch of this public works firm was responsible for the construction of the Glovelier-Saignelégier line.

Gauge: 900 mm   Locn: ?   Date: 05.2008

0-4-0T	Bn2t	Heilbronn	413	1902	new	(1)
0-4-0T	Bn2t	Heilbronn III a	430	1903	new	(2)

(1) 1903 returned to Heilbronn; 1903 resold as Heilbronn 433.
(2) 1905 to Jos. Hoffmann & Söhne, Bauunternehmung, Ludwigshafen [D].

PLACE D'ARMES FÉDÉRALE DE BURE, Bure   JU09

This establishment includes a tank testing ground. A steeply graded 4.72 kilometre long branch from Courtemaîche was opened on 19 February 1968. The station lies just outside the military area and the CFF are responsible for trains to this point. The locomotive is available for shunting within the station as required.

Gauge: 1435 mm   Locn: 212 $ 5679/2561   Date: 08.1988

          0-4-0WE   Bw2   SLM,MFO Te^{III} 4583, ?   1965   new

PORTLANDCEMENTFABRIK LAUFEN AG, Soyhières-Bellerive
JU10

later: Entrepôt de Bellerive-Delémont SA

This cement works lay just south of the older station of Soyhières-Bellerive (closed 1986), itself south of the later station of Soyhières (closed 25.03.1993). The quarries were to the west of the railway and the works to east. After closure (1977?) the majority of the buildings and sidings were reused by Entrepôt de Bellerive-Delémont SA, but without the need for their own locomotive.

Gauge: 1435 mm   Locn: 223 $ 5943/2482   Date: 08.1990

          4wPM   Bbm   Breuer III   1096   192x   new   (1)

(1)   after 1977 to Portland Cementfabrik Laufen AG, Werk Liesberg.

vonRoll casting (rondez) sa, Delémont   JU11

formerly: von Roll Fonderie des Rondez SA, Delémont (till 2/5/2003)
             von Roll SA, Usine de Rondez, Delémont (till 12/6/1995)

A foundry and machinery-making factory, acquired by von Roll in 1883 and rail connected from 1888. See also under von Roll, Gerlafingen for changes in the company name. Since about 1996, METSO Papier SA has used some of the site.

Gauge: 1435 mm   Locn: 223 $ 5942/2460   Date: 06.2008

          7 GNOM   2-2-0RT   1Azn2t   Olten   20   1870   (a)   (1)
          4   0-4-0WT   Bn2t   SLM   1267   1900   (b)   (2)
          5   0-4-0WT   Bn2t   SLM   1670   1905   (c)   (2)
          4wDM   Bdm   RACO 85 LA7   1727   1965   (d)

(a)   new to Steinbruchgesellschaft Ostermundigen, Ostermundigen "1"; ex Nestlé, Neuenegg /1907; rack gear removed /19xx.
(b)   ex von Roll, Gerlafingen /1964.
(c)   ex von Roll, Choindez /1940.
(d)   ex CFF "Tm^I 502" /1996.

(1)   1940 withdrawn; 1979 display Olten; 2000 to VHS.
(2)   2000 donated to La Traction, Pré-Petitjean.

VON ROLL HYDROTEC SA, Choindez   JU12

formerly: von Roll SA, Usine de Choindez

A steelworks, foundry and rolling mills that was established in 1846. It initially specialised in cast iron pipes, later in plastic materials. It was rail connected from 1876, at the same time as the construction of the Delémont-Moutier line. See also under von Roll, Gerlafingen for changes in the company name. A narrow gauge system was used in the casting area till around 1988, but appears earlier to have been a railway system in the associated chalk quarry; this was derelict by 1968.

SLM 1088 of 1898 performed some shunting for the benefit of a visiting party on 29 May 1973. The narrow gauge system in use at the time was to right of the main building in the background.

Left: "1" Jenbacher Werke 2450 of 1965 (type DM20/1) pulled a train of the foundry on this day. Right: A derelict system serving a chalk quarry existed a little further beyond in 1968.

Gauge: 1435 mm				Locn: 223 $ 5954/2413			Date: 05.1998	
11	0-4-0WT	Bn2t		SLM E 2/2	236	1881	(a)	(1)
3	0-4-0WT	Bn2t		SLM E 2/2	726	1892	(b)	(2)
5	0-4-0WT	Bn2t		SLM	1670	1905	new	(3)
9	0-4-0WT	Bn2t		SLM E 2/2	2168	1911	(d)	(4)
12	0-6-0WT	Cn2t		SLM E 3/3	1088	1898	(e)	(4)
	4wDM	Bdm		Breuer V	3049	1952	(f)	(6)
Tm 2/2 No 3	4wDH	Bdh		RACO 300 MA7 HT	1635	1964	(g)	
	4wDM	Bdm		RACO 85 DA7	1811	1972	(h)	
Gauge: 500 mm								
1	4wDM	Bdm		JW DM20/1	2450	1965	new?	(9)
2	4wDM	Bdm		JW DM20/1	2379	1962	new?	(10)

(a) purchased from GB "11" /1890.
(b) purchased from Sihltalbahn (SiTB) "3" /1897.

(d)	transferred from von Roll, Gerlafingen /1928 (before /1918 according to OFT records).
(e)	purchased from HWB "8" /1930.
(f)	new via RACO who allocated their number 1424.
(g)	hydrostatic transmission (Hydro-Titan); ex von Roll, Klus /1996.
(h)	new; 1996 to von Roll , Klus and returned 1/1997-11/2007.
(1)	1901 to von Roll, Klus "1".
(2)	1912 to von Roll, Klus "1".
(3)	1940 to von Roll, Rondez.
(4)	1984 to store; 1998 donated to La Traction, Pré-Petitjean.
(6)	1972-82 to von Roll, Klus.
(9)	1988 to von Roll, Gerlafingen.
(10)	1988 to von Roll, Gerlafingen for scrap.

Luzern   LU

AD. AEBERLI, Reiden   LU01 C
Probably a public works concern.
Gauge: ? mm   Locn: ?   Date: 05.2002
                4wDM   Bdm   MBA   1926   19xx   (a)   s/s
(a)   via O&K/MBA, Dübendorf.

ALCOSUISSE AG, Schachen LU   LU02
formerly: Eidgenössische Alkoholverwaltung (EAV) (till 1996-9)
An alcohol depot. It has been transferred to the sales organisation of the EAV.
Gauge: 1435 mm   Locn: 234 $ 6538/2099   Date: 08.1986
    No. 1   4wPM   Bbm   Kronenberg DT 45P   104   1941   (a)
    No. 3   4wDH   Bdh   O&K MB 125 N   26696 1970   new
(a)   new; permitted to carry 5t; alcohol fuel.

ARGE GÜTSCHSTOLLEN LOS V.1, Kriens   LU03 C
A consortium formed by Theiler & Kalbermatter, Kopp and Brun. Theiler & Kalbermatter provided the locomotives.
Gauge: 750 mm   Locn: ?   Date: 08.1992
'281.24.102'   4wD   Bd   Ageve DMD   x779   19xx   (a)   s/s
'281.24.104'   4wDH   Bdh   Schöma CFL 100DCL   5242   1991   (b)   (2)
(a)   via P. Schlatter, Münchwilen AG.
(b)   ex Theiler & Kalbermatter, Luzern "281.24.104".
(2)   by 1995 to Konsortium Bärengraben, Sanierung Deponie Bärengraben, Würenlingen.

ARGE TUNNEL SÖRENBERG, Luzern   LU04 C
Part of a gas pipe-line linking the Netherlands with Italy passes through this tunnel. It was rebuilt between August 2000 and February 2001. Ilbau of Wien [A] provided the locomotives.
Gauge: 750 mm   Locn: ?   Date: 05.2009
                 4wDH   Bdh   Schöma CHL 60G   5647   2000   new   s/s
                 4wDH   Bdh   Schöma CHL 60Tandem   5648   2000   new   s/s

BRAUEREI EICHHOF, Kriens   LU05
A brewery. The locomotive was used for steam generation - not traction.
Gauge: 1435 mm   Locn: 235 $ 6643/2103   Date: 01.1980
                 0-6-0WT Cn2t   SLM E 3/3   1971   1909   (a)   (1)
(a)   new to SBB "8491"; ex Papierfabrik Perlen, Gisikon-Root "3" /1977.
(1)   1986 to Oswald-Steam, Samstagern ZH; 1996 scrapped; boiler to Sursee-Triengen (ST) "5".

BRAUNKOHLENWERK AG, Zell LU — LU06

Lignite mines exploited during World War II.

Gauge: 750 mm   Locn: ?   Date: 02.2002

	0-4-0T	Bn2t	O&K 30 PS	847	1901	(a)	(1)
	0-4-0T	Bn2t	O&K 30 PS	985	1902	(a)	(2)
	0-4-0T	Bn2t	Jung 50 PS	1684	1911	(c)	(3)
FLÜELA	0-4-0T	B?2t	Maffei	3624	1910	(d)	1948 s/s

(a) made available by F. Stirnimann, Bauunternehmer, Olten.
(c) made available by AG Heinrich Hatt-Haller, Zürich "1".
(d) ex Entreprise Erdigt, Russin /1942.

(1) 1949 removed from boiler register without having been used.
(2) 1955 removed from boiler register.
(3) 1948 removed from boiler register; returned to AG Heinrich Hatt-Haller; 1966 to Technorama, Winterthur.

BRUN & CIE AG, Nebikon — LU07 C

A company specialising in machinery construction, including internal-combustion locomotives and also a stockist for machinery required on construction sites, including locomotives. Evidence is that the company stocked, repaired, reconditioned and copied Diema and Jung diesel locomotives.

Gauge: 600 mm   Locn: ?   Date: 02.2002

0-6-0T	Cn2t	Henschel Brigadelok	5672	1901	(a)	s/s
0-6-0T	Cn2t	Henschel Brigadelok	6335	1903	(b)	s/s

(a) new to Deutsche Feldbahnen; ex Heinrich Hatt-Haller, Hoch- & Tiefbau AG (HHH), Zürich /1948.
(b) new to Deutsche Feldbahnen "113"; ex Heinrich Hatt-Haller, Hoch- & Tiefbau AG (HHH), Zürich /1948.

DIE SCHWEIZERISCHE POST — LU08

formerly: (Schweizerische) Post-, Telephon- und Telegraphenbetriebe (PTT) (1985 till 31/12/1997)

Luzern

The mail handling section at the SBB station. A reorganisation of packet handling reduced the shunting requirements. Since about 1999-2000 this service has been provided by the SBB.

Gauge: 1435 mm   Locn: 235 $ 6663/2112   Date: 06.1991

4	0-6-0WE	Cw1	SLM,BBC Ee 3/3	4360,6147	1959	(a)	(1)
14	6wWE	Cw3	SLM,BBC Ee 3/3	5467, ?	1992	new	(1)

Kriens

Gauge: 1435 mm   Locn: 235 $ 6657 2103   Date: 03.2008

13	4wDM R/R	Bdm	Mercedes Unimog	?	1978

(a) ex PTT, Zürich-Sihlpost /1985.
(1) 2000 to Die Schweizerische Post, Zürich-Mülligen.

EGLI-MÜHLEN AG, Nebikon — LU09

The history of this flourmill dates from 1891, but the first silo was not built till 1960. The third silo was built, in 1980, on a new site where a rail connection was possible. Address: Schürmatte 4.

	Gauge: 1435 mm			Locn: 234 $ 6407/2273				Date: 01.2008
		4wDM	Bdm	RACO RA11	1377	1948	(a)	1992 scr
		4wDM	Bdm	RACO 95 SA3 RS	1667	1963	(b)	

(a) new to SBB "Tm 301"; ex SBB "Tm 542" 7/1980; a building date of 1949 is also quoted.
(b) ex SBB "Tm^{II} 724" /1992.

see: www.egli-muehlen.ch/index.php?option=com_content&task=view&id=6&Itemid=7

GALLIKER TRANSPORT AG   LU10
A transport/logistics company with several depots in the area.

Altishofen
This is the principle base of the company. A locomotive has been placed on display and can be seen from passing trains. No company locomotives are known. Address: Kantonsstrasse 2, 6246 Altishofen

	Gauge: 1435 mm			Locn: 234 $ 6403 2280				Date: 01.2008
	Ae 6/6 11418 'ST. GALLEN'							
		CoCoWE	CCw6	SLM,BBC Ae 6/6	4236,6053	1957	(a)	display

Schachen
formerly: Imbach Logistik AG (till 31/12/2007)

	Gauge: 1435 mm			Locn: 234 $				Date: 09.2008
		4wDH	Bdh	Jung RC 24 B	14025	1969	(b)	

(a) ex SBB "Ae 6/6 11418" 3/2007.
(b) ex Avia AG, Birsfelden "Tm 2/2 TONY" by /2004.

JOH. GOETZ HOCH- & TIEFBAU, Sursee   LU11 C
A public works concern from Strasbourg [F] that participated in the construction of the Sursee Triengen Bahn.

	Gauge: 600 mm			Locn: ?				Date: 08.1992
7 THERESE		0-4-0T	Bn2t	Märkische 20 PS	*181*	1897	(a)	s/s
		0-4-0T	Bn2t	O&K 20 PS	455	1900	(b)	(2)

(a) imported for the contract /1911; number not confirmed, mistake for 187?
(b) new to Hartwich, Berlin [D]; imported for the contract /1911.
(2) ? to kuk. Heeresfeldbahn [A] "IIb 822" possibly via Smoschewer.

KOPP AG LUZERN, Zweigniederlassung Horw   LU12 C
This was one depot of this construction firm. The following locomotives were seen in store in 1967. The SIG list of references gives eleven ETB 70, four ETS 100, and four EIM 100 locomotives delivered here. Five were seen in 1976. This branch went into liquidation in 1998.

Gauge: 600 mm   Locn: 235 $ 6660/2079   Date: 10.1976

MAKIES AG   LU13
By 2007 there were two gravel pits, about five kilometres apart, each with its own separation plant and rail connection. The locomotives may exchange locations. The motor coaches are

intended for trains on the lines of the SBB and BLS. These trains are operated by Swiss Rail Traffic.

Gettnau

These pits lie between Gettnau and Willisau and were rail connected in 1983.

Gauge: 1435 mm Locn: 234 $ 6416/2211 Date: 12.2002

1	0-6-0DH	Cdh	Henschel DH 500Ca	30590	1964	(a)

Zell LU

These pits were rail-connected in 1991, but are not always in use. In the absence of activity here, the Henschel locomotive is regularly rented to third parties.

Gauge: 1435 mm Locn: 234 $ 6369/2210 Date: 12.2002

	BoBoWERC		SIG,BBC	?	1959	(b)
	BoBoWERC		SIG,BBC	?	1978	(c)
	BoBoWERC		SIG,BBC	?	1979	(d)
	BoBoWERC		SIG,BBC	?	1979	(e)
2	0-6-0DH	Cdh	Henschel DHG 500C	30858	1965	(f)

(a) new to Thyssen-Edelstahl AG, Witten [D] "10"; ex Monteforno, Bodio, /1983 (on hired from Asper); in use on the "Rail 2000" project at Langenthal /2000-03; also recorded as DH 500C.
(b) new to SOB "BDe 4/4 81"; ex SOB "576 049" /2008.
(c) new to SOB "BDe 4/4 83"; ex SOB "576 055" /2008; to be dismantled for spares.
(d) new to SOB "BDe 4/4 84"; ex SOB "576 056" /2008.
(e) new to SOB "BDe 4/4 85"; ex SOB "576 057" /2008.
(f) new to Rheinstahl, Essen [D], ex WLH "52" /1987 via OnRail/Asper; rented to CIBA, Basel Kleinhüningen /1988-90; Migros, Neuendorf /1990-91; Amsteg power station reconstruction /1993-96; also recorded as DH 500C.

see: de.wikipedia.org/wiki/Hochleistungstriebwagen

MÜLLER AG VERPACKUNG, Reiden   LU14

A branch of a firm involved in the packing industry. Address: Industriestrasse, 6260 Reiden.

Gauge: 1435 mm Locn: 224 $ 6397/2336 Date: 05.2002

	4wD R/R	Bd	unknown R/R	?	19xx

PANLOG AG, Emmenbrücke   LU15

A firm providing logistics: for locomotive details see Swiss Stahl AG, Emmenbrücke. Address: Emmenweidstrasse 74, 6021 Emmenbrücke.

PAPIERFABRIK PERLEN, Gisikon-Root   LU16

A paper factory established in 1873 and rail connected in 1883. The railway installations were extended between 1885 and 1888. In 1992 the internal network had fourteen kilometres of track.

"5", a Tm^{III} returns along the branch line from the SBB station on 25 August 1967.

Gauge: 1435 mm   Locn: 235 $ 6696/2178   Date: 05.2005

	No.							
		0-4-0T	Bn2t	Krauss-M XVIII a	188	1872	(a)	(1)
	1	0-4-0T	Bn2t	O&K 100 PS	3120	1908	new	(2)
No.	2	0-4-0F	Bf2	O&K	8783	1919	new	(3)
	3	0-4-0T	Bn2t	Henschel Riebeck	20593	1925	(d)	(4)
	3	0-6-0WT	Cn2t	SLM E 3/3	1971	1909	(e)	(5)
	4	0-4-0F	Bf2	SLM	3299	1929	(f)	
	5	0-4-0DE	Bde1	SLM,BBC Tm^{III}	4158,6026	1956	new	(7)
	6	0-6-0WT	Cn2t	SLM E 3/3	1972	1909	(h)	
	7	0-6-0DH	Cdh	Henschel DH 500Ca	31193	1966	(i)	
		4wPM	Bbm	Breuer II, III or IV	?	c193x	(j)	s/s
		4wDM	Bdm	Kronenberg DT 50	101	1938	(k)	1988 scr
		4wDM	Bdm	Gmeinder 130 PS	5331	1964	(l)	
	9	6wDH	Cdh	KrMa M 500 C ex	19405	1968	(m)	
Tm 2/2	10	4wDE	Bde2	Moyse BN40EE250D	1211	1971	(n)	
		BBDH	BBdh	MaK G 1204 BB	1000819	1983	(o)	

(a) purchased from Emmentalbahn /1884.
(d) into service /1929.
(e) purchased from SBB "E 3/3 8491" /1962.
(f) purchased from Brown Boveri 3, Baden "3" /1955.
(h) purchased from SBB "E 3/3 8492" /1962.
(i) ex Raffinerie de Cornaux "4" /1976.
(j) features in the Breuer list of references of about 1930, but not that of 1928.
(k) new; 45 kW until /1979.
(l) ex Bosch-Siemens, Traunreut [D] /1988, via Asper.
(m) ex Maximilianshütte, Haidhof [D] "MH 1" /1991 via MF/Asper.
(n) ex Rhône-Poulenc Filtec, Rothenburg /2000.
(o) new to Verkehrsbetriebe Kreis Plön GmbH, Kiel "V 154" [D]; ex Häfen und Güterverkehr Köln AG, Köln [D] "V 26" /2007 via railimpex/LSB.

(1) by 1918 no longer in OFT records (withdrawn?).
(2) by 1962 stored; 1986 display in Großseifen [D] (Westerwald); 4/8/2004 Westerwälder Eisenbahnfreunde, Westerburg [D].
(3) by 1977 stored; ?1990 s/s.

(4) 1962 sold to Papierfabrik Cham, Cham "2".
(5) 1977 sold to Brauerei Eichhof, Kriens; 1986 to Oswald-Steam (OSS), Samstagern; 1996 scrapped; boiler to Sursee-Triengen (ST) "5".
(7) 1991 to Asper; 1993 to Aciers Schmutz, Orbe; ? to CJ "237 482"; 2007 to La Traction, Pré-Petitjean.

PETROPLUS AG, Rothenburg   LU17

formerly: Shell (Switzerland) (19xx-2000)

A fuel depot.

Gauge: 1435 mm   Locn: 235 $ 6611/2163   Date. 08.1994

	4wDH	Bdh	Gmeinder 190 PS	5246	1960	(a) (1)

(a) new to Südzucker, Werk Plattling [D] "1"; previously on loan from Shell, Veyrier /1988-89; purchased /1992-3.
(1) before 1994 to Stauffer, Schlieren.

W. & J. RAPP AG, Various Locations   LU18 C

A construction firm: in Switzerland the head office was in Basel.

Gauge: 600 mm   Locn: ?   Date: 03.2002

0-4-0T	Bn2t	Maffei 40 PS	2833	1908	(a)	(1)
0-4-0T	Bn2t	O&K 30 PS	995	1902	(b)	s/s

(a) used during the construction of the second track between Nebikon and Sursee /1908.
(b) used during the construction of the second track between Dagmarsellen and Wauwil /1907.
(1) by 1913 used on construction of the Hauenstein line.

RHODIA INDUSTRIAL YARNS AG   LU19

formerly: Rhône-Poulenc Filtec AG
            Rhône-Poulenc Viscosuisse SA (from 1993)
            Viscosuisse SA (till 1993)
            Société de la Vicose Suisse (SVS) (from 1906)
later: Nexis Fibers AG (from 2007)

Emmenbrücke

A plant to produce artificial textiles (rayon, nylon). It was rail connected in 1906. This plant was closed progressively from 1980. Latterly a road-rail vehicle handled rail traffic.

Gauge: 1435 mm   Locn: 235 $ 6634/2139   Date: 08.1994

	A1BE	A1a	Oehler, BBC	?	1912		(1)
	0-4-0F	Bf2	Jung	911	1905	(b)	(2)
Tm 2/2 1	4wDE	Bde2	Moyse BN40EE250D	1211	1971	new	(3)
LU 10398	4wDM R/R	Bdm	Mercedes Unimog	?	19xx	new	?

Walingen, Rothenburg

A stocking area for the parent plant at Emmenbrücke. It commenced operations in 1973 and was sold about 2000.

Gauge: 1435 mm   Locn: 235 $ 6610/2170   Date: 08.1994

1	4wDE	Bde2	Moyse BN40EE250D	1211	1971	(e)	(5)

(b) ex Saurer AG, Arbon /1948.
(e) ex Viscosuisse SA, Emmenbrücke "1" /1973.

(1) 1941 to Viscosuisse, Widnau.
(2) 1970 to private ownership, Alpnach Dorf; 1976 to Hotel Stalden, Berikon "HEINRICH".
(3) 1973 to Viscosuisse SA, Rothenburg.

RIGI BAHNEN, Vitznau — LU20 M

The company retains heritage locomotives listed. Further SLM 1 of 1873 from the Verkehrshaus, Luzern (VHS) operates here from time to time.

Gauge: 1435 mm + Riggenbach rack   Locn: 235 $ 6795/2070   Date: 01.2008

16	0-4-2RT	3bh2t	SLM H		2871	1923
17	0-4-2RT	3bh2t	SLM H		3043	1925
CFhe 2/3 6		A1ARRC		SWS,MFO	?	1911

SALON TRAIN SPECIAL, Rothenburg — LU21 M

This was a short lived operation with one vehicle for hire.

Gauge: 1435 mm   Locn: ?   Date: 05.2009

TAe 5		0-4-0WERC	Bw1	SLM,MFO	3991	1949	(a) (1)

(a) new as SBB "Te^1 260"; later "Te^1 960" (withdrawn 6/19189); rebuilt as a "Salon Traktor" and available for hire /1990.
(1) xxxx to Galleria Baumgartner, Mendrisio.

SCHERER & BÜHLER AG, Meggen — LU22

Wine merchants, rail connected since 1897.

Gauge: 1435 mm   Locn: 235 $ 6718/2116   Date: 08.1999

	4wPM	Bbm	Kronenberg DT 60P	112	1948	(a) (1)
	4wDM	Bdm	RACO 85 LA7	1577	1960	(b)

(a) ex Régie des Alcools, Delémont "Tm 2" /1973.
(b) ex SBB "Tm^l 414" /1998.
(1) 2/2001 to Draisine Sammlung Fricktal (DSF).

SCHINDLER BAUUNTERNEHMUNG AG, Horw — LU23 C

This civil engineering firm has a yard between the lake and railway. Locomotives have been stored there between contracts.

Gauge: 1000 mm   Locn: 235 $ 6660/2079   Date: 07.1967

152	4wBE	Ba	SIG	?	19xx s/s
433 ROSA	4wBE	Ba	unknown	?	19xx s/s

SCHWEIZEISCHE UNTERNEHMUNG FÜR FLUGZEUGE UND SYSTEME, Emmen — LU24

formerly: Eidgenössisches Flugzeugwerk (F+WE)

Originally a concern making and maintaining military aircraft, but latterly also handling civil aircraft.

Gauge: 1435 mm   Locn: 235 $ 6654/2144   Date: 10.1998

E1	4wDE	Bde2	Stadler	118	1965	new (1)

(1) 2000 to Sika Norm AG, Düdingen.

STÖCKLI AG, Sursee — LU27

A scrap merchant (Entsorgung & Recycling). A new plant was established in 1998 on the edge of the Sursee industrial estate and was rail connected at that time. Address: Allmendstrasse 17, 6210 Sursee.

Gauge: 1435 mm   Locn: 234 $ 650485,226345   Date: 12.2008

4wDM	Bdm	KHD A4L514 R	56810	1957	(a)

(a) new to Berkenhoff & Drebes (later Thyssen-Draht), Asslar/Wetzlar [D]; ex LSB /1999; carries oval Deutz plates 55126 (see Ing. Greuter AG, Hochfelden); also recorded as K55B.

SURSEE TRIENGEN (ST), Triengen — LU28 M

This public railway terminated its passenger service in 1971. Freight traffic continued initially worked by the ST and latterly by the SBB. Some heritage stock is retained for use on special occasions. Operational motive power is not listed.

Gauge: 1435 mm   Locn: 224 $ 6480/2317   Date: 09.2007

E 3/3	5		0-6-0WT	Cn2t	SLM E 3/3	1810	1907	(a)
	8522		0-6-0WT	Cn2t	SLM E 3/3	2345	1913	(b)
Em 2/2	1	Lisi	4wDE	Bde	SIG,BBC	? ,6406	1964	(c)

(a) ex SBB "8479" /1963; fitted with boiler from "E 3/3 8491" (SLM 1971) /1996.
(b) ex SBB "8522" /1964, fitted with electrical steam raising equipment from 1942-43 till after World War II.
(c) new; may not have been delivered till /1965.

see: www.dampfzug.ch/rollmaterial/home_inhalt.htm

SCHNYDER GOTTHARD AG, Emmen — LU29

A metal recovery firm. The site is connected to that part of the Seetalbahn (SeTB) now bypassed by the new route into Emmenbrücke from Waldibrücke. This section west of Waldibrücke has been retained to serve this firm and RUAG. Address: Seetalstrasse 195.

Gauge: 1435 mm   Locn: 235 $ 6662/2157   Date: 03.2002

	4wDM	Bdm	Kronenberg DT 180/250	141	1956	(a)	2001 scr
Em 837 815-0	0-4-0DH	Bdh	Henschel DH 240B	30594	1964	(b)	(2)
Em 837 904-2 JOSEFA	0-6-0DH	Cdh	Jung R 42 C	13290	1961	(c)	(3)
Em 837 818-4	6wDH	Cdh	Jung RC 43 C	14163	1974	(d)	
Em 837 901-9	6wDH	Cdh	Henschel DHG 500C	31184	1967	(e)	

(a) ex Holderbank Cement und Beton (HCB), Rekingen AG via LSB.
(b) new to Thyssen Edelstahl, Krefeld [D] "8"; ex EH [D] "410" via OnRail/LSB by /1999.
(c) hired from LSB /2002.
(d) new to Holzmüller Seehafenbetrieb KGaA, Hamburg-Waltersdorf [D] "1"; ex Eurokai KGaA, Hamburg-Finkenwerder, Werk Walterhof [D] /2002 via PACTON/LSB.
(e) hired from LSB /2008.
(2) 2003 to Ing. Greuter AG, Hochfelden.
(3) 2002 returned to LSB.

SCHUMACHER & CO, Gisikon — LU30

Ziegelei Körbligen

formerly: Ziegelei Chörbligen

This brickworks was built in 1860. In 1908 an animal-worked 1.2 kilometre long 500 mm gauge tramway was installed to connect the pits at Pfaffwil with the brickworks. Locomotives were introduced in 1926 and the line converted to 600 mm gauge in 1989. A guest train is available for this line, but in addition, the firm maintains a private collection of locomotives.

Gauge: 500 mm Locn: 235 $ 6725/2205 Date: 05.2009

		4wDM	Bdm	Brun B12 1HKG65	41	19xx	(a)	(4)	
		0-4-0DM	Bdm	Deutz OME117 F	10282	1932	(b)	s/s	
		4wDM	Bdm	Diema LR	479	1929	(c)		
	rebuilt	4wBE	Ba1	Stadler/BBC	?	1944		1946	wdn
		4wD	Bd	Diema DS20	2638	1963	(d)	(5)	
		4wDM	Bdm	Schöma CDL 15	1499	1953	(e)	s/s	
		4wDM	Bdm	Schöma CDL 20	2683	1963	(f)	s/s	
		4wDH	Bdh	Schöma CHL 20G	4451	1981	new	(5)	

Private collection

Gauge: 600 mm Locn: 235 $ 6725 2205 Date: 05.2005

	4wD	Bd	Diema DS20	2638	1963	(h)	
	4wD	Bd	Diema DFL 60/11	2803	1965	(i)	
	4wDH	Bdh	Schöma CHL 20G	4451	1981	(h)	
TITANUS	4wDH	Bdh	Schöma CHL 30G	5140	1990	new	
	4wDM	Bdm	O&K MV0	25102	1951	(l)	

Gauge: 500 mm and others Locn: 235 $ 6725/2205 Date: 01.1984

	4wDM	Bdm	Brun FL 12 10/12 PS	36	19xx	(m)	
	4wDM	Bdm	Brun B12 1HKG65	41	19xx	(n)	
	4wDM	Bdm	O&K MD 2b	25205	1951	(o)	
	4wDM	Bdm	O&K MV0a	25685	1955	(p)	
	0-4-0DM	Bdm	Deutz OME117 F	10282	1932	(n)	
14	4wDM	Bdm	Jung EL 110	7974	1938	(r)	
	4wDM	*Bdm*	*Hatlapa Junior II*	*5097*	*1951*	*(s)*	*(19)*
	4wDM	Bdm	Kröhnke Lorenknecht	298	1957	(t)	
	4wDM	Bdm	Kröhnke Lorenknecht	300	1958	(u)	
	4wDM	Bdm	Diema DL6	1797	1955	(v)	
	4wDH	Bdh	Schöma CHL 20G	4451	1981	(w)	

(a) ex ? /1960, Fabr. No. 41983.
(b) new via Robert Aebi & Cie. AG, Zürich.
(c) Stadler records indicate that a Diema of 1926 was rebuilt. This one was delivered to Gisikon and the assumption has been made that the two are the same.
(d) new to Flemmingsche Ziegelei, Hannover-Iserhagen [D]; obtained via Diema /1979. It may have been upgraded to 4wDH Bdh at that time - to be confirmed.
(e) new via Hans F. Würgler, Zürich; was reported as being of 22PS in /1974. The original 15 PS plates were still mounted /1976.
(f) new via Hans F. Würgler, Zürich, but never reported here.
(h) ex 500 mm /1989.
(i) new to IVB NV, Zwolle [NL]; ex NV Stenbakkerijen van Biervliet, Zonnebeke [B] via Diema /1984 as DFL60/11. It may have been upgraded to 4wDH Bdh at that time - to be confirmed.
(l) ex M. H. Bezzola AG, Bauunternehmung, Biel by 5/2005.
(m) ex Frutiger Söhne, Thun/1983; (n° 95913).
(n) ex operations; but different type recorded - to be clarified, Fabr. No. 41983.
(o) ex Frutiger Söhne, Thun "281.24.001" /1983 via Wander-Wendel (750 mm).
(p) new to Jost Hinrich Havemann & Sohn, Lübeck [D] (785 mm); via Eilers /1987 (gauge uncertain).
(r) new to Jost Hinrich Havemann & Sohn, Lübeck [D]; via Eilers /1987 (600 mm).
(s) if correctly identified ex Dampfziegelei Puchner, Regenstauf [D]; the number has been quoted as 50097 of 1956 which is too high for a Hatlapa, while known Strüver numbers start at 60001.
(t) new to Max Hohl, Stuttgart [D].
(u) new to Briquetterie de la Barrière, Bury [B].
(v) ex Matador-Kalkwerke, Springe/Deister [D] 11/2001 (gauge 600 mm).
(w) ex operations (600 mm).
(4) 1989 to private collection.
(5) 1989 converted to 600 mm.
(19) if correctly identified 1984 to Feldbahn-Museum 500 eV, Nürnberg [D].

see: www.ziegelei-schumacher.ch/Backstein.asp

SEEVERLAD + KIESHANDELS AG (SEEKAG), Luzern   LU31

A firm that extracts sand and gravel from the Vierwaldstättersee and uses the gravel to produce concrete. Office address: Landenbergstrasse 41, 6005 Luzern.

The Vollert shunting locomotive and train was photographed on 6 June 2008.

Gauge: 1435 mm   Locn: 235 $ 6569 2110   Date: 03.2008

                 4wDE   Bde   Vollert DER 3000   02/003   2002   new

SWISS STEEL AG, Emmenbrücke   LU32

formerly: Schmolz+Bickenbach AG (2003-2007), name change only from 2006
            Swiss Stahl AG (1996-2003)
            von Moos Holding AG (1988-1996)
            von Moos Stahl (?-1988)
            A.G. der von Moos'schen Eisenwerke
            Eisenwerk von Moos (1853-?)
            ? von Moos (1842-1853)

A wire-works was established in 1842. This developed to include an ironworks and rolling mill, and later a Siemens-Martin smelter in 1889. Rail connection was provided in 1891. An electric steel works followed in 1939. It now has a large internal network with a fleet of internal wagons. From about 1999 the railway operations were contracted out to Panlog, the firm currently responsible for the logistics of Swiss Steel.

Gauge: 1435 mm   Locn: 235 $ 6635/2139   Date: 09.2001

		0-4-0DM	Bdm	Deutz R IV 30-35 PS	1633	1915	new	s/s
		0-4-0PM	Bbm	Deutz R III	2760	1918	(b)	s/s
2		0-4-0T	Bn2t	SLM E 2/2	1534	1903	(c)	1966 scr
3		0-6-0WT	Cn2t	SLM E 3/3	1359	1900	(d)	(4)
4		0-6-0WT	Cn2t	SLM E 3/3	1014	1896	(e)	1957 scr
4		0-6-0WT	Cn2t	SLM E 3/3	1532	1903	(f)	19xx scr
5		0-6-0WT	Cn2t	SLM E 3/3	900	1894	(g)	(7)
6		0-6-0WT	Cn2t	SLM E 3/3	2075	1910	(h)	(8)

	7		0-6-0WT	Cn2t	SLM E 3/3	1879	1907	(i)	(8)
Tm 3/3	8		6wDE	Cde2	Moyse CN52EE500D	3512	1970	new	(10)
			6wDE	Cde2	Moyse CN52EE500D	3515	1970	(k)	
837 805-1			6wDE	Cde2	Moyse CN52EE500D	3524	1972	(l)	
837 806-9			6wDE	Cde2	Moyse CN52EE500D	3525	1972	(m)	(13)
			4wBE	B?a	von Roll	-	19xx	(n)	
	rebuilt		4wDE	B?de			197x		(14)
Tm 3/3 12 later 413			6wDH	Cdh	MaK 600 C	600340	1960	(o)	1991 scr
837 804-4			6wDE	Cde3	MaK,BBC DE502 C	700081, ?	1986	(p)	
	6		4wDH	Bdh	Vollert Diesel, FFS	86/010	1986	new	
847 852-1			BBDH	BBdh	MaK G 1206 BB	1001019	1999	(r)	

(b) new to Motorwagenfabrik Arbenz, Zürich-Albisrieden, ZU; ex BLS /19xx.
(c) new to Seetalbahn (SeTB) "21"; purchased from Elektrochemische Fabrik Kallnach /1928.
(d) new to SCB "41"; purchased from SBB "E 3/3 8410" /1941.
(e) new to Nordostbahn "461"; purchased from SBB Werkstätte Zürich "E 3/3 8559" /1948.
(f) purchased from SBB "E 3/3 8455" /1957; rebuilt to oil firing /1965.
(g) new to Nordostbahn "256"; purchased from Société Industrielle Sébeillon, Lausanne /1949.
(h) purchased from SBB "E 3/3 8500" /1960.
(i) purchased from SBB "E 3/3 8483" /1962.
(k) new, originally "Tm 3/3 9", later "410".
(l) new, originally "Tm 3/3 10", later "411".
(m) new, originally "Tm 3/3 11", later "412".
(n) constructed by von Roll, Bern as an electrically propelled crane. It may only have been driven on one, not both, axles; later rebuilt as a diesel-electric locomotive.
(o) new to Dortmunder Eisenbahn [D] "D5" /1980 via HIS/Asper.
(p) electrical equipment fitted by BBC-Mannheim [D]; purchased /1987; originally "Tm 3/3 414".
(r) new to PANLOG AG.
(4) 1973 to private ownership; by 2004 stored at Zürcher Freilager AG (ZF), Zürich-Albisrieden.
(7) 1973 to private ownership; 1986 to Modell Bahn Club Dietikon (MBCD), Dietikon.
(8) 1987 to EUROVAPOR, Sektion Haltingen (Kandertalbahn) [D].
(10) 1980 destroyed in an accident.
(13) 1999-2001 to Stahl Schweiz AG, Gerlafingen.
(14) 1986-7 s/s.

THEILER & KALBERMATTER AG, Buchrain   LU33 C

also: Batigroup, Schachen

The works yard of this construction firm is called "Werkhof Schachen" and is located at Sagenwald. The SIG list of references gives two ET 15KL locomotives delivered here.

Gauge: 750 and 600 mm   Locn: 235 $ 6687/2178   Date: 03.2002

281.5	4wBE	Ba2	SIG ETB 70	669 214	1967	(a)	
281.6	4wBE	Ba	SSW EL 7	6190	1966	(b)	
281.21.101	4wDM	Bdm	Schöma KDL 8	3295	1970	(c)	(3)
281.24.102	4wD	Bd	Ageve DMD	x779	19xx	(d)	
281.24.103?	4wDH	Bdh	Schöma CFL 100DCL	5241	1991	new	
281.24.104	4wDH	Bdh	Schöma CFL 100DCL	5242	1991	new	(6)
281.34.101	4wBE	Ba2	SIG ETB 70	649 211	1965	(g)	
281.34.103	4wBE	Ba2	SIG ETB 50	618 211	1961	(h)	
281.34.104	4wBE	Ba2	SIG ETB 70	588 102	1958	(i)	

(a) 600 mm, stored, partially dismantled.
(b) supplied to MBA, Dübendorf as 750 mm; 600 mm, stored, partially dismantled.
(c) ex Holzimprägnierungs GmbH, Laufenburg [D] (600 mm); stored, partially dismantled; may be 750 mm; if correctly identified.
(d) ex P. Schlatter, Münchwilen AG.
(g) 600 mm, stored for repair.
(h) 750 or 600 mm, stored, partially dismantled.
(i) 750 mm; in use on "Bärengraben" drainage works at Würenlingen /1995.
(3) by 6/1999 to FWF, Otelfingen.
(6) to ARGE Gütschstollen Los V.1, Kriens.

TOTTOLI & MÜLLER, Sursee — LU34 C

Construction of the station at Sursee.

Gauge: 600 mm   Locn: ?   Date: 02.2008

| SURSEE | | 0-4-0T | B?2t | Maffei | 3564 | 1911 | (a) | (1) |

(a) new via Robert Aebi & Cie AG, Zürich.
(1) 19xx to construction of the goods station at Pratteln.

UNKNOWN OWNER, Emmenbrücke — LU35 C

Gauge: 850 mm   Locn: ?   Date: 02.2002

| ELSA | | 0-4-0T | Bn2t | Krauss-S LXXII I | 2836 | 1895 | (a) | s/s |

(a) new via A. Maffei, Zürich.

C. VANOLI AG, Rothenburg — LU36 C

see under C. Vanoli AG, Samstagern.

VEREIN PENDELZUG MIRAGE, Zell LU — LU37 M

Gauge: 1435 mm   Locn: ?   Date: 02.2009

	BoBoWERC	SIG,BBC	?	1966	(a)	2008 scr
	BoBoWERC	SIG,BBC	?	1966	(b)	
	BoBoWERC	SIG,BBC	? ,6117	1959(79)	(c)	

(a) new to EBT "ABDe 4/4 201"; ex RM "BDe 4/4 250" /2007.
(b) new to SOB "ABDe 4/4 82"; ex SOB "576 054" /2008.
(c) new to SOB "ABe 4/4 71"; rebuilt as "BDe 4/4 80" /1979; later SOB "576 048"; planned to arrive 3/2009 for spare parts.

VEREIN BRÜNIG NOSTALGIE BAHN, Luzern — LU38 M

The society is amassing a collection of heritage stock from the Brünig line of the SBB. Some of it is kept at Alpnachstad.

Gauge: 1000 mm   Locn: ?   Date: 05.2008

Te¹ 198	0-4-0WE	Bw1	SLM,MFO Te¹	3770, ?	1941	
Te¹ 199	0-4-0WE	Bw1	SLM,MFO Te¹	3758, ?	1941	
Tm 491	4wDM	Bdm	RACO 40 PS	993	1931	(c)
120 009	Bo2BoRWE	B2bBw6	SLM,SAAS Fhe 4/6	3732, ?	1941	(d)
HGe 4/4 1992	BoBoRWE	BbBbw4	SLM,BBC HGe 4/4	4079,5464	1953	

(c) new to SBB "Tm 491"; ex LSE "101".
(d) new to SBB "Fhe 4/6 909".

VERKEHRSHAUS DER SCHWEIZ (VHS), Luzern — LU39 M

The museum lies on the north shore of the Vierwaldstättersee, east of the centre of Luzern. It encompasses all forms of transport. The museum changes the locomotives on display continuously; some locomotives from SBB-Historic are present on long-term loan. Operational SBB locomotives, locomotives from SBB-Historic, the museum's own store or other preservation concerns may be present. The museum is connected to, and has a station on, the

Luzern-Immensee line. This list concentrates on the details of those locomotives with an industrial connection. Address: Lidostrasse 5.

Gauge: 1435 mm   Locn: 235 $ 6682/2118   Date: 03.2008

HG 1/2 GNOM	2-2-0RT	1Azn2t	Olten	20	1870	(a)		
RB 7	2-2-0RT	2an2t	SLM H	1	1873		(V)	
HG 2/2 ELFE	0-4-0RT	Baz2t	Aarau	10	1876	(c)	(3)	
E 2/2 11	0-4-0WT	Bn2t	SLM E 2/2	236	1881	(d)		
EB 3 LANGNAU	0-6-0T	Cn2t	SLM Ed 3/3	229	1881			
OC CFe 2/2 11	4wERC		SIG,SAAS	N7798, ?	1894			
BTB E^2 E2	0-4-0WE	Bed1	SLM,BBC De 2/2	1197, ?	1899	(g)		
8512	0-6-0WT	Cn2t	SLM E 3/3	2135	1911			
BLS 151	2-10-2WE	1E1w2	SLM,MFO Be 5/7	2304, ?	1912	(i)		
RVT AB^{M}2/5 9	3BoDERC		Sulzer	MT 7	1922	(j)		
Tm 891	4wPM	Bbm	Breuer III	?	1931	(k)		
BN Ce 2/4 727	2BoWERC		SIG,MFO	?	1935	(l)		
BLS Ae 4/4 258	BoBoWE	BBw4	SLM,BBC Ae 4/4	4128,6017	1955			
Ae 6/6 11413 SCHAFFHAUSEN								
	CoCoWE	CCw6	SLM,BBC Ae 6/6	4148,6014	1955	(n)		
SBB Xrot m 100	4-4w	22	Henschel	4309	1895	(o)		

Gauge: 1000 mm   Locn: 235 $ 6682/2118   Date: 03.2008

JB He 2/2 1	0-4w+4RWE						
		2deb2	SIG,BBC	?	1897	(ba)	
SEB HGe 2/2 No 1	0-4-0WRE	Baed2	SLM,BBC HGe 2/2	1139, ?	1898		
HG 3/3 1063	0-6-0RT 4cc						
		Can4vt	SLM HG 3/3	1993	1909		
SGA BDeh 2/3 '3'	A1ARWE		SIG,EGA	?	1911	(bd)	
RhB Ge2/4 207	2-4-2WE	1B1w1	SLM,BBC Ge 2/4	2310,729	1913		
RhB Ge 6/6^I 402	0-6-6-0WE	CCw2	SLM,BBC Ge 6/6	2754,1546	1920,21		
KWO BFa2/2 4	4wBERC	Ba	SIG,SAAS	?	1939		

Gauge: 800 mm + Locher rack   Locn: 235 $ 6682/2118   Date: 05.1997

PB 9	0-2-2RRC	2an2t	SLM Dampftriebwagen	563	1889	(ca)	

Gauge: 750 mm   Locn: 235 $ 6682/2118   Date: 05.1997

WB 6 Waldenburg	0-6-0T	Cn2t	SLM G 3/3	2276	1912	

Gauge: 500 mm   Locn: 235 $ 6682/2118   Date: 05.1997

	4wBERC	Ba2	SIG,SAAS	?	1948	(ea)

(a) new to Steinbruchgesellschaft Ostermundigen, Ostermundigen; in use at Nestlé, Neuenegg /1904; to von Roll, Usine de Rondez, Delémont "7" /1907; withdrawn /1940.
(c) new to Steinbruchgesellschaft Ostermundigen, Ostermundigen; ex von Roll, Gerlafingen "6" /1941.
(d) new to Gotthardbahn "11"; to von Roll, Choindez /1890; to von Roll, Klus /1901 ; to von Roll, Gerlafingen /1934 ; withdrawn /1950 .
(g) may have exchanged identity with Bergdorf-Thun Bahn (BTB) "1".
(i) new to BLS "161", renumbered "151" /1943; with plates SLM 2265 1912 (the original "151").
(j) new as 3BoPERC /1914.
(k) new to SBB "Tm 464"; renumbered "Tm 891" /1963; withdrawn 12/1965.
(l) new to BLS "727"; renumbered "722" 1956.
(n) ex SBB "Ae 6/6 11413 SCHAFFHAUSEN" 9/1/2009.
(o) rotary snow plough with limited self-propulsion capability (not for use during operations), new to Gotthardbahn.
(ba) adhesion capability removed.
(bd) this motor coach is SGA "BDeh 2/3 1", but is restored as SGA "BDeh 2/3 3".
(ca) plates of "PB 6" SLM 514 1888.
(ea) ex Kraftwerke Oberhasli (KWO), Innertkirchen by /1997.
(3) 1981 to Gemeinde Ostermundigen, Bernstrasse Schule, Ostermundigen.
(V) operates on the Rigi Bahnen from time to time.

VAPEUR VAL-DE-TRAVERS (VVT), St. Sulpice, NE — NE15 M

This preservation group has gradually moved away from small shunting locomotives to larger steam locomotives.

Gauge: 1435 mm   Locn: 241 $ 5332/1957   Date: 11.2008

11	0-6-0WT	Cn2t	SLM E 3/3	631	1890	(a)	(1)
2	0-4-0T	Bn2t	SLM E 2/2	917	1895	(b)	
8511	0-6-0WT	Cn2t	SLM E 3/3	2134	1911	(c)	
	0-4-0TVB	Bn2t	Cockerill IV	2951	1920	(d)	
Papierfabrik Cham No 2							
	0-4-0T	Bn2t	Henschel Riebeck	20593	1925	(e)	
	4wPM	Bbm	Breuer III	?	1928	(f)	19xx scr
	4wPM	Bbm	Breuer III	1159	1929	(g)	19xx scr
2002	4wDM	Bdm	O&K RL 4	20162	1931	(h)	
BDZ 01.22	4-8-2	1D1h2	SLM	3592	1935	(i)	(9)
	4wD	Bd	Moyse 7T5	42-17	1939(51)	(j)	(10)
ÖBB 52 221	2-10-0	1Eh2	Škoda BR52	1584	1944	(k)	
	4wDH	Bdh	KHD A6M517 R	47304	1944	(l)	
	0-6-0T	Cn2t	KrMa 40 PS	16388	1945	(m)	
SNCF 241.P.30	4-6-2 4cc	2D1h4v	Schneider 195	4932	1951	(n)	
	4wDM	Bdm	Breuer V	3054	1952	(o)	(15)
Tkp 16	0-8-0T	Dn2t	Chrzanów Slask	2667	1952	(p)	
PKP TKt48-188	2-8-2T	1D1n2t	Chrzanów TKt48	4778	1956	(q)	
	4wDH	Bdh	Jung Köf II	13150	1959	(r)	
9	0-4-0WT	Bn2t	SLM E 2/2	2168	1911	(s)	
12	0-6-0WT	Cn2t	SLM E 3/3	1088	1898	(t)	
	4wDM	Bdm	Breuer V	3049	1952	(u)	1991 scr.

(a) new to JS "855"; later SBB "8575"; ex Von Roll, Gerlafingen "11" /1988 via VHS/DBB.
(b) new to Gebrüder Sulzer, Winterthur "1"; later Monteforno, Bodio and „GR "2"; ex Stiftung Zweiacher, Toffen, Belp /1988.
(c) new to CFF "8511"; ex Service du Gaz, Genève 14/6/1987.
(d) ex Usines E. Henricourt, Court-St. Etienne [F] /1984.
(e) new to Papierfabrik Perlen, Gisikon-Root "3", ex Cham-Tenero AG, Cham "No 2" /1988 (privately owned at the time but still stored at Papierfabrik Cham).
(f) originally CFF "Tm 438"; ex Lagerhaus Kloten AG, Kloten /1996.
(g) originally CFF "Tm 899"; ex Lagerhaus Kloten AG, Kloten /1996.
(h) ex HOLCIM SA, Eclépens "2002" (originally "2") /2000; arrived 1/11/2008.
(i) ex BDZ 1/4/2006; the locomotive is privately owned, the class has the nickname "Tabaklok"
(j) ex Aciéries de Champignole, Jura [F].
(k) new to DR "52 7486"; ÖBB mounted boiler of "52 221 BMAG 12226 1943" and numbered the result "52 221"; ex Oswald Steam, Samstagern 7/1992.
(l) new to DR "Kbf 5208" (4wPH Bbh); later DB "321 105-9"; later "323 967-0"; ex Schmutz Aciers SA, Fleurier /2003.
(m) ex Vereinigten Aluminium und Metallwerke Ranshofen, Braunau am Inn [A]; ex Königsfelder Eisenbahn, Österreich [A] 25/7/1986. The time between commencing construction and delivery appears to been longer than usual and various dates between 1942 and 1945 have been quoted.
(n) new to SNCF; later displayed in Vallorbe; ex Verein TMC 21/8/2003.
(o) new to Holzindustrie AG, St. Margrethen; ex Zollfreilager, St. Margrethen /1990; also reported as 4wPM Bbm.
(p) new to a coal mine in Częstochowa, Poland; ex PKP /1996.
(q) new to PKP; carries maker's plates and boiler from Chrzanów 4731 1956 from Tkt48-141; ex Firma Wagon-AW LAPY [PL] /2002.
(r) new to DB "Köf 6712"; later "323 782-3"; ex Oswald Steam, Samstagern /1993, via Vanoli.
(s) owned by La Traction, Pré-Petitjean; taken into custody /2008.
(t) purchased from La Traction, Pré-Petitjean /2008.
(u) also RACO 1424; ex von Roll, Klus /1991.

(1) 3/2008 to Bahn Museum Kerzers (BMK), Kerzers.
(9) to Betrieb Historische Schienen Fahrzeug (BHS).
(10) by 5/1997 dismantled.
(15) to Betrieb Historische Schienen Fahrzeug (BHS); 1996 to Swisstrain, Liesberg; later Bodio; 2007 to Le Locle.

Unterwalden (Obwalden + Nidwalden) OW + NW

Nidwalden NW

AG FRANZ MURER, Beckenried NW1 C
This firm was founded in 1897 and specialises in drilling techniques.
Gauge: 600 mm Locn: ? Date: 05.2002
    4wDM   Bdm   O&K(M) MD 1   11413   1939   (a) s/s
(a) ex O&K/MBA, Dübendorf for a contract at Andermatt (Kanton Uri).

Obwalden OW

BATIGROUP, Lungern OW1 C
A road tunnel, three kilometres long, was built ca. 1994. This group provided a railway and two locomotives for personnel transport at the northern end.
Gauge: 900 mm Locn: 245 $ 6559/1836 Date: 01.2001
    4wD   Bd   GIA D12   1201   1994   s/s
    4wD   Bd   GIA D12   1202   1994   s/s

EICHI, Alpnach Dorf OW2 C
A civil engineering concern or project.
Gauge: 750 mm Locn: ? Date: 05.2002
    4wDM   Bdm   O&K MD 2b   25231   1952   (a) (1)
    4wDM   Bdm   O&K MD 2b   25232   1952   (b) (2)
(a) built as Schöma 1308, new via O&K/MBA, Dübendorf, other sources quote the locomotive as going directly to Hinteregg.
(b) built as Schöma 1309, new via O&K/MBA, Dübendorf.
(1) to a firm in Hinteregg (perhaps via Neuhaus).
(2) 1984 to FWF, Otelfingen (600 mm) via Asper.

OBERKRIEGSKOMMISSARIAT (OKK), Giswil OW3
An arsenal and other military installations.
Gauge: 1000 mm Locn: 245 $ 6571/1875 Date: 12.2008
    5   4wDH   Bdh   RACO 80 ST4   1423   1953   (a)
(a) new; for disposal /2008.

SCHLIEREN, Alpnach Dorf OW4 M
Gauge: 1435 mm Locn: 245 $ 6630/1975 Date: 09.1973
    0-4-0T   Bf2   Jung 80 PS   911   1905   (a) (1)
(a) ex Viscosuisse SA, Emmenbrücke /1972.

(1) 1976 to Hotel Stalden, Berikon "HEINRICH".

VEREIN 11406, Alpnachstad OW5 M
The locomotive was, by 2008, in a secure location and being prepared for future display.

Gauge: 1435 mm Locn: ? Date: 03.2008

Ae 6/6 11406 OBWALDEN
                  CoCoWE   CCw6   SLM,BBC Ae 6/6   4141,6007   1955   store

see: www.11406.ch

AKTIENGESELLSCHAFT STEINBRUCH GUBER, Alpnach Dorf
OW6

A stone quarry and dressing area once connected by an aerial ropeway, three kilometres long, to the outskirts of Alpnach. The railway served the dressing area and connected it with the aerial ropeway. there was also an internal funicular on which the wagons of stone travelled two at a time, side by side across the direction of travel. No locomotives were seen in 8/1974, two in 3/1991 and a different one in 12/1999. By this time the locomotive was stored.

The locomotive shed and its contents (2 Jenbach JW10a Ponies) on 11 March 1991.

The dressing area on 11 August 1974

A loading point for two wagons in parallel onto the incline on 11 August 1974

Gauge: 500 mm Locn: 245 $ 6605/1980 Date: 12.1999

4wDM	Bdm	JW JW8	238	1954	(a)	OOU
4wDM	Bdm	JW JW10a	319	1957	(b)	s/s
4wDM	Bdm	JW JW10a	539	1983	(c)	s/s

(a) seen 12/1999.
(b) new to ? [I], seen 3/1991; but not seen 12/1999.
(c) seen 3/1991; but not seen 12/1999.

see: www.film-schlumpf.ch/GU/GU.html

WILH. WÄLTI, Giswil   OW7 C

A public works concern.

Gauge: 600 mm   Locn: ?   Date: 05.2002

4wDM	Bdm	O&K(M) RL 1a	4593	1931	(a)	(1)
4wDM	Bdm	O&K(M) RL 1c	8102	1937	(a)	

(a) via O&K/MBA, Dübendorf.
(1) by 4/1996 to FWF, Otelfingen.

Sankt Gallen (St. Gallen)   SG

ACKERMANN, BÄRTSCH & CIE, Mels   SG01 C
formerly: Entreprise de construction, J. Ackermann
A construction company.

Gauge: 600 mm   Locn: ?   Date: 02.2002

| | | 0-4-0T | Bn2t | Heilbronn II | 293 | 1893 | (a) | (1) |
| | | 0-4-0T | Bn2t | Zobel | 551 | 1907 | (b) | (2) |

(a)   ex ? /1905 (to Ackermann, Bärtsch & Cie).
(b)   via Fritz Marti, Bern /1910 (to Ackermann).
(1)   to Robert Aebi, Zürich; repurchased (by J. Ackermann) /1922; 1943 removed from boiler register; 1943-46 to F. Stirnimann, Olten.
(2)   1919 to Fietz & Leuthold AG, Baugeschäft, Zürich (600 mm?).

AGIP (SUISSE) SA, RAFFINERIE RHEINTAL, Sennwald   SG02
formerly: ESSO Switzerland SA, Sennwald (from 1989 till ?)
             Raffinerie Rheintal AG, Sennwald (RRAG) (19xx-1989)
             AGIP Suisse SA, Sennwald (1974-19xx)

Fuel refineries installed and rail connected with a branch from Salez-Sennwald station in 1974. Later the site was extended to include storage facilities and by 1976 had 4.9 kilometres of internal track.

Gauge: 1435 mm   Locn: 227 $ 7575/2354   Date: 07.1995

| | 4wDH | Bdh | Henschel DHG 300 | 31556 | 1973 | (a) |

(a)   new via Asper.

APPENZELLER BAHNEN (AB), Rorschach   SG03 M
formerly: Rorschach-Heiden-Bergbahn (RHB)

The last two locomotives from the erstwhile Maschinenfabrik Rüti, Rüti ZH have found a place on this line.

Gauge: 1435 mm + rack   Locn: ?   Date: 04.2008

| RHB 3 | Rosa | 0-4-0RT | Ban2t | SLM | 4046 | 1951 | (a) |
| Thm 237 916-2 | Chappi | 4wRDE | Bbde | SLM,BBC Tmh,Thm | 4400,6184 | 1962 | (b) |

(a)   ex Sulzer Rüti, Rüti ZH /1997 to EUROVAPOR, Sektion Sulgen/Rorschach which operates it on this line.
(b)   ex Sulzer Rüti, Rüti ZH /2002 hired to Kummler & Matter, Zürich which operated it on this line; 2006 sold to Appenzeller Bahnen (Rorschach-Heiden-Bergbahn).

JOSEPH AUER, Sennwald   SG04 C
later: ?Hs Auer Bagger-Trax- Unternehmen

If correctly identified a concern performing excavations and similar work. Address: Hauptstrasse 8, 9466 Sennwald.

Gauge: 600 mm   Locn: 227 $ 7562/2364   Date: 10.2007

| | 4wDM | Bdm | O&K(M) RL 1a | 4652 | 1932 | (a) | (1) |

(a)   new via O&K, Zürich.

(1)   by 2003 on display at an adjacent private address; 2004 Helmuth Lampeitl, Muntlix [A].

AVO WIEDERKEHR RECYCLING AG, Schwarzenbach SG   SG05

formerly: Autoverwertung Ostschweiz AG (AVO)

The company performs vehicle demolition and metal recovery. Address: Industriegebiet Salen, 9536 Schwarzenbach SG.

Gauge: 1435 mm   Locn: 216 $ 7230/2563   Date: 07.2002

	1	0-4-0DM	Bdm	SLM Tm 2/2	3221	1927	(a)	(1)
		4wDH	Bdh	O&K MV6B	25813	1958	(b)	(2)
Tm 2/2		4wDE	Bde1	Stadler,BBC	111, ?	1960	(c)	

(a)   ex Gebrüder Bühler, Uzwil /1979.
(b)   new to Hackethal Draht- und Kabelwerke AG, Hannover [D]: ex Kali-Chemie AG, Heilbronn [D] "2" /1983, via WBB/Asper.
(c)   ex Chemie Reichhold, Lupfig /1997, via Stadler.
(1)   1983-6 scrapped.
(2)   1997-9 scrapped.

BASALTSTEIN AG, Buchs SG   SG06

alternative: Basaltstein Aktiengesellschaft

Stone and ballast quarries. The material is brought down to a loading hopper adjacent to SBB line. The quarries lie about 1.5 kilometres away and were connected to the area below by an aerial ropeway. The skips that travelled on this also fit onto the narrow gauge chassis rail wagons. There used to be an inclined plane in the quarry. Currently rail traffic is restricted to the standard gauge.

This shunting device used the same weight transfer principles that are found on the more well-known Breuer locomotives. It differs however in having a permanently coupled wagon. It was seen here on 24 December 1968.

Gauge: 1435 mm			Locn: 237 $ 7548/2254				Date: 05.2002	
	4wPH	Bbh	Stemag V		1	19xx	(a)	OOU
	4wDH	Bdh	Diema DVL90/1.1		3154	1970	(b)	
Gauge: 600 mm			Locn: 237 $ 7531 2250				Date: 05.2002	
	4wPM	Bbm	O&K(M) RL 2		2614	1927	new	s/s
	4wPM	Bbm	O&K(M) M		2743	1928	(d)	(4)
	4wDM	Bdm	O&K(M) RL 2		2858	1928	(d)	s/s
	4wDM	Bdm	O&K(S) RL 1c		1639	1948	(f)	OOU

(a) ex Bahnhofkühlhaus, Basel-Wolf by /1967 (/1958 also quoted). The locomotive is construction from a small chassis and a VW engine. To improve braking characteristics it is permanently attached to a wagon; OOU since 1988-89.
(b) ex Felix Fiand KG, Hagen-Haspe [D] /1988-89.
(d) via O&K/MBA, Dübendorf.
(f) ex ? /1950; since ca. /1970 OOU and placed in store.
(4) 19xx to Zementfabrik Därligen ?via O&K, Zürich.

BAUUNTERNEHMUNG KÜHNIS AG, Oberriet   SG07 C

formerly: Frei & Kühnis
            Kühnis & Lüchinger

A public works concern.

Gauge: 600 mm			Locn: 228 $ 7602/2435				Date: 02.1988	
	4wDM	Bdm	O&K(M) RL 1		3645	1929	(a)	(1)
	4wDM	Bdm	O&K(M) RL 1		3661	1929	(b)	(2)
	4wDM	Bdm	O&K(M) RL 1a		4332	1930	(a)	(1)

(a) owned by Kühnis & Lüchinger; later Kühnis, Oberriet SG; purchased from A. Kiesel, Winterthur.
(b) ex Schmidheiny, Heerbrugg (500 mm).
(1) 1969 to either Elkuch & Co, Eschen [FL] (a scrap dealer) and scrapped 198x or to Ziegelei Rehag AG, Bern.
(2) 1986 to Oliver Weder, Diepoldsau.

BAUUNTERNEHMER PETER BÜCHEL AG, Gossau SG   SG08 M

This civil engineering company has a collection of machinery.

Gauge: 600 mm			Locn: ?				Date: 08.1995	
	4wDM	Bdm	Diema DS28		2580	1962	(a)	(1)
	4wBE	Ba	unknown		?	19xx		

(a) ex W. Schlittler, Niederurnen.
(1) 1999 to Nutzfahrzeug Altstätten SG, for Nuevo Arenal Tilaran [Costa Rica].

BODENSEE-TOGGENBURG BAHN (BT), St. Gallen   SG09 C

For the construction of that part of the Bodensee-Toggenburg railway line between Herisau and Wattwil in 1906 the following lots and locomotives are known:

Lot n° 2, between Herisau and Weissenbach, awarded to a consortium formed of Locher & Cie, Müller, Zeerleder & Gobat, Ritter-Egger, L. Kürsteiner and P. Rossi-Zweifel

Lot n° 3, between Weissenbach and Mogelsberg, awarded to Buchser & Broggi, Degersheim

Lot n° 4, between Mogelsberg and Lichtensteig, awarded to Favetto & Gatella, Brunnen, and later performed directly by the BT

Lot n° 5, between Lichtensteig and Wattwil, awarded to Baumann & Stiefenhofer, Wädenswil

Gauge: 750 mm				Locn: ?					Date: 03.2002
0		0-4-0T	Bn2t	SLM	?	18xx		s/s	
1		0-4-0T	Bn2t	Heilbronn II	46	1876	(b)	s/s	
2	ALBIS	0-4-0T	Bn2t	Krauss-S IV cd	2647	1893	(c)	s/s	
3	CALANDA	0-4-0T	Bn2t	O&K 30 PS	700	1900	(d)	s/s	
4		0-4-0T	Bn2t	Krauss-S IV tu	3185	1895	(e)	(5)	
5	SPERANZA	0-4-0T	Bn2t	Krauss-S XXVII vv	3254	1895	(f)	(6)	
6	IMPAVIDA	0-4-0T	Bn2t	Krauss-S IV pp	2098	1889	(g)	s/s	
7	SÄNTIS	0-4-0T	Bn2t	Krauss	?	18xx		s/s	
8		0-4-0T	Bn2t	Heilbronn III	64	1877	(i)	s/s	
9		0-4-0T	Bn2t	Krauss-S IV bv	5757	1908	(j)	s/s	
		0-4-0T	Bn2t	O&K 50 PS	391	1900	(k)	s/s	
	HELVETIA	0-4-0T	Bn2t	Krauss-S IV kl	2960	1893	(l)	(12)	
	THUR	0-4-0T	Bn2t	O&K 50 PS	1898	1906	(m)	s/s	
3		0-4-0T	Bn2t	O&K	?	19xx	(n)	s/s	
4		0-4-0T	Bn2t	O&K	?	19xx	(n)	s/s	
5	SPEER	0-4-0T	Bn2t	O&K 50 PS	1899	1906	(m)	s/s	
6	LICHTENSTEIG	0-4-0T	Bn2t	Krauss-S XXVII bg	6173	1908	new	(17)	
7	SIHL	0-4-0T	Bn2t	Krauss-S IV cd	2648	1893		(18)	
8		0-4-0T	Bn2t	Krauss	?	18xx		s/s	
9		?	?	unknown	?	1xxx		s/s	

(b) made available by Rossi-Zweifel, St. Gallen.
(c) made available by A. Baumann, Bauunternehmer, Wädenswil by /1908.
(d) new to Galli & Co, (which branch?).
(e) made available by E. Ritter-Egger, Bauunternehmer, Zürich by /1907.
(f) ex Gribi & Hassler, Burgdorf /1905-08 made available by Buchser & Broggi, Degersheim.
(g) ex Fischer & Schmutziger, Zürich /1899-1908.
(i) originally 730 mm; ex Bauunternehmer, Beuthen. Oberschliessen (PL); belonged to Favetto & Catella, Brunnen.
(k) ex Favetto & Catella, Brunnen /1906-08.
(l) made available by Baumann & Stiefenhofer, Bauunternehmer, Wädenswil by /1908.
(m) ex RTU Rickentunnel-Unternehmung AG, Wattwil /1908; made available by Regiebau Bodensee-Toggenburg, Lichtensteig.
(n) ex RTU Rickentunnel-Unternehmung AG, Wattwil /1908.
(5) before 1991 returned to E. Ritter-Egger, Bauunternehmer, Zürich.
(6) 1909 made available to Buchser & Broggi, Degersheim; 1911 to Robert Aebi & Co, Zürich.
(12) 1912 to Bau RhB, Somvix.
(17) 1911 to Robert Aebi & Co, Zürich.
(18) before 1908 to Favetto & Catella, Brunnen; possibly sold in 1912 to Julius Berger, Olten for the building of the south side of the Hauenstein tunnel.

ALFRED BONARIA, St. Gallen   SG10 C

Gauge: 600 mm			Locn: ?				Date: 04.1999
	4wDM	Bdm	O&K(M) RL 1a	4627	1931	(a)	s/s

(a) via O&K/MBA, Dübendorf.

BUCHSER & BROGGI, Degersheim   SG11 C

A public works concern that participated in the building of the Bodensee-Toggenburg railway.

Gauge: 750 mm			Locn: ?				Date: 02.1988
IMPAVIDA	0-4-0T	Bn2t	Krauss-S IV pp	2098	1889	(a)	(1)
SPERANZA	0-4-0T	Bn2t	Krauss-S IV tu	3190	1895	(b)	(1)
SÄNTIS?	0-4-0T	Bn2t	Krauss-S IV bv	5757	1908	new	(3)
ROMA	0-4-0T	Bn2t	O&K	?	19xx	(d)	s/s

(a) ex Fischer & Schmutziger, Zürich /1889-1908.
(b) new to Ing. Galli & Co, Willisau; ex Gribi & Hassler, Burgdorf /1905-08.
(d) ex ? /1900-08.

(1) 1910-31 to Schafir & Mugglin, Klingnau.
(3) 1908-11 to A. Baumann (& Stiefenhofer), Wädenswil.

BÜHLER AG, Uzwil   SG12
formerly: Gebrüder Bühler AG (1860-198x)

An undertaking, founded in 1860, that specialises in constructing machines for the food industry; in particular flourmills. From 1926 it has been connected to Uzwil station by a line that descends under the railway viaduct and then through the streets.

The Henschel locomotive stands ready to take a load to the SBB station on 14 October 2008.

Gauge: 1435 mm			Locn: 217 $ 7283/2559			Date: 10.1991	
1	0-4-0PM	Bbm	SLM Tm 2/2	3221	1927	New	(1)
Tm 237 915-4	4wDH	Bdh	Henschel DHG 300B	31988	1978	(b)	

(b) new via Asper; into service /1979.
(1) 1979 to Autoverwertung Ostschweiz AG, Schwarzenbach SG.

AG ALB. BUSS & CIE, St. Gallen-St. Fiden   SG13 C

A company charged from 1906 with building Lots 1 and 2 of the Bodensee-Toggenburg railway line from St. Fiden to Romanshorn.

Gauge: 900 mm			Locn: ?			Date: 03.2002	
BASILEA 4	0-4-0T	Bn2t	Henschel	5408or9	1900	(a)	(1)
	0-4-0T	Bn2t	unknown	?	1xxx	(b)	(1)
	0-4-0T	Bn2t	Heilbronn	?	18xx	s/s	
	0-4-0T	Bn2t	Heilbronn III alt	197	1884	(d)	s/s

Gauge: 750 mm			Locn: ?			Date: 03.2002	
WEISSENSTEIN	0-4-0T	Bn2t	Krauss-S IV bh	5295	1905	(e)	(5)
OBERDORF	0-4-0T	Bn2t	Krauss-S LXV n	5492	1906	(e)	(5)

Gauge: 600 mm			Locn: ?					Date: 03.2002
	0-4-0T	Bn2t	O&K 40 PS	551	1900	(g)	(1)	
	0-4-0T	Bn2t	O&K 30 PS	517	1900	(a)	(1)	
	0-4-0T	Bn2t	unknown	?	1xxx	(g)	(1)	

(a)	transferred to this contract /1906; locomotive details not confirmed.
(b)	transferred to this contract /1906; possibly ex Schmidlin, Basel (in which case a Krauss of 1874)
(d)	new to Jos. Messing, Rümelingen [L]; in use here /1909.
(e)	transferred to this contract from the contract for the tunnel at Weissenstein, Oberdorf /1906.
(g)	transferred to this contract /1906.
(1)	1908 returned.
(5)	1908 returned to the contract for the tunnel at Weissenstein, Oberdorf.

CELLUX AG, Rorschach   SG14

formerly: Feldmühle AG (till 1979)

A plant making artificial silk and transparent sheets. It was closed and then re-instated by Cellux AG. It has been rail connected since 1951 and is located at Goldach SBB station.

Gauge: 1435 mm				Locn: 217 $ 7547/2602			Date: 05.1988
Tm	1	4wDM	Bdm	Kronenberg DT 60	118	1951	(1)
Tm	2	4wDH	Bdh	Kronenberg DL 200	147	1964	(2)

(1)	1978 to Halter, Biel Mett.
(2)	1983 to SBB "Tm 940", via Schweizer, Zofingen.

CEMENTWERK RÜTHI, Rüthi SG   SG15

later: Gips-Union AG

A cement works constructed 1906-07. It passed to Holderbank in 1914 and was closed. A narrow gauge line ran between the quarry (7579/2391) and the cement works serving additionally the SBB station of Rüthi. It was restarted in 1929 as a gypsum works, with the narrow gauge remaining in the works and manually operated. The works was derelict by 1976.

Gauge: 500 mm			Locn: 227 $ 7589/2398					Date: 02.1988
	0-4-0T	Bn2t	Maffei 20/25 PS	2954	1910	(a)	(1)	
	0-4-0T	B?2t	Maffei	3505	1909	(b)	(2)	

(a)	purchased or hired from Robert Aebi & Cie, Zürich.
(b)	new via Robert Aebi & Cie AG, Zürich.
(1)	c1910 to Robert Aebi & Cie AG, Zürich.
(2)	1910 to Robert Aebi & Cie AG, Zürich; 1919 to Keller & Cie, Pfungen (600 mm).

DEBRUNNER ACIFER AG, St. Gallen   SG16

formerly: Debrunner AG
             Debrunner & Co. AG
also: Debrunner & Acifer

A metal dealer, rail connected to St. Gallen-Haggen SOB station. Address: Hechtackerstrasse 33, 9014 St. Gallen.

Jung 12883 of 1957 (type R 30 B) was photographed on its way to the station on 13 October 2008.

Gauge: 1435 mm Locn: 227 $ 7436/2527 Date: 03.1987

	4wPM	Bbm	Breuer III	?	1929	(a)	1968 scr.
	4wPM	Bbm	Breuer III	?	1930	(b)	(2)
	4wPM	Bbm	Breuer III	?	1930	(c)	(3)
	4wDM	Bdm	RACO 45 PS	1246	1936	(d)	(4)
	4wDM	Bdm	RACO 45 PS	1250	1937	(e)	(5)
402	4wDH	Bdh	Henschel DH 160B	26924	1960	(f)	(6)
	0-4-0DH	Bdh	Jung R 30 B	12883	1957	(g)	

(a) ex SBB "Tm 402" /1963.
(b) ex SBB "Tm 404" /1963.
(a) ex SBB "Tm 406" /1962.
(d) new to SBB "Tm 323"; ex SBB "Tm 533" /1963.
(e) new to SBB "Tm 327"; ex SBB "Tm 529" /1963.
(f) ex Gaswerk St. Gallen-Rietli "402" /1973; Henschel records indicate the locomotive as type DH240.
(g) ex Industrieterrains Düsseldorf-Reisholz AG [D] "3" /1990, via MF/Asper.

(2) 1964 to BT "Tm 7".
(3) 1964 to BT "Tm 6".
(4) 1964 to BT "Tm 8".
(5) after 1981 s/s.
(6) 1991 to Asper; 1992 to Scintilla AG, Derendingen.

EISENBAHNFREUNDE ZÜRICHSEE RECHTES UFER, Rapperswil SG17 M

Gauge: 1435 mm Locn: ? Date: 03.2009

| Ae 3/6^I | 10664 | 2Co1WE | 2C1w3 | SLM,BBC Ae 3/6^I | 3088,2125 | 1925 | (a) |

(a) on long term loan from SBB-Historic.

EISENBERGWERK GONZEN AG (EGAG), Sargans — SG18

This underground ironstone mining operation commenced in 1919, though there had been many earlier workings in the mountain. There were several underground mining areas, each with its own railway and also one line which brought ore out over a viaduct to Malerva, where a transhipment point to standard gauge wagons existed. Working ceased in 1966 but the 600 mm line continued in use for another three or four years for service purposes. At the same time ore was brought in by rail and added to the stockpiles to give the desired mix for use in the von Roll smelting plant at Gerlafingen. Since then a preservation scheme has developed to provide a tourist attraction.

Malerva

Gauge: 1435 mm   Locn: 237 $ 7524/2132   Date: 02.1994

1	4wDM	Bdm	Kronenberg DT 60	102	1940	new	(1)

Basisstollen

alternative: Vild

In use from 1951 to 1966.

Gauge: 600 mm   Locn: 237 $ 7524/2132   Date: 02.1994

	4wDM	Bdm	O&K(M)	?	1918	(b)	(2)
X 15	4wDM	Bdm	R&H 30DL	256190	1948	(c)	(3)
	4wDM	Bdm	KHD A4M517 G	47169	1951	(d)	(4)
	4wDH	Bdh	KHD A4M517 G	55442	1952	(e)	(5)
	4wDH	Bdh	KHD A6M517 G	56209	1955	(e)	(6)
2-30	4wDM	Bdm	KHD A2M517 G	56506	1957	(g)	a1992 s/s

Galerie 5

Gauge: 500 mm   Locn: ?   Date: 02.1994

	4wDM	Bdm	Brun FL 12	34	1938	(8)

Galerie 12

alternative: Naus

In use from 1928 with one diesel locomotive, details required.

Gauge: ? mm   Locn: ?   Date: 02.1994

	4wDM	Bdm	unknown	?	19xx	s/s

Gesenk

In use from 1957 to 1966.

Gauge: 600 mm   Locn: ?   Date: 02.1994

	4wDM	Bdm	Brun	33	1938		s/s
'4'	4wGBE	Bga	MFO ElectroGyro	?	1954	(k)	(11)

Wolfsloch

This working was in use from 1938, full details of locomotives used are unknown. In 1967 the miners reported one locomotive outside and five in the mine; presumed to be all from this working.

Gauge: 585? mm   Locn: ?   Date: 08.2008

	0-4-0DM	Bdm	Deutz	?	1920		(2)
	0-4-0DM	Bdm	Deutz	?	1921		(2)
	4wPE	Bbe	Deutz MLH222 G	8402	1928	(n)	(14)
	4wDM	Bdm	Brun	?	1939	(o)	s/s

		4wDM	Bdm	Brun	35	1940	(o)	s/s	
		4wDM	Bdm	Brun 2 DS 90E	?	1940		s/s	
	1-30	4wDM	Bdm	KHD A2M517 G	56507	1957	(r)	(18)	

Unknown Galerie

Gauge: 500 mm			Locn: ?				Date: 12.2008	
		4wBE	Ba	SSW		1922	new	s/s

Gauge: ? mm			Locn: ?				Date: 02.1994	
		4wDM	Bdm	O&K	?	19xx	(s)	s/s

(b)	quoted as 'Montania' type, so from O&K, Nordhausen [D].
(c)	replacement Deutz 25.8 kW motor /1959.
(d)	new via Hans Würgel, Zürich.
(e)	new via HWZ.
(g)	still present in Museums-Eisenbergwerk Gonzen, Sargans /1983.
(k)	new to Mines St. Pierremont [F] (as A1GBEM A1gam) ; acquired from MFO /1956 and purchased /1958.
(n)	new; possibly one of the two locomotives listed immediately above.
(o)	in service from /1959.
(r)	via O&K, Zürich.
(s)	in service from /1959.
(1)	1966 to A. Käppeli's Söhne AG, Sargans as part of the Malerva installations.
(2)	1959 authorised for scrap.
(3)	after 1966 to Ziegelei Landquart, Landquart (750 mm); later to Museums-Eisenbergwerk Gonzen, Sargans (600 mm).
(4)	1966 to St. Gallische Rheinkorrektion "KAMOR" (750 mm); later to Museums-Eisenbergwerk Gonzen, Sargans (600 mm).
(5)	1966? to St. Gallische Rheinkorrektion "ALVIER" (750 mm).
(6)	1966 to St. Gallische Rheinkorrektion "GONZEN" (750 mm).
(8)	2000 to FWF, Otelfingen, Fabr. No. 93913.
(11)	1956 withdrawn; 1981 donated to Technorama, Winterthur; 1986 to Museums-Eisenbergwerk Gonzen, Sargans (600 mm).
(14)	gauge approximately 500 mm; 1967 seen off the track and outside the mine; 19xx to .
(18)	In 1983 and 1988 noted as 600 mm; to Museums-Eisenbergwerk Gonzen, Sargans "BARBARA II" (600 mm).

see: www.rail.lu/materiel/mfogyrolok.html

ENTREPRISE P. ROSSI-ZWEIFEL, St. Gallen   SG19 C

formerly: Entreprise St. Rossi (till 1902)

A civil engineering concern involved in reconstructing the station at St. Gallen.

Gauge: 700? mm			Locn: ?				Date: 02.2002	
		0-4-0T	Bn2t	Heilbronn II	46	1876	(a)	s/s
		0-4-0T	Bn2t	Krauss-S IV kl	2960	1893	(b)	(2)
		0-4-0T	Bn2t	SLM	526	1888	(c)	(3)

(a)	new to Strauss & Bleiber, Backnang [D]; ex Castelli & Rüttimann for RhB construction /1900.
(b)	previously used for the building of the Fussacher Durchstich, in service /1900.
(c)	ex Jakob Mast, Zürich (construction of Landquart-Davos line of the RhB) /1888-1906.
(2)	1900-03 to Fritz Marti, Winterthur.
(3)	after 1911 s/s.

EUROPÄISCHE VEREINGUNG VON EISENBAHNFREUNDEN FÜR DEN ERHALT VON DAMPLOK (EUROVAPOR), Sektion Sulgen/Rorschach   SG20 M

see entry under Sulgen in Canton Thurgau.

JEAN FISCHELET, St. Gallen   SG21 C

Probably a construction concern.

Gauge: 790 mm   Locn: ?   Date: 02.2002

	0-4-0T	Bn2t	Hohenzollern	120	1879	new	s/s

GANTENBEIN & LAZZARINI, Buchs SG   SG22 C

alternative: Firma Gantenbein

A public works concern. The exact title and any relation to L. Gantenbein are unknown.

Gauge: 600 mm   Locn: ?   Date: 04.1999

	4wDM	Bdm	O&K(M) RL 1	3840	1929	s/s
	4wDM	Bdm	O&K(M) RL 1a	4650	1932	(2)
	4wDM	Bdm	O&K(M) RL 1c	8371	1937	(3)

(2) 1994 donated to FWF, Otelfingen.
(3) demolished and parts used in O&K 4650; at FWF, Otelfingen the resulting locomotive has the plates of O&K 8371.

LEONHARDT GANTENBEIN & CO, Werdenberg   SG23 C

A company engaged in structural and civil engineering with head offices in Werdenberg, but with several depots in the area. These two locomotives were seen in a yard visible from the line into Buchs SG from Austria in 1973; but the exact ownership then was not clear.

Gauge: 600 mm   Locn: 237: ?   Date: 07.1999

6	4wDM	Bdm	O&K(M) RL 1a	4291	1930	(a)	s/s
	4wDM	Bdm	O&K MV0	25042	1951	new	(2)

(a) ex Schmidheiny, Heerbrugg /19xx.
(2) 1987-94 to Museums-Eisenbergwerk Gonzen, Sargans and then private preservation.

GASWERK ST. GALLEN, Goldach-Rietli   SG24

A gasworks, rail connected in 1903 from Horn station and demolished ca. 1970. The site lies nearer Rorschach than St. Gallen town. The site passed to the SBB and later to the town. The rail connection is retained in connection with decontaminating the soil. The works is also referred to as Gaswerk St. Gallen-Rietli and the locality as Riet.

Gauge: 1435 mm   Locn: 217 $ 7535/2612   Date: 02.1988

1	0-4-0WT	Bn2t	Krauss-S XVIII f	485	1876	(a)	1948 scr
2	0-4-0WT	Bn2t	SLM E 2/2	686	1891	(b)	(2)
401	0-6-0WT	Cn2t	SLM E 3/3	1387	1901	(c)	(3)
402	4wDH	Bdh	Henschel DH 160B	26924	1960	(d)	(4)
2	4wBE	Ba2	SIG,MFO	?	1939	(e)	(5)

(a) new to VSB "A"; purchased from SBB "E2/2 8197" /1902.
(b) new to SOB "51"; purchased from Kriens-Luzern-Bahn "51" /1912.
(c) purchased from Uerikon-Bauma-Bahn "Ed 3/3 401" /1944.
(d) new, Henschel records give the locomotive as type DH240B.

(e)	privately owned and stored here; ex Maggi AG, Kemptthal.
(2)	1961 display Schönaustrasse, St. Gallen; 1967-71 scrapped.
(3)	1971 to private ownership; 1979 DVZO.
(4)	ca. 1970 stored at SBB, Rorschach; 1973 to Debrunner & Co AG, St. Gallen-Haggen; 1991 to Asper.
(5)	by 6/2008 to ?

GEBERIT AG, Rapperswil-Blumenau   SG27

formerly: Geberit & Co, Rapperswil

A factory making sanitary installations.

Gauge: 1435 mm   Locn: 226 $ 7066/2313   Date: 02.2003

		4wDH	Bdh	Schöma CHL 20GR	2168	1960	(a)	(1)
		4wDM	Bdm	O&K MV4a	25845	1958	(b)	(2)
237 902-2		4wDH	Bdh	O&K MV9	25826	1958	(c)	

(a) new to Alfred Teves GmbH, Gifhorn [D] (as CHL30G); ex Schöma after reconditioning /1971 via Asper.
(b) delivered new to O&K Amsterdam [NL]; ex Glunz Beropan, Berlin-Marienfelde [D] via MF/LSB /1989.
(c) new to Westfalenhütte, Dortmund [D] "48"; later Bergrohr GmbH, Herne (D); via Kai Winkler/SERSA/METRAG/WBB /1995.
(1) 1990 to Jean Trottet SA, Collombey-Muraz /1989 via LSB.
(2) 1996 to NStCM.

GEMEINDE BÜTSCHWIL, Grämigen Spielplatz   SG28 M

The locomotive is in a playground opposite the Anker Gasthof.

Gauge: 600 mm   Locn: 226 $ 7224/2491   Date: 05.2009

	201	4wDM	Bdm	O&K(M) MD 2r	12004	1944	(a)	plygrd

(a) ex Widmer AG "201" /19xx.

GLEIS- UND TIEFBAU AG (GLEISAG), Rorschach   SG29 C

This infrastructure concern has its head office in Goldach. Office address: Mühlegutstrasse 6, 9403 Goldach.

Gauge: 1435 mm   Locn: ?   Date: 04.2008

Bm 847 960-2 Iris	0-8-0DH	Ddh	MaK 1000 D	1000009	1958	(a)	(1)

(a) ex Rinteln-Stadthagener Eisenbahn [D] "V101" /1997 via WBB and Walo Bertschinger, Bern.
(1) 1998 to Stauffer.

HARTSCHOTTERWERK SEVELEN, Sevelen   SG30

The title applies from 1927. A quarry for 'Campion' stone, worked by the Hans Rüesch enterprise. The line connected the quarry with the SBB station (7558/2217).

Gauge: probably 500 or 600 mm   Locn: 237 $ 7548/2232   Date: 02.2002

0-4-0T	Bn2t	Freudenstein	106	1901	(a)	(1)
4wDM	Bdm	unknown	?	1xxx		s/s

(a) ex Entreprise Couchepin, Dubuis & Cie, Mex sur Evionnaz (Mex VS) /1932.
(1) ca. 1945 used in the construction of the Rupperswil-Auenstein hydroelectric power station; 1951 removed from boiler register.

HOLCIM KIES UND BETON AG   SG31

Werk St. Gallen, St. Gallen-Haggen

formerly: Fertigbeton AG (till 2007)

A concrete prefabrication works, located near St. Gallen-Haggen BT station. Address: Walenbüchelstrasse 17, 9000 St. Gallen.

Schöma 2381 of 1960 (type CFL 80DR) positions some cement wagons on 13 October 2008.

Gauge: 1435 mm   Locn: 227 $ 7439/2529   Date: 08.2002

	4wDH	Bdh	RACO 30 VA/HT	1549	1958	(a)	1981 scr
	4wDH	Bdh	Schöma CFL 80DR	2381	1960	(b)	
	4wDH	Bdh	RACO 70 DA4 H	1817	1974	(c)	(3)
	4wDH	Bdh	KHD A6M517 R	55126	1952	(d)	(4)

Werk St. Margrethen, St. Margrethen-Bruggerhorn

formerly: Fertigbeton AG (199x-2000)
: Interbeton AG
Gautschi AG, Bauunternehmer (1902-9x)

The Gautschi company started in 1902 by operating a quarry. It still exists, but no longer operates quarries. This site, installed in 1992, prepares and distributes sand, gravel and concrete.

Gauge: 1435 mm   Locn: 218 $ 7668/2580   Date: 07.2002

| 'SANDY' | 0-4-0DH | Bdh | Jung R 30 B | 12994 | 1959 | (e) |

Werk Schwarzenbach SG

formerly: Beton AG (till 16/08/01)

This works produces cold concrete and is located to the north-east of Schwarzenbach SG station.

After serving for many years with the PTT at Ostermundigen, Schöma 2151 of 1958 (type CFL 80DR) is now employed by HOLCIM at Schwarzenbach SG.

Gauge: 1435 mm   Locn: 226 $ 723480,256620   Date: 05.2009

| | 4wDH | Bdh | RACO 70 DA4 H | 1817 | 1974 | (f) | (6) |
| Tm 237 937-8 Alex | 4wDH | Bdh | Schöma CFL 80DR | 2151 | 1958 | (g) | |

(a) ex Shell, Cornaux NE ca. /1965; it is also reported as a diesel-mechanical locomotive (clarification required).
(b) ex Hamburger Gaswerke, Kokerei Kattwyk [D] "2" /1982 via Schöma/Asper.
(c) hired from LSB /1994.
(d) hired from LSB /1995.

(e) new to Industrieterrains Düsseldorf-Reisholz AG [D] "DL VI" /1991 via MF/Ferriere Belloli, Grono.
(f) new to Cece Graphitwerk, Zürich-Affoltern; ex LSB /1995.
(g) ex DHL Solution, Ostermundigen /2007 via Stauffer.
(3) 1995 returned to LSB.
(4) 199x returned to LSB.
(6) by 2009 s/s.

HOLZINDUSTRIE AG (HIAG), St. Margrethen　　　　　　　　SG32

The company deals in timber and has its own sawmill. It was originally established in Rorschach in 1876 and moved to St. Margrethen in 1911. Address: Grenzstrasse 24, 9430 St. Margrethen.

Gauge: 1435 mm　　　　　Locn: 218 $ 7661/2582　　　　　　Date: 06.1995

4wPM	Bbm	Breuer V	3054	1952	(a)	(1)
0-4-0DM	Bdm	KHD A4L514 R	56893	1959	(b)	

(a) new via RACO.
(b) originally Deutsche Philips GmbH, Krefeld-Linn [D] "1"; ex Vorholt & Schega, Haltern (D) /1987.
(1) 1987 to Zollfreilager, St. Margrethen; 1990 donated to VVT, St. Sulpice NE.

HOLZ STÜRM AG, Goldach　　　　　　　　　　　　　　　　　SG33

This wood yard is located by the rail entrance to Gaswerk St. Gallen and remains rail-connected from Horn station. It still uses a non-locomotive narrow-gauge rail system to serve the impregnation chambers and adjacent buildings. This includes a wagon traverser and the system is mentioned in this Handbook as a representative of several others throughout Switzerland. Address: 9403 Goldach.

Gauge: 750 mm　　　　　Locn: 217 $ 7534/2614　　　　　　Date: 08.2008

INTERNATIONALE RHEINREGULIERUNG (IRR)　　　　　　SG34

A joint Swiss/Austrian project for modifying and grading of the course of the R. Rhein over its last twenty kilometres before the Bodensee. It was created after the authorities on both sides had each established their own organisations. These national bodies continue to co-exist. Offices were located in Rorschach and Bregenz [A]. Workshops were established at Widnau (228 $ 7667/2522) and Lustenau [A] (218 $ 7675/2575). Other depots were located at least in Haag (227 $ 7561/2311), Kreissern (228 $ 7640/2475), Montlingen (228 $ 7622/2452), Sennwald (227 $ 7580/2377) and Fussach [A] (218 $ 7680/2607). Stone quarries were established at Montlingen, Koblach [A] and Unterklein [A]. There were two bridges for rail traffic over the R. Rhein. The railways were used in the quarries and in the new bed of R. Rhein for its preparation as well as bringing stone for reinforcement purposes later. In the lists that follow no distinction is made between any Swiss or Austrian allocation. A R&H engine number 285901 has been seen here at various dates from about 1960, but R&H never fitted this engine into a locomotive. Interchange of locomotives between the various bodies took place as different phases of the work proceeded. The disposal of four locomotives is unclear: all of them were seen on the Austrian section ca. 1992; they may still be there. On 29/8/2008 the use of the railway on an industrial basis formally ceased by international agreement; all remaining use of the railway from that date is solely for the Verein Rhein-schauen Museum und Rheinbähnle, Lustenau Markt [A]. Locomotives new to system after ca. 2000 were intended for the museum and are excluded from this Handbook.

Gauge: 750 mm　　　　　Locn: see text　　　　　　　　　　Date: 09.2008

3	0-4-0T	Bn2t	unknown	?	19xx	(a)	s/s
LÖTSCHBERG 28	0-6-0T	Cn2t	Borsig	6991	1908	(b)	s/s
27	0-6-0T	Cn2t	Borsig	6992	1908	(c)	(3)
WIDNAU	0-4-0T	Bn2t	Heilbronn II	228	1886	(d)	
RHEIN rebuilt	4wDM	Bdm	IRR	-	1950		(4)
SCHMITTER	0-4-0T	Bn2t	Humboldt	?	1906	(e)	1917 sold

Name		Wheel	Type	Builder	Works №	Year	Status	Notes
DIEPOLDSAU		0-4-0T	Bn2t	Jung 40 PS	98	1891	(d)	(6)
SCHWEIZ		0-4-0T	Bn2t	Jung 80 PS	1515	1910	new	(7)
ST. GALLEN		0-4-0T	Bn2t	Jung 80 PS	1516	1910	new	(8)
SÄNTIS		0-4-0T	Bn2t	Jung 80 PS	1622	1911	new	(7)
ALVIER		0-4-0T	Bn2t	Jung 80 PS	1682	1911	new	(7)
FALKNIS		0-4-0T	Bn2t	Jung 80 PS	1683	1911	new	(7)
ALFENZ		0-4-0T	Bn2t	Krauss-S LXV	1190	1882	(l)	s/s
ALBONE		0-4-0T	Bn2t	Krauss-S LXXI	1284	1883	(l)	s/s
ZÜRS		0-4-0T	Bn2t	Krauss-S LXXI	1285	1883	(l)	s/s
STUBEN		0-4-0T	Bn2t	Krauss-S LXXI a	1430	1883	(l)	s/s
HELVETIA		0-4-0T	Bn2t	Krauss-S IV kl	2960	1893	(p)	(16)
RHEIN		0-4-0T	Bn2t	Krauss-S XXXV nn	3566	1897	new	1916 sold
LUSTENAU		0-4-0T	Bn2t	Krauss-S XXXV qq	3683	1897	new	1916 sold
HÖCHST		0-4-0T	Bn2t	Krauss-S XXXV qq	3684	1897	new	1916 sold
HOHENEMS		0-4-0T	Bn2t	Krauss-S XXXVV ss	3699	1897	new	1916 sold
MAFFEI		0-4-0T	Bn2t	Maffei	4124	1921	new	(21)
STEFFI		0-4-0T	Bn2t	O&K 70 PS	2641	1908	(v)	(22)
'BRUN'		4wDM	Bdm	Brun	78	1946	(w)	(23)
'BENZINER'		4wPM	Bbm	Deutz C XIV F	983	1913	new	s/s
ELFI		4wW+DE	Bg+de	Gebus EDL 109	569	1958	new	(24)
'MAIKÄFER'		4wDM	Bdm	JW JW20	2019	1950	(aa)	(24)
JUNO		4wDM	Bdm	MR 32/42HP	10091	1950	new	(24)
MIKI		4wDM	Bdm	MR 32/42HP	10151	1951	new	(23)
SUSI		4wDM	Bdm	MR	10xxx	1952	new	(30)
SUSI		4wDM	Bdm	MR 50HP	12043	1960	new	(23)
WALD		4wDM	Bdm	O&K(M) MD 2	8476	1940	(af)	(24)
ROSL		4wDM	Bdm	O&K(M) MD 2	9400	1939	(ag)	1975 scr
'MONTANIA'		0-4-0DM	Bdm	O&K(M) MD 3	11713	1943	(ah)	(34)
'RACO 2' 2		4wDM	Bdm	RACO 50 PS	1312	1945	(ai)	(35)
'RACO 1' 1		4wDM	Bdm	RACO 50 PS	1313	1945	(ai)	(36)
		4wDM	Bdm	R&H 30DLU	256189	1948	(ak)	(37)
HEIDI		4wBE	Ba	Stadler	22	1946	new	
	rebuilt	4wW+DE	Bg+de	Stadler	109	1959		(24)
URS		4wBE	Ba	Stadler	43	1949	new	
	rebuilt	4wW+DE	Bg+de	Stadler	108	1958		(23)
SÄNTIS		4wBE	Ba	Stadler	105	1953	new	
	rebuilt	4wW+DE	Bg+de	Stadler	?	1954		(23)
		2-2-0PMRC1AP		unknown	?	19xx		(24)
		2-2-0PMRC1AP		Asper	398	1948		s/s
		2-2-0PMRC1AP		Asper type 14	402	1949		(24)
		2-2-0BERC 1AP		Stadler	?	1956		(24)

Gauge: 500 mm   Locn: ?   Date: 06.2002

Name	Wheel	Type	Builder	Works №	Year	Status	Notes
FRUTZ	0-4-0T	Bn2t	Krauss-S LXXII f	3562	1896	new	s/s
ILL	0-4-0T	Bn2t	Krauss-S LXXII f	3589	1897	new	1955 scr
LUTZ	0-4-0T	Bn2t	Krauss-S LXXII f	3754	1897	new	1955 scr
SEELACHE	0-4-0T	Bn2t	Krauss-S LXXII g	3839	1898	new	s/s
DORNBIRNER ACH	0-4-0T	Bn2t	Krauss-S LXXII g	3878	1897	new	s/s

(a) existence to be confirmed.
(b) ex Entreprise générale du chemin de fer des Alpes Bernoises "7"; purchased from RUBAG /1921.
(c) ex Entreprise générale du chemin de fer des Alpes Bernoises "8"; hired from RUBAG /c1921.
(d) ex RTU Rickentunnel-Unternehmung AG, Kaltbrunn /1909.
(e) ex ? /1913; if the date is correct then Humboldt 317, 319 or 322 are likely candidates.
(l) ex construction of the Arlberg tunnel [A] /1894; via SGRK.
(p) provided by Galli & Cie, Meggen c/1897.
(v) ex Union Bergbau Gesellschaft, Wien [D] /1936.
(w) into service /1947.
(aa) purchased /1950 (or /1953?).
(af) ex Landwasserbauamt Vorarlberg [A] /1957 (or Bundes-Wasser-Verwaltung [A] /1956?).
(ag) ex RAD Aufbaugruppe Innsbruck/Vorarlberger Landesregierung [D] /1957.
(ah) via MBA, Dübendorf /1948.
(ai) ex Conrad Zschokke, Näfels /19xx.
(ak) new as type 30DL; Robert Aebi & Cie, Zürich obtained the necessary parts from R&H to effect the conversion; via Robert Aebi & Cie, Zürich (probably as second-hand).

(3) by 1928 probably returned to RUBAG.
(4) 1945 removed from boiler register 1945; to St. Gallische Rheinkorrektion, Rorschach (Widnau?).
(6) 1952 removed from boiler register.
(7) c1924 to IRR; 1931 to Schafir & Mugglin for the construction of the hydroelectric power station at Klingnau.
(8) 1944-4x hired for the construction of the hydroelectric power station at Rupperswil-Auenstein; 1962 removed from boiler register en 1962; 1972? placed on display in Widnau; 200x operational with Verein Rhein-schauen Museum und Rheinbähnle, Lustenau Markt [A]; the old boiler is also present.
(16) by 1900 at Rossi, St. Gallen.
(21) 1949 wdn; 1969 placed on display in Lustenau [D]; 199x operational with Verein Rhein-schauen Museum und Rheinbähnle, Lustenau Markt [A] "200-90 LIESL".
(22) 1967 donated to Technisches Museum, Wien [A]; leased to Gurkthalbahn (Treibach-Althofen) [A].
(23) 2008 disposal uncertain.
(24) to Verein Rhein-schauen Museum und Rheinbähnle, Lustenau Markt [A].
(30) 1951? damaged by a grass fire; 1960 scrapped.
(34) 197x to P. Schlatter, Münchwilen AG; later Feldbahnmuseum Gross-Schwechat [A].
(35) 1969 (or 1959) to SGRK.
(36) 1970 to SGRK.
(37) reported as before 1950 scrapped.

see: Eisenbahn Amateur 5/1997

ISELI & CIE, St. Gallen — SG35 C

A public works concern (Hoch- & Tiefbau).

Gauge: 600 mm   Locn: ?   Date: 04.1999

| | | 4wDM | Bdm | O&K(M) RL 1a | 4209 | 1930 | (a) | (1) |

(a) via MBA Zürich /?1951.
(1) to Müller-Knecht, Kieswerk Riedikon /?1951.

A. KÄPPELI'S SÖHNE AG, Sargans — SG36

formerly: A. Käppeli & Co (till 1930)

The ore from the ironstone mine of Eisenbergwerk Gonzen AG within the Gonzen Mountain was brought to a storage and transhipment point at Malerva. When the mine closed in 1966, this part was sold to A. Käppeli's Söhne AG for use in conjunction with their stone quarrying activities.

Gauge: 1435 mm   Locn: 237 $ 7524/2132   Date: 01.2008

1		4wDM	Bdm	Kronenberg DT 60	102	1940	(a)	(1)
		0-4-0DM	Bdm	SLM Tm	3599	1935	(b)	(2)
		0-4-0DM	Bdm	SLM Em 2/2	3412	1930	(c)	(3)

(a) acquired with site from Eisenbergwerk Gonzen AG (EGAG) /1966.
(b) new to SBB "Tm 710"; ex SBB "Tm 870" 1/1968.
(c) ex PTT, Lausanne "1", via Karl Kaufmann AG, Thörishaus /1969.

(1) 1992-5 scrapped after a long period OOU.
(2) 1970 to Prodo SA, Domdidier.
(3) 1987 returned to PTT; 1997 to Swisstrain.

K. KELLER, Rorschach — SG37 M

A collector.

Gauge: 750 mm   Locn: ?   Date: 03.2002

| | | 0-4-0T | Bn2t | Märkische 50 PS | 302 | 1898 | (a) |

(a) ex VEBA AG, Zürich-Niederhasli /197x.

see: Eisenbahn Amateur 1/93

KERENZERBERGTUNNEL, Weesen — SG37a C

This four kilometre long rail tunnel between Sargans and Ziegelbrücke first opened in 1960, but has been modified since. The SIG list of references gives five ETB 70 locomotives delivered here.

KIESWERK ESPEL, Gossau SG — SG38

A gravel works; the railway has been abandoned, but two locomotives still exist.

Gauge: 600 mm Locn: 227 $ 7345/2523 Date: 02.1988

4wPM	Bbm	Austro-Daimler PS 6	259	19xx	(a)	(1)	
4wPM	Bbm	Austro-Daimler PS 6	541	19xx	(b)	(2)	
4wDM	Bdm	RACO	?	1944	(c)	s/s	

(a) probably supplied via Robert Aebi & Co, Zürich; carries plate "Robert Aebi & Co, Inginierbureau-Maschinenfabrik Regensdorf, Zürich and the given works number; motor type AD 8199 3.
(b) probably supplied via Robert Aebi & Co, Zürich; engine casing stamped "541" in several places; motor type AD 8199; motor number "10255"; 184 cm long, 100 cm wide.
(c) motor type Weber 2DS90E n° 2275.
(1) 19xx to playground attached to the "Lindenhof" restaurant north of Gossau SG.
(2) 19xx to Gemeinde Gossau; 1996 display Stadtwerke, Gossau SG.

KNIE KINDERZOO, Rapperswil SG — SG39 M

The locomotive has been placed on display as one of the attractions. These also a "Pony-tram" — a pony powered 600 mm tramway. Address: Oberseestrasse, 8640 Rapperswil.

Gauge: 1435 mm Locn: 226 $ 7047/2312 Date: 10.1967

0-4-0RT	Ban2t	SLM Eh 2/2	2126	1910	(a)	playgrd

(a) ex Maschinenfabrik Rüti AG, Rüti "2" /1962.

ROBERT KÖNIG AG, Oberriet — SG40

Stone quarry and ballast works (Steinbruch & Hartschotterwerk).

Gauge: 600 mm Locn: 228 $ 7612/2447 Date: 05.2002

4wDM	Bdm	O&K(M) RL 1a	5644	1934	(a)	

(a) ex Sigrist & Merz, St. Gallen /1956.

AG KOHLENWERKE, Uznach — SG41

Owned by Rickli "Im Hof". An opencast working for coal operating between 1918 and 1920.

Gauge: 600 or 750 mm Locn: 226 $ 786/2322 Date: 07.1988

0-4-0T	Bn2t	unknown	?	1xxx	(a)	s/s

(a) belonged to a sub-contractor.

LANDVERBAND ST. GALLEN "LANDI", St. Margrethen — SG42

A mill and a supermarket for farmers and gardeners. In 2007 the site was being remodelled and the railway lines temporarily blocked. Address: Industriestrasse.

Gauge: 1435 mm Locn: 218 $ 7668/2576 Date: 04.2002

4wDH	Bdh	Diema DVL60/1.3	3328	1973	(a)	OOU

(a) new to Eisen- und Metallwerke Bender, Kreuztal-Ferndorf [D]; ex Schrotthandlung Allgaier, Uhingen [D] /1986, via Asper.

LINDENHOF, Gossau SG — SG43 M

A restaurant with a children's playground situated to the north of Gossau SG. The locomotive is visible from trains on the Gossau AG-Bischofszell line. Address: Bischofszellerstrasse 133, 9200 Gossau SG.

Gauge: 600 mm   Locn: 217 $ 736000,254510   Date: 10.2008

| | 4wPM | Bbm | Austro-Daimler PS 6 | 259 | 19xx | (a) | playgrd |

(a) ex Kieswerk Espel, Gossau SG /19xx.

LOGISTIKBASIS DER ARMEE (LBA), Bronschhofen — SG44

formerly: Bundesamt für Betriebe des Heeres (BABHE) (till 31.12.2003)
Oberkriegskommissariat (OKK)

Military vehicle pool (PAA/AMP) and fuel reservoirs, which was connected to Wil-Weinfelden line of the MThB in 1965 at Bronschhofen AMP station.

One of Stauffer's hire fleet, Stadler 119 of 1966 shunts at Bronschhofen on 5 June 2008.

Gauge: 1435 mm   Locn: 216 $ 7199/2603   Date: 02.2003

Em 836 908-4	6wDE	Cde3	Stadler	113	1961	(a)	
12	6wDE	Cde3	Stadler	121	1967	new	(2)
Tm 237 933-7 Seppli	4wDE	Bde	Stadler	119	1966	(c)	(3)

(a) ex OKK, Oberburg /1979; numbered "9" till 200x.
(c) hired from Stauffer /2008.
(2) 1979 to OKK Interlaken West; according to OKK records it has 150 PS.
(3) 2008 returned toStauffer.

MIGROS OSTSCHWEIZ, Gossau SG   SG45

Regional warehouse and distribution centre for the Migros chain of stores.

Gauge: 1435 mm   Locn: 227 $ 738100,252800   Date: 10.2001

Tm 237 927-9   4wDH   Bdh   O&K MB 10 N   26679   1970   (a)

(a)   ex Migros Schönbühl /2001.

MODELLEISENBAHN CLUB DES BEZIRKS HORGEN (MECH), Rapperswil SG   SG46 M

For details please refer to the same club under Horgen (Kanton Zürich).

MODELLEISENBAHN KLUB WIL (MEKW), Wil SG   SG47 M

The locomotive is kept in a cage near the north-eastern corner of the Stadtweier. Address: Krebsbachweg, 9500 Wil.

Gauge: 1000 mm   Locn: 216 $ 7213/2588   Date: 06. 1997

    2 WYL   0-6-0T   Cn2t   SLM G 3/3   462   1897   (a)

(a)   ex FW "2 HÖRNLI", originally "2 WYL", OOU /1946; acquired /1965; display /1973.

JEAN MÜLLER & CIE, St. Gallen   SG48 C

Gauge: ? mm   Locn: ?   Date: 04.1999

            4wDM   Bdm   O&K(M) RL 1   3011   1928   (a)   s/s
            4wDM   Bdm   O&K(M) RL 2   12043   194x   (a)   s/s

(a)   via O&K/MBA, Dübendorf.

MUSEUMS-EISENBERGWERK GONZEN, Sargans   SG49 M

A museum was founded in 1983 by Pro Gonzen Bergwerk eV to present the history and activity of the Gonzen iron mine. The railway is used to transport visitors from the lecture hall at the entrance through the Vild or Basis Tunnel to a demonstration area. More detailed foot tours of the mine sometimes start from this point. Initially visitors were conveyed from an area around the locomotive depot over a viaduct and into the mountain. However this outer area has now been abandoned and trains now never come out into the open. One other Deutz locomotive from the mining days was present in 1983 and 1997.

Gauge: 600 mm   Locn: 237 $ 7528/2138 (ex 7525/2132)   Date: 04.1997

BARBARA I   4wDM   Bdm   Brun FD 16   ?   1938   (a)
BARBARA II   4wDM   Bdm   KHD A2M517 G   56507   1957   (b)
'BARBARA III'   4wDM   Bdm   R&H 30DL   256190   1948   (c)
'BARBARA IV'   4wDM   Bdm   KHD A4M517 G   47169   1951   (d)
    '4'   A1GBE   A1gb   MFO ElectroGyro   ?   1954   (e)
                 4wDM   Bdm   KHD A2M517 G   56506   1957   (f)   a1992 scr
                 4wDM   Bdm   O&K(M) LD 2   8586   1937   (g)   (7)
                 4wDM   Bdm   O&K(M) LD 2   10678   1939   (g)   (7)

(a)   ex Eisenbergwerk Gonzen AG, Sargans /1983. The locomotive exhibits a number of RACO features despite the Brun label.
(b)   ex Eisenbergwerk Gonzen AG, Sargans /1983 "1-30".
(c)   used at Eisenbergwerk Gonzen AG, Sargans till /1966; ex Ziegelei Landquart, Landquart (750 mm) /1986.
(d)   used at Eisenbergwerk Gonzen AG, Sargans till /1996; ex P. Schlatter, Münchwilen AG/198x; but reported previously at Locher AG, Stettbach and Gäbris (1000 mm).

(e)		new to Mines de St. Pierremont [F] "4"; used at Eisenbergwerk Gonzen AG, Sargans (as 4wBM Bb) till /1996; ex Technorama, Winterthur with chain to rear axle removed again via "Arbeitsgruppe Gyro Nr. 4"/1994.
(f)		used at Eisenbergwerk Gonzen AG, Sargans "2-30" till /1966 ; acquired for use as spare parts.
(g)		ex Steinbruch Lochezen, Walenstadt; stored.
(7)		1995 to a private individual, Gossau SG; scrapped without being restored.

(e) new to Mines de St. Pierremont [F] "4"; used at Eisenbergwerk Gonzen AG, Sargans (as 4wBM Bb) till /1996; ex Technorama, Winterthur with chain to rear axle removed again via "Arbeitsgruppe Gyro Nr. 4"/1994.
(f) used at Eisenbergwerk Gonzen AG, Sargans "2-30" till /1966 ; acquired for use as spare parts.
(g) ex Steinbruch Lochezen, Walenstadt; stored.
(7) 1995 to a private individual, Gossau SG; scrapped without being restored.

see: Die Elektrogyro-Lokomotive des Gonzenbergwerks, Norman Lang, 1994.

NIEDERER AG, Altstätten SG   SG52 C

formerly: Niederer Erben

A public works concern (Hoch- & Tiefbau).

Gauge: ? mm   Locn: ?   Date: 04.1999

	4wDM	Bdm	O&K(S) RL 1c	1502	1947	s/s

RTU RICKENTUNNEL UNTERNEHMUNG AG   SG53 C

Wattwil

Construction site for building the railway line through the Ricken rail tunnel between Uznach and Wattwil 1904-10. The company was based in Lausanne and this sometimes features in the quoted title and gives rise to the speculation that the locomotives passed through Lausanne.

Gauge: 750 mm   Locn: ?   Date: 03.2002

ST. GALLEN	0-4-0T	Bn2t	Freudenstein	229	1905	new	(1)
	0-4-0T	Bn2t	Heilbronn II ur	14	1872	(b)	s/s
SANTERNO	0-4-0T	Bn2t	Heilbronn II	227	1886	(c)	(3)
RENO	0-4-0T	Bn2t	Heilbronn II	228	1886	(c)	(4)
ORNE	0-4-0T	Bn2t	Krauss-S LXV s	3048	1894	(e)	s/s
ENGADIN	0-4-0T	Bn2t	Krauss-S IV zs	4589	1901	(c)	(6)

Kaltbrunn

Construction site for boring the Ricken rail tunnel between 1906 and 1910.

Gauge: 750 mm (or 720 mm)   Locn: ?   Date: 05.2002

	0-4-0T	Bn2t	O&K 50 PS	1460	1904	new	(7)
	0-4-0T	Bn2t	O&K 50 PS	1461	1904	new	(7)
	0-4-0T	Bn2t	O&K 50 PS	1462	1904	new	(7)
RICKEN	0-4-0T	Bn2t	O&K 50 PS	1896	1906	new	s/s
TOGGENBURG	0-4-0T	Bn2t	O&K 50 PS	1897	1906	new	(1)
THUR	0-4-0T	Bn2t	O&K 50 PS	1898	1906	new	(12)
SPEER	0-4-0T	Bn2t	O&K 50 PS	1899	1906	new	(12)
SÄNTIS	0-6-0T	Cn2t	O&K 140 PS	1935	1906	new	s/s
	0-4-0T	Bn2t	Jung 40 PS	98	1891	(o)	(15)

(b) ex Müller & Stähle, Weinheim [D] (810 mm).
(c) originally 700 mm; ex construction of the Albula line /1904.
(e) new to Hüttengesellschaft Novéant, Moyeuvre [F] "5" (700 mm).
(o) ex RhB construction /1904.
(1) 1909 to Joh. Rüesch, St. Gallen.
(3) 1906-18 to Fietz & Leuthold AG, Zürich.
(4) 1906-09 to Internationale Rheinregulierung (IRR), Rorschach.
(6) 1909 to Robert Aebi & Cie, Zürich.
(7) 19xx to Vereinigte Dampfziegeleien und Industrie AG, Halle/Saale [D].
(12) 1908 to Regiebau Bodensee-Toggenburg, Lichtensteig.
(15) 1909 to Internationale Rheinregulierung (IRR), Rorschach.

HANS ROTH, Haag, Gams SG54 C

Gauge: 600 mm Locn: ? Date: 03.2008

| | 0-4-0DM | Bdm | Deutz OME117 F | 11008 | 1933 | (a) |

(a) new to Robert Aebi & Cie. AG, Zürich: ex ? /19xx.

HANS RÜESCH ING, St. Gallen SG55 C

formerly: Johannes Rüesch (till 1929)

A building concern managed first by father (Johannes Rüesch) and then son (Hans Rüesch) till closure in 1954. Address: Krügerstrasse 1, 9000 St. Gallen.

Gauge: 750 mm Locn: ? Date: 02.2002

ENGADIN	0-4-0T	Bn2t	Krauss-S IV zs	4589	1901	(a)	(1)
ST. GALLEN	0-4-0T	Bn2t	Freudenstein	229	1905	(b)	s/s
TOGGENBURG	0-4-0T	Bn2t	O&K 50 PS	1897	1906	(b)	(3)

Gauge: 600 mm Locn: ? Date: 02.2002

	4wDM	Bdm	O&K(M) RL 1	3012	1928	(d)	(4)
	4wDM	Bdm	O&K(M) LD 2	8191	1937	(e)	(5)

(a) ex Robert Aebi & Cie, Zürich /1912.
(b) ex RTU Rickentunnel Unternehmung AG, Wattwil /1909.
(d) may have come from Oskar Reutimann, Guntalingen.
(e) via O&K/MBA, Dübendorf.
(1) 1912-13 to Favetto, Bosshard, Steiner & Cie, Brienz BE.
(3) either 1913 to Robert Aebi & Cie AG, Zürich; or 1913-28 to Steiner & Cie, Ingenieurbureau, Zürich.
(4) may have gone to Oskar Reutimann, Guntalingen.
(5) to Rathgeber, Zürich (source data gives Rathgeb).

A. SAXER, St. Gallen SG56 C

A construction concern.

Gauge: 600 mm Locn: ? Date: 02.2002

| | 0-4-0T | Bn2t | Krauss-S XIV u | 2615 | 1892 | (a) | (1) |

(a) new to H. Lehmann, Guben [D]; acquired /1929.
(1) 1931 removed from boiler register.

SCHLÄPFER ALTMETALL AG, St. Gallen-Winkeln SG57

formerly: Robert Schläpfer AG, St. Gallen

A metal recovery firm rail-connected to the west of Winkeln station. Adddress: Martinsbruggstrasse 111a, 9016 St. Gallen.

Gauge: 1435 mm Locn: 227 $ 740000,252150 Date: 10.2008

| | 4wDM R/R | Bdm | Mercedes/Zagro Unimog | | ? | 06/1984 |

GEBHARD SCHMITTER AG, Widnau SG58 C

A road building firm that changed to the recovery business in 2003.

Gauge: 750 mm Locn: ? Date: 07.1999

| | 4wDM | Bdm | RACO 50 PS | 1313 | 1945 | (a) | (1) |

(a) probably new to Entreprises Électriques Fribourgeoises (EEF), Rossens FR; later with Conrad Zschokke, Näfels; ex SGRK /1972.

(1)   1999 to Nutzfahrzeug Altstätten; for Nuevo Arenal Tilaran [Costa Rica].

SIGRIST-MERZ & CIE, St. Gallen — SG59 C
A public works company.

Gauge: 600 mm   Locn: ?   Date: 05.2002
                4wDM   Bdm   O&K(M) RL 1a   5644   1934   (a)   (1)
(a)   via O&K/MBA, Dübendorf.
(1)   1956 to König, Oberriet.

SIGRIST-MERZ & GRÜEBLER, St. Gallen — SG60 C
A public works concern. The last name might correctly be Gräbler and the relation to the similarly named company also from St. Gallen has not been established.

Gauge: 600 mm   Locn: ?   Date: 08.2002
                4wBE   Ba   SSW EL 9   5470   1953   (a)   s/s
                4wDM   Bdm   O&K MV0   25043   1951   (b)   s/s
(a)   via MBA, Dübendorf.
(b)   new via MBA, Dübendorf for the "Reuss" project at Wasen im Emmental (600 mm).

WERNER SOLENTHALER ALTMETALLE, St. Margrethen — SG61
A metal recovery firm.

Gauge: 1435 mm   Locn: 218 $ 7665/2582   Date: 07.1992
                0-4-0DM Bdm   SLM Tm   3608   1936   (a)   (1)
(a)   rebuild of SLM 2853 of 1932; ex SAIS, Arbon "2" /1986.
(1)   1991 scrapped without having been used.

SOLENTHALER RECYCLING AG, Gossau SG — SG62
formerly: Walter Solenthaler, Winkeln AG (SOW AG)
            Ammann & Co AG

A material recovery firm with locations in St. Gallen and Gossau SG. This one is rail connected between Gossau SG and Arnegg. Address: Moosburgstrasse, 9200 Gossau SG.

Many SBB Tm^{II} that have entered industrial service. This one - Tm^{II} 643 (RACO 1520 of 1958) was seen inside the company's warehouse on 14 October 2008.

Gauge: 1435 mm				Locn: 227 $ 735950,253550				Date: 10.2008	
		4wDM	Bdm	RACO 80 SA3	1397	1950	(a)	(1)	
Tm^{II}	643	4wDM	Bdm	RACO 95 SA3 RS	1520	1958	(b)		

(a) ex Saurer, Arbon /1982.
(b) withdrawn 31/8/2006; ex SBB /2006.
(1) 2006 wdn; 2007 to Schweizerische Militär Museum (SMM), Full, Full-Reuenthal.

ST. GALLER ZOLLFREILAGER UND LAGERHAUS AG, St. Margrethen   SG63

later: SFL AG
A duty free area and warehouses: Address: Grenzstrasse 24a.

Gauge: 1435 mm			Locn: 218 $ 7659/2582			Date: 03.1990	
	4wPM	Bbm	Breuer V	3054	1952	(a)	(1)

(a) ex Holzindustrie AG (HIAG), St. Margrethen /1987.
(1) 1990 to VVT, St. Sulpice NE.

ST. GALLISCHE RHEINKORREKTION (SGRK), Rorschach   SG64

Widnau

The Swiss base for work in connection with the maintenance of the course of R. Rhein on its way into the Bodensee was located here. From 1892, this work was performed in conjunction with the Vorarlberg [A] authorities under a joint state agreement (see Internationale Rheinregulierung (IRR)). At least seven distinct phases in the requirements over the last 120 years can be distinguished; some locomotives passed between SGRK and IRR and vice versa in connection with these new requirements. The SGRK workshops were located at Widnau and other depots from the IRR scheme were used as required. A sister organisation existed on the other bank of the river, based at Lustenau in Austria and trains of stone from the quarry at Koblach in Austria had to cross the R. Rhein and pass through Switzerland to reach destinations in Austria.

Gauge: 750 mm   Locn: 228 $ 7667/2522   Date: 06.2002

	0-4-0T	Bt	unknown	?	18xx	(a)	c1923 s/s
	0-4-0T	Bt	unknown	?	18xx	(a)	c1923 s/s
MONTLINGEN	0-4-0T	Bn2t	Heilbronn II	69	1878	(c)	(3)
'KRAUSS'	0-4-0T	Bn2t	Krauss-S IV o	430	1873	new	(4)
'KRIESSERN'	0-4-0T	Bn2t	Krauss-S IV w	1292	1883	(e)	s/s
FAXE	0-4-0T	Bt	Karlsruhe	284	1866	(f)	c1923 s/s
	0-4-0T	Bn2t	O&K 60 PS	4281	1911	(g)	(7)
CHURFIRSTEN	0-6-0DH	Cdh	Gmeinder HF 130 C	4005	1943	(h)	(8)
FALKNIS	0-6-0DH	Cdh	Gmeinder HF 130 C	4233	1946	(i)	(9)
PIZOL	0-6-0DH	Cdh	Gmeinder HF 130 C	4234	1946	(i)	(10)
KAMOR	4wDM	Bdm	KHD A4M517 G	47169	1951	(k)	(11)
ALVIER	4wDH	Bdh	KHD A4M517 G	55442	1952	(l)	(12)
GONZEN	4wDH	Bdh	KHD A6M517 G	56209	1955	(k)	(11)

Gauge: 500 mm (or 600 mm)   Locn: ?   Date: 06.2002

'KLEINE MAFFEI'	0-4-0T	Bn2t	Maffei	3505	1909	(n)	(14)

(a) ex ? c/1894 (existence not confirmed by all sources).
(c) new to Rommel & Schoch, Stuttgart, Stuttgart-Heslach [D] (800 mm); ex Kettner Bauunternehmer, Muri/Rotkreuz /1883 (see Kettner AG, Muri AG).
(e) new to Ing. Wey, Buchs SG for St. Gallische Rhein-Korrektion.
(f) ex Faxe Jernbane [DK] (785 mm) /1882.
(g) ex Eisenbergwerk Gonzen, Sargans /1966?
(h) new to Heeresfeldbahn [D] "M 13808"; ex P. Schlatter, Münchwilen AG /1968.
(i) new to Glasser & Pflaum, Mannheim (D); ex P. Schlatter, Münchwilen AG /1968.
(j) originally Heilbronn 228 of 1886 "WIDNAU"; ex IRR.
(k) ex Eisenbergwerk Gonzen, Sargans /1966.
(l) hire from Robert Aebi & Co, Zürich for Diepoldsauer Durchstrich /1936.
(n) ex RUBAG, Zürich /1922 (500 mm).

(3)   1918 to Aargauische Torfgesellschaft, Müri AG; 1922 to RUBAG.
(4)   c1909 to IRR; 1913 to Jura-Cement-Fabriken (JCF).
(7)   1942 to Schafir & Mugglin, Liestal.
(8)   1976 to P. Schlatter, Münchwilen AG; 1991 to Sächsisches Schmalspurbahn-Museum Rittersgrün (D).
(9)   1976 to P. Schlatter, Münchwilen AG; 1991 to privately ownership (first in Oberrittersgrün, later on the island of Rügen [D]).
(10) 1976 to P. Schlatter, Münchwilen AG; 1991 to IG Preßnitztalbahn, Jöhstadt [D].
(11) 197x to P. Schlatter, Münchwilen AG.
(12) 197x to Pitsch, Thusis.
(14) 194x hired for work in connection with the building of the Rupperswil-Auenstein hydroelectric power station; 1958 withdrawn. The boiler was still at Widnau 5/1976.

SCHWEIZERISCHE BUNDESBAHNEN (SBB)   SG65 M

Bahnhof, Buchs SG

Gauge: 1435 mm   Locn: 237 $ 7546/2261   Date: 06.2007

8487	0-6-0WT	Cn2t	SLM E 3/3	1967	1909	(a)	display

Depot, Sargans

The depot is no longer used for operational purposes but, among others, SBB Bau Management, Zürich is present.

Gauge: 1435 mm			Locn: 237 $ 7529/2120				Date: 01.2008	
Tm^{II} 631		4wDM	Bdm	RACO 95 SA3 RS	1489	1957	(b)	display
(a)	ex SBB "8487" /1965.							
(b)	ex SBB "Tm^{II} 631" /2007.							

SILO AG, Wil SG   SG66

This grain storage silo lies to the west of the SBB station. Address: Silostrasse 6, 9500 Wil.

Gauge: 1435 mm			Locn: 216 $ 7207/2580			Date: 12.2007	
	4wDM	Bdm	RACO 85 LA7	1641	1962	(a)	s/s?
	4wDM	Bdm	RACO 85 LA7	1608	1961	(b)	

(a)   new to SBB "Tm^I 340"; ex SBB "Tm^I 440" /2004.
(b)   new to SBB "Tm^I 328; later 428"; ex COOP Neuchâtel-Jura, La Chaux-de-Fonds /2006.

STADLER ALTENRHEIN AG, Staad SG   SG67

formerly: Schindler Waggon (SWP) (1991-97)
            Schindler Waggon Altenrhein (SWA) (1987-91)
            Flug- und Fahrzeugwerk Altenrhein (FFA) (194x-87)
            Dornier Flugzeuge AG (1928-4x)

This factory originally made aeroplanes and later railway vehicles. Altenrhein lies just outside Staad, north along the shore of the Bodensee. Address: Dorfstrasse 1, 9423 Altenrhein.

Gauge: 1435 mm			Locn: 217 $ 7588/2618			Date: 07.1999
Tm 239 929-5	4wDH	Bdh	Jung RK 11 B	14038	1969	(a)

(a)   ex Hille & Müller, Düsseldorf-Reisholz [D] "1" /1988 via OR/LSB.
Address: Bischofszellerstrasse 90, 9201 Gossau.

STADTWERKE, Gossau SG   SG68 M

Address: Bischofszellerstrasse 90, 9201 Gossau.

After closure of Kieswerk Espel (gravel works) two Austro-Daimler locomotives have been retained in the town. This one has not been adapted for use in a playground, but has been placed outside the town offices and workshops. It appears to be Austro-Daimler 541.

Gauge: 600 mm		Locn: 217 $ 736050,253970				Date: 10.2008
	4wPM	Bbm	Austro-Daimler PS 6	541	19xx	(a) display
(a) ex Kieswerk Espel, Gossau SG /19xx.						

EUGEN STEINMANN, St. Gallen SG69

later: Osterwalder St. Gallen AG
A dealer in timber and coal (Kohle- und Holzhandlung) rail connected in 1905.

Gauge: 1435 mm		Locn: 227 $ 7447/2532				Date: 03.1987
	A1BE	A1a	unknown,MFO	?	1916	(a) s/s
(a) The builder who fitted MFO electrical equipment into this locomotive is not known.						

ANDREAS STIHL AG & CO., Wil SG70

A branch of a chain saw and other equipment manufacturer. Address: Hubstrasse 100, 9500 Wil.

Gauge: 1435 mm		Locn: 216 $ 7199/2579				Date: 12.2008
	4wDM	Bdm	RACO 85 LA7	1732	1965	(a)
(a) ex SBB "Tm^1 505" /2008 (withdrawn 9/2008).						

TORFWERK NEUMEIER (?), Oberriet SG71

This peat works never had locomotives, only animal traction.

Gauge: 600 mm   Locn: 228 $ 76xx/24xx   Date: 05.1988

UNKNOWN CONCERN, Altstätten SG   SG72 C

A locomotive was used in the construction of the railway from Altstätten to Gais between 1909 and 1911.

Gauge: 750 mm (gauge not confirmed) Locn: 227 $ 759x/248x   Date: 02.2002

                  0-6-0T   Cn2t   Maffei   3508   1909   (a)   (1)

(a)   see note for this locomotive under Robert Aebi & Cie AG, Regensdorf, new as 750 mm; hired from Robert Aebi & Cie AG, Zürich /1909.
(1)   c1911 returned to Robert Aebi & Cie AG, Zürich.

VANOLI AG, Wattwil   SG73 C

As part of a scheme to upgrade the Wil-Nesslau line and to operate it remotely from St. Gallen, Wattwil station is to be rebuilt.

Gauge: 1435 mm   Locn: ?   Date: 05.2009

Tm 237 959-2   4wDM   Bdm   RACO   184x   1980   (a)

(a)   new to BLS "Tm 85"; on hire from Stauffer /2009.

VERSUCHSSTOLLEN HAGERBACH AG, Flums Hochwiese SG74 M

This is a research and testing laboratory. Public tours are run using a railway system. Address: Polistrasse 1, 8893 Flums Hochwiese. This location is nearer Mels than Flums.

Gauge: 600 mm   Locn: 237 $ 7482/2158   Date: 07.2008

                  4wBE   Ba   unknown   ?   xxxx

Gauge: 500 mm   Locn: 227 $ 7482/2158   Date: 05.2009

                  4wPE   Bbe   Deutz MLH222 G   8402   1928   (b)   (2)

(b)   ex Eisenbergwerk Gonzen AG (EGAG), Sargans /19xx.
(2)   on display till 2008; 2009 disposal uncertain.

VISCOSUISSE WIDNAU AG, Widnau   SG77

formerly: Setila AG (till Autumn 2005)
            Rhône-Poulenc Setila AG (from before 1996 till before 1999)
            Rhône-Poulenc Viscosuisse SA (from 1993)
            Viscosuisse SA (till 1993)
            Société de la Viscose Suisse, Widnau (from 1924)

The artificial silk manufacturers, Société de la Viscose Suisse of Emmenbrücke, set up a new branch here in 1924. Several changes of owner after 1993 attempted to keep it running but it finally closed under Setila AG in the autumn of 2005. A new start was made on 26/7/2007. The rail access is formed with a branch across fields from Heerbrugg SBB station nearly a kilometre away.

Gauge: 1435 mm   Locn: 228 $ 7665/2537   Date: 12.2002

Name		Type	Class	Builder	Works No.	Year	Notes
'BÜGELEISEN' 3	A1BE	A1a	Oehler,BBC	?	1912	(a)	1969 scr
'TRAM'	4wBE	Ba2	SIG,MFO	?	1924	new?	1969 scr
E 2/2 2 'OSTERHASE'	0-4-0F	Bf2	Jung Flink	11795	1953	new	(3)
RIBEL 3	0-6-0DH	Cdh	Henschel DH 360C	30704	1963	(d)	

(a)   ex Viscosuisse SA, Emmenbrücke /1941.
(d)   ex hire locomotive /1969 via Asper.
(3)   1993-4 sold to private ownership (to be confirmed).

VOGEL & FREI, Widnau   SG78 C

Probably a public works concern.

Gauge: 750 mm   Locn: ?   Date: 04.2002

	0-4-0T	Bn2t	Krauss-S IV kl	2960	1893	(a)	(1)

(a) new via E. Ritter-Egger, Stäfa.
(1) 1893-97 to Galli & Cie of Meggen; but used on SGRK Fussacher Durchstich.

OLIVER WEDER, Diepoldsau   SG79 M

A collector of machinery, including locomotives from building sites. The railway collection was closed after c2003. Other locomotives have been associated in the literature with this collection, but details are not confirmed. As the owner was also a member of the Feld- und Werkbahnfreunde, Otelfingen (FWF) locomotives from the collection could also be present at times in Otelfingen.

Gauge: 600 mm   Locn: ?   Date: 07.2008

4wDM	Bdm	Diema DS28	2702	1964	(a)	(1)
4wDM	Bdm	Jung ZL 233	8457	1939	(b)	(2)
4wDM	Bdm	Jung ZL 105	8208	1938	(c)	(3)
4wDM	Bdm	Jung ZL 114	8860	1939	(d)	(4)
4wDM	Bdm	Kröhnke Lorenknecht	313	1959	(e)	(5)
0-4-0DM	Bdm	LKM Ns2f	248665	1955	(f)	(6)
4wDM	Bdm	O&K(M) RL 1	3661	1929	(g)	(7)
4wDM	Bdm	O&K MV0	25072	1951	(h)	(8)
4wDM	*Bdm*	*O&K(M) LD 16*	*4609*	*1932*	*(i)*	*s/s*

(a) new to J. Schmidheiny & Co. AG, Ziegelei, Heerbrugg, ,SG (500 mm); ex Belloli SA, Grono.
(b) new to L. Baumann, Ansbach [D]; ex Joh. Baumman, Ansbach [D] /1981; exchanged cab parts and plates with Jung 8860.
(c) ex Schinznacher Baumschulbahn (SchBB), Schinznach Dorf.
(d) ex Joh. Baummann, Ansbach [D] "3" /1980; exchanged cab parts and plates with Jung 8457.
(e) ex Torfwerk Friedrich Meiners, Gnarrenburg [D] /198x.
(f) ex ? [D] /1995.
(g) ex Baufirma Kühnis, Oberriet /19.
(h) ex Energie Ouest Suisse (EOS), Salanfe, Vernayaz (800 mm).
(i) ex Toneatti, Bilten /19xx (entry not confirmed).

(1) by 6/1999 to FWF "6 SUSI".
(2) 1997-98 donated to FWF.
(3) 1995 to B. Könen, Losheim [D], via Bernhard Schramm, Weiskirchen [D].
(4) 1995 to Bernhard Schramm, Weiskirchen [D].
(5) 2004 to Christian Felten, Kelsterbach [D].
(6) 1998 to private ownership, Zürich.
(7) 2005 to Christian Felten, Kelsterbach [D].
(8) by 6/1999 to FWF.

WIDMER AG, Grämigen   SG80 C

formerly: Erwin Widmer

This civil engineering concern is located in the centre of Grämigen and is 15 minutes by foot from Lütisburg SOB Bahnhof.

Gauge: 600 mm   Locn: 226 $ 7224/2491   Date: 05.1988

201	4wDM	Bdm	O&K(M) MD 2r	12004	1944	(a)	(1)
	4wDM	Bdm	RACO 16PS	1264	1944	(b)	(2)

(a) ex Bless & Co, Zürich "201".
(b) ex Bless & Co, Zürich "inventory number 1-9-5 (281.21.105)".

(1) to playground, Grämigen.

(2) 1994 donated to FWF, Otelfingen.

WUNDERLI AG, Rapperswil-Blumenau   SG81

A dealer in building materials.

Gauge: 1435 mm   Locn: 226 $ 7060/2313   Date: 09.1996

    4wDM   Bdm   Jung ZN 133   7863   1938   (a)

(a)   ex Eisen- & Metallwerke Bender, Kreuztal-Ferndorf (D), via Asper by /1984; originally 36-40 PS, 11.8 t, 11.5 km/h.

ZEMENT- UND KALKFABRIK UNTERTERZEN   SG82

Unterterzen

The cement works are situated on the south shore of the Walensee. Raw material has been mined or quarried on the opposite shore are brought across to the works by ship.

Gauge: 600 mm   Locn: ?   Date: 04.1999

    4wBE   Ba1   Stadler   39   1950   (a)   s/s

Steinbruch Lochezen, Walenstadt

Raw material for chalk, cement and gravel has been mined here for centuries. On closure part of the cave system was reused for an underground wartime hospital. Nowadays the caverns are part of a Geo-park tour.

Gauge: 600 mm   Locn: 237 $ 7400/2214   Date: 07.1999

4wPM	Bbm	Ruhrthaler 11/13 PS	543	1922	(b)	(1)
4wDM	Bdm	O&K(M) LD 2	8586	1937	(c)	(2)
4wDM	Bdm	O&K(M) LD 2	10678	1939	(c)	(2)

(a)   conversion of a diesel locomotive.
(b)   via Robert Aebi AG.
(c)   via O&K/MBA, Dübendorf.
(1)   198x to Montan- und Werksbahnmuseum, Graz, Graz [A].
(2)   198x-88 to Museums-Eisenbergwerk Gonzen, Sargans; later scrapped.

ZIEGELEI BRUGGWALD, St. Gallen   SG83

A Schmidheiny brickworks brought into service in 1904 and closed in 1974.

Gauge: 500 mm   Locn: 217 $ 7476/2574   Date: 02.1988

    4wDM   Bdm   O&K RL 1a   ?   19xx   (a)   (1)

(a)   probably came from Schmidheiny, Heerbrugg.
(1)   ca. 1974 withdrawn.

ZIEGELEI LÜCHINGER, Oberriet   SG84

A brickworks railway operated 1928-69. The gauge has been quoted as 500 mm: this appears to be an error.

Gauge: 600 mm		Locn: 228 $ 7602/2435				Date: 07.2008	
	4wDM	Bdm	O&K(M) RL 1	3645	1929	(a)	(1)
	4wDM	Bdm	O&K MV0	25175	1952	(b)	(2)

(a) owned by Kühnis & Lüchinger, Oberriet.
(b) ex Gips Union, Zürich /19xx (identity to be confirmed).

(1) 19xx to Kühnis, Oberriet SG.
(2) 19xx either s/s or via Schmidheiny to Oliver Weder, Diepoldsau (not confirmed).

ZÜRCHER ZIEGELEIEN AG   SG85

formerly: Schmidheiny & Cie.
alternative: Ziegelei Schmidheiny

Altstätten

A railway was used to transport clay from several clay pits to a transhipment point on the SBB for onward transport to Heerbrugg.

Gauge: ? mm   Locn: 227 $ 759x/248x   Date: 02.1989

Heerbrugg

Brickworks acquired by Schmidheiny in 1870 and abandoned in 1971 after a serious fire. A railway system was used both internally at the Heerbrugg works and between the clay pits and a transhipment point for transferring clay to Heerbrugg over the SBB. The locomotives were moved to whichever system needed them for collecting clay and to other works of the Schmidheiny & Cie. as required. Use of rail ceased in the period 1966-67. Not all the locomotives can be safely assigned to a particular gauge.

Gauge: 600 mm		Locn: 228 $ 7647/2532				Date: 08.2002	
	4wDM	Bdm	O&K(M) RL 1	3907	1929	(a)	s/s
	4wDM	Bdm	O&K(M) RL 1a	4023	1930	(a)	s/s
	4wDM	Bdm	O&K(M) LD 2	4883	1933	(a)	s/s
	4wDM	Bdm	O&K MV0	25042	1951	(d)	(4)

Gauge: 500 mm		Locn: 228 $ 7647/2532				Date: 08.2002	
	4wDM	Bdm	RACO	?	19xx		
rebuilt	4wBE	Ba	Stadler/BBC	?	1944		s/s
	4wPM	Bpm	O&K(M) S10	1891	1924	(g)	s/s
	4wPM	Bpm	O&K(M) H 1	2182	192x	(g)	s/s
	4wDM	Bdm	O&K(M) H 1	2899	1928	(a)	s/s
	4wDM	Bdm	O&K RL 1	3661	1929	(j)	(10)
6	4wDM	Bdm	O&K RL 1a	4291	1930	new	(4)
	4wDM	Bdm	O&K MV1a	25351	1955	(a)	s/s
	0-4-0DM	Bdm	KHD A2L514 F	56281	1956	(m)	(13)

Montlingen, Oberriet

The railway system carried clay from the "Hilpert" clay pits to a loading point alongside the SBB for onward transport to Heerbrugg.

Gauge: 600 mm   Locn: 228 $ 7611/2447   Date: 02.1988

Oberriet

Acquired by Ziegelei Schmidheiny in 1925 and abandoned by 1974. The railway conveyed clay between the clay pits, the brickworks and the SBB. The line crossed that of the Lüchinger brickworks.

Gauge: 600 mm		Locn: 228 $ 7600/2443				Date: 02.1989	
	4wDM	Bdm	RACO		?	1923	s/s

4wDM	Bdm	O&K MV0	?	c1950	(o)	s/s
4wDM	Bdm	O&K RL 1a	?	19xx		s/s
4wDM	Bdm	Kromag	?	19xx	(q)	s/s

(a) via MBA, Dübendorf.
(d) via Wander-Wendel.
(g) gauge not confirmed, via MBA, Dübendorf.
(j) via O&K, Zürich.
(m) new via HWZ.
(o) in use from 1950 till ca. 1968.
(q) ex Wehrmacht, used a gas generator.
(4) 19xx to L. Gantenbein & Co, Werdenberg.
(10) 19xx to Baufirma Kühnis, Oberriet SG (600 mm).
(13) 1990 to Niedersächsische Bergwerks- & Hüttenschau, Langelsheim-Lautenthal [D].

Schaffhausen

SH

AEBLI, ROSSI & KRIEGER BAUUNTERNEHMUNG, Schaffhausen
SH01 C

A public works concern.

Gauge: 750 mm Locn: ? Date: 01.2002

0-4-0T	Bn2t	Heilbronn II	274	1891	(a)	(1)
0-4-0T	Bn2t	Krauss-S IV fg	2811	1893	new	(2)

(a) 1894 ex Angelo Castelli, Rheinfelden AG "ANGELO"
(1) before 1902 to Minder & Galli, Seftigen BE.
(2) to J. Wampfler SA, Bienne BE.

BAUVERWALTUNG, Neuhausen am Rheinfall
SH02

What the involvement the local government building department had with the locomotive is not known.

Gauge: 600 mm Locn: ? Date: 05.2009

4wDM	Bdm	Schöma CDL 10	1533	1954	(a)	s/s

(a) new via Hans F. Würgler, Zürich.

CLUB Bm 22-70, Schaffhausen
SH03 M

An artist's group that make "art" from scrap materials. Club address: Rikonerstrasse 17, 8310 Grafstal ZH.

Gauge: 1435 mm Locn: ? Date: 05.2009

Tem^{III} 334

0-4-0W+DEBw2	SLM,SAAS Tem^{III}		4200, ?	1956	(a)

(a) new to SBB "Tem^{III} 34"; ex SBB-Cargo "Tm^{III} 334" by /2009.

GEORG FISCHER AG (+GF+), Schaffhausen
SH04

Foundries and ironworks first established in 1802

Werk Herblingertal

A cast iron foundry and a plant to construct foundries both set up in 1969. The foundry closed in 1989.

Gauge: 1435 mm Locn: 206 $ 6904/2849 Date: 02.1995

1	4wDE	Bde1	Moyse BN30E210B	1212	1971	new	(1)

(1) 1994 to Asper.

Werke Mühlethal (I, II, III, IV)

The factories were built between 1802 and 1906 and the railway network installed in 1913 by using the existing metre gauge street tramway line for linking the various factories. The network was reduced and de-electrified in 1980, by which time the use of the tramlines for public transport had long ceased. It was finally abandoned when the steelworks ceased operation in 1993.

In 1929 the rolling stock included 4 electric locomotives, 16 transporter wagons, 15 goods wagons, 1 bogie tank, 1 bogie flat and 74 tipper wagons.

Two other locomotives probably "82 and 83" may also have come from the Compagnie Genevoise des Tramways Électriques (CGTE), but that is not confirmed.

Gauge: 1000 mm   Locn: 205 $ 6895/2854   Date: 10.1998

	75	4w+4wWE	BBg4	SLM,MFO Ge 4/4	2327, ?	1913	new	(1)	
	76	4w+4wWE	BBg4	SLM,MFO Ge 4/4	2328, ?	1913	new	(2)	
	77	4wWE	Bg2	SIG,MFO	?	1915	new	1968 scr	
	78	4wWE	Bg2	SIG,MFO	?	1921	new	(4)	
	79	0-6-0T	Cn2t	SLM G 3/3	635	1890	(e)	1921 sold	
	80	0-6-0T	Cn2t	SLM G 3/3	474	1887	(f)	19xx sold	
	81	0-6-0T	Cn2t	SLM G 3/3	624	1890	(g)	b1925 s/s	
	GEORG	4wDH	Bdh	Schöma CFL 200DCL	3325	1972	(h)	(8)	

(e) new to Birsigtalbahn (BTB) "G 3/3 4"; ex Frauenfeld-Wil (FW) "G 3/3 5" /1917.
(f) ex Birsigtalbahn (BTB) "G 3/3 2" /1913.
(g) ex Compagnie Genevoise des Tramways Électriques (CGTE) "G 3/3 15" /1913.
(h) ex construction of the Arlberg tunnel [A] /1980.
(1) 1980 donated to Blonay-Chamby (BC).
(2) 1980 donated to Blonay-Chamby (BC); 1981 scrapped.
(4) 1980 donated to Blonay-Chamby (BC); 1980-81 to M. Léchelle, Rouillac, Charentes [F].
(8) 1993 OOU; 1998 to Appenzeller Bahnen "98".

HÖHENER TRANSPORT AG, Ramsen   SH05

A transport firm with a locomotive stored. Address: Petersburg 341, 8262 Ramsen

Gauge: 1435 mm   Locn: 206 $ 703800,285000   Date: 02.2009

4wDM	Bdm	RACO 95 SA3 RS	1707	1964	(a)

(a) ex SBB "Tm^{II} 801" /2009; supplied to Arnold Schmid Recycling AG, Schaffhausen.

KIESWERK, Neuhausen am Rheinfall   SH06

Gravel and sand pit. Address: Birchstrasse 8, 8212 Neuhausen am Rheinfall.

Gauge: 600 mm   Locn: 205 $ 6870/2815   Date: 09.2007

4wDM	Bdm	unknown	?	1xxx	s/s

JOSEPH KODE, Stein am Rhein   SH07 C

Probably a public works concern.

Gauge: 720 mm   Locn: ?   Date: 01.2002

0-4-0T	Bn2t	Heilbronn II	42	1875	new	(1)

(1) to Ruttimann & Riedlinger, Pfullendorf [D] /1878.

SCHWEIZERISCHE INDUSTRIE-GESELLSCHAFT (SIG),
Neuhausen am Rheinfall   SH08

A factory for making railway rolling stock was set up in 1853 and rail connected to Neuhausen am Rheinfall station from 1897. The siding was about 1500 m long and was electrified from 1897-1986.

Gauge: 1435 mm 250V AC   Locn: 205 $ 6887/2814   Date: 08.1992

	xxWE	xgx	SIG,MFO		?	1897	(a)
rebuilt	xxBE	xax	SIG,MFO			(b)	(2)
1	4wWE	Bg2	SIG,MFO		?	1920	(c) (3)

(a) new, it is not clear if one or two axles were driven.

(b)	rebuild of previous locomotive; it probably never worked here in this form.				
(c)	new; a date of 1919 is also quoted.				
(2)	1920 to Holzlager und Sägewerk (SIG), Rafz.				
(3)	1986 OOU; preserved in a hall of an old +GF+ factory in Schaffhausen; 1998 display outside SIGG Modell, Lokdepot, Winterthur.				

VEREIN ZUR ERHALTUNG DER DAMPFLOK MUNI (VDM), Ramsen
SH09 M

A group founded in 2001 and located on the closed Etzwilen-Singen (Hohentwiel) [D] international railway. Stock is kept at Hemishofen and Etzwilen. Address: Postfach 14, 8262 Ramsen.

Gauge: 1435 mm Locn: 206 $ 703730,279920 Date: 06.2008

1 MUNI	0-6-0T	Cn2t	Hohenzollern Leverkusen	4267	1922	(a)
Tm 2/2 2 Grizzly	4wDH	Bdh	KHD A8L614 R	56900	1958	(b)
Tm 92 GIRAFF	4wDM	Bdm	RACO 100 DA4 H	1818	1975	(c)
2	0-6-0WT	Cn2t	SLM E 3/3	1220	1899	(d)
RM/EBT 80 62 97 <u>12</u> 100-6						
			ex 2BoDMRC			
			ex 2Bdm SIG,MFO (,SAAS)		?1930(37,61)	(e)

(a) new to Badische Anilin und Sodafabrik (BASF), Ludwigshafen [D]; fitted with boiler Hohenzollern 4268 (BASF "60") /19xx; to Rheinische Braunkohlenwerke (RBW) "6" (later "327") 5/10/1959; ex private ownership (Herr Rüesch) 27/4/2002; fitted with a plate reading "Hohenzollern 4268 1922". .
(b) ex Bülachguss AG, Bülach /2005.
(c) new to SZU; purchased from DVZO /2008.
(d) ex Oensingen Balsthal Bahn, Balsthal /2008.
(e) new as "Fm 2/4 18601"; "Fm 891" /1947; "Dm 2/4 1692" /1961; Sulzer diesel motor; SAAS parts on rebuilding; withdrawn 31/12/1971; used as a coach by EBT as "X2 80 62 97-<u>12</u> 100"; ex EBT /2000.

see: www.muni-dampflok.ch

VEREIN ZUR ERHALTUNG DER EISENBAHNLINIE ETZWILEN-SINGEN (VES), Ramsen
SH10 M

At the start of 2007 authority was granted to run museum trains on Swiss section of this closed international line, including the bridge over the R. Rhein.

WILLIAM COOK RAIL GMBH, Schaffhausen
SH11 M

formerly: Urs Rüesch

Initially owned by a collector and engineer the locomotive is now kept in the old SBB depot at Schaffhausen, together with the locomotive from DLM. It is intended that in the future the locomotive will be maintained by the VVT at St. Sulpice.

Gauge: 1435 mm Locn: 206 $ 6998/2841 Date: 09.2007

SNCF 141.R.568	2-8-2	141h2	BLW 141.R	72381	1945

Solothurn   SO

ALTOLA AG, Olten   SO01
The company is active in the materials recovery field. Address: Gösgerstrasse 154, 4600 Olten.

Gauge: 1435 mm   Locn: 224 $ 636350,246400   Date: 02.2009

Tm 237 871-9	4wDM	Bdm	O&K MV3	26596	1966	(a)	
Tm 237 920-4 Felix	4wDM	Bdm	RACO 80 SA3	1417	1953	(b)	(2)

(a)   ex MIGROS Genossenschaft Basel, Münchenstein /2003 via Stauffer.
(b)   hired from Stauffer /2008.
(2)   2009 returned to Stauffer.

BAUUNTERNEHMUNG DÜNNERNKORREKTION LOS D, Olten
SO02
One of the contracts(Los D) for grading the R. Dünnern around Olten.

Gauge: 750 mm   Locn: ?   Date: 03.2002

WEISSENSTEIN	0-4-0T	Bn2t	Krauss-S IV bh	5295	1905	(a)	(1)
	0-4-0T	Bn2t	O&K 30 PS	847	1901	(b)	(2)
	0-4-0T	Bn2t	O&K 30 PS	985	1902	(a)	(2)
TOGGENBURG	0-4-0T	Bn2t	O&K 50 PS	1897	1906	(a)	(4)

(a)   ex Steiner & Cie, Ingenieurbureau, Zürich /1936.
(b)   ex Ad. Baumann (& Stiefenhofer), Wädenswil /1936.
(1)   1936-39 to Constantin von Arx, Olten.
(2)   1938 to Mangold & Cie, Zürich.
(4)   1936 to F. Stirnimann, Olten.

BELART(?) & CIE, BAUMANN & STIEFENHOFER, Olten-Klingnau
SO03
Probably one or more civil engineering firms. The exact location of Belart & Cie has not been determined, but it seems likely that it was related to the above firm.

Gauge: 750 mm   Locn: ?   Date: 02.2002

	0-4-0T	Bn2t	O&K 80 PS	1622	1905	(a)	(1)
BERNA	0-4-0T	B?2t	O&K 70 PS	6572	1913	(b)	(2)

(a)   purchased from O&K, Zürich /1916-19.
(b)   ex Ernst & Hammann, Laufenburg /1913-19.
(1)   1919 to Schafir & Mugglin during the construction of water courses for the Klingnau hydroelectric power station.
(2)   1913-19 to Schafir & Mugglin, Muri AG.

BELART & CIE, Unknown location   SO04
A public works firm?

Gauge: 750 mm   Locn: ?   Date: 05.2002

BERNA	0-4-0T	B?2t	O&K 70 PS	6572	1913	(a)	(1)

(a)   ex Ernst Hammann, Laufenburg before /1919.

(1)   1919 to Schafir & Mugglin, Muri bei Bern.

JULIUS BERGER, Olten   SO05

A concern chiefly involved in boring of the southern end of the Hauenstein tunnel, though there appears to have been some involvement with the northern end as well, see Tecknau.

Gauge: 750 mm (unless specified)   Locn: ?   Date: 02.2002

	0-4-0T	Bn2t	Hanomag	4767	1907	(a)	(1)
	0-4-0T	Bn2t	Hanomag	4925	1907	(b)	s/s
	0-4-0T	Bn2t	Freudenstein	157	1904	(c)	s/s
	0-4-0T	Bn2t	Freudenstein	205	1905	(c)	s/s
	0-4-0T	Bn2t	Freudenstein	211	1905	(c)	s/s
	0-4-0T	Bn2t	Henschel	6612	1905	(f)	s/s
	0-4-0T	Bn2t	Krauss-S IV cd	2648	1893	(g)	s/s
	0-4-0T	Bn2t	O&K 50 PS	2951	1908	(f)	s/s
	0-4-0T	Bn2t	Borsig	6850	1909	(i)	s/s
	0-4-0CA2cc	Bp2v	Borsig DrL	8376	1912	(j)	s/s
	0-4-0CA2cc	Bp2v	Borsig DrL	8377	1912	(j)	s/s
	0-4-0CA2cc	Bp2v	Borsig DrL	8378	1912	(j)	s/s
	0-8-0CA2cc	Dp2v	Borsig DrL	8379	1912	(j)	s/s
	0-8-0CA2cc	Dp2v	Borsig DrL	8380	1912	(j)	s/s
REHAG	0-4-0T	Bn2t	SLM G 2/2	181	1880	(o)	s/s

(a)   new to Julius Berger, Bromberg [D]; imported from Germany /1912 (600 mm).
(b)   imported from Germany /1912, possibly for the northern end of the tunnel; (gauge not specified).
(c)   imported from Germany /1913 (?750 mm).
(f)   imported from Germany /1912.
(g)   a locomotive with works number 2648 was imported from Germany in 1912 and could have been of 600 mm gauge. The builder of Krauss and the details above are not confirmed. The locomotive shown corresponds to a locomotive delivered new to F. Lusser & Co, Baar.
(i)   new to Julius Berger, Bromberg [D]; to Olten for this project /1912 (gauge not specified).
(j)   ordered by Julius Berger AG Berlin [D] for this project.
(o)   ex Waldenburgerbahn /1913.
(1)   19xx returned to Germany; 1964 reported at VEB Grube "Glückauf", Olbersdorf/Kr. Zittau (DDR) (750 mm).

TH. BERTSCHINGER, Olten-Gösgen   SO06

Work for the hydroelectric scheme there.

Gauge: 600 mm   Locn: ?   Date: 02.2008

	0-4-0T	Bn2t	O&K 30 PS	2213	1906	(a)	(1)

(a)   made available by Th. Bertschinger, Lenzburg /1932.
(1)   19xx returned to Th. Bertschinger, Lenzburg.

BORNER AG, Trimbach   SO07

alternative: Borner AG, Transporte, Olten
               Gebr. E. + R. Borner AG

A firm offering warehousing and transport services from a site within the Olten-Trimbach industrial estate and on the other bank of the river to the SBB establishments. Address: Industriestrasse 11, 4632 Trimbach.

				Gauge: 1435 mm	Locn: 224 $ 635625,246100				Date: 06.2008
Tm¹			4wDM	Bdm	Kronenberg DT 90	145	1962	(a)	(1)
			4wDM	Bdm	RACO 85 LA7	1676	1963	(b)	

(a) ex Firestone, Pratteln "1" via Giezendanner, Rothrist et Meier & Jäggi, Reiden.
(b) ex SBB "Tm¹ 483" /1994.
(1) 1995 donated to Betriebsgesellschaft Historischer Schienenfahrzeuge (BHS); 1998 to Swisstrain, Bodio; by 2007 to Le Locle.

BORREGAARD SCHWEIZ AG, Riedholz   SO08

formerly: Atisholz AG (4/2000-3/2003)
            Cellulose Attisholz AG, Riedholz (from ?2/1996)
            Cellulose Attisholz AG, Attisholz (till ?2/1996)
            Cellulosefabrik Attisholz AG (CFA)
            D. Sieber, Luterbach

A cellulose plant created in 1882 and from 1890 rail connected to Luterbach-Attisholz station. Attisholz lies in Gemeinde Riedholz. All these names and the version of Attisholz with only one "t" can be found in the literature referring to this one site. The locomotive depot dates from 1917. There is an extensive internal railway system including a bridge over the R. Aare. The locomotive running numbers are usually made from the last two digits of the year in which they were acquired. The product range has been extended to include other chemicals. The plant closed on 4 November 2008. Address: Attisholzstrasse 10, 4533 Riedholz.

A power parade on 29 May 1973. 26 (SLM 1332 of 1901), 71+65 Jung 13869 of 1965 and 14136 of 1972), 58 (Jung 13008 of 1958) and 54+64 (Jung 12126 of 1954 and 13874 of 1965).

					Gauge: 1435 mm		Locn: 223 $ 6108/2304				Date: 01.1991
				?	?	unknown	?	18xx	(a)	s/s	
				4wBE	Ba2	Rastatt,MFO	-,21	1910	(b)	(2)	
		16		0-4-0T	Bn2t	O&K 110 PS	8089	1916	new	(3)	
		26		0-6-0WT	Cn2t	SLM E 3/3	1332	1901	(d)	(4)	
		43		0-6-0WT	Cn2t	SLM E 3/3	1361	1901	(e)	1958 scr	
		46		0-6-0WT	Cn2t	SLM E 3/3	1457	1902	(f)	1965 scr	
		54		0-4-0F	Bf2	Jung Flink	12126	1954	(g)	(7)	
		58		0-4-0F	Bf2	Jung Freia	13008	1958	new	(7)	
		64		0-4-0F	Bf2	Jung Flink	13874	1965	new	(9)	
Tm		65		4wDH	Bdh	Jung RC 24 B	13869	1965	new		
		71		4wDH	Bdh	Jung RC 24 B	14136	1972	new		
				0-4-0F	Bf2	Jung Flink	12999	1958	(l)	1984 scr	
				4wDH	Bdh	Jung Köf II	13178	1960	(m)		
		89		4wDH	Bdh	Schöma CFL 350DCLR	4942	1987	(n)		
				4wDM	Bdm	RACO 85 LA7	1576	1960	(o)		
Tm 2/2	94			4wD R/R	Bdm	Zephir Lok 10.170	?	1994	new		
Tm 2/2	95			4wD R/R	Bdm	Zephir Lok 10.170	?	1995	new		

(a) A locomotive was sought in an announcement of 1889. Several (?) engines are mentioned in the railway statistics of 1900. But it appears that the firm did not have a locomotive before 1918.
(b) ex Nestlé and Anglo-Swiss Condensed Milk Co, Cham ca. /1934.
(d) new to Gürbetalbahn "3"; purchased from BLS "77" /1926.
(e) new to SCB "43"; purchased from SBB "8412" /1943.
(f) ordered as SCB "48"; purchased from SBB "8452" /1946.
(g) new, the Jung lists from Merte give it as a Freia and of 1955 - this data appears to be incorrect.
(l) purchased from Papierfabrik Biberist "4" /1980 as a source of spare parts.
(m) new to DB "Köf 6740"; ex DB "323 810-2" /1984, brought into service /1985; also reported as being allocated number "85".
(n) ex ARGE Freudenstein Tunnel, Oberderdingen [D] "3" /1989 via Schöma (reconditioning) and Asper.
(o) ex SBB "Tm^l 413" /1996 via Betriebsgesellschaft Historischer Schienenfahrzeuge (BHS).

(2) after 1952 to Portland Cementfabrik Laufen AG, Werk Liesberg.
(3) 1963 withdrawn; 1964 scrapped at von Roll, Gerlafingen.
(4) 1972 withdrawn; 1975 to a private individual; 1977 (also reported as 1988) BLS "3".
(7) 1992 withdrawn; 1995 scrapped.
(9) 1992 withdrawn; 1995 to Ateliers SNCF, Saintes [F] for display.

A. BUSS & CO   SO09

alternative: Albert Buss & Cie

A public works concern.

Oberdorf SO

Construction site for building of the Weissenstein tunnel on the Solothurn-Moutier railway line.

Gauge: 750 mm   Locn: ?   Date: 04.2002

WEISSENSTEIN	0-4-0T	Bn2t	Krauss-S IV bh	5295	1905	new	(1)
OBERDORF	0-4-0T	Bn2t	Krauss-S LXV n	5492	1905	new	(1)

Olten-Gösgen

Construction site for building the hydroelectric plant.

Gauge: 900 mm   Locn: ?   Date: 04.2002

0-4-0T	Bn2t	Jung 125 PS	785	1904	(c)	(3)
0-4-0T	Bn2t	O&K 125 PS	1268	1904	(d)	(3)
0-4-0T	Bn2t	O&K 140 PS	2328	1907	(e)	(3)
0-4-0T	Bn2t	O&K 140 PS	3065	1908	(f)	(3)
0-4-0T	Bn2t	O&K 140 PS	3941	1910	(g)	(3)

(c) new to M. Brenner, Magdeburg [D], hired from O&K, Zürich /1916.
(d) new to H. Reifenrath, Strassbourg branch [F], hired from O&K, Zürich /1915.
(e) new to Justus Kranz, Hemer, Westfalen [D]; hired from O&K, Zürich /1915.
(f) new to Heinrich Weber, Bauunternehmung, Unna, Westfalen [D]; hired from O&K, Zürich /1915.
(g) new to Steven Arntz, Milligen [NL]; hired from O&K, Zürich /1915.

(1) 1906-09 transferred for the building of the St. Finden-Wittenbach line.
(3) either returned to O&K, Zürich or s/s.

COOP, Olten   SO10

A distribution centre for the Coop chain of shops.

Gauge: 1435 mm   Locn: 224 $ 6351/2461   Date: 09.1995

4wDM	Bdm	Breuer III	?	1931(69)	(a) (1)

(a) new to EBT "Tm 10" as 4wPM Bbm; rebuilt /1969 and fitted with Ford diesel engine, strengthening bars and curved cab roof; withdrawn /1978; ex EBT /1979.

(1) 11/1999 donated to Betriebsgesellschaft Historischer Schienenfahrzeuge (BHS); later Swisstrain, Bodio; by 10/2007 to Bahn Museum Kerzers (BMK), Kerzers.

DIE SCHWEIZERISCHE POST   SO11
formerly: (Schweizerische) Post-, Telephon- und Telegraphenbetriebe (PTT) (1985 till 31/12/1997)

Däniken SO
A postal sorting centre; the activity was reduced in 1999 and the SBB have since performed the necessary shunting.

Gauge: 1435 mm   Locn: 224 $ 6398/2446   Date: 01.2008

7	0-6-0WE	Cw1	SLM,BBC Ee 3/3	3188,2533	1926,7	(a)	(1)
10	6wWE	Cw3	SLM,BBC Ee 3/3	5288, ?	1985	new	(2)

Härkingen
Located near Egerkingen this sorting office and centre for packet distribution opened in 1999.

Gauge: 1435 mm   Locn: 224 $ 6263/2399   Date: 06.2000

6	837 814-3	0-6-0DH	Cdh	Henschel DH 500Ca	30710	1964	(c)	(3)
17	237 917-0	4wDE	Bde1	Stadler	523	1999	new	

(a)   ex SBB "Ee 3/3 16311" /1977.
(c)   ex Die Schweizerische Post, Zürich-Mülligen /1999.

(1)   1999 donated to Swisstrain, Bodio.
(2)   2000 to Die Schweizerische Post, Bern.
(3)   by 2006 to EUROVAPOR, Sektion Balsthal, Balsthal (to be confirmed).

DISPLAYED LOCOMOTIVES, Gerlafingen   SO12
Three locomotives were brought to Gerlafingen for scrap. They were retained and placed on display in various locations. The authorities objected and the locomotives were scrapped. The Maffei locomotives were part of a fleet of similar machines and the identity of the two that came here is not completely clear. It would appear that parts may have been interchanged.

For a short while three locomotives were displayed around the town. The two Maffei locomotives had once worked on the Chantier de Barberine, Le Châtelard VS contract. A photograph of one at work is now on a wall in the museum at the power station at Le Châtelard. Here one (Maffei 4208 of 1922) is seen in the Grüttstrasse, Gerlafingen on 29 October 1982.

Gauge: 750 mm	Locn: 233 $ *see notes*: (x) to (z)	Date: 02.2002

'3'	0-4-0WT	B?2t	Maffei	4206	1921	(a, X)	(1)
'4'	0-4-0WT	B?2t	Maffei	4208	1922	(a, Y)	(1)
'ROSINA'	0-4-0WT	Bn2t	O&K 50 PS	2391	1907	(a, Z)	(1)

(a) ex J. Seeberger, Frutigen to von Roll, Gerlafingen /1952-73.

(X) playground in Wilerstrasse (6096/2243).
(Y) playground in Grüttstrasse/Nordringstrasse (6105/2250); also reported as Maffei 4203 of 1921.
(Z) outside Lehrlingsschule (6098/2247).

(1) 1982-84 scrapped.

EMMENHOF IMMOBILIEN AG, Derendingen　　　　　　　　SO13

formerly:　G. Scolari (from 1977)
　　　　　　Baumwollspinnerei Emmenhof (from 1862)

Originally a weaving factory constructed in 1862-63, and later a food plant for G. Scolari pasta. From 1884 it used a section, 1021 metres long, of the early line between Derendingen and Biberist as a private siding.

Gauge: 1435 mm　　　　　　Locn: 233 $ 6109/2271　　　　　　　　Date: 07.1988

	1ABE	1Aa1	Oehler,BBC Akku-Fahrz. ? ,1099		1917	(a)	
rebuilt	1APH	1Abh			1972-3		(1)

(a) new; rebuilt /1972-73 with petrol engine and hydrostatic drive.
(1) 1976 OOU, 1982-88 scrapped.
　　see: Bulletin du Personnel des CFF 89/71 (May).

FELDSCHLÖSSCHEN GETRÄNKE AG, Depot Gurten, Solothurn
　　　　　　　　　　　　　　　　　　　　　　　　　　　　SO14

This distribution centre has a cable operated Windhoff mule for moving wagons alongside the loading dock.

Gauge: 1435 mm　　　　　　Locn: ?　　　　　　　　　　　　Date: 03.1991

GEMEINSCHAFTSUNTERNEHMUNG BANNWART & CO,
F. RENFER, Solothurn　　　　　　　　　　　　　　　　SO15

A consortium of public works concerns.

Gauge: 600 mm　　　　　　Locn: ?　　　　　　　　　　　　Date: 05.2002

	4wDM	Bdm	O&K(M) RL 1a	4331	1930	(a)	s/s

(a) ex MBA, Dübendorf /1949.

GRUBITZ & ZIEGLER, Solothurn　　　　　　　　　　　　SO16

Probably a civil engineering firm.

Gauge: 800 mm　　　　　　Locn: ?　　　　　　　　　　　　Date: 02.2002

BERTHA	0-4-0T	Bn2t	Heilbronn II alt	30	1874	(a)	s/s
CLARA	0-4-0T	Bn2t	Heilbronn II alt	31	1874	(a)	s/s
SZIGET	0-4-0T	Bn2t	Heilbronn III	48	1875	new	(3)

Gauge: 700 mm　　　　　*Locn: ?*　　　　　　　　　　　*Date: 05.2008*

	0-4-0T	Bn2t	Heilbronn II ur	12	1870	(d)	(4)

(a) ex Wolf & Grafs, Zürich (796 mm).

(d)	ex Felmaier & Cie, Wildberg /18xx (presence here not confirmed).	
(3)	18xx sold to Castelli & Rüttimann for the construction of the RhB.	
(4)	1876 to Jardini & Co, Basel (destination not confirmed).	

J.H. HEUSSER, Riedholz — SO17 M

A private collector, this locomotive has resided at Cellulose Fabrik Attisholz.

Gauge: 1435 mm   Locn: 223 $ 6108/2304   Date: 05.2009

       0-4-0WT Bn2t Hohenzollern 2015 1911 (a) (1)

(a) ex Kreis Jülicher Zuckerfabrik [D] /1972.
(1) 2004 to Niederlausitzer Museumseisenbahn, Finsterwalde [D] via Zürcher Museumsbahn (ZMB).

HOLZKONTOR AG, Däniken SO — SO18

Gauge: 1435 mm   Locn: ?   Date: 01.1992

       4wPM Bbm Breuer ? c192x (a) s/s

(a) mentioned in the Breuer list of references of 1928 as being with Comptoir de Bois SA, Yverdon and transferred here in the list of references of about 1930.

HUNZIKER & CIE, PORTLANDZEMENTWERK AG, Olten — SO19

A cement works; was it perhaps in Olten-Hammer?

Gauge: 750 mm   Locn: ?   Date: 02.2002

	0-4-0WT	Bn2t	Borsig	6485	1908	(a)	(1)
	0-4-0T	Bn2t	Krauss-S IV o	430	1874	(b)	(2)
HELVETIA	0-4-0T	Bn2t	Krauss-S IV kl	2960	1893	(b)	(3)
ENGADIN	0-4-0T	Bn2t	Krauss-S IV zs	4589	1901	(d)	(4)
	0-4-0T	Bn2t	Krauss-S IV tu	3190	1895	(e)	(5)
	0-4-0T	Bn2t	Jung 30 PS	69	1889	(b)	(6)

(a) ex RUBAG /1929.
(b) ex Jura-Cement-Fabrik (JCF), Aarau /1932.
(d) ex ? /1929.
(e) ex ? (probably RUBAG) /1928.
(1) 1939-41 to Kies AG Bollenberg, Tuggen.
(2) to Th. Bertschinger, Lenzburg after /1932 (to be confirmed).
(3) 1939 transferred to Brugg AG; ?1953 to Bern/Fischermättli; 1953 removed from boiler register.
(4) 1949 removed from boiler register.
(5) 1932 removed from boiler register.
(6) 1938 to Kies AG Bollenberg, Tuggen.

IDEAL-STANDARD, Dulliken — SO20

formerly: Radiatorenfabrik Dulliken

A factory making radiators that was taken over by Ideal-Standard. It was closed and sold in 1971 becoming the SBB stores warehouse.

Gauge: 1435 mm   Locn: 224 $ 6379/2452   Date: 12.1988

       4wPM Bbm RACO RA7 1215 1933 new b1971 s/s

KEHRICHTBESEITIGUNG AG (KEBAG), Zuchwil — SO21

A rubbish disposal plant, rail connected midway between Solothurn and Luterbach-Attisholz stations in 1993. It has a yard containing 1.5 kilometres of track.

Gauge: 1435 mm   Locn: 233 $ 6100/2294   Date: 03 2002

4wDH	Bdh	Jung RK 12 B	11509	1956	(a)	(1)
4wDE	Bde1	Moyse BN30E210B	1212	1971	(b)	(2)
4wDH	Bdh	Schöma CFL 200DCLR	5362	1994	new	

(a) new to Hoesch Walzwerke AG, Hohenlimburg-Oege [D] "1" ; 1992 Trans-Manche Link (TML) [GB] "73"; hired from Schöma /1993.
(b) hired from Asper /1994.
(1) 1994 returned to Schöma; 1997 Betonwerk Neu-Ulm GmbH, Neu-Ulm [D].
(2) 1995 returned to Asper.

LEHMANN & VON ARX, Wangen an der Aare — SO22

Probably a civil engineering firm.

Gauge: 800 mm   Locn: ?   Date: 02.2002

0-4-0T	Bn2t	Krauss-S IV o	425	1874	(1)

(1) sold; later used by Th. Bertschinger /1901.

MEIER & JÄGGI AG, Breitenbach — SO23

A depot of a public works company based in Zofingen.

Gauge: 600 mm   Locn: ?   Date: 02.1991

0-4-0DM	Bdm	Deutz OMZ117 F	10876	1933	(a)	(1)
4wDM	Bdm	Gmeinder 20/24 PS	1557	1936	(b)	(1)

(a) new to Robert Aebi & Cie. Co. AG, Zürich; it is not known if the locomotive came directly to Meier & Jäggi or not.
(b) via Robert Aebi & Cie, Zürich.
(1) by 2006 to FWF, Otelfingen.

MIGROS VERTEILBETRIEB NEUENDORF AG, Oberbuchsiten — SO24

A depot and distribution centre of the Migros group of co-operatives. It was constructed 1974-75. By 2007 the locomotive listed is quoted as being the reserve one. Details of the operational one are not known.

Gauge: 1435 mm   Locn: 224 $ 6260/2397   Date: 03.2000

Tm 237 825 1	4wDH	Bdh	O&K MB 280 N	26791	1974	new

M-REAL, Biberist — SO26

formerly: Papierfabrik, Biberist

In the 1860s, the firm of Locher (of Zürich) built a canal from the R. Emme to the R. Aare to provide power for a chain of factories. This particular factory was built 1863-65. Along the eastern bank of the canal, a horse worked railway connected the factory with the SCB at what became the station of Derendingen in 1863. The route became part of the EB in 1872.

Gauge: 1435 mm   Locn: 233 $ 6100/2260   Date: 07.1995

BIBER	0-4-0T	Bn2t	SLM E 2/2	922	1895	new	1930 scr

2	BIBER	0-4-0T	Bn2t	SLM E 2/2	2592	1917	(b)	(2)
3	BIBER	0-6-0T	Cn2t	SLM E 3/3	1194	1899	(c)	1965 scr
	4wDM		Bdm	Kronenberg DT 50	119	1950	new	(4)
4	BIBER	0-4-0F	Bf2	Jung Flink	12999	1958	new	(5)
5		0-4-0DH	Bdh	Henschel DH 240B	30585	1963	new	(6)
1		4wDH	Bdh	MaK G 320 B	220079	1965	(g)	
2 Em 837 820-0		0-6-0DH	Cdh	Henschel DH 500Ca	31309	1968	(h)	
402		4wDH	Bdh	Henschel DH 160B	26924	1960	(i)	(9)

(b) ex Dreispitz-Verwaltung, Basel /1930.
(c) ex Rondchâtel SA, Frinvillier /1940.
(g) new to Jülicher Kreisbahn [D] "34"; ex KLB "1" /1979.
(h) ex Monteforno, Acciaierie e Laminatori SA, Giornico, Bodio /?1996.
(i) hired from Asper /1991; Henschel records indicate the locomotive as type DH240.

(2) 1958 stored; 1959 scrapped.
(4) c1962 converted to non self-propelling flat wagon for internal use; s/s?
(5) 1980 sold to Cellulosefabrik Attisholz AG, Luterbach-Attisholz for spare parts.
(6) 1991-92 sold to Asper; though remaining on site for some time afterwards.
(9) 1991 returned to Asper.

MÜLLER, ZEERLEDER, GOBAT, Olten-Gösgen   SO27

A construction company for the hydroelectric plant at Olten-Gösgen.

Gauge: 900 mm   Locn: ?   Date: 04.2002

0-4-0T	Bn2t	Henschel	6659	1904	(a)
0-4-0T	Bn2t	Henschel	6838	1904	(a)
0-4-0T	Bn2t	O&K 125 PS	1321	1904	(c)
0-4-0T	Bn2t	O&K 140 PS	4107	1910	(d)
0-4-0T	Bn2t	O&K 140 PS	4127	1911	(e)

(a) hired and then purchased from Robert Aebi & Co, Zürich /1914.
(c) hired from H. Reifenrath & J. Christ /1915.
(d) new to Gebr. Kratz, Baugeschäft, Ludwigshafen [D]; hired from O&K, Zürich /1915.
(e) new to Josef Ell, Heidelberg [D]; hired from O&K, Zürich /1915.

NÄFF & ZSCHOKKE, AARAU, Derendingen   SO28

A public works concern using the Derendingen-Biberist paper factory industrial line. This line was brought into service on 1 April 1864, initially using animal traction. In 1869-70 it was extended to Gerlafingen. In turn, it passed on 1 January 1873 to the Emmentalbahn; this took over the operation in 1875. From 1884, it reverted to being an industrial siding again.

Gauge: 1435 mm   Locn: ?   Date: 11.1993

0-4-0T	Bn2t	Krauss-M XVIII a	188	1872	new	(1)

(1) ?1873 to Emmentalbahn; 1880 wdn; 1884 to Papierfabrik Perlen, Gisikon-Root.

OENSINGEN BALSTHAL BAHN, Balsthal   SO29

alternative: Verein Interkantonales Kulturprojekt

The company owns some heritage stock and participates in a Kulturprojekt. The depot area is also used by Meier und Jägi AG, EUROVAPOR, Sektion Balsthal and SBB-Historic for their locomotives. Some items from Swisstrain are present here on loan.

Gauge: 1435 mm   Locn: 223 $ 6195/2406   Date: 03.2008

1	0-6-0T	Cn2t	Maffei	2983	1909	(a)	
2	0-6-0WT	Cn2t	SLM E 3/3	1220	1899	(b)	(2)
De 6/6 15301	0-6-6-0WE	CCw2	SLM/BBC De 6/6	3056/2140	1926		
6 837 814-3	0-6-0DH	Cdh	Henschel DH 500Ca	30710	1964	(f)	

(a)	ex von Roll, Klus "1".
(b)	ex von Roll, Gerlafingen "16".
(f)	*EUROVAPOR; ex Die Schweizerische Post, Härkingen by /2006 (to be confirmed).*
(2)	2008 to Verein zur Erhaltung der Dampflok Muni (VDM)
	see: www.kulturprojekt.ch/index.php

PATRONENFABRIK AG, Solothurn   SO30

A factory opened in 1923.

Gauge: 1435 mm   Locn: ?   Date: 05.1989

| | | 4wPM | Bbm | Breuer I, II or III | ? | 192x | (a) | s/s |

(a) mentioned in the Breuer list of references of 1928, but not the one of 1926.

PORTLANDCEMENTWERK AG, Olten-Hammer   SO31

later: PCO Olten AG, Olten

Cement works near Olten Hammer station. It was demolished in 2004. Address: Cementweg 30, 4600 Olten.

Gauge: 1435 mm   Locn: 224 $ 6344/2439   Date: 10.2007

| | 4wCE | B-el | Vollert Kabeltrommel | 81/009 | 1981 | new | s/s |

REXROTH AG, Oensingen   SO32

formerly: Hydraulik AG (1983-1994)
          von Roll (Secteur Robinetterie) (till 1983)

The hydraulics section of von Roll. The title of Rexroth AG applied from 1994. Production ceased in 1997.

Gauge: 1435 mm   Locn: 224 $ 6205/2374   Date: 06.1990

		0-6-0PM	Cbm	von Roll	-	1946	(a)	(1)
Tm	21	0-4-0DH	Bdh	SLM Tm^{III}	4218	1956	(b)	(2)
		4w-4wBEcrane			unknown	?	19xx (c)	s/s
		4wDM R/R	Bdm	unknown R/R	?	19xx	(d)	s/s

(a) ex von Roll, Klus.
(b) ex von Roll, Klus; hydrostatic transmission (Hydro-Titan); "20" till /1958.
(c) fitted with rail buffers and apparently used for moving and unloading wagons
(d) in use from /1988.
(1) after 1968 (an unconfirmed report suggests 2/1970) s/s.
(2) 1988 to E. Flückiger, Rothrist; 1989 scrapped.

SCHAFIR & MUGGLIN, Solothurn   SO33

Construction site in the grounds of the gas works seen in March 1967.

Gauge: 600 mm   Locn: 233 $ 6068/2285   Date: 02.2002

| 2.20.01 | 4wBE | Ba | SIG | 563 507 | 19xx | s/s? |

SCINTILLA AG, ZUCHWIL, Werk Derendingen — SO34

formerly: Schoeller Textilwerk (19xx till 1987)
Kammgarnspinnerei Derendingen (1872 till 19xx)

The spinning mill was opened in 1862-3 and closed in 1987. The site was then re-used for the construction of electrical equipment by a member of the Bosch group.

Gauge: 1435 mm Locn: 233 $ 6106 2280 Date: 08.1995

Ta	1	4wBE	Ba	MFO	?	1928	(a)	1986 scr
	402	4wDH	Bdh	Henschel DH 160B	26924	1960	(b)	

(a) constructed on the frame of a freight wagon.
(b) new to Gaswerk St. Gallen-Rietli "402"; ex Asper /1992; Henschel records indicate the locomotive as type DH240.

Schweizerische Bundesbahnen (SBB), Hauptwerkstatt, Olten — SO35

alternative: Hauptwerkstatt SBB
formerly: Hauptwerkstätte SCB

Various locomotives have been displayed at various places around the works and the adjacent depot. The depot is used by SBB-Historic for some of its fleet.

Gauge: 1435 mm Locn: 224 $ 6360/2460 and others Date: 11.2007

C 5/6 2958		2-10-0 4cc	1Eh4v	SLM C 5/6	2495	1914	(a)	(1)
	1 GNOM	2-2-0RT	1Azn2t	Olten	20	1870	(b)	(2)

(a) ex SBB "C 5/6 2958" /1973.
(b) new to Steinbruchgesellschaft Ostermundigen, Ostermundigen; ex von Roll, Usine des Rondez, Delémont on loan from VHS /19xx.
(1) 1996 to EUROVAPOR, Sulgen; later dismantled.
(2) 2002 returned to VHS.

SILO OLTEN AG (SOAG), Olten — SO36

Grain silos in the goods yard.

Gauge: 1435 mm Locn: 224 $ 6361/2458 Date: 01.2008

72	0-4-0DM	SLM Em 2/2	3464	1930	(a)	(1)
	4wDM	RACO 85 LA7	1697	1964	(b)	

(a) ex CIBA, Monthey /1970.
(b) ex SBB "Tm^I 479" /1996.
(1) 1997 to Historische Eisenbahn Gesellschaft, Biel/Bienne and kept at Delémont.

STAHL GERLAFINGEN AG, Gerlafingen — SO37

formerly: Gesellschaft der Ludwig von Roll'schen Eisenwerke (1823-1996)
von Roll (1813-1823)

A steel works and rolling mill established in 1813. Rail connected from 1870 to the industrial line Derendingen-Gerlafingen and from 1875 to the EB. In 1996 the Gerlafingen factory joined with von Moos, Emmenbrücke to form Stahl Schweiz AG.

Gauge: 1435 mm Locn: 233 $ 6094/2244 Date: 05.2008

1		0-4-0T	Bn2t	SLM E 2/2	236	1881	(a)	(1)	
2		0-4-0T	Bn2t	SLM	377	1884	new	(2)	
4		0-4-0WT	Bn2t	SLM	1267	1900	new	(3)	
73		0-6-0WT	Cn2t	SLM E 3/3	38	1874	(d)	(4)	
6	ELFE	0-4-0RT	Baz2t	Aarau	10	1876	(e)	(5)	
8		0-4-0Tm	Bn2t	Krauss-S LVIII i	1719	1886	(f)	1928 scr	
9		0-4-0WT	Bn2t	SLM E 2/2	2168	1911	new	(7)	

10	0-6-0WT	Cn2t	SLM E 3/3	629	1890	(h)	(8)	
11	0-6-0WT	Cn2t	SLM E 3/3	631	1890	(i)	(9)	
14	0-6-0WT	Cn2t	SLM E 3/3	955	1896	(j)	1966 scr	
16	0-6-0WT	Cn2t	SLM E 3/3	1220	1899	(k)	(11)	
17	2-6-0WT	1Ch2t	SLM Ed 3/4	1904	1908	(l)	(12)	
8	4wDH	Bdh	ABL IV NHT	4702	1956	(m)	(13)	
22	4wDH	Bdh	RACO 300 PS HT	1634	1964	new	(14)	
24	4wDH	Bdh	RACO 200 PS HT	1699	1966	new	(15)	
5525	6wDE	Cde	Moyse CN52EE500D	3501	1972	(p)		
26	4wDH	Bdh	SLM Tm^{IV}	4952	1972	new	(17)	
5527	6wDE	Cde	Moyse CN52EE500D	3526	1972	(r)		
28	4wDH	Bdh	SLM Tm^{IV}	4984	1973	new	(19)	
5529	6wDE	Cde	Moyse CN52EE500D	3535	1973	(t)		
	0-6-0DH	Cdh	Henschel DH 500Ca	30709	1964	(u)		
11	6wDE	Cde2	Moyse CN52EE500D	3525	1972	(v)		
	0-4-4-0DE	BBde	B+L	?	19xx	(w)		
846 350-7	BBDH	BBdh	MaK G 1202 BB	1000781	1978	(x)		
	6wDH	Cdh	Henschel DHG 500C	31236	1968	(y)		
	BBDH	BBdh	MaK G 1202 BB	1000804	1983	(z)	(26)	
Em 847 904-0	BBDH	BBdh	Henschel DHG 1200 BB	31575	1973	(aa)	(27)	

Additionally there were at least two self-propelling cranes that could be used for shunting:

2	4w-4wD	
3	4w-4wD	

Gauge: 500 mm   Locn: 233 $ 6094/2244   Date: 12.1999

	4wBE	Ba	BBC	861	1918	new	s/s
	0-4-0DM	Bdm	Deutz OME117 F	10296	1932	(bb)	s/s
	0-4-0DM	Bdm	Deutz OME117 F	11069	1933	(bc)	s/s
	0-4-0DM	Bdm	Deutz MLH714 F	21114	1937	(bc)	s/s
129	4wDM	Bdm	RACO 12/16PS	1304	1945	(be)	(57)
	4wDM	Bdm	RACO 12/16PS	1349	1947	(be)	(58)
145	4wDM	Bdm	R&H 30DL	323589	1952	(bg)	a1982 s/s
146	4wDM	Bdm	R&H 30DL	323590	1952	(bg)	a1982 s/s
147	4wDM	Bdm	R&H 30DL	323591	1952	(bg)	a1978 s/s
154	4wDM	Bdm	R&H 30DL	371380	1954	(bg)	a1976 s/s
165	4wDM	Bdm	R&H 30DLU	285356	1950	(bg)	a1982 s/s
166	4wDM	Bdm	R&H 30DLU	285351	1950	(bg)	(64)
187	4wDM	Bdm	R&H LBT	476114	1962	(bg)	a1982 s/s
195	4wDM	Bdm	R&H LBT	497713	1963	(bg)	a1982 s/s
216	4wDH	Bdh	R&H LFT	518189	1965	(bg)	(67)
217	4wDM	Bdm	R&H LFT	518188	1965	(bg)	a1971 scr
	4wDM	Bdm	JW DM20/1	2450	1965	(bq)	(69)

(a)   new to GB "11"; ex von Roll, Klus "1" /1934.
(d)   new to ARB "E 3/3 11"; ex Gaswerk Basel /1934.
(e)   ex Steinbruch Ostermundigen "2" /1907.
(f)   ex KLB "1" /1905; worked at von Roll, Bern (Fonderie de Bern) at an undetermined time.
(h)   new to JS "853"; ex von Roll, Klus /1941.
(i)   new to JS "855"; ex RVT "8" /1928.
(j)   new to SCB "78"; ex von Roll, Klus /1964.
(k)   ex OeBB "2" /1943.
(l)   new as saturated locomotive (1Cn2t) ex VHB "11" /1949.
(m)   Breuer licence; may have been used elsewhere at a von Roll plant.
(p)   new as "25"; renumbered after 1982.
(r)   new as "27"; renumbered ca. 1982.
(t)   new as "29"; renumbered before 1982.
(u)   ex Monteforno, Bodio "8" /1996.
(v)   ex PANLOG, Emmenbrücke "11" /1999-2001.
(w)   ex ?HBNPC /2001 via CFD. A wheel arrangement of BoBoDE would appear more likely.
(x)   new to NIAG "5"; hire from Crossrail /2005.
(y)   hire from LSB /200x.
(z)   new to Regionalverkehr Ruhr-Lippe GmbH [D] "66"; on hire from Voith Turbo Lokomotivtechnik GmbH & Co. KG, Kiel [D] 7/2007.
(aa)   hire from LSB by 7/2007.
(bb)   data from Deutz records where gauge is shown as 600 mm; locomotive not confirmed on-site.

(bc)	data from Deutz records; locomotives not confirmed on site.
(be)	new; von Roll records indicate a building date of 1946.
(bg)	new via R. Aebi AG, Regensdorf.
(bq)	ex Von Roll, Choindez /1988.
(1)	1950 stored at Le Locle; 1957 donated to VHS, Luzern.
(2)	1949 stored; 1960 scrapped.
(3)	1964 to von Roll, Usine des Rondez, Delémont.
(4)	1946 rebuilt as diesel locomotive and transferred to von Roll, Klus "4".
(5)	1941 OOU; 1955 stored; 19xx VHS; 1981 displayed Schule, Ostermundigen.
(7)	?1928 to von Roll, Choindez.
(8)	1973 donated to DBB.
(9)	1973 donated to VHS, Luzern; 1988 to VVT, St. Sulpice NE via DBB.
(11)	1967 donated to OeBB "2".
(12)	1973 donated to EUROVAPOR, Sektion Emmental; thence Verein Historische Eisenbahn Emmental (VHE), Huttwil, BE.
(13)	1973 to Centrale Pelli & Fonditoio Grassi SA, Bellinzona.
(14)	1973 to GFM "82".
(15)	1974 to von Roll, Klus.
(17)	1977 to BT "6".
(19)	1977 to BT "7".
(26)	2007-8 returned to Voith Turbo Lokomotivtechnik GmbH & Co. KG, Kiel [D].
(27)	2007-8 returned to LSB.
(57)	1/1978 to CFTT "7".
(58)	1976 to CFTT "5".
(64)	1996 donated to FWF, Otelfingen; by 1999 to Dampf Express Alpenhof, Bülach. Some reports exchange the identities of "165" and "166". The version given here is based on details observed on one cabside and one R&H plate.
(67)	1981 sold to CFTT "10".
(69)	1995 sold to A. Schwarz, Sumiswald.

FRANZ STIRNIMANN, Olten-Hammer   SO38
A civil engineering firm.

Gauge: 750 mm   Locn: ?   Date: 04.2002

	0-4-0T	Bn2t	Henschel	6525	1903	(a)	(1)
	0-4-0T	Bn2t	O&K 30 PS	847	1901	(b)	(2)
	0-4-0T	Bn2t	O&K 30 PS	985	1902	(b)	(2)
TOGGENBURG	0-4-0T	Bn2t	O&K 50 PS	1897	1906	(d)	(4)

Gauge: 600 mm   Locn: ?   Date: 04.2002

	0-6-0T	Cn2t	Krauss-S Zwilling n	4132A	1899	(e)	(5)
	0-4-0DM	Bdm	Deutz PME117 F	10207	1931	(f)	s/s
	4wDM	Bdm	O&K(M) RL 1	3490	1929	(g)	(7)
	4wDM	Bdm	O&K(M) RL 1	3491	1929	(g)	(8)

Gauge: 750 or 600 mm   Locn: ?   Date: 04.2002

	0-4-0T	Bn2t	Heilbronn II	293	1893	(i)	(9)
	0-4-0T	Bn2t	Zobel	569	1909	(j)	(10)

(a)	owned by Impresa costruzioni Ing. Bianchi, Lugano/1928 "MELIDE"; purchased before /1947.
(b)	ex Mangold & Cie, Zürich /1939.
(d)	ex Steiner & Cie, Ingenieurbureau, Zürich /1936.
(e)	ex ? /1936.
(f)	via Würgler.
(g)	via O&K/MBA, Dübendorf.
(i)	ex J. Ackermann, Baugeschäft, Mels /1943-46.
(j)	ex Martin Eichelgrün & Cie, Feldbahnfabrik, Frankfurt [D] /1936 (Moser refers to 589).
(1)	1949 removed from boiler register; after 1959 scrapped.
(2)	1949 removed from boiler register; ? sent to Braunkohle AG, Zell LU but not used there.
(4)	from 1936 used for grading the R. Dünnern, Olten; 1936-39 to Constantin von Arx, Olten.
(5)	before 1939 to Losinger & Cie, Bern '1'.
(7)	1939-48 to VEBA Vereinigte Bauunternehmer GmbH, Wildegg.

(8) 19xx to J. Freienmuth, Frauenfeld.
(9) 1949 removed from boiler register.
(10) 19xx to Franz Vago AG, Müllheim.

SWISSMETAL- UMS SCHWEIZERISCHEMETALLWERKE AG, Dornach   SO39

UMS = Usines Métallurgiques Suisses
formerly: Metallwerke AG
            Metallwerke Dornach AG (MD)

A non-ferrous metals plant built in 1885 and rail connected to Dornach-Arlesheim via a short tunnel in 1912. Address: Weidenstrasse 50, 4143 Dornach.

Gauge: 1435 mm   Locn: 213 $ 6128/2587   Date: 06.2000

1 'DIE BLAUE'	4wBE	Ba	SIG,MFO	?	1912	(a)	(1)
2 ZEPHIR	0-4-0WT	Bn2t	Krauss-S XIII d	290	1874	(b)	(2)
Tm 2/2	4wDE	Bde	Stadler,BBC	107,1422	1958	(c)	

(a) constructed using the motors from BDB "TM 8".
(b) new to BB "3 Zephir"; ex TSB "72" /1916-17; via RUBAG.
(c) new; originally referred to as '1 DER GELBE' but the present colour scheme is green and orange. The alleged BBC number is that allocated to the electrical gear fitted to SBB "Ae 4/8 11001"; perhaps some of that equipment was recovered and re-used in this locomotive.

(1) 1958 to Stadler; frame reused for Stadler 131 (Schlachthof, Basel and later Henkel AG, Pratteln).
(2) 1970 donated to MEFEZ, Zweilütschinen; 1970 to Verkehrshaus, Luzern; 1986 stored Spiez; 1995 to EUROVAPOR, Sektion Emmental, Huttwil for restoration to operational status; 1996 to Verkehrshaus Luzern (VHS); by 2007 to Historische Eisenbahn Gesellschaft, Delémont.

see: Eisenbahn Amateur Jan. 1971

H. TAUFER, KORROSIONSSCHUTZ, Olten   SO40

formerly: von Roll (1866-198x)

Breuer 3039 of 1951 (type V) acts as advertising material rather than a working locomotive on 1 April 1997.

Originally a foundry of the von Roll enterprises. The works were rail connected to the mainline 1897-8 and also had a manually operated internal narrow gauge system. On closure the firm of H. Taufer was established in a section and acquired the locomotive, though little or no use was made of it.

Gauge: 1435 mm   Locn: 224 $ 6351/2460   Date: 02.1999

| | 4wPM | Bbm | Breuer | ? | c193x | (a) | s/s |
| | 4wDM | Bdm | Breuer V | 3039 | 1951 | new | (2) |

(a) mentioned in the Breuer list of references of about 1930.
(2) 1999 to Dampfbahn Bern (DBB).

TERRAZZO UND JURASITWERK AG, Bärschwil   SO41
alternative: Usine de Terrazzo et de Jurasite SA

Gauge: 1435 mm   Locn: 223 $ 6029/2500   Date: 01.2008

| | 4wPM | Bbm | Breuer III | ? | 1929 | (a) | s/s |

(a) ex SBB "Tm 457" 1/1962.

VEREIN DS BLAUE BÄHNLI, Solothurn   SO42
One motor coach is under restoration. Address: Postfach 544, 3076 Worb.

Gauge: 1000 mm   Locn: ?   Date: 01.2008

| BDe 4/4 36 | BoBoWERC | SIG,BBC | ? | 1913 | (a) |

(a) new to WTB "CFe 4/4 101"; VBW "36" /1927; rebuilt /1951; ex MOB /2007.

CONSTANTIN VON ARX, Olten   SO43
A civil engineering firm.

Gauge: 750 mm   Locn: ?   Date: 02.2002

| TOGGENBURG | 0-4-0T | Bn2t | O&K 50 PS | 1897 | 1906 | (a) | (1) |
| WEISSENSTEIN | 0-4-0T | Bn2t | Krauss-S IV bh | 5295 | 1906 | (b) | (1) |

(a) originally 600 mm; ex Franz Stirnimann, Olten /1936-39.
(b) ex Bauunternehmung Dünnernkorrektion Los D, Olten /1936-39.
(1) from 1944 used for grading water courses in connection with the Rupperswil-Auenstein hydroelectric power station; 1948 to VEBA Vereinigte Bauunternehmer GmbH, Wildegg.

VON ROLL UMWELTTECHNIK, Klus   SO44
formerly: von Roll

This plant was originally a steelworks (1813-77), and then a foundry and machine factory. It was rail connected, to the OeBB, from 1899. A locomotive shed was built in 1909 at the southern end of the tracks. Production ceased in 1982 and the site was let to von Roll Umwelttechnik and other users. OeBB has performed shunting operations on the site since 1994.

Gauge: 1435 mm   Locn: 223 $ 6196/2388   Date: 07.1999

1	0-4-0WT	Bn2t	SLM E 2/2	236	1881	(a)	(1)
3	0-4-0WT	Bn2t	SLM E 2/2	726	1892	(b)	(2)
10	0-6-0WT	Cn2t	SLM E 3/3	629	1890	(c)	(3)
1	0-6-0WT	Cn2t	Maffei	2983	1909	(d)	(4)
3	0-6-0T	Cn2t	SLM Ec 3/3	1075	1897	(e)	1960 scr
2	0-6-0WT	Cn2t	SLM E 3/3	955	1896	(f)	(6)
2	0-6-0WT	Cn2t	SLM E 3/3	2507	1914	(g)	(7)

	4	0-6-0WT	Cn2t	SLM E 3/3	38	1874	(h)	
		rebuilt	0-6-0DM	Cdm	von Roll		1946		(9)
Tm	3	4wDH	Bdh	RACO 300 MA7 HT	1635	1964	new	(10)
	24	4wDH	Bdh	RACO 200 PS HT	1699	1966	(k)	(11)
		4wDM	Bdm	Breuer V	3049	1952	(l)	(12)
		4wDM	Bdm	RACO 85 DA7	1811	1972	(m)	(13)

(a) new to GB "11"; ex von Roll, Choindez "11" /1901.
(b) ex von Roll, Choindez "3" /1912.
(c) ordered as JBL 203; new to JS "853"; purchased from RVT "7" /1928.
(d) new to KLB "1"; purchased from SBB "E 3/3 8651" /1933.
(e) new to GB "303"; later SBB "Ed 3/3 6403"; purchased from SBB Werkstätte Olten /1941.
(f) new to SCB "78"; purchased from SBB "E 3/3 8423" /1941.
(g) purchased from SBB "E 3/3 8527" /1963.
(h) ex von Roll, Gerlafingen /1946 on completion of conversion to diesel.
(k) ex von Roll, Gerlafingen /1974.
(l) ex von Roll, Choindez /1972-82; also RACO 1424.
(m) ex Von Roll, Choindez /1996.

(1) 1934 to von Roll, Gerlafingen.
(2) 1941 sold to Gebrüder Sulzer, Ludwigshafen [D].
(3) 1941 to von Roll, Gerlafingen.
(4) 1975 sold to OeBB "1".
(6) 1964 to von Roll, Gerlafingen "14".
(7) 1981 to Messerli AG, Kaufdorf; by 2006 to Mainstation 1901, Chur.
(9) before 1968 to von Roll, Oensingen.
(10) 1984 to von Roll, Choindez.
(11) 1997 to OeBB.
(12) 1978-198x loan to von Roll, Subingen; 1991 donated to VVT, St. Sulpice NE; 1991 scrapped.
(13) 1/1997 seen at Balsthal; by 2007 returned to von Roll, Choindez; s/s.

WAVIN AG, Subingen SO45

formerly: von Roll (till 1992)
This factory makes units in plastic materials. The siding was lifted in 1995 or 1996.

Gauge: 1435 mm   Locn: 233 $ 6128/2269   Date: 05.1999

		4wDM	Bdm	Breuer V	3049	1952	(a)	(1)
	20	4wDH	Bdh	ABL IV NHT	4707	1957	new	c1995 s/s

(a) ex von Roll, Klus about /1978; seen at Subingen /1982; also RACO 1424.
(1) 198x returned to von Roll, Klus.

Schwyz   SZ

AUF DER MAUR BAU COMPANY, Steinen   SZ01
A building concern. Address: Frauholzstr. 10, 6422 Steinen.

Gauge: 600 mm   Locn: ?   Date: 02.2008
          4wDM   Bdm   Brun 2 DS 90E   67   19xx   (1)

(1)   Masch. Nr. 26483, PS 16; by 2003 to FWF, Otelfingen.

SERGE BOURGINET, Seewen SZ   SZ02
This private individual provides workshop and storage facilities for locomotives as required. At the time of closing this handbook it has been proposed that SBB-Historic "Ae 6/6 11403 SCHWYZ" comes here for restoration.

Gauge: 1435 mm + rack   Locn: ?   Date: 11.2007
          1   2-2-0RT   1Aan2t   Aarau   11   1876   new
          1 rebuilt   2-2-0RT   1Aan2t   SLM   -   1893
          1 rebuilt   0-4-0RT   Ban2t   SLM   -   1925   (a)
Ae 4/7 10997   2D1WE   2D1w4   SLM,MFO Ae 4/7 3536,? 1932 (b)

(a)   new to Caspar Honneger, Rüti; on loan from Technorama, Winterthur by /2006.
(b)   ex SBB "Ae 4/7 10997" by /200x.

DRUCKSTOLLEN HINTERTHAL, Muotathal   SZ03
The power station of Hinterthal near Muota is fed by water in a pressure tunnel 3746 metres long. It was brought into use in 1960. A railway was used during the construction.

Gauge: 600 mm   Locn: ?   Date: 08.2002
          4wDM   Bdm   O&K MV2   25340   1954   (a)   s/s

(a)   purchased new from MBA, Dübendorf.

HANS EGLI, Steinen   SZ04
A private collector resident in Besazio.

Gauge: 600 mm   Locn: ?   Date: 01.2002
          4wDM   Bdm   Gmeinder   2954   1940   (a)   (1)

(a)   1992 stored in the open in Goldau; c2000 transferred to Steinen.
(1)   by 2006 not located.

HEINRICH HATT-HALLER & ED. ZÜBLIN AG   SZ05

Etzel
Construction of the reservoir for hydroelectric scheme at Etzel.

Gauge: 600 mm   Locn: ?   Date: 01.2002
          0-6-0T   Cn2t   Henschel Brigadelok   6335   1903   (a)   (1)

Wäggital

Building of the dam for the hydroelectric works in the Wäggital carried out 1921-24. The AG Kraftwerk Wäggital is a jointly owned firm of the power supply for the city (EWZ) and the district of Zürich (NOK).

Gauge: 750 mm   Locn: ?   Date: 03.2002

	0-4-0T	Bn2t	Hanomag	8009	1920	(b)	(2)
	0-4-0WT	Bn2t	O&K 90 PS	3216	1908	(c)	(3)
	4.M	B.m	O&K(M)	?	19xx		s/s

(a) ex Hoch- & Tiefbau AG, Zürich /1922; used 1934 till 193x.
(b) ex Rollmaterial AG, Zürich /1922.
(c) ex J.A. Reif & Kröll, Baugeschäft, Koblenz [D] (900 mm) /1922.

(1) 1948 to Brun & Cie AG, Nebikon.
(2) 1922-29 to AG Heinrich Hatt-Haller (HHH), Hoch- & Tiefbau AG, Zürich.
(3) sold; by 1955 in use by Kieswerk Hardwald, Dietikon (900 mm).

HOLCIM ZEMENT AG, Brunnen   SZ06

formerly: Zementfabrik K. Hürlimann Söhne AG, Brunnen (till 2001)

The cement works was founded in 1883 and rail connected to Brunnen station in 1888. The narrow gauge system links a conveyor belt from the quarries at Schönenbach to the works and was originally worked by horses. It passes under the Gotthard main line at right angles. Locomotives may be stabled at either end of the line. The last pair of locomotives were used in conjunction with a driving trailer. Quarrying ceased at the end of 2007 and stocks were cleared in the first part of 2008; disposal continues at the time of closure for publication.

Gauge: 1435 mm   Locn: 235 $ 6893/2065   Date: 09.2001

	4wPM	Bbm	Breuer III	1201	1928	(a)	(1)
	4wDM	Bdm	Gebus Lokomotor	567	1957-58	(b)	(1)
	4wDH	Bdh	KHD KG230 B ex	58175	1967	(c)	

Gauge: 750 mm   Locn: 235 $ 6893/2065   Date: 09.2008

	0-4-0T	Bn2t	Heilbronn	?	1888		(4)
	0-4-0T	Bn2t	Maffei	3732	1912	(e)	c1955 scr
	0-4-0PM	Bbm	Deutz C XIV F	1583	1915	(f)	(6)
	0-4-0DM	Bdm	Deutz OMZ117	10830	1932	new	s/s
	1ABE	1Aa1	Stadler, BBC	17, ?	1945	(h)	(8)
	4wDM	Bdm	O&K MV2a	25323	1954	new	(8)
	4wDM	Bdm	O&K MV4a	26245	1963	new	(8)
	4wDH	Bdh	Diema DFL60/1.6	4525	1981	(k)	
	4wDH	Bdh	Diema DFL150/1.2	5146	1991	new	(12)
	4wDH	Bdh	Diema DFL150/1.2	5147	1991	new	(12)

(a) new; the Breuer works list shows Lagerhäuser, Brunnen – probably the delivery address for railway purposes.
(b) new to Ing. Struppe, Wien [A] (dealer?); purchased via RACO.
(c) ex Gewerkschaft Victor, Castrop-Rauxel [D] "3" /1989 via MF.
(e) new via Robert Aebi & Cie AG, Zürich; spare boiler Maffei 4402 of 1928 fitted /1939.
(f) new to HFB (600 mm); ex Deutz /1920.
(h) rebuilt from a Deutz 21PS locomotive, probably C XIV F 1583 1915.
(k) new via Asper.

(1) 1989 donated to Museum für Verkehr und Technik (MVT), Berlin [D].
(4) 1901 offered for sale due to lack of use.
(6) 1945 probably converted by Stadler to become Stadler 17.
(8) 1994 to FWF, Otelfingen.
(12) 2008 to Tafag AG.

MARTIN HORATH, Goldau SZ07

Gauge: 1435 mm			Locn: ?					Date: 11.2007
T 2/2	1	0-4-0Fic	Bf2	SLM		1141	1898	(a)

(a) new to Gaswerk Winterthur; on loan for overhaul from Technorama, Winterthur /2007.

A. KAUFMANN AG, Goldau SZ08

A firm founded in 1946 and active in metallic constructions and electrical equipment. From about 1990 the field extended to mounting catenary.

Gauge: 1435 mm			Locn: ?					Date: 12.2002
92	TIRAMISU	6wDH	Cdh	Henschel DHG 500C	30926	1964	(a)	OOU

(a) ex Ruhrchemie GmbH, Oberhausen [D] "11" via OnRail/LSB.

KIBAG SZ09

A gravel and cement concern with headquarters in Zürich.

Bäch SZ

Gravel works near Freienbach. Address: Bächaustrasse 73, 8806 Bäch SZ.

Gauge: 750 mm			Locn: 236 $ 6984/2293				Date: 01.2002	
5	IDA	0-4-0T	Bn2t	Hanomag	8009	1920	(a)	(1)
9	MARTHA	0-4-0T	Bn2t	Märkische 50 PS	296	1898	(b)	(1)
		4wBE	Ba1	Stadler	27	1948	(c)	

Nuolen

One locomotive is recorded as delivered here. It is not sure that it worked here though there is a quarry to the southeast of the village, which lies near Wangen SZ on the Obere Zürichsee (Obere Zürisee). Address: Seestrasse 95, 8855 Nuolen.

Gauge: 900 mm		Locn: 236 $ 7102/2287			Date: 03.2008	
	4wBE	Ba1	Stadler	27	1948	new (4)

Seewen SZ

formerly: Walter Weber, Steinbruch Zingel AG, Seewen
Alois Weber

Quarrying began in 1902. Initially a roadside line connected the quarry to the works alongside the SBB. An underground conveyor belt replaced the narrow gauge line in 1974. One locomotive was then displayed on the roof of a building there. Before the purchase of the standard gauge locomotives an electric crane by Oehler was used to position wagons.

The last narrow gauge locomotive acquired, Ruston & Hornsby 386851 of 1955 (type 48DL with frame extensions) still has its original engine 4VRHL 384746 and has now been placed on the roof of a building. It is visible from passing trains. The photograph was taken on 5 April 2005.

Gauge: 1435 mm			Locn: 237 $ 6904/2094				Date: 12.2002
	0-6-0DH	Cdh	KrMa ML 500 C	18320	1958	(e)	s/s
202	4wDE	Bde	Moyse BN32E150B	1067	1964	(f)	
Gauge: 600 mm			Locn: 237 $ 6904/2094				Date: 12.2002
	0-4-0T	Bn2t	O&K 20 PS	10391	1923	new	(7)
	4wDM	Bdm	O&K(M) RL 1	3839	1929		(8)
	0-4-0DM	Bdm	Deutz MLH332 F	9718	1931	(i)	s/s
	4wDM	Bdm	R&H 48DL	386851	1955	new	display

Zürich

This section is reserved for locomotives delivered to the company, but for which no record of the place of use has been found.

		Gauge: 600 mm		Locn: ?			Date: 04.2009	
		4wDM	Bdm	O&K(M) RL 1	3615	1929	(i)	s/s
		4wDM	Bdm	O&K(M) RL 1	3710	1929	(j)	s/s

(a) ex Heinrich Hatt-Haller, Hoch- & Tiefbau AG, Zürich /1931 via Kibag AG, Zürich.
(b) fitted with replacement boiler O&K 344 of 1905; ex Robert Aebi & Co, Zürich /1943.
(c) ex Kibag AG, Nuolen (900 mm).
(e) new to Westfalenhütte, Dortmund [D] "70"; ex Dortmunder Eisenbahn GmbH, Dortmund "525" /1985 via NEWAG.
(f) ex SA des Ciments Francais, Canton, Nord [F] /1992.
(i) new via Robert Aebi & Cie. AG.
(j) new via Robert Aebi & Cie. AG; no gauge given.
(1) 1960-62 to Dietiker, Zürich-Seebach.
(2) 1956 removed from boiler register; 1960 -62 to Dietiker, Zürich-Seebach.
(4) to Kibag AG, Bäch SZ /19xx (750 mm).
(7) 1931 to Robert Aebi & Cie, Zürich.
(8) after 4/1974 scrapped.

KIES AG BOLLENBERG, Tuggen   SZ10

Gravel pits. The company was later resident in Pfäffikon SZ and ceased to exist as a consequence of a merger in 2006. There is a HOLCIM in Girendorf: the same location?

		Gauge: 750 mm		Locn: ?			Date: 01.2002	
		0-4-0T	Bn2t	Krauss-S IV tu	3185	1895	(a)	(1)
		0-4-0T	Bn2t	O&K 40 PS	1086	1903	(b)	(2)
		0-4-0T	Bn2t	Jung 30 PS	69	1889	(c)	(2)
		0-4-0T	Bn2t	O&K 40 PS	1106	1903	(d)	s/s
		0-4-0WT	Bn2t	Borsig	6485	1908	(e)	(2)
		0-4-0T	Bn2t	O&K 50 PS	4347	1911	(f)	(6)

(a) ex Rubag, Basel before or in /1928.
(b) ex Hunziker & Co, Zürich /1931.
(c) ex Hunziker & Co, Baustoff-Fabriken, Olten /1938.
(d) ex Arbeitsgemeinschaft Stauwehrbau Klingnau before or in /1939 (originally 600 mm).
(e) ex Hunziker & Cie/ Portlandzementwerk AG, Olten /1941.
(f) ex Robert Aebi & Co, Zürich /1950 (other sources quote 1946).
(1) 1939 removed from boiler register.
(2) 1945 removed from boiler register.
(6) 1963 removed from boiler register.

G. LEIMBACHER, Lachen SZ   SZ12

formerly: Eidg. Dipl. Baumeister
later: Batigroup AG (from 1995)

A construction firm. Address: Spitalweg 6, 8853 Lachen.

		Gauge: 600 mm		Locn: 236 $ 7077/2277			Date: 04.1999	
		4wDM	Bdm	O&K(M) RL 1c	8372	1937	(a)	s/s
		4wDM	Bdm	O&K(M) RL 1c	11428	1939	(b)	s/s
Gauge: 900 mm				Locn: 236 $ 7077/2277			Date: 04.1999	
		4wDH	Bdh	Schöma CFL 180DCL	4504	1982	(c)	(3)
	DL1	4wDH	Bdh	Schöma CFL 180DCL	5177	1990	(d)	(4)
	DL2	4wDH	Bdh	Schöma CFL 180DCL	5176	1990	(d)	
		4wDH	Bdh	GIA DHD35	3502	1993	(f)	

(a) via O&K/MBA, Dübendorf.
(b) via O&K/MBA, Lager Zürich; recorded as type RL 2.
(c) new to Trans-Manche Link (TML), Cheriton [GB], ex Schöma, /1990.
(d) new to Trans-Manche Link (TML), Sandgatte [F].
(f) hired from Schlattler AG, Rheinsulz.

(3)   to Porr. Technobau, Stollen Semmering, Murzzuschlag [A] /1996.
(4)   to Porr. Technobau, Stollen Semmering, Murzzuschlag [A] /1998.

Logistikbasis der Armee (LBA), Schwyz   SZ13

formerly: Bundesamt für Betriebe des Heeres (BABHE) (till 31.12.2003)
            Oberkriegskommissariat (OKK), Seewen
includes: Armeeverflegungsmagazin (AVM), Seewen-Schwyz

Federal arsenal constructed in 1888 and rail connected from 1892. In 1902 a large grain store (AVM) was added and also rail connected.

Gauge: 1435 mm   Locn: 236 $ 6907/2081   Date: 09.2008

		4wPM	Bbm	RACO RA7	103	1933	new	(1)
Tm 2/2	1	4wDM	Bdm	Kronenberg DT 90	120	1950	(b)	(2)
	4	4wDM	Bdm	Kronenberg DT 100/220	142	1957	new	(3)
Em 836 908-4		6wDE	Cde3	Stadler	112	1961	(d)	(4)

(b)   ex OKK, Ostermundigen after /1962; seen at AVM, Altdorf UR/ 2004.
(d)   ex OKK, Göschenen "8" /1996.
(1)   1941 transferred to OKK, Ostermundigen.
(2)   a2004 to Rynächt Arsenal or Zeughaus Eyschächen, Altdorf UR (to be confirmed).
(3)   19xx transferred to OKK, Ostermundigen; 1983 transferred to BABHE, Brenzikofen-Herbligen.
(4)   2007-9 to LBA, Brenzikofen-Herbligen via Stauffer.

Losinger AG, Einsiedeln   SZ14 C

This civil engineering firm has sites throughout Switzerland. One locomotive is recorded as delivered new here.

Gauge: 600 mm   Locn: ?   Date: 04.2009

| | 4wDM | Bdm | O&K(M) MD 1 | 10503 | 1940 | (a) | (1) |

(a)   (purchased from MBA, Dübendorf.
(1)   19xx to Losinger AG, Bern.

METRAG AG, Pfäffikon SZ   SZ15

This company specialises in mechanised track maintenance and automation in locomotives and has a registered address in Freienbach. Principle workshop Address: Churerstrasse 80, 8808 Pfäffikon. Other work is performed in Uznach and Rümlang.

Gauge: ? mm   Locn: ?   Date: 01.2008

| | 4wDH | Bdh | GIA BS2 | 204 | 1996 | new |

MINARINI & BERTON, Goldau   SZ16

Probably a public works concern.

Gauge: 750 mm   Locn: ?   Date: 01.2002

| | 0-4-0T | Bn2t | Krauss-S IV fg | 2808 | 1893 | (a) | s/s |

(a)   ex Ing. Silvio Viglino, where? (Chavornay?) before or in /1896.

MODELLEISENBAHN-CLUB EINSIEDELN (MECE), Einsiedeln SZ17

Gauge: 1000 mm   Locn: ?   Date: 01.2008

| | 0-4-0WT | Bn2t | O&K 90 PS | 3216 | 1908 | (a) | (1) |

(a) new to J.A. Reif & Kroll, Coblenz [D] (900 mm); ex Oswald Steam Samstagern (1000 mm) via private ownership.
(1) returned to Kieswerk Hardwald, Dietikon.

OTT'S ERBEN, Schwyz   SZ18

Probably a public works concern.

Gauge: 600 mm   Locn: ?   Date: 08.2002

| | 4wDM | Bdm | O&K MV0 | 25173 | 1952 | (a) |

(a) new via MBA, Dübendorf; seen at Goldau /19xx "55".

PRADER AG TUNNELBAU, Siebnen   SZ19

formerly: Prader AG, Filial Siebnen

A civil engineering firm, specialising in tunnel construction. Address: Kreuzstrasse 30, 8854 Siebnen.

Gauge: 750 mm   Locn: 236 $ 7104/2268   Date: 08.1995

		4wBE	Ba	SIG	?	19xx	(a)
	281 2595 111	4wBE	Ba	Ageve	813	1976	(b)
1	281 2595 411	4wBE	Ba	Ageve	865	1980	(b)
		4wBE	Ba	Ageve	?	19xx	(d)
		4wBE	Ba	Ageve	?	19xx	(d)

Gauge: 600 mm   Locn: 236 $ 7104/2268   Date: 08.1995

4wDM	Bdm	O&K(M) RL 1c	8422	1937	(f)	(6)
4wDM	Bdm	O&K MD 2b	?	19xx		display
4wBE	Ba	SSW	?	19xx		(8)
4wBE	Ba	SIG	?	19xx		in use

(a) in use on work for the Wasserwerk, Zürich, Galerie Lyrenweg-Handhof /1995.
(b) in use by Arge Anschluss Stollen, Glatt, Zürich-Oerlikon c. /1995-98; the numbers given are inventory numbers.
(d) in use on works at Oerlikon station /?1995.
(f) transferred from Prader & Co., Zürich /19xx.
(6) 1989-90 to Hans Egli, Besazio at Arth-Goldau; c2000 to Steinen.
(8) 1996 donated to FWF, Otelfingen.

REIS MÜHLE, Brunnen   SZ20

Three interested groups founded this mill in 1957. Since 2004, it has been a division of the COOP. Address: Industriestrasse 1, 6440 Brunnen.

Gauge: 1435 mm   Locn: 235 $ 6894/2065   Date: 01.2009

| Tm 2/2 | 756 | 4wDM | Bdm | RACO 80 SA3 | 1405 | 1951 | (a) |
| Tm | 9587 | 4wDH | Bdh | RACO 225 SV4 H | 1867 | 1981 | (b) |

(a) ex SBB "Tm^{II} 756" /2004.
(b) ex SBB "Tm^{III} 9587" /2007; a report indicates that it may have been rebuilt to 4wDE Bde.

BAUUNTERNEHMUNG SENN, Brunnen   SZ21

A civil engineering firm.

Gauge: 600? mm   Locn: 236/ 6893/2065   Date: 03.1996

| 4wDM | Bdm | O&K MV2 | 25801 | 1957 | (a) | (1) |

(a) ex Ziegelei auf der Maur GmbH, Einsiedeln.

(1)   1989 placed on a plinth in Franzosenstrasse, Seewen SZ; since vanished and whereabouts unknown.

STEINFABRIK ZÜRICHSEE, Pfäffikon SZ — SZ22

A dependency of AG Hunziker & Cie, Zürich, rail connected since 1899. Address: Unterdorfstrasse 12, 8808 Pfäffikon SZ.

Gauge: 1435 mm   Locn: ?   Date: 07.1998

	1APM	1Abm	Oberursel	?	1911	(a)	s/s

(a)   also quoted as 4wPM, Bbm.

TAFAG AG, Goldau — SZ23

An engineering firm that modifies, repairs and deals in locomotives. One of the several workshops of the Dampf Furka Bahn is located in the same complex. Address: Chräbelstrasse 3, 6410 Goldau.

Gauge: various   Locn: 235 $ 6842/2107   Date: 07.1996

TSCHÜMPERLIN, *Küssnacht am Rigi* — SZ24

Probably a public works firm. The original source (MBA listing) gives only Küssnacht, which suggests that the firm may have been resident in Küssnacht-am-Rigi. An alternative might be Küsnacht ZH, but this cannot be verified. A company with the title A. Tschümperlin AG provides building materials in Lüsslingen.

Gauge: ? mm   Locn: ?   Date: 05.2002

	4wDM	Bdm	MBA MD 1	2001	c1950	(a)	s/s?

(a)   second-hand from MBA, Dübendorf /1951; the type is not given in the MBA records.

C. VANOLI AG, Immensee — SZ26

see under C. Vanoli AG, Samstagern.

VEREIN HISTORISCHER TRIEBWAGEN 5, Einsiedeln — SZ27

The motorcoach is kept in the SOB depot. Address: Postfach 359, 8840 Einsiedeln.

Gauge: 1435 mm   Locn: ?   Date: 10.2008

SOB ABDe 4/4 5	BoBoWERC BBw4	SWS,MFO,BBC,SAAS	?	1939	(a)	

(a)   new as SOB "CFZe 4/4 12"; reclassified "Ce 4/4 12" /1945; renumbered "BCe 4/4 5" /1949; reclassified "ABe 4/4 5" /1956.

JAK. WILD, Pfäffikon SZ — SZ28

A gravel quarry. It is not sure whether Pfäffikon SZ or Pfäffikon ZH is correct.

Gauge: ? mm   Locn: ?   Date: 05.2002

	4wDM	Bdm	MBA MD 1	2002	c1950	(a)

(a)   purchased via O&K/MBA, Dübendorf.

ZIEGELEI AUF DER MAUR GmbH, Einsiedeln — SZ29

A brickworks operated by this engineering company.

Gauge: 600 mm   Locn: 236 $ 6992/2211   Date: 08.1990

| | 4wDM | Bdm | O&K MV2 | 25801 | 1957 | (a) | (1) |
| | 4wDM | Bdm | Diema DS11/2 | 2790 | 1965 | new | 19xx scr |

(a) via MBA, Zürich.
(1) to Bauunternehmung Senn, Brunnen.

(a)	purchased from Heine AG, Arbon /1913.
(1)	1948 sold to Viscose Suisse SA, Emmenbrücke.
(2)	1982 to Amman AG, Gossau SG; later renamed Solenthaler Recycling AG, Gossau SG.

F.T. SONDEREGGER AG, Egnach   TG18

A household and dairy farming appliances dealer, founded in 1956 and since moved to Herisau.

Gauge: 1435 mm   Locn: 217 $ 746x/267x   Date: 01.2008

		4wPM	Bbm	Breuer III		?	1930	(a)	(1)
(a)	ex SBB "Tm 459" 3/1963.
(1)	1980 OOU; s/s.

STADLER BUSSNANG AG, Bussnang   TG19

formerly: Stadler-Fahrzeuge AG (1950-2000)

Mechanical engineering workshops established and rail connected from 1962/63. Previously located in Zürich the Stadler firm makes, amongst other items, shunting locomotives and special railway equipment. It also produces railcars for the worldwide market and has a growing number of subsidiaries.

A small number of Tm^{III} went directly into industrial service, rather than to the SBB. This one (SLM,BBC 4439,6180 of 1962) went initially to Dreispitz-Verwaltung, Basel as their "11"; in 1999 it was acquired by Stadler. On 20 April 2000 it was photographed during a visit to the station.

Gauge: 1435 mm   Locn: 217 $ 7240/2685   Date: 02.2002

'BARRY'		4wBE	Ba	Stadler	155	1981	(a)	
		4wDE	Bde2	Stadler,BBC	111, ?	1960	(b)	(2)
	11	0-4-0DE	Bde1	SLM,BBC Tm^{III}	4439,6180	1962	(c)	
(a)	made from two construction locomotives.

(b)   ex Wander AG, Neuenegg /1989.
(c)   ex Dreispitz-Verwaltung, Basel "11" /1999.
(2)   1990 to Chemie Reichhold, Hausen.

STAUFFER SCHIENEN-UND SPEZIALFAHRZEUGE, Frauenfeld
TG20 C

A firm specialising in locomotive repair and providing maintenance services at customer sites. It was first established on the old Gaswerk der Stadt Zürich site in Schlieren and moved to Frauenfeld in 2007-8. There are other offices and locomotives may be located elsewhere, such as Pratteln, Ristet and Roggwil BE. Address: Langdorfstrasse 16, 8500 Frauenfeld.

ex-Bundeswehr Jung 12348 of 1956 (a R 40C) was stored amongst other stored stock at Ristet (Kanton Zürich) on 3 June 2008.

Gauge: 1435 mm			Locn: 216 $ 7105/2694				Date: 05.2009	
	4wDM	Bdm	RACO 95 SA3 RS	1461	1955	(a)	(1)
237 931	4wDM	Bdm	RACO 95 SA3 RS	1664	1963	(b)	
	0-4-0DE	Bde1	SLM,BBC Tm^{III}	4564,7516	1965	(c)	
Bm 847 960-2 Iris	0-8-0DH	Ddh	MaK 1000 D	1000009	1958	(d)	(S)
237 866-9	0-4-0DE	Bde1	SLM,BBC Tm^{III}	4586,7539	1965,66	(e)	
133	4wDH	Bdh	KHD A6M617 R	57338	1960	(g)	(7)
Am 847 900-8	BBDE	BBde	MTE BB800H	6174	1982	(h)	
rebuilt			CFD BB800	851	2000		
	0-6-0DH	Cdh	Jung R 40 C	12348	1956	(i)	(S)
Tm 236 314	4wDH	Bdh	RACO 100 DA3 H	1826	1975	(j)	(R)
Tm 237 920-4 Felix	4wDM	Bdm	RACO 80 SA3	1417	1953	(k)	
Tm 237 933-7 Seppli	4wDE	Bde	Stadler	119	1966	(l)	
	4wDM	Bdm	RACO 95 SA3 RS	1783	1967	(m)	(R)
	0-4-0DE	Bde1	SLM,BBC Tm^{III}	4158,6026	1956	(n)	
	4wDH	Bdh	Gmeinder 190 PS	5246	1960	(o)	
Tm^{III} 922	0-4-0DE	Bde1	SLM,BBC Tm^{III}	4579,7531	1965	(p)	
Tm 236 313-3	4wDM	Bdm	RACO 95 MA3 RS	1798	1971	(q)	
Tm 237 959-2	4wDM	Bdm	RACO	184x	1980	(s)	

(a) ex SBB "Tm^{II} 605" /2004.
(b) ex SBB "Tm^{II} 726" /2004.
(c) ex SBB "Tm^{III} 907" /2005.
(d) new Rinteln-Stadthagener Eisenbahn [D] "V101"; ex Gleisag, Rorschach.
(e) ex Dreispitz-Verwaltung, Basel "237 866-9" /2006.
(g) new to DB "DB "Köf 6480"; later DB "323 235-2"; ex Furrer & Frey, Gwatt (Thun) /2003 (to be confirmed here).
(h) ex Elf, Lacq [F] via CFD.
(i) new to Bundeswehr [D], für Meppen, Erprobungsstelle 91; ex Unirail 8/2006; still carries plate "2210-12-10.1639".
(j) ex EBT "Tm^{III} 14" 5/2008.
(k) ex SBB "Tm^{III} 760" by 6/2008 (withdrawn 5/2007).
(l) new to MThB "51"; later "62", to SBB "Tm 236 642"; withdrawn 8/2007; ex SBB 7/07.
(m) ex SBB "Tm^{II} 844" by /2008 (withdrawn 31/8/06).
(n) ex Schmutz Aciers SA, Orbe /2008; after a period in store on the premises of La Traction, Pré-Petitjean.
(o) before 1994 from Petroplus AG, Rothenburg.
(p) ex SBB "Tm^{III} 922" 1/2009.
(q) new to EBT "Tm 13", ex BLS 3/2009.
(s) new to BLS "Tm 85"; ex BLS "235 085" /2009 and hired to Vanoli.

(1) to be stripped for parts.
(7) 10/2007 to Unirail [D]; (to be confirmed - seen at RMM, Full 10/2007).

(R) at Roggwil BE /2008.
(S) at Ristet /2008.

TANKLAGER ALTISHAUSEN AG, Berg TG — TG21

A fuel depot constructed and rail connected north of Berg TG in 1967.

				Locn: 217 $ 7306/2715			Date: 06.2008	
		4wPM R/R	Bbm	Unilok B6000S	A208	1966	(a)	?
Tm	6	4wDM	Bdm	Kronenberg DT 90/120	125	1952	(b)	

(a) new; constructed by Jung with number 13951.
(b) transferred from Bundestankanlagen (BTA), Boudry /2000-01.

TORFWERKE AG (TOWAG), Märwil — TG22

A peat extraction scheme operated during World War II.

Gauge: 600 mm		Locn: ?				Date: 02.2002	
	0-4-0T	Bn2t	O&K 40 PS	550	1900	(a)	(1)
	0-6-0T	Cn2t	Maffei	3508	1909	(b)	(1)

(a) see note for this locomotive under Maschinen- und Bahnbedarf (MBA), Dübendorf; ex Steinemann & Cie, Bauunternehmung, St Gallen /1942.
(b) see note for this locomotive under Robert Aebi & Cie AG, Regensdorf; new as 750 mm; ex Gobba & Bondietti, Impresa costruzioni, Locarno /1943.

(1)   1943 to Locher & Cie, Zürich.

GEBRÜDER TUCHSCHMID AG, Frauenfeld   TG23 C
This firm constructs locomotive shells for, among others, the SBB, but does not own a shunting tractor. Address: Kehlhofstrasse 54, 8501 Frauenfeld.
Gauge: 1435 mm   Locn: 216 $ 7105/2695   Date: 11.2007

FRANZ VAGO AG, Müllheim-Wigoltingen   TG24 C
A public works and road building concern (Strassen- und Tiefbauunternehmung) with a (now closed) subsidiary in Stein am Rhein. A Maffei steam road roller is preserved outside the offices, which are located adjacent to Müllheim-Wigoltingen station. The fate of the locomotive has not been determined. Address: Hasli, 8554 Müllheim-Wigoltingen.
Gauge: 600 mm   Locn: ?   Date: 06.2008

		4wDM	Bdm	O&K(M) RL 1	3490	1929	(a)	?
(a)	ex Franz Stirnimann, Olten /19xx.							

see: www.parlament.ch/cv-biografie?biografie_id=227
see: www.moneyhouse.ch/u/vago_ag_filiale_stein_am_rhein_CH-290.9.004.489-6.htm

VEREIN HISTORISCHE MITTELTHURGAU BAHN, Romanshorn
TG25 M
see: Locorama, Romanshorn

GEBRÜDER WEIBEL, Eschlikon   TG26
An automated brickworks.
Gauge: 600? mm   Locn: 216 $ 7153/2578   Date: 02.1988

		4wDM	Bdm	O&K?	?	19xx	(a)	s/s
(a)	in service till about the 1960s.							

WELLAUER AG, Frauenfeld   TG27 C
formerly: H. Wellauer
A public works concern (Tiefbau) that owned four or five locomotives. Address: Zürcherstrasse 354, 8500 Frauenfeld.
Gauge: 600 mm   Locn: 216 $ 7103/2697   Date: 06.1991

		4wDM	Bdm	O&K(M) LD 16	4684	1932	(a)	s/s
		4wDM	Bdm	O&K(M) RL 1a	4881	1933	(a)	s/s
		4wDM	Bdm	O&K(M) LD 2	5691	1934	new	display
(a)	via O&K/MBA, Dübendorf.							

ZIEGELEI EHRAI AG, Frauenfeld   TG28
A brickworks with a short line that passed under the SBB line to the east of Frauenfeld station. The line was abandoned in 1971.

Gauge: 600 mm		Locn: 216 $ 7105/2691				Date: 06.1991	
	4wDM	Bdm	MBA MD 1		2008	c1950	(a) (1)
(a)	ex Ad. Merk, Pfyn.						
(1)	1971 wdn; to Plättli-Zoo, Frauenfeld for display; 1991 donated to FWF, Otelfingen.						

ZUCKERFABRIK FRAUENFELD (ZFF), Frauenfeld   TG29

This sugar factory was constructed 1961-63 and has a large yard between Frauenfeld and Islikon.

"Tm 237 904-8" a Gmeinder with the memorable number of 5678 of 1989 (type DHS 20 B) undergoes modification at the nearby works of Stauffer Schienen-und Spezialfahrzeuge on 4 June 1980.

Gauge: 1435 mm			Locn: 216 $ 7080/2680				Date: 12.2002
		4wDH	Bdh	Gmeinder 190 PS	5253	1961	(a)
Tm 237 904-8		4wDH	Bdh	Gmeinder DHS 20 B	5678	1989	(b)
82		4wCE	B-el	Vollert Kabeltrommel	82/045	1982	new
84		4wCE	B-el	Vollert Kabeltrommel	83/092	1983	new
(a)	new, into service 1962.						
(b)	new, has carried the name SUGAR LADY.						

E. ZWICKY AG, Müllheim-Wigoltingen   TG30

Grain mills connected to Müllheim-Wigoltingen station and situated in Hasli. Address: Hasli, 8554 Müllheim-Wigoltingen.

Gauge: 1435 mm		Locn: 216 $ 7183/2720				Date: 01.2008
	4wDM	Bdm	RACO 85 LA7	1654	1963	(a)
(a)	new to CFF "Tm^I 348"; ex CFF "Tm^I 448" /2004.					

Ticino (Tessin) TI

AARE-TESSIN AG FÜR ELEKTRIZITÄT (ATEL), Lavorgo TI01

One of several rail-connected power plants operated by this company is that of 'Centrale Piottino'. This had a connection from the FFS station of Lavorgo to the turbine hall situated towards the south of the site. The presence of a locomotive is indicated in an FFS report of 1977, but without specifically stating that it worked this connection. On a visit in May 2005 a rail-borne battery-operated shunting device was found on a separate system within the switching station towards the north of the site. Whether or not that is what is meant in the FFS report is not clear.

Gauge: 1435 mm   Locn: 266 $ 7239/1441   Date: 12.2007

AEBISCHER & GUSCETTI, Ambri TI02 C

Probably a public works consortium. The SIG list of references gives one ETS 35 locomotive delivered here.

Gauge: 600 mm   Locn: ?   Date: 12.2008

	4wBE	Ba	SSW EL 9	5568	1954	(a)

(a) new to Losinger & Cie AG, Bern; purchased via MBA, Dübendorf /?1954, who have recorded the number as 5868 which was built in 1957.

ALBARELLI SPA VERONA, Ponte Chiasso [I] TI03

A forwarding agent located on Italian soil, but rail connected on the Swiss side of the border to Chiasso station. Activity ceased 1995-96.

Gauge: 1435 mm   Locn: 296 $ 7239/0766   Date: 07.1993

	4wDH	Bdh	ABL	?	19xx		(1)
	4wDH	Bdh	ABL V-A	5232	1969	new	(2)
	4wDH	Bdh	Diema DVL150/1.2	3238	1972	(c)	s/s

(c) ex Filipini & Figlio, Airolo /1987.
(1) 1969-70 s/s.
(2) 1993-97 s/s.
(3) 1997 to Fincantieri, Sestri Ponente [I]; /1997 via Nuova Ralfo s.r.l. Olginate [I].

ALBERGO LOSONE, Losone TI04 M

The locomotive has been placed in the children's playground of the hotel in the company of a road roller and what appears to be the old station building from "Solduno". The current FART station serving that village is underground on the other bank of the Maggia river. Address: Via dei Pioppi 14, 6616 Losone.

Gauge: 1000 mm   Locn: 276 $ 7229/1139   Date: 08.2000

'MAGGIA'	0-4-0WT	Bn2t	Jung 50 PS	104	1891	(a)	playgrd

(a) ex Consorzio Correzione Maggia, Locarno.

ANTOGNINI & NOLI, Bellinzona TI06 C

A civil engineering firm.

Gauge: 750 mm			Locn: ?			Date: 01.01.1994	
SAN PAOLO	0-4-0T	Bn2t	O&K 50 PS	4347	1911	(a)	(1)
VALLONE	0-4-0T	B?2t	Maffei	3609	1911	(b)	(2)

(a) purchased from Bernische Braunkohlengesellschaft AG, Gondiswil /1920.
(b) see note for this locomotive under Robert Aebi & Cie AG, Regensdorf; purchased from Robert Aebi & Co, Zürich /1919.

(1) 1920 to Impresa costruzioni Ing. Bianchi, Lugano.
(2) 1927 sold to Grimsel-Staumauer AG, Meiringen.

ANTONIETTI & BRAUCHI, Lugano — TI07 C

A construction concern.

Gauge: 750 mm			Locn: ?			Date: 01.01.1994	
ROSINA	0-4-0T	Bn2t	O&K 50 PS	2391	1907	(a)	(1)

(a) new to Hermann Sohn, Thorn; ex O&K, Zürich /1922.

(1) in or before 1926 sold to Robert Aebi, Zürich.

ASSOCIAZIONE LAVORI PIORA, Faido — TI08 C

Ca. 1996 there was a project to investigate the stability of the rocks under the Piora Mulde through which the Gotthardbasis tunnel was to be driven. Murer AG, Erstfeld provided the locomotives. An exploratory tunnel was driven with an exit nearer Polmengo than Faido. On completion of the work the tunnel was retained for access purposes, but without a railway.

"1" Schöma 5189 of 1991 (type CFL 180DCL) waits outside the tunnel on 22 October 1996.

Gauge: 750 mm				Locn: 266 $ 7027/1494			Date: 12.2007	
1		4wDH	Bdh	Schöma CFL 180DCL	5189	1991	(a)	(1)
2		4wDH	Bdh	Schöma CFL 180DCL	5190	1991	(a)	(1)
	1674	4wDH	Bdh	Diema DS40	2391	1960	(c)	s/s?

3	1675	4wDH	Bdh	Schöma CFL 180DCL	4999	1989	(d)	s/s?	

(a) ex ArGe EKW, Ramosch.
(c) ex C. Zschokke AG, Genève.
(d) ex ArGe Pierre Pertuis, Sonceboz.
(1) returned to Schöma.

A. BACCIARINI & CIE   TI09 C
alternative: Bacciarini & Co
A civil engineering firm.

Ambri
Gauge: 600 mm   Locn: ?   Date: 01.1994
RITOM   0-4-0T   Bn2t   Maffei 40 PS   2920   1909   (a)   (1)

(a) Maffei lists show it as 1000 mm; new via Robert Aebi & Co, Zürich.
(1) from 1915 used by L. Boretto & Maffiotti; 1927 to Impresa costruzioni Taddei Domenico, Castagnola.

Unknown location
Gauge: 750 mm   Locn: ?   Date: 01.1994
CERESIO   0-4-0T   Bn2t   O&K 50 PS   2110   1906   (a)   (1)

(a) ex Pandolfi & Co, Lugano /1917-20.
(1) by 1929 to Fasoletti & Malfanti Impresa costruzioni, Lugano-Viganello.

O. BETTELINI, Bellinzona   TI10 C
A public works contractor ?
Gauge: ? mm   Locn: ?   Date: 11.2003
          4wBE   Ba   Oehler G3053   1136   1958   new   s/s

BIANCHI & HYOS, Airolo   TI11 C
A civil engineering firm ?
Gauge: 600 mm   Locn: ?   Date: 05.2002
          0-6-0T   Cn2t   O&K 20 PS   3720   1910   new   s/s

G. CAPRIOGLIO E. C., Magadino   TI12 C
A civil engineering firm (?).
Gauge: 700 mm   Locn: ?   Date: 04.2002
          0-4-0T   Bn2t   Krauss-S IV q   799   1879   new   s/s

CARAVATI (?) & BIANCHI, ENTREPRENEURS, Melide   TI13 C
A civil engineering firm.
Gauge: 750 mm   Locn: ?   Date: 1994
MELIDE   0-4-0T   Bn2t   Henschel   6525   1903   (a)   (1)

(a) new to M. Brenner, Magdeburg [D]; purchased ex ? /1912.
(1) in or before 1928 to Impresa costruzioni Ing. Bianchi, Lugano.

CARTIERE CHAM-TENERO AG, Tenero　　　　　　　　　　TI14

formerly: Cartiera di Locarno SA, Tenero (till 1978)
alternative: Fabbrica di Cartone duro Zurigo SA

These paper mills were founded in 1854. They became part of the Papierfabrik Cham concern in 1978 and was closed in 2007 (locomotive status in 2008 unknown). A narrow gauge steam locomotive was hired, probably for use as a stationary boiler.

Gauge: 1435 mm　　　　　　　　Locn: 276 $ 7094/1149　　　　　　　Date: 02.2005

	4wPM	Bbm	RACO 40 PS	11	1930	(a)	(1)
	4wDH	Bdh	KHD A6M617 R	57906	1965	(b)	(2)
	4wDM	Bdm	RACO 85 LA7	1723	1965	(c)	

Gauge: 1000 mm　　　　　　　　Locn: ?　　　　　　　　　　　　　Date: 02.2008

	0-4-0T	Bn2t	Jung 50 PS	920	1905	(d)	(4)

(a) ex Gaswerk der Stadt Zürich, Schlieren /19xx.
(b) new to DB "Köf 6806"; ex DB "323 326-9" /1985.
(c) ex FFS "Tm^l 499" /1994.
(d) hired from Robert Aebi & Cie AG /1944.
(1) 1985 stored; 1994-95 s/s.
(2) 1994-98 s/s.
(4) returned to Robert Aebi & Cie AG.

SOCIETÀ ANOMINA CEMENTI ARMATI CENTIFUGATI (SACAC), Bodio　　TI15

By 1987 this cement company (as Sacac Schleuderbetonwerk SA) was located in via ai Salici, 6514 Sementina to the south west of Bellinzona, just across the river Ticino from Giubiasco. The history of the Bodio operation has not been determined; as the time between arrival and departure of the locomotive is very short the planned operation may not have come to fruition.

Gauge: 1435 mm　　　　　　　　Locn: ?　　　　　　　　　　　　　Date: 04.1999

	4wDM	Bdm	O&K MV2a	25728	1957	(a)	(1)

(a) new via MBA, Dübendorf.
(1) 1957 to Interfrigo (IF), Basel Badischer Bahnhof "270".

CENTRALE PELLI & FONDITOIO GRASSI SA, Bellinzona　　TI16

This factory was a subsidiary of Häute- und Fettwerk AG (HFZ), Zürich. The factory prepared skins and fats and was rail connected to the FFS marshalling yard. Other variations of the title are known in railway enthusiast literature, but the above corresponds to the text on the building in the two references below.

Gauge: 1435 mm　　　　　　　　Locn: ?　　　　　　　　　　　　　Date: 06.2008

	4wDH	Bdh	ABL IV NHT	4702	1956	(a)	(1)

(a) Breuer style with HydroTitan transmission; ex Von Roll, Gerlafingen "8" /1973.
(1) 1980-86 to Häute und Fettwerk AG, Zürich.

　　see: Lok Report 11/2007 page 7
　　see: www.seak.ch

CLUB DEL SAN GOTTARDO (CSG), Mendrisio — TI17 M

alternative: Associazione Club del San Gottardo

This preservation group operates steam and diesel hauled trains on the otherwise closed international route from Mendrisio via Stabio to Valmorea-Ródera in Italy. It also runs electrically hauled trains to local destinations on the FFS. It is intended that the line to Stabio should re-open in 2013; this will affect the present operation.

Gauge: 1435 mm   Locn: 296 $ 7194/8008   Date: 01.2008

8463	0-6-0WT	Cn2t	SLM E 3/3	1623	1904	(a)
8501	0-6-0WT	Cn2t	SLM E 3/3	2076	1910	(b)
Ce 6/8 ^{II} 14276	2-6-6-2WE	1CC1w2				
			SLM,MFO Ce 6/8^{II}	2773, ?	1922	(c)
42	0-4-4-0WE	BBw2	SLM,BBC ex Ce 4/6	2694,1320	1919,20	(d)
6	2-6-2T	1C1h2t	Maffei	3126	1910	(e)
Tm^{II} 725	4wDM	Bdm	RACO 95 SA3 RS	1662	1963	(g)
Tm^{II} 769	4wDM	Bdm	RACO 95 SA3	1431	1953	(h)
Tm^{II} 785	4wDM	Bdm	RACO 95 SA3	1439	1955	(i)
836 506-6 LEU	0-6-0DE	Cde	SLM,BBC Em 3/3	4361, ?	1959	(j)
DE.500-2	BoBoDE	BBde	unknown	?	19xx	(k)
	BoBoWERC					
		BBw4	SIG,BBC	?	1966	(l)
	4wDM	Bdm	RACO 85 LA7	1599	1961	(m)
	0-4-0WE	Bw1	SLM Te^{I}	3987	1949	(n) b2008 s/s
Ae 6/6 11401 TICINO	CoCoWE	CCw6	SLM,BBC Ae 6/6	4050,5429	1951	(o)

(a) new to SBB "8463"; ex Gotthardwerke AG, Bodio /1995.
(b) new to SBB "8501"; ex HOVAG, Ems via Oswaldsteam, Samstagern /1993.
(c) ex SBB "Ce 6/8"14276" /1986.
(d) new to GBS "Ce 4/6 312"; ex SZU "Ce 4/4 312" /1995.
(e) new to BT "6"; later SBB "5886"; from Degersheim /2004; at Mendrisio /2007.
(g) ex FFS "Tm ^{II} 725" /2007.
(h) ex FFS "Tm ^{II} 769" /200x.
(i) ex FFS "Tm ^{II} 785" /200x.
(j) ex SZU "Em 3/3 11" /2006.
(k) ex FNM /200x.
(l) new to VHB "ABDe 4/4 251"; later RM "251"; SOB "BDe 576 060" 3/2004; to CSG 25/2/2005.
(m) new to SBB "Tm^{I} 325"; purchased from SBB "Tm^{I} 425" /2008.
(n) new as SBB "Te^{I} 254"; later "Te^{I} 47", ex SBB by 8/2000 (loan or purchase?).
(o) ex FFS "Ae 6/6 11401 TICINO" expected /2009.

see: www.clubsangottardo.ch/default.asp?nid=4&lid=5

COMBONI & FELTRINELLI, Cadenazzo — TI18

Comboni & Feltrinelli is an Italian firm. The locomotive was probably exported directly to Italy through Cadenazzo.

Gauge: 750 mm   Locn: ?   Date: 01.1994

0-4-0T	Bn2t	SLM	84	1881	(a) ?

(a) presented at the "Guidovie a vapore" in Mantua [I] /1902.

CONSORZIO BRONTALLO, Menzonio — TI19 C

The work performed by this consortium is not known. The location is near Bignasco.

Gauge: 600 mm   Locn: ?   Date: 08.2002

4wDM	Bdm	O&K MV0	25162	1952	(a) (1)
4wDM	Bdm	O&K MV1	25228	1952	(a) s/s

(a) purchased new from MBA, Dübendorf.
(1) 19xx reported at Mannesmann Handel, Oberhausen [D].

CONSORZIO CORREZIONE MAGGIA, Locarno   TI20 C

A consortium responsible for modifying the course of the river Maggia.

Gauge: 1000 mm   Locn: ?   Date: 04.2002

MAGGIA   0-4-0WT   Bn2t   Jung 50 PS   104   1891   (a)   (1)

(a)   new via F. Marti, Winterthur.
(1)   1953 removed from boiler register; to children's play area at "Albergo Losone", Losone.

CONSORZIO GOTTARDO SUD (GCS), Airolo   TI21 C

This consortium constructed the southern end of the Gotthard road tunnel. One of the participants was Walo Bertschinger of Zürich. This firm was responsible for the railway operation: the equipment used came almost entirely from German open-cast brown-coal mining operations. The BoBo electric locomotives (ex 1200V DC) were prepared on the site of Schlieren gasworks. The equipment chosen came almost entirely from open cast brown coal mining operations. The larger electric locomotives were coupled to a generator wagon as required.

The standard gauge Breuer locomotive (1235) in company of ex-RBW 165 (Krupp 3464 1955, type 200 PS) on 30 May 1976.

On the same day CGS "1" (Schalke, BBC 57626/5740 of 1959) stands near the tunnel entrance with its generator wagon and in the company of overhead maintenance equipment.

Gauge: 1435 mm				Locn: 265 $ 6891/1536				Date: 05.1976
ATEL No. 1		4wPM	Bbm	Breuer 40 PS		1235	c1929	(a) 19xx scr
Gauge: 900 mm				Locn: 265 $ 6895/1535				Date: 06.1988
	1	BoBoWE	BBg4	Schalke,BBC Abraumlok				
					57626/5740	1959	(b)	(2)
FAVRE	2	BoBoWE	BBg4	Schalke,BBC Abraumlok				
					57625/5739	1959	(c)	(2)
	3	BoBoWE	BBg4	Schalke,BBC Abraumlok				
					57624/5738	1959	(d)	(2)
		BoBoWE	BBg4	Schalke,BBC Abraumlok				
					57623/5737	1959	(e)	19xx scr
BIRICCHINO	4	0-6-0DH	Cdh	Henschel DH 360C		26759	1961	(f) (2)
LISA	11	4wW+DE	Bg+de	Henschel,SSW		28318,5722	1955,6	(g) 19xx scr
	12	4wW+DE	Bg+de	Henschel,SSW		28319,5725	1955,6	(h) 197x scr
CRISTALLINA	21	0-4-0DH	Bdh	O&K MV9		25833	1958	(i) (9)
	23	0-4-0DH	Bdh	O&K MV9		25834	1958	(j) 1977 scr
	22	0-4-0DH	Bdh	O&K MV9		25892	1959	(k) (11)
'24 or 25'		4wDH	Bdh	Krupp 200 PS		3464	1955	(l) (12)
'25 or 24'		4wDH	Bdh	Krupp 200 PS		3465	1955	(m) (12)
	71	4wDRC	Bd	unknown		?	19xx	(a) s/s
		4wBE	Ba	SIG				(o) s/s

(a) ex? /19xx. It would be logical for this locomotive to have come from Fratelli Tenconi SA and thus be ex FFS "Tm 449". However it has been recorded as having a plate "ATEL No. 1" (ATEL = Aare-Tessin AG für Elektrizität) and so possibly the locomotive from ATEL, Lavorgo. But the plate may have had nothing to do with the locomotive.
(b) ex Rheinische Braunkohlenwerke (RBW) [D] "245" /1971.
(c) ex RBW "244" /1971.
(d) ex RBW "243" /1971.
(e) ex RBW "242" /1971; used as a spare parts bank and never came to Airolo; remains scrapped at Schlieren.

(f)	ex Braunkohlen- und Brikett-Industrie Aktiengesellschaft (BUBIAG) - Braunkohlenwerk Frielendorf [D]; later running number "003".	
(g)	ex RBW "268" /1970.	
(h)	ex RBW "269" /1970.	
(i)	ex RBW "151" /1969.	
(j)	ex RBW "162" /1969; some reports exchange the identities of "22" and "23".	
(k)	ex RBW "159" /1971; some reports exchange the identities of "22" and "23".	
(l)	ex RBW "164" /1973; there is no evidence to indicate which of the pair was allocated "24" or "25".	
(m)	ex RBW "165" /1973; there is no evidence to indicate which of the pair was allocated "24" or "25"; one retained the RBW running number "165" in 6/1976.	
(o)	the frame of a SIG battery locomotive was seen here in 5/1976; details required.	
(2)	1978 to Consortium BEY, Chavornay.	
(9)	1979-82 to Kummler + Matter, Zürich (1000 mm) "Tm 2/2 21" via RhB/LSB/Tafag & used on AB and RhB; some reports suggest that this one did not go further than Tafag.	
(11)	1979-82 to RhB/LSB/Tafag; some reports suggest that this one became Kummler + Matter, Zürich (1000 mm) "Tm 2/2 21".	
(12)	197x to Consortium BEY, Chavornay.	

CONSORZIO MORROBIA, Bellinzona   TI22 C

A consortium of public works contractors?

Gauge: ? mm   Locn: ?   Date: 11.2003

    4wBE   Ba   Oehler G2049   779   1950

CONSORZIO TAT, Pollegio   TI23 C

alternative: ArGe TAT
TAT = Tunnel AlpTransit

The work of constructing the southern end of the Gotthardbasis tunnel for AlpTransit Gotthard AG has been awarded to a consortium formed by Zschokke Locher AG, Zürich, Alpine Mayreder GmbH, Salzburg [A], CSC Impresa Costruzioni SA, Lugano, Impregilo S.p.A., S. Giovanni [I] and Hochtief AG, Essen [D]. This includes both the Faido and Pollegio work sites.

CH 263 TI

"581 002" Schöma 5736 2002 (type CFL 350DCL) heads a service train on 8 June 2007.

"582 009" Schöma 5884 of 2004 (type CFL 200DCL) carries a group of tourists along the Umgehungsstollen on 25 September 2004.

Gauge: 900 mm   Locn: 266 $ 7149/1364   Date: 06.2008

581 001	4wDH	Bdh	Schöma CFL 350DCL	5734	2002	new
581 002	4wDH	Bdh	Schöma CFL 350DCL	5736	2002	new
581 003	4wDH	Bdh	Schöma CFL 350DCL	5735	2002	new
581 004	4wDH	Bdh	Schöma CFL 350DCL	5737	2002	new
581 005	4wDH	Bdh	Schöma CFL 350DCL	5738	2002	new
581 006	4wDH	Bdh	Schöma CFL 350DCL	5740	2002	new
581 007	4wDH	Bdh	Schöma CFL 350DCL	5739	2002	new
581 008	4wDH	Bdh	Schöma CFL 350DCL	5741	2002	new
581 009	4wDH	Bdh	Schöma CFL 350DCL	5742	2002	new
581 010	4wDH	Bdh	Schöma CFL 350DCL	5743	2002	new
581 011	4wDH	Bdh	Schöma CFL 350DCL	5744	2002	new
581 012	4wDH	Bdh	Schöma CFL 350DCL	5745	2002	new
581 013	4wDH	Bdh	Schöma CFL 350DCL	5787	2003	new
581 014	4wDH	Bdh	Schöma CFL 350DCL	5788	2003	new
581 015	4wDH	Bdh	Schöma CFL 350DCL	5789	2003	new
581 016	4wDH	Bdh	Schöma CFL 350DCL	5790	2003	new
581 017	4wDH	Bdh	Schöma CFL 350DCL	5791	2003	new
581 018	4wDH	Bdh	Schöma CFL 350DCL	5792	2003	new
581 019	4wDH	Bdh	Schöma CFL 350DCL	5793	2003	new
581 020	4wDH	Bdh	Schöma CFL 350DCL	5794	2003	new
581 021	4wDH	Bdh	Schöma CFL 350DCL	5795	2003	new
581 022	4wDH	Bdh	Schöma CFL 350DCL	5796	2003	new
581 023	4wDH	Bdh	Schöma CFL 350DCL	5797	2003	new
581 024	4wDH	Bdh	Schöma CFL 350DCL	5798	2003	new
581 025	4wDH	Bdh	Schöma CFL 350DCL	5876	2004	new
581 026	4wDH	Bdh	Schöma CFL 350DCL	5877	2004	new
581 027	4wDH	Bdh	Schöma CFL 350DCL	5878	2004	new
581 028	4wDH	Bdh	Schöma CFL 350DCL	5879	2004	new
581 029	4wDH	Bdh	Schöma CFL 350DCL	5880	2004	new
581 030	4wDH	Bdh	Schöma CFL 350DCL	5881	2004	new
581 031	4wDH	Bdh	Schöma CFL 350DCL	6038	2005	new
581 032	4wDH	Bdh	Schöma CFL 350DCL	6039	2005	new
581 033	4wDH	Bdh	Schöma CFL 350DCL	6041	2005	new

581 034	4wDH	Bdh	Schöma CFL 200G	6070	2006	new
581 035	4wDH	Bdh	Schöma CFL 350DCL	6040	2005	new
581 036	4wDH	Bdh	Schöma CFL 200G	6071	2006	new
581 037	4wDH	Bdh	Schöma CFL 350DCL	6068	2005	new
581 038	4wDH	Bdh	Schöma CFL 350DCL	6183	2007	new
581 039	4wDH	Bdh	Schöma CFL 350DCL	6184	2007	new
581 040	4wDH	Bdh	Schöma CFL 350DCL	6185	2007	new
	4wDH	Bdh	Schöma CFL 350DCL	6321	2008	new
	4wDH	Bdh	Schöma CFL 350DCL	6322	2008	new
	4wDH	Bdh	Schöma CFL 350DCL	6323	2008	new
	4wDH	Bdh	Schöma CFL 350DCL	6324	2008	new
	4wDH	Bdh	Schöma CFL 350DCL	6325	2008	new
	4wDH	Bdh	Schöma CFL 350DCL	6326	2008	new
	4wDH	Bdh	Schöma CFL 350DCL	6327	2008	new
582 001	4wDH	Bdh	Schöma CFL 200DCL	5730	2002	new
582 002	4wDH	Bdh	Schöma CFL 200DCL	5733	2002	new
582 003	4wDH	Bdh	Schöma CFL 200DCL	5731	2002	new
582 004	4wDH	Bdh	Schöma CFL 200DCL	5732	2002	new
582 005	4wDH	Bdh	Schöma CFL 200DCL	5799	2003	new
582 006	4wDH	Bdh	Schöma CFL 200DCL	5800	2003	new
582 007	4wDH	Bdh	Schöma CFL 200DCL	5801	2003	new
582 008	4wDH	Bdh	Schöma CFL 200DCL	5882	2004	new
582 009	4wDH	Bdh	Schöma CFL 200DCL	5883	2004	new
582 010	4wDH	Bdh	Schöma CFL 200DCL	6186	2007	new
582 011	4wDH	Bdh	Schöma CFL 200DCL	6187	2007	new
584 001	4wD	Bd	Schöma D60-20	5766	2003	new
584 002	4wD	Bd	Schöma D60-20	5767	2003	new

see: www.alptransit.ch/pages/d/aktuell/presseitem.php?id=27

CONSORZIO TICINESE TI24 C

A consortium of public works contractors?

Airolo

Gauge: ? mm Locn: ? Date: 11.2003

	4wBE	Ba	Oehler 1042	492	1942	new

Corcapolo (Intragna)

Gauge: 750 mm Locn: ? Date: 08.2002

	4wDM	Bdm	O&K MD 2b	25208	1951	(a) s/s

(a) new via MBA, Dübendorf.

CONSORZIO UTA, Bodio TI27 C

A consortium of public works contractors?

Gauge: ? mm Locn: ? Date: 11.2003

	4wBE	Ba	Oehler G2050	958	1951	new s/s

CONSORZIO VERBANO, Locarno TI28 C

A consortium of public works contractors?

Gauge: ? mm				Locn: ?					Date: 05.2002
		4wDM	Bdm	O&K(S) MD 2		1517	194x	(a)	s/s?
(a)	1952 ex MBA, Dübendorf.								

CORREZIONE DEL FIUME TICINO, Bellinzona — TI29 C

A consortium of civil engineering contractors modifying the course of the river Ticino; the work commenced in 1886.

Gauge: 1000 mm				Locn: ?					Date: 1994
2	TICINO	0-4-0WT	Bn2t	Jung 50 PS	59	1889	(a)	(1)	
	BRENNO	0-4-0T	Bn2t	Jung 50 PS	60	1889	(b)	s/s	
	MOESA	0-4-0T	Bn2t	Jung 50 PS	178	1893	(c)	(3)	
	MOROBBIA	0-4-0T	Bn2t	Jung 50 PS	1310	1908	(d)	(4)	

(a) purchased new via F. Marti, Winterthur. The following details are recorded; height 2800 mm, width 1900 mm, wheel diameter 600 mm, cylinders 200*300 mm, 68 tubes 1960 mm long, 11bar pressure.
(b) new via F. Marti, Winterthur; still in use /1905.
(c) new via F. Marti, Winterthur.
(d) new to H. von Arx, Zürich; possibly for Italy.
(1) 1941 removed from boiler register; donated to Scuola di Arti e Mestieri, Bellinzona; 1961 to Guido Travaini, Mendrisio.
(3) 1943 to Losinger & Cie, Bern.
(4) 1941 removed from boiler register; 1943 to Losinger & Cie, Bern.

CORREZIONE DEL FIUME VEDEGGIO, Taverne — TI30 C

The firm of Maspoli & Cie, Taverne modified the course of the river Vedeggio.

Gauge: 1000 mm			Locn: ?					Date: 1994
VEDEGGIO	0-4-0T	Bn2t	Jung 50 PS	920	1905	(a)	(1)	
CERESIO	0-4-0T	Bn2t	O&K 50 PS	2110	1906	(a)	(2)	

(a) new via F. Marti, Winterthur.
(1) 1912 to Robert Aebi & Co, Zürich.
(2) 1917 to Pandolfi & Co, unknown location (750 mm).

COSTRUZIONI STRADALI E CIVILI SA, Lugano — TI30a C

The SIG list of references gives 15 ETB 70 locomotives delivered here.

CSC IMPRESA COSTRUZIONI SA, Lugano — TI31 C

The following locomotives were delivered to this construction company, but where they were employed is not known. Office address: Via Pioda 5, 6900 Lugano.

Gauge: 750 mm			Locn: ?				Date: 05.2009
	4wDH	Bdh	Schöma CFL 180DCL	5530	1997	new	
	4wDH	Bdh	Schöma CFL 180DCL	5531	1997	new	

ENTREPRISE LOUIS FAVRE, Airolo — TI32 C

This engineering firm had a base here 1871-83 and was responsible for boring the first Gotthard tunnel linking Göschenen with Airolo (opened 1882).

Gauge: 1000 mm			Locn: 265 $ 6893/1536				Date: 06.1988	
	0-4-0	Bn2	Canada	114	1863	(a)	s/s	
1 REUSS	0-4-0T	Bn2t	Schneider 50 bis	1559	1873	(b)	s/s	
2 TESSIN	0-4-0T	Bn2t	Schneider 50 bis	1560	1873	(b)	s/s	
3	0-4-0CA	Bp2	Schneider 50 ter	1715	1874	new	s/s	
4	0-4-0CA	Bp2	Schneider 50 ter	1716	1874	new	s/s	
5	0-4-0CA	Bp2	Schneider 50 ter	1863	1876	new	s/s	
6	0-4-0CA	Bp2	Schneider 50 ter	1864	1876	new	s/s	
7	0-4-0CA	Bp2	Schneider 50 ter	2000	1879	(h)	s/s	
8	0-4-0CA	Bp2	Schneider 50 ter	2001	1879	(h)	s/s	

(a) this locomotive was tested on a section of the Cromford and High Peak Railway in England; operated on the temporary Mt. Cenis railway between France and Italy; became LEB "1 LAUSANNE"; and was then apparently used on this contract though no photographic evidence has emerged.
(b) new; also stated to have been of type 50 ter which used compressed air.
(h) according to the manufacturer's records supplied as type 50 bis which had a firebox; other sources quote these two as Schneider 2001 and 2002.

FABBRICA DI CARTONE DURO ZURIGO SA, Tenero — TI33

A cardboard making factory. The relation to Cartiera di Locarno SA, Tenero, if any, is not known.

Gauge: 1000 mm			Locn: 276 $ 7237/1201 and elsewhere				Date: 04.2005
VEDEGGIO	0-4-0T	Bn2t	Jung 50 PS	920	1905	(a)	(1)

(a) ex Robert Aebi & Co, Zürich /1944; probably obtained for use as a stationary boiler.
(1) 1947 removed from boiler register.

FERROVIE E AUTOLINEE REGIONALE TICINESSE (FART), Locarno-S. Antonio — TI34 M

One locomotive from the construction period has been identified. Heritage stock from this company may be present. Tramcar "Ce 2/2 7" is also usually present.

Gauge: 1000 mm			Locn: 276 $ 7033/1141				Date: 09.2007
	0-6-0Tm	Cn2tk	SLM G 3/3	456	1887	(a)	s/s
BCFe 4/4 17	BoBoWERC BBg4		CEM,BBC BCFe 4/4	? ,1968	1923-4	(b)	

(a) provided by Sutter, Lugano for both the FRT (Centovalli) and the LPB (Maggiatalbahn) after /1915.
(b) new to SSIF "BCFe 4/4 17".

FERRIERE CATTENEO SA, Giubiasco — TI35

A workshops making metallic structures and also railway freight vehicles. The firm was founded in 1932 and established on the site of "Acciaierie Elettriche del San Gottardo" that closed in 1925.

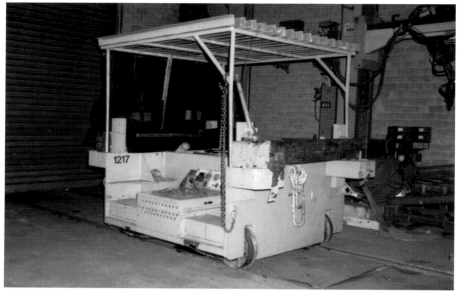

ABL (Badoni) 279 of 1962 (type 1L) inside the works on 3 August 2000.

Gauge: 1435 mm   Locn: 276 $ 7213/1151   Date: 02.2003

1215	4wPM	Bbm	Unilok A2500	A122	19xx	(a)	s/s
1216	4wDM	Bdm	Diema DVL15/3	3229	1971	(b)	2000 scr
1217	4wD	Bd	ABL 1L	279	1962	(c)	
1218	4wDM	Bdm	CMR Tm^l	'Tm^l 318'	1960	(d)	

(a) built by Jung.
(b) new on 18/1/1972 (what did the plate show?); ex Dätwyler AG, Altdorf UR /1988 via LSB.
(c) ex ? /199x-96.
(d) new to FFS "Tm^l 318"; ex FFS "Tm^l 418" /1996.

FERROVIA DEL MONTE GENEROSO (FMG), Capolago   TI36 M

Gauge: 800 mm + Abt rack   Locn: 296 $ 7196/0848   Date: 09.2007

H	2	0-4-2T	B/a1n2t SLM H	604	1890	(a)

(a) new; locomotive was on static display at Capolago and possibly elsewhere from 19xx till 199x.

FERROVIA MESOLCINESE, Castione-Arbedo   TI37 M

operated by: Società Esercizio Ferroviario Turistico (SEFT)

The RhB withdrew services from the remaining part of the isolated metre gauge line between (Bellinzona-) Castione-Arbedo and Mesocco as from 31 December 2003. Since then SEFT have operated a tourist service over the line. Address: Casella Postale 2612, 6500 Bellinzona.

Gauge: 1000 mm   Locn: 276 $ 7237/1201 and elsewhere   Date: 04.2005

ABe 4/4	1	BoBoWERC	SIG,MFO	?	1933	(a)
ABe 4/4	2	BoBoWERC	SIG,MFO	?	1933	(b)
BCFe 4/4	4	BoBoWERC	Ringhoffer,Rieter	?	1909	(c)
ABe 4/4	5	BoBoWERC	SWS,SAAS	?	1963	(d)
BDe 4/4	6	BoBoWERC	SWS,BBC	?	1958	(e)

(a) ex AB "ABe 4/4 41".

(b) ex AB "ABe 4/4 42".
(c) new to BM "BCe 4/4 4"; ex RhB "ABDe 4/4 454".
(d) new to BA "ABe 4/4 5"; later OJB "Be 4/4 80"; ex ASm (ex BTI) "Be 4/4 521".
(e) ex RhB "BDe 4/4 491"

see: www.seft-fm.ch/index99.html

FERROVIE LUGANESI SA (FLP), Agno　　　　　　　　　　　　TI38 M

When the Tramvie elettriche Lugano (TEL) closed some of its lines in 1954 one of the trams (Ce 2/2 3 SWS,EGA of 1910) passed to the Ferrovia Lugano-Ponte Tresa for use on service trains. Eventually it became their "Ce 2/2 6" and was withdrawn and sold for scrap in 1970. Nevertheless it has survived inside what has since become a bus depot. Address: Via Stazione 8, 6982 Agno.

FERT SA, Castione-Arbedo　　　　　　　　　　　　　　　　TI39

formerly:　Lorenzo Cattori

A builders' merchant that closed in 2000.

Gauge: 1435 mm　　　　　　　Locn: 276 $ 7236/1202　　　　　　　　Date: 03.2003

| | | 4wPM | Bbm | RACO RA11 | 1257 | 1939 | (a) | (1) |

(a) ex Alusuisse, Chippis, VS /1952 via K. Kaufmann, Thörishaus.
(1) 1980 OOU; 2003-05 scrapped.

FILIPINI & FILIO, Airolo　　　　　　　　　　　　　　　　TI40

formerly:　Fratelli Tenconi SA (till 1982)

A mechanical engineering workshop founded in 1871. At least the last locomotive is stated to have been the private property of the owner and was disposed of after his death in 1982. A road tractor equipped with rail buffers has performed shunting operations since then.

Gauge: 1435 mm　　　　　　　Locn: 266 $ 6902/1536　　　　　　　　Date: 01.2008

	Bbm	4wPM	Breuer III	?	1929	(a)	s/s
388	4wDH	Bdh	O&K MV2a	26191	1962	(b)	(2)
	4wDH	Bdh	Diema DVL150/1.2	3238	1972	(c)	(3)

(a) ex FFS "Tm 449" 10/1961.
(b) new via MBA, Dübendorf; it is not clear when the running number "388" was applied.
(c) new via Asper.
(2) 1972-77 to Thévenaz-Leduc, Ecublens VD.
(3) 1987 to Albarelli, Chiasso [I] (but listed here under TI).

FISCHER & RECHSTEINER, Chiasso　　　　　　　　　　　TI41 C

A public works contractor?

Gauge: 600 mm　　　　　　　Locn: ?　　　　　　　　　　　　　　Date: 05.2002

| | 4wDM | Bdh | O&K(M) RL 1a | 4109 | 1930 | (a) | s/s |

(a) purchased via O&K/MBA, Dübendorf.

GALLERIA BAUMGARTNER, Mendrisio　　　　　　　　　　TI42 M

This model railway shop has a rebuilt Traktor on display. Address: Via S. Franscini 24, 6850 Mendrisio.

Gauge: 1435 mm		Locn: ?					Date: 05.2009	
TAe 5		0-4-0WERC						
			Bw1	SLM,MFO	3991	1949	(a)	display
(a)	new as FFS "Te^1 260"; later "Te^1 960"; ex Salon Train Special, Rothenburg "TAe 5".							

GALLERIA DI BASE DEL FURKA, lotto 63, Bedretto — TI43 C

As part of the construction of the Furka-Basistunnel (1973-82) a 5.2 kilometre long side tunnel was built on the south side, leading towards Airolo. This was for access purposes only. The same contractors of Evéquoz & Cie SA, Gebrüder Arnold AG and Gebrüder Bonetti AG held the contact for both Los 61 and Lotto 63. Railway details are not known.

HOTEL CORONADO, Mendrisio — TI44 M

Tramcar "Ce 2/2 3" from the Tram elettrici mendrisiensi (TEM) was displayed on the Campo Sportivo 1968-76 and stored in Stabio 1980-2006. In both cases it was labelled "2". It has been restored to its original identity and been put on display in Mendrisio 10/2006.

IMPRESA COSTRUZIONI FASOLETTI & MALFANTI, Lugano-Viganello — TI45 C

A public works contractor.

Gauge: 750 mm		Locn: ?					Date: 1994	
CERESIO		0-4-0T	Bn2t	O&K 50 PS	2110	1906	(a)	(1)
(a)	ex Bacciarini & Co, unknown location /1920-29.							
(1)	1931 removed from boiler register; 1931-43 to Hoch- & Tiefbau AG, Aarau.							

IMPRESA COSTRUZIONI GOBBA & BONDIETTI, Locarno — TI46 C

A construction concern.

Gauge: 600 mm (gauge not confirmed)		Locn: ?					Date: 01.1994	
		0-6-0T	Cn2t	Maffei	3508	1909	(a)	(1)
(a)	see note for this locomotive under Robert Aebi & Cie AG, Regensdorf, new as 750 mm; ex Robert Aebi & Co /1932 (or 1942).							
(1)	in or before 1942 sold to Torfwerke AG (TOWAG), Märwil TG.							

IMPRESA COSTRUZIONI H. HATT-HALLER, Giornico — TI47 C

This firm of construction engineers (based in Neuhausen am Rheinfall and Zürich) had a yard near Giornico. One locomotive was seen there after a contract, details not available.

Gauge: 600 mm		Locn: 266 $ 7113/1393				Date: 10.1967
	4wBE	Ba	unknown	?	19xx	OOU

IMPRESA COSTRUZIONI ING. BIANCHI, LUGANO, Lugano — TI48 C

A civil works contractor.

Gauge: 750 mm			Locn: ?				Date: 01.1994		
MELIDE	0-4-0T	Bn2t	Henschel		6525	1903	(a)	(1)	
SAN PAOLO	0-4-0T	Bn2t	O&K 50 PS		4347	1911	(b)	(2)	

(a) new to M. Brenner, Magdeburg [D]; ex Caravati (?) & Bianchi, Melide in or before /1928.
(b) ex Antognini & Noli, Bellinzona /1920; belonged to the firm (Bianchi &) Pagani in or before /1942.
(1) in or before 1947 to Franz Stirnimann, Olten-Hammer.
(2) 1946 to Robert Aebi & Co, Zürich.

IMPRESA COSTRUZIONI MUTTONI & CATTANEO   TI49 C

A civil engineering firm.

Biasca

Gauge: 1000 mm			Locn: ?				Date: 01.1994	
VERBANO	0-4-0T	Bn2t	Jung 50 PS		105	1891	(a)	(1)

Faido

Gauge: ? mm			Locn: ?				Date: 05.2002	
	4wPM	Bpm	O&K(M) S10		1969	c1924	(b)	s/s

(a) formerly J. Rieter & Cie AG, Winterthur probably during the construction of the Bellinzona-Mesocco railway; ex ? /1923.
(b) new via MBA, Dübendorf.
(1) 1932 removed from boiler register; s/s.

IMPRESA COSTRUZIONI TADDEI DOMENICO, Castagnola   TI52 C

A civil engineering firm.

Gauge: 600 mm			Locn: ?				Date: 1994	
RITOM	0-4-0T	Bn2t	Maffei 40 PS		2920	1909	(a)	(1)
	4wDM	Bdm	O&K(M) RL 1a		3946	1930	(b)	s/s
	4wDM	Bdm	O&K(M) RL 1a		3947	1930	(b)	s/s

(a) ex A. Bacciarini & Cie, Ambri /1927.
(b) ex O&K/MBA, Dübendorf.
(1) 1940 to Gebrüder Caprez Erben, Chur.

IMPRESE TICINESE, Bellinzona   TI53 C

A consortium of public works contractors.

Gauge: ? mm			Locn: ?			Date: 11.2003	
	4wBE	Ba	Oehler G2050		998	1954	new

INDUSTRIA TICINESE LATERIZI, Balerna   TI54

A brickworks.

Gauge: 600 mm			Locn: ?				Date: 05.2002	
	4wDM	Bdm	O&K(M) RL 1		3616	1929	(a)	s/s
	4wDM	Bdm	O&K(M) RL 1		3763	1929	(b)	s/s
	4wDM	Bdm	O&K(M) RL 1a		5680	1934	(c)	s/s

(a) ex Civeli, Castione.

(b) ex O&K/MBA, Dübendorf.
(c) ex Jean Chiavazza, Saint-Prex.

LGV IMPRESA COSTRUZIONE, Osogna — TI54a C

The yard of this Bellinzona based construction firm is located in the Industrie Sud of Osogna. The SIG list of references gives one ETS 35 locomotive delivered to Empresa Costruzioni SA, Bellinzona – probably the same concern. Office address: viale Officina 6, 6500 Bellinzona.

Gauge: 750 mm   Locn: 276 $ 719300,129370   Date: 05.2005

1	4wBE	Ba	SIG B 20 MG	807 809	1987	(a)	str
2	4wBE	Ba	SIG B 20 MG	806 308	1986	(b)	str

(a) new as 600 mm gauge.
(b) crash damage.

LOGISTIKBASIS DER ARMEE (LBA) — TI55

formerly: Bundesamt für Betriebe des Heeres (BABHE) (till 31.12.2003)

Bodio

formerly: Bundestankanlagen (BTA)
Fuel reservoirs run by Carbura SA.

Gauge: 1435 mm   Locn: 266 $ 7127/1377   Date: 12.2008

	4wDM	Bdm	RACO 30 PS	1259	1940	new	(1)
2	4wDM	Bdm	Kronenberg DT 90/120	128	1953	(b)	(2)

Claro

Military stores.

Gauge: 1435 mm   Locn: 276 $ 7217/1240   Date: 06.2007

Tm 98 85 5237 942-8	4wDE	Bde2	Stadler	117	1965	(c)	(3)
	4wD R/R	Bd	UCA UCA-Trac	?	2008	new	

(b) new; "21" in BABHE inventory.
(c) ex BABHE, Göschenen /2001-02 as "Tm 236 910-6", renumbered /2006-7.
(1) 1954 transferred to BTA, Zollikofen.
(2) 200x transferred to LBA, Munchenbuchsee.
(3) 2008-9 to OKK, Zollikofen (Münchenbuchsee) via Stauffer.

LONATI & CAVADINI, Lugano — TI56 C

A public works contractor?

Gauge: 600 mm   Locn: ?   Date: 05.2002

4wDM	Bdm	O&K(M) RL 1a	?	19xx	(a)	s/s
4wDM	Bdm	O&K(M) RL 1a	7858	1937	(a)	s/s

(a) ex O&K/MBA, Dübendorf.

ARTH. LUCHESSA, Bellinzona — TI57 C

A public works contractor?

Gauge: ? mm   Locn: ?   Date: 05.2002

4wDM	Bdm	O&K(M) RL 1a	4378	1931	(a)	s/s

a)   purchased via O&K/MBA, Dübendorf.

MAGGIAWERKE, Locarno   TI58 C

A consortium of public works concerns (?) formed to build hydroelectric plants.

Gauge: ? mm   Locn: ?   Date: 11.2003

4wBE	Ba	Oehler D1158	1163	1958	new	
4wBERC	Ba	Oehler	2139	1988	new	(1)

(1)   19xx for the construction site at Bavona (north east of Bignasco).

SOCIETÀ MARSAGLIA, Faido   TI59

Details unknown; data from SLM lists.

Gauge: 1435 mm   Locn: ?   Date: 11.2007

0-6-0WT	Cn2t	SLM		68	1880	new (1)

(1)   1882 to GB, Bellinzona "13".

MASPOLI & CIE, Taverne   TI60 C

A civil engineering concern? See also Correzione del Fiume Vedeggio, Taverne.

Gauge: 1000 mm   Locn: ?   Date: 0.2002

CERESIO	0-4-0T	Bn2t	O&K 50 PS	2110	1906	(a)	(1)

(a)   new via F. Marti, Winterthur.

(1)   1908-17 to Pandolfi & Co, unknown location (750 mm).

MATUCCI, Bellinzona   TI61 C

A civil engineering firm ?

Gauge: 600 mm   Locn: ?   Date: 05.2002

0-4-0T	Bn2t	O&K 30 PS	9001	1919	new	s/s

MIGROS TICINO SpA   TI62

S. Antonino

A distribution centre built in 1970 and rail connected to the Locarno line between S. Antonino and Cadenazzo.

Gauge: 1435 mm   Locn: 276 $ 7175/1127   Date: 02.2003

4wDM	Bdm	Kronenberg DT 90	110	1947	(a)	(1)
4wDH	Bdh	RACO 100 SA7 HT	1595	1961	(b)	(2)
4wDM	Bdm	RACO 85 LA7	1659	1963	(c)	

Taverne

A distribution centre connected to the FFS at Taverne-Torricella. The activity was transferred to S. Antonino in 1970.

Gauge: 1435 mm   Locn: 286 $ 7156/1022   Date: 04.1988

4wDM	Bdm	Kronenberg DT 90	110	1947	(d)	(1)

(a)	ex Migros Ticino SpA, Taverne /1970.
(b)	ex Weiacher Kies, Weiach /1975 via Asper.
(c)	new to FFS "Tm^I 360"; ex FFS "Tm^I 460" /2001.
(d)	ex Neptun AG, Basel ca. /1960.
(1)	1976 OOU; 1992 scrapped.
(2)	2001 sold to Fondeca SA, Cadenazzo for scrap.
(4)	1970 to Migros Ticino SpA, S. Antonino.

MONTEFORNO ACCIAIERIE E LAMINATORI SA (MF), Bodio   TI64

later: Parco Industriale e Immobiliare SA, Giornico (1994-2004)

These steelworks and rolling mills were built under the direction of von Roll in 1946, were taken into the group in 1977 and closed in 1996. In 1986 they had five locomotives, about 50 internal wagons and twenty-five kilometres of track.

From 1994 to 2004 Parco Industriale e Immobiliare SA, a subsidiary of von Roll, was responsible for the services for the companies established in the former steelworks.

During this period rolling stock from Classic Rail, Swisstrain and SBB-Historic was stored on and around the site.

Gauge: 1435 mm   Locn: 266 $ 7120/1379   Date: 06.1998

1	4wDM	Bdm	Kronenberg DT 90	115	1949	new	(1)	
2	0-4-0WT	Bn2t	SLM E 2/2	917	1895	(b)	(2)	
	4wDM	Bdm	Kronenberg DT 180/250	130	1953	new	(3)	
4	0-6-0WT	Cn2t	SLM E 3/3	1456	1902	(d)	1964 scr	
5	0-4-0DH	Bdh	Henschel DH 240B	30335	1962	new	(5)	
6	0-6-0DH	Cdh	Henschel DH 500Ca	30504	1962	new	(6)	
7	0-6-0DH	Cdh	Henschel DH 500Ca	30506	1962	new	(7)	
8	0-6-0DH	Cdh	Henschel DH 500Ca	30709	1964	new	(8)	
9	0-6-0DH	Cdh	Henschel DH 500Ca	31309	1968	new	a1996 s/s	
10	0-6-0DH	Cdh	Henschel DH 500Ca	30590	1965	(j)	(10)	
10	0-6-0DH	Cdh	Henschel DH 500Ca	31078	1964	(k)	(11)	
	4wDH	Bdh	KHD KG230 B	57869	1965	(l)	(12)	

(b)	purchased from Sulzer, Winterthur /1953.
(d)	purchased from FFS "E 3/3 8451" /1955.
(j)	Asper hire locomotive from /1981.
(k)	ex Rheinische Kalksteinwerke GmbH & Co KG (RKW), Wülfrath-Flandersbach [D] "5" /198x via WBB and Asper.
(l)	new to Vereinigte Tanklager und Transportmittel GmbH, Hamburg [D] "AE 3005" ; ex Preussag Metall AG, Hüttenwerk Harz, Goslar-Oker [D] /1981 via WBB/Layritz.
(1)	?1960 to Papierfabrik Ziegler, Grellingen.
(2)	19xx sold to Valmoesa SA, San Vittore.
(3)	after 1970 to Valmoesa SA, San Vittore.
(5)	198x to Nuova Ralfo s.r.l, Olginate [I]; later xx "T 503".
(6)	stored; ?1998 to Papierfabrik Biberist.
(7)	stored; 2002 to Tensol Rail SA, Bodio.
(8)	1996 to Stahl- und Walzwerke, Gerlafingen.
(9)	by 1996 to Papierfabrik, Biberist "2".
(10)	1983 to Makies, Zell LU.
(11)	1985 OOU; by 8/1990 to IPE, Pradelle di Nogarole Rocca [I].
(12)	198x to Asper.

PANDOLFI & CO, Unknown location   TI65 C

A civil engineering firm.

Gauge: 750 mm   Locn: ?   Date: 01.1994

CERESIO	0-4-0T	Bn2t	O&K 50 PS	2110	1906	(a)	(1)

(a)   ex A. Maspoli, Taverne (Correzione del Fiume Vedeggio) (1000 mm) /1917.

Gauge: 900 mm   Locn: 256 $ 6942/1815   Date: 06.2007

L01	4wDH	Bdh	Schöma CFL 180DCL	5275	1992	(a)
L02	4wDH	Bdh	Schöma CFL 180DCL	5276	1992	(a)
L03	4wDH	Bdh	Schöma CFL 180DCL	5803	2003	new
L04	4wDH	Bdh	Schöma CFL 180DCL	5804	2003	new
L05	4wDH	Bdh	Schöma CFL 180DCL	5805	2003	new
L06	4wDH	Bdh	Schöma CFL 180DCL	5806	2003	new
L07	4wDH	Bdh	Schöma CFL 180DCL	5807	2003	new
L08	4wDH	Bdh	Schöma CFL 180DCL	5808	2003	new
L09	4wDH	Bdh	Schöma CFL 180DCL	5809	2003	new
L10	4wDH	Bdh	Schöma CFL 180DCL	5810	2003	new
L11	4wDH	Bdh	Schöma CFL 180DCL	5811	2003	new
L12	4wDH	Bdh	Schöma CFL 180DCL	5812	2003	new
	4wDH	Bdh	Schöma CFL 180DCL	5134	1990	(m)
	4wDH	Bdh	Schöma CFL 180DCL	5933	2004	new
L13	4wDH	Bdh	Schöma CFL 180DCL	5934	2004	new
L16	4wDH	Bdh	Schöma CFL 180DCL	5935	2004	new
L17	4wDH	Bdh	Schöma CFL 180DCL	5966	2005	(q)
L18	4wDH	Bdh	Schöma CFL 180DCL	5967	2005	(q)
	4wDH	Bdh	Schöma CFL 180DCL	5968	2005	(q)
L20	4wDH	Bdh	Schöma CFL 180DCL	5969	2005	(q)
PL01	4wD RC	B	Schöma D60-20	5802	2003	new

(a) new to Ilbau (Isola & Lerchbaumer), Spittal [A] (750 mm); ex Schöma /1993.
(m) new to Trans-Manche Link (TML) [GB] "RS 032 MAREEN"; ex Schöma by 5/2005.
(q) new, Schöma works lists shows date as 2004.

ARBEITSGEMEINSCHAFT AMA, Amsteg   UR03 C

formerly: SBB - Kraftwerk Amsteg (1992-1999)

The original SBB power station was put into service in 1922. It was enhanced 1992-99 by the construction of Amsteg II. To facilitate this work a six kilometre long branch from Erstfeld was inaugurated on 18 March 1994. From 2001, the first 4.95 kilometres of this line was reused to serve a new installation for handling the spoil from the north end of Gotthardbasistunnel being built for AlpTransit Gotthard AG (ATG). The consortium AMA (Arnold & Co, Flüelen, Aggregat AG Erstfeld, Mattli Beton AG Wassen, Neue Agir AG Affoltern and Niederberger AG Stans) is performing the work. For details of the narrow gauge system in use in the tunnel see AGN, Los 252, Amsteg.

Gauge: 1435 mm   Locn: 246 $ 6940/1815   Date: 06.2002

1	0-6-0DH	Cdh	Henschel DH 500Ca	30590	1964	(a)	(1)
2	0-6-0DH	Cdh	Henschel DHG 500C	30858	1965	(b)	(2)
	0-6-0DH	Cdh	Jung R 40 C	12041	1961	(c)	(3)
837 905-9	6wDE	Cde	Moyse CN60EE500D	3548	1975	(d)	
837 953-9	6wDE	Cde	Moyse CN54EE500C	3523	1972	(e)	
837 954-7	6wDE	Cde	Moyse CN60EE500Ba	3546	1974	(f)	

(a) hired from Makies, Zell LU /1994; also recorded as DHG 500C.
(b) hired from Makies, Zell LU /1993.
(c) hired from LSB /1993.
(d) ex HOLCIM, Rekingen AG /2001; into service after overhaul by CFD Industries [F] /2002.
(e) ex Lorraine Charbonnière, Laneuveville devant Nancy (F) /2001 via CFD Industries [F].
(f) ex Cargotrans, Duisburg [D], via CFD Industries [F].
(1) 1994 returned to Makies, Zell LU.
(2) 1998-2000 returned to Makies, Zell LU.
(3) 199x returned to LSB.

see: KM Info (Kraftwerk Amsteg AG) Mai 1994

ARGE DRUCKSTOLLEN AMSTEG LOS 4, Amsteg   UR04 C

The contract was for the construction of a pressure tunnel as part of the upgrade of the SBB Amsteg hydroelectric plant (east of Amsteg) 1993-98. Walker-Porr AG, Altdorf UR were also involved in this work and some sources assign at least one of the Schöma locomotives to this company.

Gauge: 900 mm   Locn: 256 $ 6943/1800   Date: 01.1996

DL1	4wDH	Bdh	Schöma CFL 180DCL	5177	1990	(a)	
DL2	4wDH	Bdh	Schöma CFL 180DCL	5176	1990	(a)	
	4wDH	Bdh	GIA DHD35	3502	1993	(c)	

(a)   purchased from Trans-Manche Link (TML) Sangatte [F] "TU 20"; via Schöma?.
(c)   hired from Schlatter AG, Rheinsulz.

ARGE FURKA BASISTUNNEL (LOS 62), Realp   UR05 C

This consortium (Schmalz, Sion + Kopp, Brig + Socosa, Lausanne) was awarded the contract for the western end of the Furka-Basistunnel (1973-82). At least the following locomotives were used.

Gauge: 750 mm   Locn: 255 $ 6815/1607   Date: 06.1978

1	4wBE	Ba	SIG	?	19xx	s/s	
3	4wBE	Ba	SIG	?	19xx	s/s	
4	4wBE	Ba	SIG	?	19xx	s/s	
5	4wBE	Ba	SIG	?	19xx	s/s	
7	4wBE	Ba	SIG	?	19xx	s/s	

ARNOLD & CO AG, Flüelen   UR06

A firm created in 1891 for extracting gravel from beds of lakes. Here the source is the Vierwaldstättersee. At the time of writing (2008) the locomotive is rarely, if ever, used. Address: Seestrasse 11, 6454 Flüelen.

Gauge: 1435 mm   Locn: 246 $ 6902/1950   Date: 06.2002

	0-4-0DH	Bdh	R&H LSSH	497746	1963	(a)

(a)   new to Raffinerie du Sud-Ouest (RSO), Collombey, VS "4"; ex Rive Bleue Express, Le Bouveret, VS /1998 via Rhôna SA, Le Bouveret, VS.

GEBR. ARNOLD AG, Bürglen UR   UR06a C

The SIG list of references gives two ETB 70 locomotives delivered to this building firm.

BAUDEPARTEMENT DES KANTONS URI, Altdorf UR   UR07 C

The public works department of Canton Uri.

Gauge: 750 mm   Locn: ?   Date: 04.2002

SCHÄCHEN	0-4-0T	B?2t	Krauss-S XIV am	6427	1911	(a)	(1)

(a)   purchased from Arx & Co, Zürich /1911-21.
(1)   1921 to Anton Bless & Co, Dübendorf.

BAUGESELLSCHAFT FLÜELEN-GÖSCHENEN, Flüelen   UR08 C

A construction consortium for building part of the northern ramp of the Gotthard railway line. The exact location of the site is not known.

Gauge: 750 mm		Locn: ?					Date: 04.2002	
	0-4-0T	Bn2t	Heilbronn		28	1873	(a)	(1)
	0-4-0T	Bn2t	SLM		40	1874	new	s/s

(a) ex Vicarino & Curty, Bötzbergbahn construction, Pratteln /18xx.
(1) 18xx to Mathier & Co, Visp, VS.

BAUMANN, EMIL AG, Altdorf UR   AG08a C

The SIG list of references gives one ET 20L5 locomotive delivered here.

DAMPFBAHN FURKA-BERGSTRECKE (DFB), Realp   UR09 M

This preservation organisation has locomotives in store at several locations and also workshops where stock is being prepared for service.

Gauge: 1000 mm + Abt Rack		Locn: 255 $ 6814/1610					Date: 03.2008	
HG 2/3	6 WEISSHORN							
		0-4-2RT4c	B1bn4t	SLM HG 2/3	1410	1902	(a)	
HG 3/4	1 FURKAHORN							
		2-6-0RT4cc	1Cbh4vt	SLM HG 3/4	2315	1913	(b)	
HG 3/4	2	2-6-0RT4cc	1Cbh4vt	SLM HG 3/4	2316	1913	(c)	
HG 3/4	4	2-6-0RT4cc	1Cbh4vt	SLM HG 3/4	2318	1913	(d)	
HG 3/4	8	2-6-0RT4cc	1Cbh4vt	SLM HG 3/4	2418	1914	(c)	
HG 3/4	9 GLETSCHHORN							
		2-6-0RT4cc	1Cbh4vt	SLM HG 3/4	2419	1914	(b)	
40 304		0-8-0RT4cc	Dbh4vt	SLM	2940	1923	(g)	
40 306		0-8-0RT4cc	Dbh4vt	Esslingen	4227	1929	(h)	
40 308		0-8-0RT4cc	Dbh4vt	SLM	3413	1930	(g)	
HGe 4/4^I	16	BoBoWER	BbBbew4	SLM,MFO HGe 4/4^I	3677, ?	1938	(j)	(10)
	rebuilt	BoBoWER	BbBbew4	SLM,MFO HGe 4/4^I	4087, ?	1952		
CFm 2/2	21	AAPMRRC	Bbbm	SLM BChm 2/2	3206	1927	(k)	
HGm 2/2	51	4wDMR	Bdm	KHD KG230 BS	57181	1961	(l)	
Tm	68	4wDMR	Bdm	RACO 45 PS	1378	1948	(m)	
Tm	91	4wDH	Bdh	RACO 65 SA3 HT	1546	1959	(n)	
Tm	92	4wDH	Bdh	RACO 65 SA3 HT	1547	1959	(n)	
Tmh	985	4wDR	Bd	RACO 140 SA3	1629	1962	(p)	
Tmh	986	4wDR	Bd	RACO 140 SA3	1630	1962	(p)	
Tm	506	4wD	Bd	Asper	?	1953	(r)	
Xrotd	9212	2-A1	22h2	SLM	2399	1913	(s)	

(a) new to BVZ.
(b) new to BFD; ex Krongh Pha-Dalat [Vietnam].
(c) new to BFD; ex Krongh Pha-Dalat [Vietnam]; parts only.
(d) new to BFD; on loan from MGB.
(g) new to Krongh Pha-Dalat [Vietnam]; in store.
(h) new to Krongh Pha-Dalat [Vietnam]; parts only.
(j) new to VZ.
(k) new to FO; on loan from VHS; being restored in Aarau /2007-8.
(l) new to ARBED, Werk Burbach, Saarbrücken [D]; ex C. Vanoli /1986 and fitted with rack gear in addition to retaining its adhesion mechanism.
(m) new to CMN (PSC); ex RhB.
(n) ex RhB.
(p) ex SBB (Brunig).
(r) new to CJ.
(s) ex RhB via BC.
(10) damaged in an accident.

see: www.furka-bergstrecke.ch/ger/technik/index.htm

DÄTWYLER AG, Altdorf UR   UR10

formerly: Schweizerische Draht- Kabel- und Gummiwerke.

A concern founded in 1917 (from a smaller operation) and making insulated cables and other rubber products.

Gauge: 1435 mm   Locn: 246 $ 6920/1925   Date: 06.2002

4wDH	Bdh	Diema DVL15/3	3229	1971	(a)	(1)
4wDH	Bdh	Gmeinder Köf II	4675	1951	(b)	(2)
4wDH	Bdh	Moyse BN24HA150D	1398	1977	(c)	

(a) new on 18/1/1972 (what did the plate show?); purchased via Asper.
(b) new to DB "Köf 6126"; ex DB "323 940-7" /1988 via OnRail/LSB.
(c) ex Migros Volketswil, Schwerzenbach /2000 via METRAG.
(1) 1988 to Ferriere Cattaneo, Giubiasco "1216".
(2) 2000 to Baldini Recycling, Altdorf UR (for scrap?).

FURKA TUNNEL, Realp   UR11 C

At least the following locomotive was used when the Furka Tunnel on the Brig-Furka-Disentis (BFD) was constructed. It is not known if the BFD or a contractor was responsible.

Gauge: 600 mm   Locn: 246 $?   Date: 05.2009

0-6-0?PM	Cpm	Ruhrthaler 16PS	201	1913	new	s/s

LOGISTIKBASIS DER ARMEE (LBA)   UR12

formerly: Bundesamt für Betriebe des Heeres (BABHE) (till 31.12.2003)
Oberkriegskommissariat (OKK) (till 1997-9)
Abteilung Waffen und Schiessplätze (AWP) (till 1997-9)
Kriegsmaterialverwaltung (KMV) (till 1997-9)

Göschenen

Military stores serving the forces stationed at Andermatt, rail connected from 1894. Locomotives from Altdorf UR and Seewen SZ have been reported here temporarily.

Gauge: 1435 mm   Locn: 255 $ 6882/1688   Date: 04.2001

7	4wDM	Bdm	Kronenberg DT 90/120	126	1952	new	(1)
8	6wDE	Cde3	Stadler	112	1961	(b)	(2)
236 910-6	4wDE	Bde2	Stadler	117	1965	(c)	(3)

Rynächt Arsenal, Altdorf UR

formerly: Kriegsmaterialverwaltung (KMV)

This arsenal is associated with the arsenal at Amsteg.

Gauge: 1435 mm   Locn: 246 $ 69xx/19xx   Date: 01.1989

	A1BE	A1a	Stadler	104	1953	(d)	(4)
10	4wDE	Bde2	Stadler	117	1965	(e)	(5)
836 914-8	4wDE	Bde1	Stadler	148	1976	(f)	

Zeughaus Eyschachen, Altdorf UR

formerly: Armee-Verpflegungsmagazin (AVM)

A military commissariat. Locomotives from Seewen SZ have been reported here from time to time.

Gauge: 1435 mm   Locn: 246 $ 6911/1915   Date: 12.2008

2	4wDM	Bdm	Breuer	?	19xx		(7)
236 903-1	4wDH	Bdh	Kronenberg DT 120	146	1963	(h)	(8)

0-6-0CA	Cp	Borsig DrL	7898	1911	new	s/s
0-6-0CA	Cp	Borsig DrL	7899	1911	new	s/s
0-8-0CA	Dp	Borsig DrL	7900	1911	new	s/s
0-8-0CA	Dp	Borsig DrL	7901	1911	new	s/s
0-6-0CA	Cp	Borsig DrL	8169	1911	new	s/s
0-6-0T	Cn2t	Borsig	8226	1912	new	s/s

BIANCHI FRÈRES, Payerne   VD09 C

Gauge: 600 mm   Locn: ?   Date: 04.1999

| | 4wDM | Bdm | O&K MV0 | 25047 | 1951 | (a) | (1) |

(a) purchased via Wander-Wendel and MBA, Dübendorf.
(1) to Tuileries et Briqueteries de Bardonnex SA, Bardonnex "29".
     see: Le Chemin de fer YStC, R. Schultz and VE 106, 3/88, p34

BLONAY-CHAMBY (BC), Chaulin   VD10 M

alternative: chemin de fer touristique Blonay-Chamby
             chemin de fer-musée Blonay-Chamby

This preservation group operates over a closed section of the CEV and owns a significant collection of tramway rolling stock (including TF "Be 2/2 7", TL "Be 2/3 28", RhStB "Ze 2/2 31", TN "Be 2/2 76, CGTE "De 4/4 151", BVB "Be 2/2 182", VB "Xe 2/2 1" and VBZ "Xe 2/2 926"). The depot is a short walk from Chamby (MOB/CEV) station.

Gauge: 1000 mm   Locn: 262 $ 5594/1448   Date: 01.2008

Mulhouse 7	2-4-2Tm	1B1n2tk	SLM	316	1882		
LEB 5 BERCHER	0-6-0T	Ch2t	SACM-G	4172	1890	(b)	
RdB 1 LE DOUBS	0-6-0T	Cn2t	SLM	618	1890		
Sernftalbahn BCFe2/2 4	4wWERC	Bg2	SIG,MFO	?	1897		
Padane 4	0-4-0Tm	Bn2tk	Krauss-S XXXV It	4278	1900		
JS 909	0-6-0T	Cn2t	SLM G 3/3	1341	1901	(f)	
GFM Be 4/4 111	BoBoWERC	BBg4	SWS,EGA	?	1903		
MOB Fe 4/4 11	BoBoWERC	BBg4	SIG,EGA	?	1905		
MCM BCFe^h 6	0-4-4-0WERC	BBg	SIG,SLM,EGA	?	1909	(9)	
RB E332	4-6-0T	2Cnt	Fives-Lille	3587	1909		
Xrotd 9214	0-6-6-0 4c	CCn4	SLM Xrotd	2299	1912		
BFD 3	2-6-0RT	1Cbh4vt	SLM HG 3/4	2317	1913		
+GF+ Ge 4/4 75	BoBoWE	BBg4	SLM,MFO Ge 4/4	2327, ?	1913	(m)	
LLB ABFe 2/4 10	0-4-4-0WERRC	BBg2	SWS,BBC BCFhe 4/4	? ,968	1914		
RhB Ge 4/4^I 181	0-4-4-0WE	BBg2	SLM,BBC Ge 4/4	3295,1054	1916(29)	(o)	
LJB Ce 2/2 12	4wWETm	Bog2	SIG,BBC	?	1917		
ZT 105	0-4-4-0T	BBn4vt	Karlsruhe	2051	1918		
ZT 104	0-6-6-0T	CCh4vt	Hanomag	10437	1925		
OG 23	2-6-2T	1C1n2t	MTM	282	1926		
99 193	0-10-0T	Eh2t	Esslingen BR 99.1	4183	1927		
	4wDM	Bdm	O&K RL 2a	20200	1931	(u)	
MOB DZe 6/6 2002	BoBoBoWERC	BBBg6	SIG,BBC Gepäck-Tw	? ,3703	1932	(v)	

Gauge: 600 mm   Locn: ?   Date: 02.1997

| | 0-4-0PM | Bbm | Oberursel | ? | c1915 | (aa) | (27) |

(b) new LEB "5"; later Grande Dixence; ex display, Feldkirch [A].

(f)	ex Holzwerk Renfer, Biel-Mett "6".
(m)	ex Georg Fischer AG, Schaffhausen "75".
(o)	built new by Berninabahn as a Ge 6/6 with BBC parts and number 1054; rebuilt by SLM as a Ge 4/4 with SLM number 3295 /1929; OOU /1965; ex RhB /1970.
(u)	new to la Grande Dixence SA (as 4wPM Bbm); used by Holzwerke Rieder, St. Stephan /1950-79; ex MOB "Tm 1" /2006.
(v)	new to MOB /1933; ex MOB 19/07/2008.
(aa)	originally Carrières d'Arvel, Villeneuve VD.
(9)	2008 on loan to TPC.
(27)	19xx on loan to Michel Crot, Vaulion.

BLONAY-LES PLÉIADES, Blonay — VD11 C

The following locomotive was used on the construction of this railway.

Gauge: ? mm   Locn: ?   Date: 01.2008

	0-4-0RT	Ban2t	SLM HG 2/2	502	1888	(a)	c1914 s/s

(a) ex construction of Monthey-Champéry-Morgins /1911.

JEAN BOLLINI SA, Baulmes — VD12 C

A public works company.

Gauge: 600 mm   Locn: 241 $ 5303/1827   Date: 08.1988

	4wDM	Bdm	O&K(M) RL 1a	4690	1932	(a)	(1)
2	4wDM	Bdm	O&K(M) RL 1a	4706	1932		(2)
	4wDM	Bdm	Ammann	-	19xx	(c)	(1)

(a) purchased from Anton Bless, Dübendorf.
(c) constructed with Jung parts; carries a plate "Gerate Nr 8456", probably Jung ZM 233 8456 1939.
(1) last used for rebuilding the electric power station at Les Clées; c2006 to FWF, Otelfingen.
(2) last used for rebuilding the electric power station at Les Clées; s/s.

BRENNTAG SCHWEIZERHALL AG, Avenches — VD13

formerly: Schweizerhalle Chimie SA

A chemical waste disposal point.

Gauge: 1435 mm   Locn: ?   Date: 01.2001

	4wDM	Bdm	RACO 85 LA7	1596	1961	(a)

(a) new to CFF "Tm^l 329"; ex CFF "Tm^l 429" /2000.

BRIQUETERIE, TUILERIE & POTERIE DE RENENS (BTR), Renens VD — VD14

A brickworks (bricks, tiles and pottery) founded in 1907 and lying near the CFF mainline. It became part of the Schmidheiny group in 1938.

Gauge: 600 mm   Locn: ?   Date: 04.1999

4wDM	Bdm	O&K(M) MD 1	11562	1940	(a)	s/s
4wDM	Bdm	O&K(S) RL 1c	1501	1947		s/s

(a) via O&K, Lager Zürich; recorded as 15562.

DR. BUGNAN, Prahins — VD15 M

To serve a scout camp and a working project (open to the public) in 1970 a kilometre long temporary line was laid down to convey visitors from the nearest bus stop to the campsite.

Gauge: 600 mm   Locn: 251 $ 5454/1763   Date: 01.1970

2230 MAGDALENA	4wDM	Bdm	Deutz OME117 F	10286	1932	(a)	(1)

(a) new to Robert Aebi & Cie AG, Zürich; ex ? /19xx.
(1) ? to Pépinière Raoul Thonney, Cheseaux-sur-Lausanne; by 2006 at FWF, Otelfingen.

BUNDESTANKANLEN (BTA) — VD16

Gland

Fuel reservoirs built and rail connected in 1953 (816 metres of track) and operated by Carbura. The installations were removed in 2002.

Gauge: 1435 mm   Locn: 260 $ 5296/1409   Date: 12.2008

2	4wPM	Bpm	RACO RA11	1260	1940	(a)	(1)
	4wDH	Bdh	O&K MB 5 N	26586	1966	(b)	(2)

Saint-Triphon

Fuel reservoirs constructed and rail connected in 1950 (860 metres of track) and operated by Carbura. The installations were removed in 2001.

Gauge: 1435 mm   Locn: 272 $ 5640/1262   Date: 05.1984

2	4wDM	Bdm	R&H 88DS	252840	1949	(c)	(3)

(a) transferred from BTA, Altdorf UR (or Ostermundigen?) /1953.
(b) ex Wollkämmerei und Wäscherei Hannover, Werk Döhren [D].
(c) new via Robert Aebi, Zürich.

(1) 1973 to Shell, Vernier.
(2) 2002 to BABHE, Thun.
(3) 2001 to Carrières du Lessus HB SA, Saint-Triphon.

CABLOFER SA, Bex — VD17

A scrap merchant specialising in metals. The firm was set-up in 1971 on the site of the previous electrochemical plant that was rail connected from 1942.

Gauge: 1435 mm   Locn: 272 $ 5664/1211   Date: 11.2000

388	4wDM	Bdm	O&K MV2a	26191	1962	(a)	1990 scr
	4wDM	Bdm	R&H 88DS	321726	1951	(b)	1996 scr
1	4wDM	Bdm	RACO 80 SA3	1388	1949	(c)	
	4wDM	Bdm	RACO 85 LA7	1731	1965	(d)	

(a) ex Thévenaz-Leduc, Ecublens VD /1980.
(b) ex Alusuisse, Chippis /1981.
(c) ex Valmetal, Martigny /1990.
(d) ex CFF "Tm^I 504" /1996.

CARRIÈRES D'ARVEL, Villeneuve VD — VD18

These quarries have been rail connected since 1891 by a roadside line 1.95 kilometres in length.

Gauge: 1435 mm   Locn: 262 $ 5614/1370   Date: 11.2000

4550	4wDE	Bde1	Moyse BN32E150B	1094	1965	(a)	

Gauge: 600 mm		Locn: 262 $ 5614/1370					Date: 02.1994	
	0-4-0PM	Bbm	Oberursel		?	c1915	(b)	(2)

- (a) ex SA Streichenberger, Vongy [F] /19xx.
- (b) purchased via Rubag.
- (2) 1936 sold to Carrières Bussien, Le Bouveret.

CARRIÈRES DU LESSUS HB SA, Saint-Triphon VD19

formerly: Sociétés Réunies des Carrières de Saint-Triphon, Saint-Triphon

The marble producing Carrières des Andonces had existed for many years before they were rail connected to the north end of St. Triphon Gare CFF in 1859. This was one of the first industrial lines in Switzerland. The railway line entered the quarry and wagons were taken directly to be loaded from the working faces with the aid of a wagon table. Later another quarry was opened at Lessus, about a kilometre further from the previous one in the direction of Aigle. This was rail connected in 1889 by extending the existing siding. The machinery in the quarry and the branch line to the Carrières des Andonces was electrified in 1898 and the electrification later extended to Lessus. The electrification was removed by 1918 and the Carrières des Andonces has now been abandoned. The present title dates from 1930 and the product is now rail ballast. In 1995 a ballast recycling plant was added. Two 600 mm skips of 0.7 m^3 capacity are displayed outside the offices; one bears plates "Motor M.V. St-Aubin, Lieferant F. Marti, Bern".

Gauge: 1435 mm 110V DC				Locn: 272 $ 5640/1271			Date: 10.2008	
		2-2-0WE	1Ag1	?,CIEM	-	1898	(a)	s/s
	7533	4wDM	Bdm	Moyse 20TD-HT	66	1958	(b)	
		4wDM	Bdm	R&H 88DS	252840	1949	(c)	(3)
		4wDH	Bdh	Jung RK 20 B	13372	1961	(d)	(3)
Tm^{III}	9551	4wDH	Bdh	RACO 210 SV4 H	1820	1976	(e)	

- (a) constructed from a flat wagon with electric motors from a tram used at the Exposition National, Genève of 1896.
- (b) new to Huilerie Lessieur-Cotelle, Croix-Sainte [F]; ex Provence Energie, Fos-sur-Mer [F] in /1995. Carries plates "No serie PE 100 1954".
- (c) ex BTA, Saint-Triphon "2" /2001, plates of Robert Aebi & Cie AG, Zürich.
- (d) ex Sihlpapierfabriken, Landquart /2001 via SERSA ; also METRAG plates.
- (e) withdrawn 31/7/2008; ex CFF "Tm^{III} 9551" /2008 via J. Müller, Effretikon.
- (3) by 10/2007 to RoutOrail SA, Satigny.

    see: Drehscheibe 187

CHAPUISAT, Lausanne VD20 C

A public works concern.

Gauge: 600 mm		Locn: ?				Date: 12.2008	
	4wBE	Ba	SSW EL 8	5857	1957	(a)	s/s

- (a) purchased new via MBA, Dübendorf, who describe it as an EL 9.

CHÂTELET ET MEUNIER, Orbe VD21 C

This concern participated in the construction of the Yverdon Sainte-Croix railway.

Gauge: ? mm		Locn: ?				Date: 02.2002	
	0-4-0T	Bn2t	Schneider	?	18xx	(a)	s/s

- (a) engaged on the construction of the Yverdon Sainte-Croix railway /1891-92.

6	0-4-0T	Bn2t	Anjubault		22	1857	new	s/s

ENTREPRISE MANIER & ERMOGLIO, Payerne — VD37 C

This consortium was formed by two public work concerns that constructed the lines of la Broye 1875-76 for the Compagnie de la Suisse Occidentale. The Ermoglio firm is from Montpellier [F].

Gauge: 1435 mm Locn: ? Date: 12.1993

	0-4-0T	Bn2t	Krauss-S XIII f	487	1875	(a)	s/s

(a) used in construction of the Yverdon-Fribourg and Palézieux-Lyss lines.

FABRIQUES DE TABAC RÉUNIES SA (FTR), Onnens — VD38

This company provides the tobacco store for the Philip Norris organisation whose headquarters are in Neuchâtel. The storage site opened in 1967 and was later extended with more warehouses. It was fully functional from 1974. The length of siding was seven hundred metres in about 1967 and 3694 metres in 1974.

Gauge: 1435 mm Locn: 241 $ 5426/1870 Date: 10.2001

Tm 2/2 MHB 101	4wDH	Bdh	O&K MHB 101	26970	1981	(a)

(a) via MBA.

FAVRE SA, Corcelles-près-Payerne — VD39

formerly: Ferraco SA (till 1974)

This company deals in metals and operates a hardware store. A 204 metre long branch line from Corcelles Nord station was brought into use in 1962.

Gauge: 1435 mm Locn: 242 $ 5633/1871 Date: 07.2002

1	0-4-0DM	Bdm	KHD KS55 B	58228	1968	(a)

(a) ex Schmutz Aciers, Orbe /1993.

FEBEX, Bex — VD40

formerly: la Fonte Électrique SA (FEBEX) (from 1917)

A concern initially specialising in electro-metallurgy, later changing to phosphoric chemistry. Rail connected in 1917; the system was extended in 1966.

Originally a CFF battery shunter "Ta 44", this locomotive was converted to diesel-electric operation, retaining the original electric motors. It is seen here on 6 Sept. 1996 before moving to the Bahn Museum Kerzers.

	Gauge: 1435 mm			Locn: 272 $ 5321/1541			Date: 01.2003	
	3	4wDE	Bde2	Olten,BBC	? ,1945	1923	(a)	(1)
		4wDE	Bde	Stadler	131	1969	(b)	c2007 s/s
Tm 2/2		4wDH	Bdh	KHD A6M617 R	55754	1955	(c)	

(a) new to CFF "Ta 44" (4wBE Ba2); later fitted with a diesel generator; ex Société des Ciments Portland de Saint-Maurice (EGT) /1986.
(b) ex Henkel, Pratteln /1998.
(c) new to DB "Köf 6207", later "323 085-1"; ex BSZ Oil AG, Tankanlage West, Muttenz-Auhafen /2007 via LSB and TAFAG /2006.
(1) 2002 to a private collector and kept at Bahn Museum Kerzers (BMK), Kerzers.

FIXIT, Bex

VD41 M

formerly: Gips-Union SA

The existing gypsum plant at Bex was acquired and developed in 1905. It had a short rail network in the quarry located on the Montet hill above Bex that ran between the underground silo (glory hole?) and an aerial ropeway leading down to the factory. Both the aerial ropeway and the railway were abandoned in 1989. After disposal of the railway items a further locomotive was obtained for display in the factory. A standard gauge railway system encircled the silo and was noteworthy for its small radius curves. The layout has been modified but is still operated using road tractors fitted with railway buffering gear.

Gauge: 600 mm			Locn: 272 $ 5667/1236			Date: 04.1997	
	4wDM	Bdm	Ammann	-	1949	(a)	(1)
	4wDM	Bdm	O&K(M) RL 1c	?	19xx	(b)	(2)
	4wDM	Bdm	O&K(M) RL 1a	4899	1933	(c)	display

(a) replacement KHD motor (A2L514, ex F2M414) /1978.
(b) ex Gips-Union; where and when ?
(c) new to Ziegelei Lufingen via O&K, Zürich; ex FWF, Otelfingen /2004.
(1) 1989 OOU; 1997 donated to FWF, Otelfingen.
(2) 1989 OOU; 19xx s/s.

GABELLA & CIE. SA, Lausanne

VD41a C

The SIG list of references gives one ETS 35 locomotive delivered here.

GAVILLET & DELESLE, Lausanne   VD42 C
A public works concern?
Gauge: 600 mm   Locn: ?   Date: 05.2002

    0-4-0T   Bn2t   O&K 30 PS   9955   1922   (a)   (1)

(a)   purchased from Entreprise Paul Juillard, Saxon /1931.
(1)   1931-39 sold to Losinger & Cie, Bern.

GEILINGER SA, Yvonand   VD43
This metal construction company (a subsidiary of Geilinger AG, Bülach) was rail connected at Yvonand CFF station. The line was constructed in 1974 and extended in 1978 (complete length 428 metres). The company closed 1996-2000.
Gauge: 1435 mm   Locn: 241 $ 5472/1834   Date: 01.2008

    4wDM   Bdm   RACO Tm11   1376   1948   (a)   (1)

(a)   new to CFF "Tm 535"; purchased from CFF "Tm 897" 3/1978.
(1)   1996-2000 sold to Jean-Jacques Bader SA, Conthey.

GOUTTE RÉCUPÉRATION SA, Lausanne   VD44
formerly: J.-M. Goutte & Cie SA
This scrap merchant started business in 1965. Address: Avenue de Sévelin 22, 1004 Lausanne.
Gauge: 1435 mm   Locn: 261 $ 5368/1527   Date: 01.2003

    4wDE   Bde1   Moyse 36TDE   140   1955   (a)

(a)   ex Société la Rochette-CENPA, Venizel [F] via Desbrugères /1992.

GRAVIÈRE DE BIOLEY-ORJULAZ, Bioley-Orjulaz   VD45
A gravel quarry originally owned by the LEB and later sold to the État de Vaud. The exact location of operation of the line is not known; there are four old workings to the west of the village.
Gauge: 600 mm   Locn: 251 $ 535x/163x   Date: 08.1989

    0-4-0DM   Bdm   Maffei   5908   1930   (a)   (1)

(a)   new to Brun & Cie AG, Nebikon (dealers).
(1)   by 1959 OOU; 1963-64 donated to CF touristique de Meyzieu (CFTM) [F] "Brouette d'Echallens"; later CF du Haut Rhône - Vallée Bleue à Montalieu (Ain) [F].

GROUPEMENT DES COPROPRIÉTAIRES DES VOIES INDUSTRIELLES DE LA GUINGUETTE, Vevey   VD46
formerly: Nestlé & Anglo-Swiss Condensed Milk Co. (1905-1953)
            Société Henri Nestlé (18xx-1905)

A factory of the Nestlé concern lying on the banks of la Veveyse at the west end of Vevey Station was connected to the rail network in 1891 via a wagon-table and a short line at right angles to what became the CFF. In 1901 the passenger station was extended and the goods station moved a little to the west leading to a complete replacement of the branch which also served other firms. The access via a wagon-table was retained, but it became mixed gauge to allow CEV wagons to use it. The line was electrified in 1921, Nestlé ceased production before 1936, retaining only a packing/storage facility. Metre gauge traffic ceased with the closure of the CEV Châtel-St. Denis line in 1970 but at that time the branch still served Nestlé, the slaughterhouses and Margot Frères. The latter became the "Union des co-operatives agricoles romandes" (UCAR/LANDI) and the sole user till closure at the end of 1997. One reason for closure was that the wagon-table was not able to handle the longer bogie hopper wagons. The locomotive was originally built with wheels of the same diameter and a cable winch for moving metre gauge wagons, but later had wheels of notably different sizes and the assistance of a Marjollet satellite shunter.

Gauge: 1435 mm   Locn: 262 $ 5544/1461   Date: 08.1997
                1AWE   1Ag1   ACMV,BBC   ? ,1338   1921   new   (1)
(1)   7/1998 donated to Swisstrain, Bodio; by 2007 at Le Locle.
       see: Voie Etroite No 169 (6/98) pp 32-3.

HOLCIM SA   VD47
formerly: Holderbank Ciments et Bétons (HCB) (1992-2001)
alternative: HCB Ciments et Bétons «Holderbank»
formerly: Société des Ciments et Bétons (SCB) (1988-1991)
alternative: Société des Ciments et Bétons de la Suisse Romande (SCB)
formerly: Société des Chaux et Ciments de la Suisse Romande (SCC) (1913-1988)
                Société des Usines de Grandchamp et de Roche (1896-1913)

Usine d'Eclépens, Eclépens
This cement works opened in 1953.

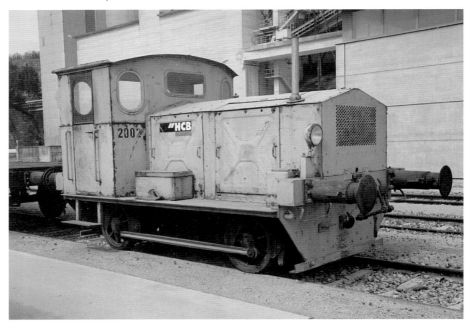

This O&K diesel from Berlin, number 20162 of 1931 (type RL 4) had already retired from active service when photographed on 21 April 1999. It has since gone into preservation.

Gauge: 1435 mm   Locn: 251 $ 5317/1679   Date: 05.2002
2001   4wBE   Ba2   SIG   520 501   1953   (a)   1998 scr
2002   0-4-0DM Bdm   O&K RL 4   20162   1931   (b)   (2)
2003   0-6-0DH Cdh   Henschel DH 500Ca   31083   1965   (c)

CH. PAYOT SA, Clarens   VD57 C

The depot of a civil engineering company.

Gauge: 600 mm   Locn: ?   Date: 04.1997

4wDM	Bdm	O&K(M) MD 1	11414	1939	(a)	(1)
4wDM	Bdm	O&K MV0	25166	1952	(b)	(2)

(a) purchased from Consorzio Presa Clancasa, Buseno /1955.
(b) used ca. 1963 for work on the Lausanne-Genève motorway at Chavannes-de-Bogis.

(1) used by Payot & Rochat at Vers l'Eglise; s/s.
(2) stored OOU at this site; by 1999 to FWF, Otelfingen.

PÉPINIÈRE RAOUL THONNEY, Cheseaux-sur-Lausanne   VD58 M

Railway equipment for a planned railway was stored at this tree nursery. In the event the project was not authorised. Address: ch. de Sous-le-Mont 1, 1033 Cheseaux-sur-Lausanne.

Gauge: 600 mm   Locn: 251 $ 5354/1595   Date: 06.1989

	0-4-0WT	Bn2t	Jung 20 PS	1693	1911	(a)	(1)
2230 MAGDALENA	0-4-0DM	Bdm	Deutz OME117 F	10286	1932	(b)	(2)

Gauge: 750 mm   Locn: ?   Date: 06.1989

4wDM	Bdm	RACO 50 PS	?	194x	(c)
4wDM	Bdm	RACO 50 PS	?	194x	(c)

(a) ex E. Bellorini & Cie, Lausanne-Malley ca. /1977 via ?, Satigny.
(b) new to Robert Aebi & Co, Zürich; in use at a scout camp at Prahins (loan from Dr. Bugnan) /1970.
(c) ex ?Barrage de Rossens FR; if correctly identified then two of the six locomotives with numbers RACO 1313-17 plus 1321 of 1945.

(1) 1999 privately owned (Thomas Brändle, Geroldswil) "LISELI".
(2) before 2006 to FWF, Otelfingen.

PERRET SA, Chavornay   VD59

This scrap merchant also operates as a machinery dealer. The site is not rail connected, though very close to the Orbe-Chavornay (OC) line. Founded in 1970, the present title dates from 1988. Address: **En Forez, 1373 Chavornay.**

Gauge: 1435 mm   Locn: 251 $ 533350,173700   Date: 01.2009

0-8-0DH	Ddh	LEW V60	12246	1968	(a)	(1)

(a) ex Conserves Estavayer SA (CESA), Estavayer-le-Lac /2002.

(1) 2005 to OeBB "22".

H. PIOT & CH. PIGUET, Bavois   VD60

Société co-operative suisse de la tourbe, Berne (1918-1921)
SA pour l'exploitation de la tourbe, Berne (1917-1918)

This was a peat extraction scheme operated near the end of World War I.

Gauge: 600 mm   Locn: ?   Date: 08.1994

	0-4-0T	Bn2t	Güstrow	?	1894	(a)	s/s
BUBI	0-4-0T	Bn2t	Jung 10PS	1349	1909	(b)	(2)
HERBERT	0-4-0WT	Bn2t	Krauss-S XIV mm	4048	1899	(c)	(3)
	0-4-0T	Bn2t	O&K 20 PS	373	1900	(d)	(4)

(a) purchased /1918.
(b) new to F. Marti à Winterthur, purchased /19xx.
(c) purchased from Robert Aebi & Cie, Zürich /1918.

(d) hired from Th. Bertschinger, Lenzburg /1917; then from F. Marti, Bern /1918-/1919; purchased /1919-1920.
(2) 1926 sold to Schafir & Mugglin, Muri bei Bern during the construction of the Klingnau hydroelectric power station.
(3) 1923 to Michel Dionisotti (entrepreneur), Saint-Maurice; later Tacot-des-Lacs [F].
(4) 1926 sold to Paul Juillard, Saxon.

PORT-FRANC ET ENTREPÔTS DE LAUSANNE-CHAVORNAY SA (PESA), Chavornay VD61

A duty free storage area containing the grain silos of André SA, connected to the Orbe-Chavornay (OC) line from 1979. Address: Place de la Gare, 1373 Chavornay.

Gauge: 1435 mm   Locn: 251 $ 533150,173710   Date: 01.2003

| | 4wDH | Bdh | Moyse BL18HS80D | 137 | 1966 | (a) |

(a) ex Tela AG, Werk Niederbipp.

PREBE, Avenches VD62

formerly: BTR-Prébéton SA (1979-3/2/2006)
            Stahlton-Prébéton SA (19xx-1979)

A factory making concrete elements rail connected from 1963 and with 195 metres of track.

Gauge: 1435 mm   Locn: 261 $ 5700/1930   Date: 07.2001

| 13 | 4wPM | Bbm | SLM Tm | 3579 | 1934 | (a) | (1) |
| | 4wDM | Bdm | RACO 60 LA7 | 1498 | 1958 | (b) | |

(a) new to CFF "Tm 214"; ex CFF "Tm 723" /1968.
(b) new to CFF "Tm^I 303"; ex CFF "Tm^I 403" /1998.
(1) OOU; 1998 donated to Swisstrain.

RÉGIE FÉDÉRALE DES ALCOOLS, Daillens VD63

Alcohol warehouses brought into service in 1970. Rail connected from 1969 to the station at Daillens which is located in the fork of the Lausanne - Vallorbe and Lausanne - Yverdon lines and was closed to passengers when no longer needed for exchange purposes. The installations included a six-track yard and had a total length of 3600 metres. The site was cleared in 1997.

Gauge: 1435 mm   Locn: 251 $ 5315/1652   Date: 10.1997

| | No.1 | 4wDH | Bdh | O&K MB 7 N | 26609 | 1969 | new | (1) |
| TM | 511 | 4wDM | Bdm | RACO 45 PS | 1248 | 1937 | (b) | (2) |

(b) new to SBB "Tm 325"; later MO "TM 511"; hired from Asper /1992.
(1) 1997 to Récupération RG SA, Sévaz.
(2) 1992 returned to Asper.

SCHEUCHZER SA, Bussigny-près-Lausanne VD64 C

formerly: les fils d'Auguste Scheuchzer SA (1947-
             Charles-Auguste Scheuchzer (1917-1947)
alternative: August Scheuchzer Söhne AG
             The Sons of August Scheuchzer Ltd

This company performs track maintenance and replacement. The rail access dates from 1943 and was extended 1950-3, and again in 1975 when it reached a length of 1974 metres. Address: Chemin du Cudrex 1-3, 1030 Bussigny-près-Lausanne.

Gauge: 1435 mm			Locn: 261 $ 5323/1552			Date: 08.2006	
BA 62		A1DM	A1dm	MATISA	?	1964	(a)	(1)
BA 84		A1DM	A1dm	Plasser	?	1967	(b)	
Tm	1	4wDE	Bde	Gmeinder DE-SF1	5466	1970	new	1991 scr
Tm	2	4wDE	Bde	Gmeinder DE-SF1	5487	1972	(d)	
Tm	3	4wDE	Bde	Gmeinder DE-SF1	5486	1972	(d)	
Tm 283 000-8 (ex 4)		4wDE	Bde	Gmeinder DE-SF3	5490	1972	new	(6)
Bm 4/4 11011 GAZELLE								
		BoBoDE	BBde	B+L 040DE600	?	1960	(g)	
BISON		BoBoDE	BBde	CFD	152-001	2004	(h)	

(a) converted from a tamper /1975.
(b) converted from a tamper /1980-86.
(d) new, also reported as type DE-SF3.
(g) new to HBNPC [F] "29"; ex CFD Industries [F].
(h) also numbered "80 85 977 70 001-8", provenance yet to be determined.
(1) 1985-6 to Thévenaz-Lèduc SA, Ecublens VD.
(6) 1989 renumbered; 9/1997 to Rhenus AG, Basel-Rheinhafen.

see: www.scheuchzer.ch

H.R. SCHMALZ SA, Joux Verte   VD65 C

This construction company has a head office in Bern and many branches throughout Switzerland. The exact purpose of the work performed in 1968 in the valley about five kilometres from Roche VD is not known (water conduit?).

Gauge: 600 mm			Locn: 262 $ 5635/1372			Date: 08.1968	
468 RITA	4wBE	Ba	SSW EL 8	6092	1961	(a)	(1)
901	4wBE	Ba	TIBB	0117	1959		(1)

(a) new to MBA, Dübendorf.
(1) probably returned to H.R. Schmalz SA, Ardon.

SCHMUTZ ACIERS SA, Orbe   VD66

The company fashioned wrought-iron products and traded in them. The workshops were established in 1987 and rail connected to Les Granges station of the Orbe-Chavornay (OC) line. Activity ceased in 1997.

Gauge: 1435 mm			Locn: 251 $ 5315/1747			Date: 08.1998	
1	0-4-0DM	Bdm	KHD KS55 B	58228	1968	(a)	(1)
	0-4-0DE	Bde1	SLM,BBC Tm^{III}	4158,6026	1956	(b)	(2)

(a) new to Thyssen Röhrenwerke AG, Düsseldorf [D] "1"; ex Mannesmann, Duisburg [D] /1987 via OnRail/LSB.
(b) new to Papierfabrik Perlen, Gisikon-Root '5'; ex via Asper/ 1993.
(1) 1993 transferred to Favre SA, Corcelles-près-Payerne.
(2) 1998 stored at la Traction, Pré-Petitjean; 2008 to Stauffer.

SOCIÉTÉ DES MINES ET SALINES DE BEX, Le Bouillet, Bex
VD68 M

Some of the output from these extensive salt mines used to travel on the BVB. The commercial entrance to the mine and silos are located near the BVB depot (siding now abandoned) and the public entrance to the part of the mine that has been converted into a tourist operation lies two kilometres away. The railway is used to convey both the tourists and the personnel in this area, but is not used for mining operations. There has been more than one generation of vehicles for this service. The present one is designed to keep the tourist enclosed within a very limited loading gauge; the locomotives have been adapted to match the coaches and to give the drivers a cab mounted forward of the frame and batteries. There have been at least twelve coaches built between 1944 and 1986. Coaches 5? (1978) and 6 (1980) are reserved for the personnel. Coaches 8 to 17 were built by Meili, Bex for the tourists (12 and 15 were not seen on a visit in 1999).

Gauge: 600 mm   Locn: 272 $ 5676/1247   Date: 07.1999

1	4wBE	Ba	Oehler	?	1942(7x)	?new
2	4wBE	Ba	home-made	-	1976	(b)
3	4wBE	Ba	home-made	-	1981	(b)
4	4wBE	Ba	home-made	-	1985	(d)

(b) built in the company workshops (from what?).
(d) also reported to have been fitted with a diesel motor /1996, and to have been provided by NEWAG: to be confirmed.
(4) rebuilt with a diesel engine (to be confirmed).

SOCIÉTÉ ROMANDE D'ÉLECTRICITÉ, Territet   VD69

The location and the reason of use of the locomotives are not known, probably for a hydroelectric scheme.

Gauge: 750 mm   Locn: ?   Date: 02.1994

	4wDM	Bdm	Deutz C V F	1462	1914	new s/s
	4wDM	Bdm	Deutz C V F	1463	1914	new s/s

TAMOIL SA, Saint-Triphon   VD70

formerly: RSO Services SA, Collombey (? till 21/12/1998)
Raffinerie du Sud-Ouest SA (RSO) (at latest 1967 till ?)
Raffineries du Rhône, SA, Collombey (1963 till at latest 1967)

This oil refinery was first operational in 1963. The offices are located in the western part of the site in Collombey while the railway installations are all across the R. Rhône to the east and so in Canton Vaud. They are connected to the CFF station of Saint-Triphon. Address: 1868 Collombey.

On 6 September 1996 the two remaining Ruston & Hornsby LSSH diesel locomotives share the shed with one of new generation of SFL locomotives. They are –left to right- "4" the old "1" R&H 497743 of 1963, "2" SFL 302 of 1994 (type 385DH44NIND) and "3" R&H 497745 of 1963. In fact the two locomotives on the left are receiving attention and the other SFL locomotive, "2" SFL 301 of 1994 is the only one working.

Gauge: 1435 mm			Locn: 272 $ 5628/1271					Date: 08.2005
4 (ex 1)	0-4-0DH	Bdh	R&H LSSH	497743	1963	(a)	(1)	
2	0-4-0DH	Bdh	R&H LSSH	497744	1963	new	(2)	
3	0-4-0DH	Bdh	R&H LSSH	497745	1963	(c)	(1)	
4	0-4-0DH	Bdh	R&H LSSH	497746	1963	new	(4)	
1	4wDH	Bdh	SFL 385DH44NIND	300	1994	new		
2	4wDH	Bdh	SFL 385DH44NIND	301	1994	new		
LSB 05	6wDH	Cdh	Henschel DHG 500C	31184	1967	(g)	(7)	
	6wDH	Cdh	Henschel DHG 500C	31236	1968	(h)	(8)	
	BBDH	BBdh	Vossloh G 1000 BB	5001533	2004	new		
Em 847 904-0	BBDH	BBdh	Henschel DHG 1200 BB	31575	1973	(j)		

(a) new as "1", renumbered "4" /1994.
(c) new; fitted with engine from "2" /1994-6.
(g) hire from LSB by /2004.
(h) hire from LSB /2003.
(j) hire from LSB by 10/2008.

(1) 2004 OOU; 2006 for scrap.
(2) 1994 dismantled for spares; engine into "3".
(4) 1995 donated to Rive-Bleue-Express, Le Bouveret; 1998 to Arnold AG, Flüelen.
(7) 9/2004 returned to LSB.
(8) 9/2004 returned to LSB.

THÉVENAZ-LEDUC SA, Ecublens VD   VD71

This scrap merchant handles paper and both ferrous and non-ferrous metals. The site is rail connected at the end of the Lausanne-Triage in Denges. Address: ch. de la Motte 5, 1024 Ecublens.

Gauge: 1435 mm			Locn: 261 $ 5321/1541				Date: 01.2008
388	4wDH	Bdh	O&K MV2a	26191	1962	(a)	(1)
	4wDM	Bdm	RACO TM11	1383	1949	(b)	1996 scr
1	4wDM	Bdm	KHD KK135 B ex	57703	1964	(c)	
3	4wDH	Bdh	O&K MB 5 N	26600	1966	(d)	
BA62	A1DM	A1dm	MATISA	?	1964	(e)	(5)
1	4wDH	Bdh	O&K MV10	26157	1962	(f)	
1912	0-6-0WT	Cn2t	SLM E 3/3	2298	1912	(g)	display

(a) purchased from Fratelli Tenconi SA, Airolo /1972-77.
(b) new to CFF "Tm 539"; ex CFF "Tm 899" 5/1976.
(c) new to BP Benzin und Petroleum AG, Tanklager Stuttgart [D]; ex Deutsche BP AG, Stuttgart Hafen [D] /1982; no running number from /1990.
(d) ex Dortmunder Union-Brauerei, Dortmund (DAB "1" /1980 via WBB/Asper.
(e) converted tamper; ex Scheuchzer SA, Bussigny-près-Lausanne /1985-86.
(f) new to Hoesch AG, Werk Westfalenhütte, Dortmund-Barop [D] "40"; ex Dortmunder Eisenbahn GmbH [D] /1989 via OR/LSB.
(g) new to Services Industriels de Genève (SIG), Usine à Gaz, Vernier; ex Alusuisse, Chippis "GERONDE 31" to private ownership /1980 and placed on display here by /1994.

(1) 1980 to Cablofer SA, Bex.
(4) 1989 to LSB à Calandabräu, Felsberg via LSB.
(5) 1987-93 scrapped.

TRUCHETET, Nyon   VD72 C

Probably related to Truchetet & Besson of Dijon [F] and Lausanne, See entry in canton Valais.

Gauge: 600 mm			Locn: ?				Date: 04.2009
	0-4-0T	Bn2t	O&K 20 PS	1194	1903	(a)	s/s

(a) recorded in O&K delivery list as new to A. Truchetet Entrepreneur, Nyon, but no reference has been found in the appropriate records.

UGINE KUHLMANN, Le Day   VD73

formerly: Société d'Électrochimie du Day
Société d'Électro-Chimie et d'Électro-Métallurgie (SECEM)
Usine de Carbure du Day

This works, located near Vallorbe, produced sodium and potassium chlorates; it was disused from 1975. The railway siding was brought into service in 1916 and connected to that of the Société électro-chimique de Vallorbe dating from 1893, making a total length of 220 metres. The system was extended and a locomotive shed built in 1930.

Gauge: 1435 mm   Locn: 251 $ 5205/1748   Date: 08.1995

         4wPM   Bbm   RACO 25 PS   977   1930   new   (1)

(1)   1970-2 s/s.

USINE DE LA PAUDÈZE, Paudex   VD74

The entry in the Deutz works list (Merte) refers to the "Usine de Paudez, Pully". It is thought that this refers to the above listed chalk works (confirmation required).

Gauge: 500 mm   Locn: ?   Date: 02.1994

        0-4-0DM   Bdm   Deutz C XIII   1477   1914   new   s/s

VETROPACK SA, Saint-Prex   VD77

The glassworks (la Verrerie de Saint-Prex) was founded in 1911 and is located next to Saint-Prex CFF station. Shunting is normally performed by the CFF. However for at least one short period a locomotive was leased for this work. Address: Rue de la Verrerie 1, 1162 Saint-Prex.

Gauge: 1435 mm   Locn: 262 $ 5245/1484   Date: 12.2007

         4wDE   Bde1   Moyse BNC   153   1963   (a)   (1)

(a)   ex SA Eaux d'Évian-les-Bains, Évian-les-Bains [F]; on loan from CFD Industries [F] /1992.
(1)   199x returned to France.

ZWAHLEN & MAYR SA (ZM), Aigle   VD78

A metalwork construction company from Lausanne that transferred to Aigle in 1966. This site lies on the Aigle industrial estate that from 1964 was rail connected to the railway line from Saint-Triphon serving the Raffineries du Sud-Ouest. There are 920 metres of dedicated track for the estate. Address: Zone Industrielle 2, 1860 Aigle.

Gauge: 1435 mm   Locn: 272 $ 5616/1279   Date: 06.1986

         1ADM   1Adm   home-made   -   197x   (a)   1984 scr

(a)   home-made; also quoted as 4wDM, Bdm.

Valais (Wallis)   VS

ALCAN ALUMINIUM VALAIS SA   VS01
formerly: Alusuisse SA (until 2000)
            Aluminium Industrie AG (AIAG) (until 1963)

Chippis
This factory produces and refines aluminium, and was built 1905-08. From 1906 it was rail connected to Sierre CFF station by a line 2.1 kilometres long and with a gradient of 1 in 50 at the main-line end. This line crosses the Rhône on a reinforced concrete bridge, a novelty at the time of construction. Later developments include a rail served stocking area situated on the station side of the river.

"RHONE" SLM 2538 of 1915 (a "Tigerli") shunts in the area outside the factory and to the north of the R. Rhône on 14 September 1974. It had just returned from a double-heading a train to the station.

Gauge: 1435 mm   Locn: 273 $ 6081/1255   Date: 03.2003

	AIAG	0-4-0Tm	Bn2tk	Krauss-S LVIII i	1720	1886	(a)	1915 wdn
4	NAVIZENCE	0-6-0WT	Cn2t	SLM E 3/3	2079	1910	new	(2)
5	RHÔNE	0-6-0WT	Cn2t	SLM E 3/3	2538	1915	(c)	(3)
	SIERRE	0-8-0T	Dh2t	SLM E 4/4	2621	1917	new	(4)
29	CHIPPIS	0-6-0WT	Cn2t	SLM E 3/3	2503	1915	(e)	(5)
30	SIERRE	0-6-0WT	Cn2t	SLM E 3/3	2130	1910	(f)	(6)
31	GERONDE	0-6-0WT	Cn2t	SLM E 3/3	2298	1912	(g)	(7)
		4wPM	Bpm	RACO RA11	1257	1939	new	(8)
		4wDM	Bdm	R&H 88DS	321726	1951	new	(9)
	GERONDE	6wDH	Cdh	Henschel DHG 700C	32479	1981	(j)	
	RHÔNE	4wDH	Bdh	Henschel DHG 300B	32475	1982	new	
837 908-3 163 RENÉ		6wDH	Cdh	Henschel DHG 500C	31239	1968	(l)	
		0-6-0DH	Cdh	Jung R 40 C	12041	1961	(m)	(13)

CH 309 VS

Steg VS

A factory producing and refining aluminium dating from 1962 and rail connected from the start to Gampel-Steg SBB station. This line also crosses the R. Rhône. The steam engines worked to Chippis for attention and were replaced by ones from Chippis for a short period. The plant closed in 2007.

"NAVIZENCE" SLM 2079 of 1910 shunts at Steg VS on 14 September 1974. It is now preserved in the Netherlands.

Gauge: 1435 mm	Locn: 274 $ 6252/1290	Date: 03.2003

27	STEG	0-6-0WT	Cn2t	SLM E 3/3	1568	1904	(n)	1972 scr
36	STEG	0-6-0WT	Cn2t	SLM E 3/3	1974	1909	(o)	(15)
	BENKEN	4wDH	Bdh	Henschel DHG 300B	31557	1973	(p)	(16)
		4wDH	Bdh	MaK G 400 B	1001305	2003	(q)	

(a) purchased from Kriens-Luzern-Bahn (KLB) "2" /1906.
(c) new as "2"; renumbered "5" ? /1915.
(e) purchased from CFF "8523" /1963.
(f) purchased from CFF "8507" /1965.
(g) purchased from Services industriels de Genève, Vernier "1" /1965.
(j) ex Henschel, Kassel [D] /1982.

(l)   hired from Lonza AG, Visp /1989.
(m) hired from LSB /1998.
(n)   purchased from SBB "8458" /1962.
(o)   ex Ciments Vigier SA, Péry "2" /1973-74.
(p)   new to Bayerische Elektro-Stahlwerke GmbH (BEST), Meitingen [D] "1"; ex Lechstahl, Meitingen [D] /1980 via WBB/Asper.
(q)   ex Vossloh Moers[D] /2005; formerly a hire locomotive.

(2)   1983 to Landesmuseum Mannheim [D]; later Museum Buurt Spoorweg (MBS), Haaksbergen [NL] "8".
(3)   1981 wdn; then donated to the Commune of Sierre; 1988 to Amicale du train à vapeur, Le Pont; 1998 to chemin de fer touristique Pontarlier-Vallorbe [F].
(4)   1961 withdrawn; 1962 scrapped.
(5)   1982 withdrawn; 1984 to Amicale du train à vapeur, Le Pont; 1998 to chemin de fer touristique Pontarlier-Vallorbe [F].
(6)   1980 withdrawn; 1981 to W. Fournier, Sierre; later on display in Sierre.
(7)   1980 withdrawn; display at Thévenaz-Leduc, Renens VD.
(8)   1952 to L. Cattori, Castione via Kaufmann, Thörishaus.
(9)   1981 to Cablofer SA, Bex.
(12) 1998 returned to LSB.
(15) 1986 to John Gaillard, Yverdon; later Amicale du train à vapeur, Le Pont.
(16) 2005 to Hafenverwaltung Kehl, Kehl [D] via LSB/railimpex.

see: Eisenbahn Amateur March 1970

ARGE BAHNTECHNIK LÖTSCHBERG, Raron   VS02 C

alternative: Totalunternehmer Arbeitsgemeinschaft Bahntechnik Lötschberg

A consortium responsible for installing the railway infrastructure necessary for the Lötschbergbasistunnel that opened in 2007. The railway locomotives were provided by SERSA, and were stationed either here or at Frutigen; Kanton Bern. On completion of the tunnel the infrastructure and the responsibility for maintaining it passed to the BLS. Brochures available before the event indicated that this maintenance work would be performed by the same consortium. As the consortium includes SERSA, the continuity would appear to be ensured.

Gauge: 1435 mm   Locn: 274 $ 6310/1284   Date: 01.2008

ARGE BALTSCHIEDER, Visp + ARGE BIETSCHTAL, Raron   VS03 C

A consortium formed by Rothpelz, Lienhard and Zschokke for the construction of double track on the BLS Bietschtal viaduct and other work. The construction tracks had been removed by the end of 1983.

Gauge: ?750 mm   Locn: 274 $ 6288/1302   Date: 03.1982

1	4wDH	Bdh	Diema DFL 90/0.2	2360	1960	(a)	(1)
2	4wDH	Bdh	Diema DFL 90/0.2	2729	1964	(a)	(2)

(a)   new to Munitions-Depot Aurich-Tannhausen [D] (as 600 mm DS90) "2" and "9" respectively; ex Diema /1981.

(1)   19xx to Conrad Zschokke, Genève.
(2)   1983-85 to construction of the Hondrich tunnel (BLS).

ARGE FURKA BASISTUNNEL (LOS 61), Oberwald   VS04 C

At least the following worked at the western end of the Furka-Basistunnel (1973-82) project. The work included the Oberwald Umfahrungstunnel as well as the end of the main tunnel. The same contractors were responsible for "Lotta 63"; the "Bedretto-fenster".

Gauge: 750 mm   Locn: 265 $ 6700/1542   Date: 10.1976

1	4wBE	Ba	SIG	?	19xx	s/s	

2	4wBE	Ba	SIG	?	19xx	s/s
3	4wBE	Ba	SIG	?	19xx	s/s
TEE 1	4wBE	Ba	SIG	?	19xx	s/s
5	4wBE	Ba	SIG	?	19xx	s/s
ROSSI	4wBE	Ba	SIG	?	19xx	s/s

see: www.csc-sa.ch/downloads/Fob-t-csc.pdf

ARGE MATRANS, Steg/Raron — VS05 C

A consortium (MaTrans) of several companies responsible for boring the Lötschbergbasistunnel. A temporary standard gauge connection from the access line leading to Alusuisse, Steg VS ran to a point near the eastern end of the Alusuisse boundary fence where the consortium had its base. The location was referred to as Steg/Raron. From there an access tunnel was bored to meet what became the future BLS tunnel. A railway used in the construction of the access tunnel was retained during the building of the main tunnel to bring in cement and other necessary material. By 2005 both standard and narrow gauge lines were partially dismantled and the locomotives of the latter stored outside. By 2007 all traces had been removed except for some narrow gauge rails on a road crossing.

The four locomotives listed without running numbers have all been reported as having worked on this project, but none have been verified there. Three have been quoted as of 750 mm gauge. Details required.

"L1" Schöma 5005 of 1989, rebuilt 1993 (type CFL 180DCL) in the service area on 9 May 2005.

Gauge: 900 mm   Locn: 274 $ 5255/1293   Date: 05.2005

L1	4wDH	Bdh	Schöma CFL 180DCL	5005	1989(93)	(a)	(1)
L2	4wDH	Bdh	Schöma CFL 180DCL	4504	1982	(b)	(1)
L3	4wDH	Bdh	Schöma CFL 200DCL	5254	1991	(c)	s/s
4	4wDH	Bdh	Schöma CFL 200DCL	5273	1992	(c)	s/s
5	4wDH	Bdh	Schöma CFL 200DCL	5255	1991	(c)	s/s

Gauge: 750? mm   Locn: ?   Date: 01.2008

| | 4wDH | Bdh | Schöma CFL 180DCL | 5060 | 1989 | (f) | (6) |

			Schöma CFL 200DCL	5089	1990	(g)	(7)
	4wDH	Bdh	Schöma CFL 200DCL	5311	1993	(h)	s/s
	4wDH	Bdh	Schöma CFL 200DCL	5316	1993	(i)	s/s

(a) new to Trans-Manche Link (TML) [GB] "RS 011"; ex Schöma after reconditioning /199x.
(b) new to ArGe Walgaustollen, Baulos Bürs [A]; reconditioned by Schöma; ex Arge KS [D] /199x.
(c) ex Lesotho Highlands Water Projekt (LHPC) [Lesotho] /200x via Schöma.
(f) new to Trans-Manche Link (TML) [GB] "RS 023 GERLINDA" ; ex Ping Ling [Taiwan] (914 mm) via Schöma.
(g) new to Trans-Manche Link (TML) [GB] "RS 030 ALISON"; verbally stated to have been on a "Grossprojekt" at Raron.
(h) new to TMT Group, Korsör, Storebaelt-Tunnel [DK] "45-35" ; ex Lesotho Highlands Water Projekt (LHPC) [Lesotho] via Schöma.
(i) new to TMT Group, Korsör, Storebaelt-Tunnel [DK] "45-40" ; ex Olympic Metro Athen "B" [GR] via Schöma.

(1) 2006 to Pajares Lote 1 [E].
(5) 2006 to ArGe Tunnel Wienerwald [A].
(7) 5/2007 seen at RhB Werkstätte, Landquart.

JEAN-JACQUES BADER SA, Conthey VS06

alternative: Vétroz (village name)

This scrap merchant specialises in vehicle demolition. Rail connected at the western end of the siding from Châteauneuf-Conthey station in 2000. Address: route de l'Industrie 2, 1964 Conthey.

Gauge: 1435 mm Locn: 273 $ 5891/1182 Date: 10.2008

Tm¹	446	4wDM	Bdm	RACO Tm11	1376	1948	(a)	(1)
		4wDM	Bdm	RACO 85 LA7	1649	1962	(b)	

(a) new to CFF "Tm 535" ex Geilinger, Yvonand /1996-2000.
(b) new to CFF "Tm¹ 346"; ex CFF "Tm¹ 446" /2006.

(1) c2006 to Asper.

M. BARATELLI & CIE, Chantier de Barberine, Le Châtelard VS
VS07 C

A Zürich and Lausanne-based construction company working on the first Barberine dam. A photograph of one of the locomotives is on display in Le Châtelard power station museum (situated at the foot of the access funicular used for the project), indicating that it was at the "carrière de Ruan". A separate report lists an SLM steam locomotive and two diesel locomotives that worked on the same project, but not who operated them.

Gauge: 750 mm Locn: ? Date: 08.1995

1	0-4-0T	B?2t	Maffei	4203	1921	(a)	(1)
2	0-4-0T	B?2t	Maffei	4204	1921	(a)	(2)
3	0-4-0T	B?2t	Maffei	4206	1921	(a)	(3)
4	0-4-0T	B?2t	Maffei	4208	1922	(a)	(4)
	0-4-0T	Bn2t	Märkische 50 PS	296	1898	(e)	(5)

(a) new via Robert Aebi & Cie AG, Zürich.
(e) made available by H. Martin & J. Baratelli, Vallorbe; by /1922; worked on the "Hohenbahn".

(1) 1929 to J. Seeberger, Frutigen "1".
(2) 1927 to J. Seeberger, Frutigen "2".
(3) 1928 to J. Seeberger, Frutigen "3".
(4) 1922 to J. Seeberger, Frutigen "4"?
(5) 19xx returned to H. Martin & J. Baratelli, Vallorbe.

see: Revue des Amis du Rail de la Suisse Romande 12/1973

BARRAGE DE LA DIXENCE, Le Chargeur   VS08 C

Work performed by SA la Dixence, Sion and Energie Ouest Suisse, Lausanne for the construction of the dam of the "la Dixence" hydroelectric scheme 1931-35.

Gauge: 1000 mm   Locn: ?   Date: 03.1989

0-6-0T	Cn2t	Corpet	1195	1908	(a)	(1)
0-6-0T	Cn2t	Pinguely	354	1919	(b)	(2)
0-6-0T	Cn2t	SACM-G	4172	1890	(c)	(2)
0-4-0DM	Bdm	Maffei	5915	1930	(d)	s/s
0-4-0DM	Bdm	Maffei	5916	1930	(d)	s/s
0-4-0DM	Bdm	Maffei	5917	1930	(d)	s/s
4wPM	Bbm	O&K RL 2a	20200	1931	new	(7)

(a) 1931 ex Brunner & Marchand [F].
(b) 1932 ex Brunner & Marchand [F] "Lv 193".
(c) 1934 ex Lausanne-Echallens-Bercher "5".
(d) new via Brun & Cie AG, Nebikon.

(1) 1935 boiler removed from boiler register; 1941 to Hilti, Feldkirch [A].
(2) 1941 to Hilti, Feldkirch [A].
(7) 1940 to MOB "Tm 1".

BAUUNTERNEHMUNG GORNERGRATBAHN, Zermatt   VS09 C

A locomotive used for the construction of the Gornergrat Bahn.

Gauge: 1000 mm   Locn: ?   Date: 01.2008

0-4-2RT	3bn2t	SLM	748	1892	(a)	(1)

(a) ordered by Chemin de Fer du Revard, Aix-les-Bains [F] "8"; new here.
(1) 1905 to Gornergratbahn; 1918 to Monistrol-Monserrat [E] "6".

BILLIEUX, PAYOT, KALBERMATTEN, Zermatt   VS10 C

The SIG list of references gives five ETM 50 locomotives delivered here.

BODENMÜLLER AG, Visp   VS11 C

The SIG list of references gives one ETB 50 locomotive delivered here.

BRANDT, BRANDAU & CIE   VS12 C

The company was the principal contractor for the construction of the first Simplon railway tunnel. There was a northern construction site in Switzerland (Brig) and a southern one in Italy (Iselle). Allocation of the locomotives to the two may not be entirely correct. The locomotives had large boilers so that the fire did not have to be used when inside the tunnel.

Brig

Gauge: 800 mm   Locn: ?   Date: 08.1995

4	2-4-0T	1Bn2t	SLM	1459	1902	new	s/s
11	0-4-0CA	Bp2	SLM	1329	1900	new	s/s
12	0-4-0CA	Bp2	SLM	1330	1900	new	s/s
13	0-4-0CA	Bp2	SLM	1331	1900	new	s/s
14	0-4-0CA	Bp2	SLM	1461	1900	(e)	s/s
15	0-4-0CA	Bp2	SLM	1462	1900	(e)	s/s

Iselle [I]

Gauge: 800 mm   Locn: ?   Date: 02.1994

		4wPM	Bbm	Deutz C I F	32	1901	new	s/s
4	ISELLE	2-4-0T	1Bn2t	SLM	1460	1902	new	s/s
	11	0-4-0CA	Bp2	SLM	1326	1900	new	s/s
	12	0-4-0CA	Bp2	SLM	1327	1900	new	s/s
	13	0-4-0CA	Bp2	SLM	1328	1900	new	s/s

(e)   possibly used at Iselle [I] rather than Brig.

BRIQUETERIE DE VERNAYAZ, Vernayaz   VS13

owned by: Dorénaz SA, Charbonnages du Valais, Genève
later: Moderna Commerce d'emballages SA

During three periods (1855-98, 1917-21 and 1941-53) anthracite was mined from the Dorénaz area. Dorénaz SA was formed by Ateliers Piccard et Pictet of Genève to work the anthracite commercially. From about 1917 it was transported by a 3.6 kilometre long aerial ropeway down to the briquetting factory at Vernayaz, built in 1918, and turned into briquettes, particularly for use by the CFF. The siding, also installed in 1918, was de-electrified in 1928. From 1932 some of the site was reused by what became Moderna Commerce d'emballages SA. The mines were operated by M. Dionisotti, from 1940-47; that is until a little after the end of World War II. Some of the anthracite produced was used by the lime works at Malévaux near Monthey. In 2007 some of the early buildings and traces of an extensive 750 mm system were still visible. It is not known if the buckets could be detached from the ropeway and transported on the narrow gauge.

Gauge: 1435 mm 250V DC   Locn: 272 $ 5692/1099   Date: 06.1988

	1AWE	1Ag	SBB(Biel)/CIEM	?	1919	(a)	s/s

(a)   built from a bogie of an old rail vehicle.
       see: la Revue Polytechnique No 1706 pp 354-6

BUNDESAMT FÜR BETRIEBE DER LUFTWAFFE (BABLW), Niedergesteln   VS14

formerly: Oberkriegskommissariat (OKK), Gampel-Steg
Fuel tanks, rail connected to Gampel-Steg SBB station.

Kronenberg126 1952 (type DT 90/120) at work in Gampel-Steg SBB station on 30 April 1978.

Gauge: 1435 mm Locn: 274 $ 6240/1316 Date: 12.2008

7	4wDM	Bdm	Kronenberg DT 90/120	126	1952	(a)	(1)
7	4wDM	Bdm	RACO 85 LA7	1682	1964	(b)	

(a) ex OKK Göschenen ca. /1975.
(b) ex CFF "Tm^I 485" /1996; offered for sale /2008.
(1) 1996 to BABHE, Oberburg; by 1998 to Zürcher Museumsbahn (ZMB), Sihlwald "9 MAX".

CHAPERY & DELORME, Visp VS15 C

A civil engineering concern?

Gauge: 750 mm Locn: ? Date: 03.2002

	0-4-0T	Bn2t	Heilbronn II alt	28	1873	(a) s/s

(a) ex Mathier & Co, Visp.

CHEMINS DE FER FÉDÉRAUX (CFF) VS16 C

This was a civil engineering project to bore the second Simplon rail tunnel. As with the first tunnel two work sites were established, one on the northern side at Brig and the other on the southern side near Iselle [I].

The Demag locomotives have also been attributed to Maschinenfabrik Rudolf Meyer - should that be Rudolf Meier Maschinenfabrik GmbH? The distribution of these locomotives given here is arbitrary and not confirmed.

The gauge of the construction railways was 750 mm, rather than the 800 mm used when constructing the first tunnel.

Brig

Gauge: 750 mm   Locn: ?   Date: 08.1995

1	0-4-0CA 2cc	Bp2v	SLM	2405	1913	new	s/s
2	0-4-0CA 2cc	Bp2v	SLM	2406	1913	new	s/s
3	0-4-0CA 2cc	Bp2v	SLM	2407	1913	new	s/s
	0-4-0CA 2cc	Bp2v	Demag	?	191x	new	s/s

Département de construction du tunnel du Simplon II

Gauge: 750 mm   Locn: ?   Date: 04.2008

2	0-4-0WT	Bn2t	O&K 50 PS	2187	1906	(e)	(5)
4	0-4-0WT	Bn2t	O&K 50 PS	2448	1908	(e)	(5)
	0-4-0WT	Bn2t	O&K 50 PS	2449	1908	(e)	(7)

Iselle [I]

Gauge: 750 mm   Locn: ?   Date: 08.1995

4	0-4-0CA 2cc	Bp2v	SLM	2408	1913	new	s/s
5	0-4-0CA 2cc	Bp2v	SLM	2409	1913	new	s/s
6	0-4-0CA 2cc	Bp2v	SLM	2410	1913	new	s/s
	0-4-0CA 2cc	Bp2v	Demag	?	191x	new	s/s

(e) ex Entreprise Générale du Lötschberg (EGL), Frutigen + Brig /1913.
(5) 1920-9 to AG Hunziker & Cie, Brugg AG.
(7) 1921 to Jaquet, Vallorbe.

COMMUNE DE SIERRE, Sierre   VS17 M

Location: Rue de l'Ile Fâcon near a car-washing establishment. This road serves the Zone Industrielle, Ile Fâcon.

Gauge: 1435 mm   Locn: 273 $ 608770,126800   Date: 03.2003

30	SIERRE	0-6-0WT	Cn2t	SLM E 3/3	2130	1910	(a) display

(a) new to CFF "8507"; ex Alusuisse, Chippis /198x via W. Fournier, Sierre.

CONSORTIUM BIPV, Fionnay   VS18 C

A consortium of civil engineering firms, in which Bellorini & Fils, Lausanne participated. Fionnay is near the dam at the end of Lac de Mauvoisin. The associated power station was brought into service in 1960 and could have been part of the project.

Gauge: 600 mm   Locn: ?   Date: 08.2002

4wDM	Bdm	O&K MV0	25166	1952	(a)	(1)

(a) new via MBA, Dübendorf.
(1) ? to Ch. Payot SA, Clarens; later FWF, Otelfingen.

CONSORTIUM CLEUSON-DIXENCE, Nendaz — VS19 C

Two separate lots for the construction of this hydroelectric scheme used railways.

Lot A/B

A consortium with the title Lochner Cleuson-Dixence, Beuson for the construction of a hydroelectric tunnel above Hérémence.

Gauge: 900 mm   Locn: ?   Date: 01.1996

1	4wDH	Bdh	Schöma CFL 180DCL	5006	1989	(a)	s/s
2	4wDH	Bdh	Schöma CFL 180DCL	5061	1989	(b)	(2)
	4wDH	Bdh	Schöma CFL 180DCL	5014	1989	(c)	s/s
4	4wDH	Bdh	Schöma CFL 180DCL	5435	1995	new	s/s
	4wDH	Bdh	GIA DHS90	980 600	19xx	(e)	
	4wDH	Bdh	GIA DHS90	?	19xx	(e)	

Lot C/D

A consortium formed of Murer/Züblin/CSC (MZC) for the construction of a hydroelectric gallery 5.5 kilometres long above Nendaz. The gauge of the three new locomotives (Schöma 5388-90) is given in the Schöma lists as both 750 and 900 mm.

Gauge: 900 mm   Locn: ?   Date: 01.1996

2	4wDH	Bdh	Schöma CFL 180DCL	5388	1994	new	s/s
16	4wDH	Bdh	Schöma CFL 150DCL	5149	1990	(h)	s/s
19	4wDH	Bdh	Schöma CFL 180DCL	5390	1994	new	s/s
22	4wDH	Bdh	Schöma CFL 180DCL	5389	1994	new	s/s

(a) new to Trans-Manche Link [GB] "RS 012 JENNY"; ex Schöma /1994.
(b) new to Trans-Manche Link [GB] "RS 024 MOLLY"; ex Schöma /1994.
(c) new to Trans-Manche Link [GB] "RS 020 TINA"; ex Schöma /1994.
(e) hired from Schlatter AG, Rheinsulz.
(h) new to Trans-Manche Link [GB] "RS 038 HANNELORE" (900 mm); ex Schöma /1994.
(2) ? to Perarolo-Calazone, Belluno [I].

JEAN-CLAUDE COQUOZ, Collombey-Muraz — VS20 M

A scrap merchant and collector with a locomotive placed outside the scrap yard. Home address: Chemin de la Ry 2, 1843 Muraz (Collombey); scrap yard: Chemin du Tonkin.

Gauge: 1435 mm   Locn: 272 $ 561725,125610   Date: 09.2008

ET^{O}B 74-Lb-00003	0-4-0DM	Bdm	BLW VM		4xxxx	c1917	(a) display

(a) new as 600 mm 0-4-0PM Bbm; ex R. Bouclier & J. Chatagnat, Thonon-les-Bains, Haute-Savoie (Carrières de Vongy) [F] by /2008.

see: Rail et Industrie No 2 (Décembre 2000)

COUCHEPIN, DUBUIS & CIE, MEX — VS21 C

later: Mex VS
alternative: Mex sur Evonniaz

This was a civil engineering company. One locomotive is recorded as delivered to Sion (office address?).

Gauge: 600 mm   Locn: ?   Date: 04.2009

4wDM	Bdm	O&K(M) RL 1a	4307	1930	(a)	(1)
4wDM	Bdm	O&K(S) L 200	1605	194x	(b)	s/s

(a) new to Sion.

(b) new via O&K/MBA, Dübendorf; the available data does not indicate if it had a petrol or diesel engine; gauge not given.
(1) 19xx to Losinger & Co., Bern.

G. DÉNÉRIAZ SA, Sion VS22 C

This was a civil engineering company. For a while it operated the coal mines on the valley side. That scheme included a téléphérique, but terminated before the locomotives listed here were available.

On 24 August 1966 four locomotives were stored in a shelter behind the cement silos. Nearest the camera is O&K 25339 of 1955 (type MV1a), followed by two cab-less Bartz battery locomotives (963 and 966 of 1952). Just visible outside is another Bartz locomotive (1114 of 1957), it appears to have been fitted with a cab from Jung diesel locomotive.

Gauge: 600 mm   Locn: 273 $ 5957/1188   Date: 02.1997

1	4wBE	Ba	Bartz	966	1952	(a)	b1978 s/s
2	4wBE	Ba	Bartz	963	1952	(b)	b1978 s/s
	4wBE	Ba	Bartz	1114	1957	(a)	b1978 s/s
6230	4wDM	Bdm	O&K MV1a	25339	1955	(d)	a1986 s/s
6231	4wBE	Ba	MFO?	?	19xx		a1979 s/s

(a) ex ? /19xx.
(b) carried a RACO plate.
(d) new via MBA, Dübendorf to Züblin & Dénériaz, Le Chargeur "1" (600 mm); reported at Losinger & Cie, Turtmann (750 mm); ex ? by /1966. Gauge at Sion not verified.

JOSEPH DIONISOTTI, Monthey VS23

A lime works situated at Malévaux. The rail network was abandoned between 197x and 198x. Joseph Dionisotti lived between 1891-1970. Later ownership is unclear (Jean Dionisotti?).

Gauge: 600 mm			Locn: 272 $ 5616/1232					Date: 02.1996	
	4wDM	Bdm	Schöma CDL 15		1642	1955	(a)	a1976 s/s	

(a) new via Hans F. Würgler, Zürich and W. Perrenoud, Constructeurs Mécaniques, Installation et Machines de Manutention, Genève.

DUPONT & SCHAFFNER, Eggerberg — VS24 C

A building contractor participating in the construction of the Brig-Goppenstein section of the BLS railway (1906-13).

Gauge: 750 mm			Locn: ?				Date: 03.2002
SAN PAOLO	0-4-0T	Bn2t	O&K 50 PS	4347	1911	new	(1)

(1) 1914 to Frutiger, Lüthi & Lanzrein, Bern.

ÉLECTRICITÉ DE LA LIENNE SA, Sion — VS27 C

One locomotive was seen at the Vatryeret (Adduction de Vatryeret, Barrage de Rawyl) water tunnel construction site above Sion. It is assumed that the other locomotive of the pair delivered to the agent was also present.

Battery locomotive Jung 12907 of 1958 (type Ez 21) at work on 25 August 1966. A sister locomotive is now with the Feld- und Werkbahn- Freunde, Otelfingen.

Gauge: 600 mm			Locn: 273 $ 6000/1332				Date: 02.1997
	4wBE	Ba	Jung Ez 21	12906	1958	(a)	(1)
	4wBE	Ba	Jung Ez 21	12907	1958	(a)	s/s

(a) new via Baumaschinen-Verkaufs AG (BMVAG), Montreux.
(1) not confirmed here; to ?; then FWF, Otelfingen.

ENTREPRISE BUSSIEN, Le Bouveret — VS28

The concern operated a quarry. The railway line connected the quarry (Carrière Bussien) via the CFF station of Bouveret to a wharf on Lac Léman and still existed as late as 1968.

Gauge: 600 mm		Locn: 262 $ 5554/1366				Date: 01.1989	
	4wPM	Bbm	Oberursel	?	c1915	(a)	(1)
	4wP/DM	Bb/dm	unknown	?	19xx	(b)	(2)

(a) supplied via Rubag, ex Carrières d'Arvel, Villeneuve VD /1936.
(b) it is not known if this was a diesel or petrol locomotive.
(1) 1974 to BC; later private collection.
(2) by 1974 sold.

ENTREPRISE DE CORRECTION DU TORRENT DE SAINT-BARTHÉLEMY, Mex/Saint-Maurice   VS29 C

later: Mex VS
alternative: Mex sur Evonniaz

Stage 1

The construction site for the first grading works on the "Torrent de Saint-Barthélemy" by the contractor Couchepin, Dubuis & Cie.

Gauge: 600 mm		Locn: ?				Date: 03.2002	
	0-4-0T	Bn2t	Freudenstein	106	1911	(a)	(1)
HERBERT	0-4-0WT	Bn2t	Krauss-S XIV mm	4048	1899	(b)	(2)

Stage 2

The construction site for the second grading works of the "Torrent de Saint-Barthélemy" in the area known as "Le Follet". This work was performed by a consortium formed of E. Bellorini & Fils + Oyex, Chessex & Cie, Mex + Conforti & Juillard.

Gauge: 600 mm		Locn: ?				Date: 08.1995	
	0-4-0WT	Bn2t	Jung 20 PS	1693	1911	(c)	(3)
	0-4-0T	Bn2t	O&K 20 PS	373	1900	(d)	(4)

(a) ex Syndicat d'assainissement de la Plaine du Rhône, Aigle /1931.
(b) ex Müller & M. Dionisotti, Entrepreneurs, Lausanne + Saint-Maurice /1931.
(c) ex O&K, Zürich /1936.
(d) ex Entreprise Paul Juillard, Saxon /1936.
(1) 1932 to Rüesch Hans, Hartschotterwerk Sevelen [D].
(2) 1933 boiler removed from boiler register; returned to M. Dionsotti.
(3) by 1939 to E. Bellorini, Lausanne-Malley.
(4) 1938 to E. Bellorini, Lausanne-Malley.

EGT CONSTRUCTION SA, Saint-Maurice   VS30 C

formerly: Entreprise de Grands Travaux (from 28/11/1936)
           Michel Dionisotti (from 1908) (born 1891, died 1970)

A civil engineering concern founded by M. Dionisotti in 1908. In 1936 this enterprise was made into a company (EGT) with the head office in Lausanne and branches in Genève and Saint-Maurice. In 1954 the head office was moved to St. Maurice (town), while in 1961 the Genève and Lausanne establishments were closed and replaced by one in Montreux. In 1986 the offices were moved out of the town to a site adjacent to cement works, the works yard and the petrol station. Among the projects were the operation of the anthracite mines at Dorénaz (1940-47), the cement works at Saint-Maurice, Genève Aeroport, the ESSO storage depot at Genève and the grading of the Torrent de Saint-Bartélemy. Address: Route d'Epinassey.

"4" Gmeinder 662 in store in 2008. Photograph: S. Jarne

Gauge: 750 mm			Locn: 272 $ 566715,117220 (plinth)				Date: 08.1995	
5		4wDH	Bdh	Gmeinder		?	19xx	(a)	display
Gauge: 600 mm			Locn: 272 $ 566765,117100 (offices)				Date: 01.2009	
1		4wDM	Bdm	O&K(M)?				(b)	s/s
2	HERBERT	0-4-0WT	Bn2t	Krauss-S XIV mm	4048	1899	(c)	(3)
3		0-4-0T	Bn2t	O&K 30 PS	9955	1922	(d)	(4)
4		4wDM	Bdm	Gmeinder	662	192x		str
		4wPM	Bpm	O&K(M) M	2231	1926	(f)	str

(a) engine (Karl Kaelble, Backnang No 3864); placed on display in the forecourt of "Station Service Horizonville" near the EGT office in Saint-Maurice /2004; the gauge is that of the internal system at the Briqueterie de Vernayaz.
(b) requires confirmation; probably from M. Dionisotti era.
(c) new to Neuchâtel Asphalt Co Ltd, Travers (610 mm) "HERBERT"; ex H. Piot & Ch. Piquet, Bavois /1923; used 1929-31 by the consortium of Müller & M. Dionisotti of Lausanne, for the first project to grade the "Torrent de Saint-Barthélemy".
(d) new; used by the consortium of Josef Zeiter & Dionisotti of Zürich for a contract at Ackersand/Stalden.
(f) new to O&K, Köln [D]; used at the Charbonnages de Dorénaz.
(3) 1931 sold to Couchepin, Dubuis & Cie, Lausanne; 1944 repurchased; 1952 boiler removed from the boiler register; before 1965 placed on display in the forecourt of "Station Service Horizonville" near the EGT office in Saint-Maurice; 2002 loaned to Tacot-des-Lacs, Grez-sur-Loing [F].
(4) 1930 sold to Entreprise Paul Juillard, Saxon.

see: http://www.egt.ch

ENTREPRISE PAUL JUILLARD, Saxon   VS31 C

A civil engineering firm.

Gauge: 600 mm		Locn: ?			Date: 08.1995	
0-4-0T	Bn2t	Jung Helikon	3025	1919	(a)	(1)
0-4-0T	Bn2t	O&K 20 PS	373	1900	(b)	(2)
0-4-0T	Bn2t	O&K 30 PS	9955	1922	(c)	(3)

	0-6-0T	Cn2t	Krauss-S Zwilling		2936B	1894	(d)	(4)
	0-6-0T	Cn2t	Krauss-S Zwilling f		3052A	1894	(e)	(5)

(a) ex Syndicat d'assainissement de la Plaine du Rhône, Aigle, bought by Juillard, Müller & Dionisotti after /1929.
(b) ex H. Piot & Ch. Piguet, Bavois /1926.
(c) ex Entreprise Josef Zeiter & Dionisotti, Zürich /1930.
(d) new to Deutsche Feldbahn "62B"; ex ? /1926.
(e) new to Eisenbahn-Bataillon Oberwiesenfeld [D] "4"; ex ? /1926.

(1) 1931 boiler removed from boiler register; s/s.
(2) 1936 to G. Bellorini & Fils, Oyex, Chessex & Cie, Mex VS, possibly with other owners in the interval.
(3) 1931 to Gavillet & Delesle, Lausanne.
(4) 1931 boiler removed from boiler register; 1943 to Schmidlin, Genève.
(5) 1933 boiler removed from boiler register; s/s.

ENTREPRISE GÉNÉRALE DU LÖTSCHBERG (EGL), Brig   VS32 C

During the construction (1907-11) of the south ramp (Goppenstein-Brig) section of the BLS a temporary 750 mm gauge line was in operation. The route roughly followed the present line, though generally below it and now forms part of the footpath between Goppenstein and Brig. The base with a two-road locomotive shed was situated at Naters just across the R. Rhône from Brig. For locomotive details see the EGL entry in the Canton Bern section.

ENTREPRISE POUR LA CONSTRUCTION DU TUNNEL DES FORCES DU RHÔNE, Chippis   VS33 C

Construction works for a hydroelectric scheme.

Gauge: 750 mm   Locn: ?   Date: 08.1995

HOHENLOHE	0-4-0T	Bn2t	Märkische 50 PS	192	1896	(a)	(1)

(a) new to Kleinbahn Landsberg-Rosenberg [D]; ex ? /1908.
(1) 1941 to Heinrich Hatt-Haller, Zürich possibly with other owners in the interval.

ENTREPRISES STUAG ET BELLORINI, Mauvoisin   VS34 C

Civil engineering firms participating in the construction of the Mauvoisin hydroelectric plant.

Gauge: 750 mm   Locn: ?   Date: 08.1995

4wDM	Bdm	RACO		?	19xx	(a)	s/s
4wBE	Ba	SSW EL 9		?	1953	(b)	s/s

(a) ex ?; overhauled /1952-53.
(b) new to Bellorini & Fils, Lausanne.

EVÉQUOZ, PIERRE-LOUIS & CIE., Pont-de-la-Morge   VS35 C

The SIG list of references gives two ETM 50 locomotives delivered to this consortium.

GIOVANOLA FRÈRES SA, Monthey   VS36

This company dealt in metal and performed mechanical construction. It declared bankruptcy in 10/2007.

Gauge: 1435 mm   Locn: 272 $ 5623/1236   Date: 03.2003

1	4wDE	Bde	Kronenberg DT 180/250	129	1952	(a)	OOU

(a)   1964 ex Grande Dixence SA, Sion /1964.

GORNERGRAT BAHN AG   VS37 M

Stalden VS
One item of heritage stock was placed on the Killerhofbrücke roundabout near Stalden VS on 23 May 2007.
Gauge: 1000 mm + Abt Rack 725V 50~ Locn: 274 $ 633000,119700   Date: 02.2009
HGe 2/2 3002   11RE   2/bd2   SLM He   1066   1898   (a)
(a)   new to Gornergratbahn "HG 2/2 2".

Zermatt
One item of heritage stock is retained.
Gauge: 1000 mm + Abt Rack 725V 50~ Locn: 284 $ 624150,096940   Date: 02.2009
Bhe 2/4 3011   1A1ARERC1A1Ag   SLM,BBC Che 2/4   3949,4535   1946   (a)
(a)   new to Gornergratbahn "Che 2/4 101"; reclassified "Bhe 2/4 3011" /1959.

GRANDE DIXENCE SA   VS38
formerly: ?
A company formed to construct and operate various hydroelectric schemes.

Ardon
A transhipment point in use 1948-49 for the cement used in constructing the Grande Dixence dam.
Gauge: 1435 mm   Locn: 273 $ 5872/1175   Date: 04.2000
               0-4-0DM   Bdm   O&K RL 4   20162   1931   (a) (1)

Sion
A siding crosses the R. Rhône and serves the Chandoline power station. It was in use to build this station and the dams of Dixence (1931-35) and Grande Dixence (1946-5x). Since 1964 it has been worked by the CFF.

The metre gauge funicular alongside the pressure pipes with the connection to the standard gauge line past the Chandolin power station visible in the background, seen on 1 October 2008.

Gauge: 1435 mm   Locn: 273 $ 5950/1191   Date: 08.1995

			4wDM	Bdm	O&K RL 4	20162	1931	new	(2)
1			4wDE	Bde	Kronenberg DT 180/250	129	1952	new	(3)

(a)   ex Grande Dixence SA, Sion ca. /1948.
(1)   c1949 to Grande Dixence SA, Sion.
(2)   1952 to Société des Chaux et Ciments de la Suisse Romande, Cimenterie Eclépens, Eclépens; 2000 to VVT, St. Sulpice NE.
(3)   1964 to Giovanola Frères SA, Monthey; possibly used during the construction of the Sembrancher-Le Châble railway line.

HOLDERBANK CIMENTS ET BÉTONS (HCB), Saint-Maurice   VS39

formerly: Société des Ciments et Bétons (SCB) (1988-1991)
            Société des Ciments et Bétons de la Suisse Romande (SCC) (1984-c1988)
            Société des Ciments Portland de Saint-Maurice SA (SCPS) (1955-c1984)

The plant was built in 1955 by Entreprise de Grands Travaux (EGT). In 1986 the ovens were stopped and the factory became a centre for grinding the clinker. The factory was closed at the end of 1994, and then demolished.

Gauge: 1435 mm   Locn: 272 $ 5666/1174   Date: 05.1995

3		0-6-0T	Cn2t	SLM E 3/3	1064	1897	(a)	(1)
3		4wBE	Ba2	Olten,BBC Ta	?,1945	1923	(b)	
	rebuilt	4wDE	Bde					(2)
2		4wBE	Ba2	Olten,BBC Ta	?,1942	1923	(c)	
	rebuilt	4wDE	Bde					(3)
		4wDM	Bdm	Breuer V	3033	1951	(d)	(4)
		4wDH	Bdh	O&K MB 7 N	26606	1966	(e)	(5)

(a)   new to Seetalbahn (SeTB) "WILDEGG 8"; later CFF "8652", ex Ciments Portland, Vernier /1953 or later via EGT, Bussigny-près-Lausanne.
(b)   ex CFF "Ta 44" /1961.

(c) new to CFF "Ta 41"; ex EGT, Bussigny-près-Lausanne /1971.
(d) new to GCG, Vevey "3"; ex EGT Bussigny-près-Lausanne /1973-80.
(e) ex Borsig-Werke, Berlin-Tegel "010" /1985 via NEWAG; short term hire to CIBA-GEIGY, Monthey /1985.

(1) to EGT Bussigny-près-Lausanne; 1968 scrapped.
(2) 1986 to la Fonte Électrique (FEBEX), Bex; 1999 to Bahn Museum Kerzers (BMK), Kerzers.
(3) 1990 to Société des Ciments et Bétons (SCB), Roche VD.
(4) 1984-5 scrapped.
(5) 1995 to Holderbank Ciments et Bétons (HCB), Roche VD.

ILLSEE TURTMANN AG (ITAG), Oberems   VS40

This company was a part of Alusuisse, then of the Forces Motrices Valaisannes and finally of the Rhônewerke Chippis.

The railway serves Alusuisse's hydroelectric plant. It runs four kilometres from Augstwaengi (6185/1243), by way of Meretschialp, to the Illsee dam (6152/1232),and was built in 1925-26. The track was renewed in 1995 using a locomotive hired from Belloli, Grono.

Gauge: 600 mm   Locn: 273 $ 6185/1243   Date: 10.1995

ILLSEE-EXPRESS	4wBE	Ba	EFAG	2121	19xx	new	
MERETSCHJ-EXPRESS							
	4wBE	Ba	EFAG	2123	19xx	new	
	4wBE	Ba	GIA BS2	204	1994	(c)	

(c) 1995 ex METRAG AG, Rümlang hired from Belloli, Grono.

LA POSTE SUISSE, Sion   VS41

formerly: Entreprise des PTT Suisses (PTT) (1985 till 31/12/1997)

A postal sorting centre dealing principally with returned packets. The installation is simple enough not to require a locomotive.

Gauge: 1435 mm   Locn: 273 $ 5397/1196   Date: 01.2003

LINDENMEYER, BOULENAZ & CIE, Sembrancher   VS42 C

A civil engineering company from Vevey involved in the construction of the Martigny-Orsières railway (1908-1910).

Gauge: 750 mm   Locn: ?   Date: 08.1995

VEVEY	0-4-0T	Bn2t	O&K 50 PS	2907	1908	new	(1)

(1) before 1917 sold to Rossi, Perusset & Cie, Ostermundigen.

LOCHER & CIE ZÜRICH, Simplon   VS43 C

One of contractors involved in boring the first Simplon rail tunnel. It is not clear whether the company worked at the Swiss (Brig) or Italian (Iselle) end of the tunnel.

Gauge: 800 mm   Locn: ?   Date: 08.1995

1	2-4-0T	1Bnt	SLM	1201	1899	new	(1)
2	2-4-0T	1Bnt	SLM	1202	1899	new	(1)
3	2-4-0T	1Bn2t	SLM	1308	1900	new	s/s

(1) 1907 to ? [I].

LOGISTIKBASIS DER ARMEE (LBA), Gamsen — VS44

formerly: Bundesamt für Betriebe des Heeres (BABHE) (till 31.12.2003)
Bundestankanlagen (BTA)

Fuel tanks, rail connected to SBB and BVZ by 1953 and managed by Carbura.

Gauge: 1435 mm   Locn: 274 $ 6397/1286   Date: 06.2008

5	4wDM	Bdm	Kronenberg DT 90/120	124	1952	(a)	(1)
	4wD R/R	Bd	UCA UCA-Trac	?	2008	(b)	
	4wD R/R	Bd	UCA UCA-Trac	?	2008	(b)	
Tm 236 342	4wDM	Bdm	RACO 95 SA3 RS	1588	1960	(d)	

(a) new; fitted with couplings for normal and metre gauge vehicles.
(b) new via Stauffer; fitted with couplings for normal and metre gauge vehicles.
(d) new to SBB "Tm^{II} 700"; withdrawn 9/2003; ex BLS "Tm 236 342" /2008.
(1) 2005 to LBA, Oberburg for scrap.

LONZA AG — VS45

Susten

To the east of the Leuk station a short narrow gauge line crossed the R. Rhône to bring stone from a quarry to staithes adjacent to the SBB. The quarry closed in 197x.

Gauge: 600 mm   Locn: 273 $ 6163/1288   Date: 02.1997

4wDM	Bdm	Gmeinder?	?	19xx	(a)	a1968 s/s
4wDM	Bdm	Gmeinder?	?	19xx	(b)	a1968 s/s

(a) new via Marti AG, Bern; Hatz engine 3020.
(b) new via Marti AG, Bern; Hatz engine 4007.

Vernayaz

An electrochemical plant making carbide was rail connected to the north end of Vernayaz station. This siding was later extended to serve the joint Lonza/Energy Ouest Suisse power station at Miéville (later Salanfe SA). There is no indication that the Lonza plant had its own locomotives. The power station serving it closed in 1992 and the plant itself was demolished in 2008. To serve the power station a 500 mm funicular ran from inside the Miéville compound. Although by 2008 the line was derelict the single vehicle for it remained on one of the two tracks at the summit.

The scene at the summit of the 500 mm gauge funicular on 30 September 2008.

Gauge: 1435 mm   Locn: 272 $ 568760,110000   Date: 12.2008
Gauge: 500 mm   Locn: 272 $ 568200,110430   Date: 12.2008

Visp

temporarily: Alusuisse-Lonza AG (1974-1997)
A works producing basic chemicals, built 1907-1909, rail connected from 1917.
Gauge: 1435 mm   Locn: 274 $ 6353/1272   Date: 09.2004

1		2-6-0T	1Cn2t	SLM Ed 3/4	1798	1907	(a)	(1)
2		0-4-0F	Bf2	SLM T 2/2	2593	1917	(b)	(2)
3		0-4-0F	Bf2	SLM	2589	1917	(c)	(3)
4		0-6-0T	Cn2t	SLM Ec 3/3	1377	1901	(d)	(4)
161	HANS	0-6-0DH	Cdh	Henschel DH 360C	30505	1962	(e)	2004 wdn
162	WILLI	0-6-0DH	Cdh	Henschel DH 360C	30703	1963	(f)	
837 908-3 163	RENÉ	6wDH	Cdh	Henschel DHG 500C	31239	1968	(g)	

(a) new to Solothurn-Münster-Bahn "Ed 3/4 1"; purchased from Dreispitz-Verwaltung, Basel /1945.
(b) new as "1"; renumbered "2" /19xx.
(c) purchased as new from SLM, Winterthur /1932; originally "2".
(d) new to Gotthardbahn "Ec 3/3 309"; ex Colormetal, Zürich /1947.
(e) new as "5".
(f) new as "6", hired to Alusuisse SA, Chippis /1989.
(g) ex Henschel, Kassel [D] (originally a demonstration locomotive) after /1971, formerly "JULES"; re-motored by LSB.
(1) 1965 to Technorama, Winterthur; 1972 to DVZO "1".
(2) c1965 wdn, 1976 to VHS, Luzern; 199x returned to Visp and placed on display inside the factory.
(3) 19xx placed on display inside the factory; 1992-93 scrapped.
(4) 19xx scrapped without being put into service.

LOSINGER & CIE, Turtmann   VS46 C
later: Losinger Sion SA
A construction site of this Sion-based engineering company.
Gauge: 750 mm   Locn: ?   Date: 08.2002

	4wDM	Bdm	O&K MV1a	25339	1955	(a)	(1)

(a) new via MBA, Dübendorf (600 mm) to Züblin & Dénériaz, Le Chargeur "1"; ex ?.
(1) by 1966 to G. Dénériaz SA, Sion.

MATHIER & Co, Visp   VS47 C
A civil engineering company?
Gauge: 750 mm   Locn: ?   Date: 03.2002

0-4-0T	Bn2t	Heilbronn II alt	28	1873	(a)	(1)

(a) ex Vicarino & Curty /1xxx.
(1) to Chapery & Delorme, Visp.

MATTERHORN GOTTHARD BAHN (MGB), Brig   VS48 M
Heritage stock includes the following.
Gauge: 1000 mm + Abt rack   Locn: 274 $ 6406/1293   Date: 06.2008

HG 2/3 7	BREITHORN	0-4-2RT 4c	B1bn4t	SLM HG 2/3	1725	1906	
HGe 4/4 15		BoBoRWE	BbBbew4	SLM,SWS,MFO HGe 4/4	3340, ?	1929	
BDeh 2/4 41		Bo2WERRC	Bb2w2	SLM,BBC	3765,4451	1941	(c)

(c) new to Schollenenbahn "BCFeh 2/4 41"; withdrawn as "FO BDeh 2/4 41" 13/12/2001.

MINERAL AG, Brig　　　　　　　　　　　　　　　　　　　　　VS49

Gauge: 600 mm　　　　　　　Locn: ?　　　　　　　　　　　　　　　Date: 03.2008
　　　　　　　　　　0-4-0T　　Bn2t　　Jung 50 PS　　　　1684　　1911　(a)　(1)
(a)　from Fritz Marti, Bern /19xx (hire or purchase?).
(1)　1936-42 sold to AG Heinrich Hatt-Haller (HHH), Zürich; 1966 to Technorama, Winterthur.

MOMECT SA, Collombey　　　　　　　　　　　　　　　　　　　VS52 C

A company set up in 1971 to repair and hire machines and machinery. The locomotives were stored awaiting sale. Address: En Bovery B, 1868 Collombey.

Gauge: 600 mm　　　　　　　Locn: 272 $ 5616/1260　　　　　　　　Date: 03.1997

　　　　　　　4wDMSO　Bdm　R&H 30DL　　256188　1948　(a)　(1)
　　　　　　　4wDMSO　Bdm　Ammann　　　　-　　　19xx　(b)　(1)

(a)　new via RACO to ?; used in the peat bogs of la Rougève near Semsales until 195x; modified with steam outline casing for use on a "Wild-west" style railway in a pleasure park, where? /1971; ex ? 19xx.
(b)　used in the peat bogs of la Rougève near Semsales until 195x; modified with steam outline casing for use on a "Wild-west" style railway in a pleasure park, where? /1971; ex ? 19xx.
(1)　1997 to FWF, Otelfingen.

ORGAMOL SA, Evionnaz　　　　　　　　　　　　　　　　　　　VS53

A fork-lift operated shunting device (possibly a Zagro) has been seen here.

Gauge: 1435 mm　　　　　　Locn: 272 $ 5680/1134　　　　　　　　　Date: 01.2003

PARC D'ATTRACTIONS DU CHÂTELARD VS SA,
Le Châtelard VS　　　　　　　　　　　　　　　　　　　　　　VS54 M

formerly:　Société Anonyme des Transports Emosson-Barberine (SATEB) (till 1997)
　　　　　Trains Touristiques d'Emosson SA

This line has been built on the track-bed of the construction line which ran from the top of the Barberine funicular (Le Châtelard VS to Château d'Eau) to the first Barberine dam, but only as far as the later and higher Emosson dam, 1.56 kilometres in all. The opening date was 12/6/1975. The views from this of Mont Blanc [F] and the surrounding country are remarkable. Address: Gare du Funiculaire, 1925 Le Châtelard VS.

Gauge: 600 mm　　　　　　　Locn: 282 $ 5618/1017　　　　　　　　Date: 10.2007

　　　　　5　　　　4wBE　　　Ba　　SSW EL 8　　　　　?　　　1952　(a)
　　　　　6　　　　4wBE　　　Ba　　SSW EL 8　　　　　?　　　1952　(b)
　　　　　7　　　　4wBE　　　Ba　　SSW EL 8　　　　　?　　　1952　(a)
　　　　　8　　　　4wBE　　　Ba　　SSW EL 8　　　　　?　　　1952　(b)
　　　　10　　　　4wDM　　　Bdm　Ammann　　　　　　-　　　194x　(e)
　　　　62　　　　4wDM　　　Bdm　R&H 48DL　　　296044　1952　(f)
　　　　63　　　　4wBE　　　Ba　　SIG ES 50　　　　1165　19xx　(g)
　　LISELI　　　0-4-0WT　　Bn2t　Jung 20 PS　　　　1693　1911　(h)

(a)　ex construction of Trient-La Fouly tunnel; possibly ex Savioz et Marti SA, Collombey /1974.
(b)　ex construction of Trient-La Fouly tunnel; possibly ex Colorama, Bulle /1977.
(e)　Lister engine, ex Conforti Frères, Martigny, VS /1975 who had used it in the construction of the Emaney-Barberine tunnel.
(f)　constructed by RACO from engine-less frame 296044 of 1950 and engine 257653 of 1952; ex Rondchâtel SA, Frinvillier /2005.
(g)　ex Rondchâtel SA, Frinvillier /2005.
(h)　made available by Thomas Brändle /2/2005.

see: www.chatelard.net/deutsch/pages/traind_3.html

SAGRAVE SA, Le Bouveret VS55

formerly: Rive-Bleue SA (till 31/12/2006)
            Rhôna SA (1975 till 31/12/2006)
            Sagrave Holding SA (from 1994)
            Sagrave SA (from 1926)

From 1 January 2007 two firms of the Arnold group merged into the already existing Sagrave SA. Arnold & Co AG, Flüelen is an even older part of the same enterprise. Rive-Bleue SA developed a tourist train operation: the "Rive-Bleue Express (RBE)" between Bouveret (CFF) and Evian-les-Bains [F] along the "Ligne du Tonkin" and on closure of this operation one locomotive was retained for transferring traffic between the CFF station and the works along a siding laid in the road. The CFF and la Poste Suisse use the appellation of Bouveret: the town and CGN prefer Le Bouveret.

Gauge: 1435 mm Locn: ? Date: 10.2007

2 HANSLI	0-6-0WT	Cn2t	SLM E 3/3	795	1893	(a)	(1)
Lok 2	0-4-0DH	Bdh	R&H LSSH	497746	1963	(b)	(2)
BB.71010 Noémie	0-4-4-0DH	BBdh	Fives-Lille-Cail	?	1965	(c)	

(a) new to Sihltalbahn (SiTB) "2"; ex Basel Gas-und Wasserwerke, Basel Kleinhüningen /1979.
(b) ex Raffinerie du Sud-Ouest SA, Raffineries du Rhône /1995
(c) ex SNCF; Poyaud motor.

(1) 2005 to Zürcher Museum-Bahn (ZMB), Sihlwald.
(2) by 2004 to Arnold AG.

see: www.sagrave.ch/index.php?goto=historique&expand=portrait

SALANFE SA, Vernayaz VS56

formerly: Energie Ouest Suisse (EOS) (till 1947)
            Elektrizitätwerk Lonza SA (1924-xx)
            la Force et Lumiere SA (192x-24)
            G. Stächelin, Basel (1902-2x)
            Société Industrielle du Valais (1897-1902)

The original project was an electrochemical (calcium carbide etc) works relying on power from the Usine de Pissevache, taking the name from a prominent waterfall almost above the site. This plant was rail connected from 1897 and operational from 1898. The railway siding was connected to the Lonza branch and enlarged 1910-28 to reach a total length of 1335 metres. It was electrified at the CFF voltage (15kV AC). This plant ceased production in 1948.

Lonza SA and Energie Ouest Suisse (EOS) then set up a joint undertaking of dating from 1949-50 to produce electric power from an underground power station, "Usine de Miéville", which lies on virtually the same site. The reservoir of Lac de Salanfe is situated near the Tour Sallière six kilometres to the west. A narrow gauge railway allows material to be taken through the mountain and up to the dam alongside the pressure pipes. The line runs from a small fan of exchange sidings into the mountain and then goes up one cable-worked incline. From the top of this another line runs in an underground tunnel some three kilometres long to another cable-worked incline at the top of which there is a final short level section with its own diesel locomotive. The intermediate level section, which has two stabling points within the mountain, is electrified with a trolley overhead. The gauge within the mountain has not yet checked – it is claimed to be 750 mm; that at the lowest level is 800 mm.

By at least 2007 the plant was being operated as Hydro Exploitation Salanfe and Lonza had pulled out leaving EOS as the sole owner of Salanfe SA. The standard gauge line was abandoned about 1995.

The BLW locomotive is now retired and displayed on a short length of track near the narrow gauge loading point. It was photographed there on 30 September 2008.

Gauge: 1435 mm			Locn: 272 $ 568390,110590				Date: 07.1988
206		0-4-0DM	Bdm	BLW VM	4xxxx	c1917	(a)	display
Gauge: 800 mm			Locn: 272 $ 568390,110590, see also in text				Date: 08.2002
		4wDM	Bdm	O&K MV0	25071	1951	(b)	s/s
		4wDM	Bdm	O&K MV0	25072	1951	(b)	(3)
		4wDM	Bdm	O&K MV0	25073	1951	(b)	s/s
		4wDM	Bdm	Brookville BMD-8	3517	1949	(e)	Miéville
		4wDM	Bdm	Brookville BMD-8	3518	1949	(f)	Salanfe
	1	4wWE	Bg1	Stadler	29	1949	new	
	2	4wWE	Bg1	Stadler	28	1949	new	

(a) new as (probably 600 mm) 0-4-0PM Bbm but now fitted with Deutz F3-6L912 motor; supplied via Val de Maizet [F] /1919 to an EOS construction site (Arbon?) before coming here. There are other versions of the history of this locomotive. In one, there were two similar locomotives. In another one locomotive was here and was later scrapped and a second was then obtained.
(b) new via MBA, Dübendorf.
(e) earlier "202"; new via International Harvester Export and Joly, Lausanne as 800 mm 4wPM Bbm.
(f) new via International Harvester Export and Joly, Lausanne as 800 mm 4wPM Bbm.
(3) to ?; to Oliver Weder, Diepoldsau; by 1999 to FWF, Otelfingen.

see: www.eosholding.ch/home/entreprise/chiffres_et_faits/historique.htm

SAUDINO, VANNI, BERTIONE, Goppenstein   VS57 C

A consortium of civil engineering firms participating in the construction of the Lötschberg line.
The name Bertione may , in fact, be Bertioni (hand-written text).

Gauge: 750 mm   Locn: ?   Date: 04.2002

           0-4-0T   Bn2t   O&K 50 PS   *2186*   1906   (a)   s/s

(a)   if the maker's number is correct then new to Deutsche Feld- und Industriebahn-Werke, Danzig [PL]; via O&K, Zürich /1911. Maybe the maker's number is in error and the locomotive was O&K 2187 or 2197 from the Entreprise Générale de Construction du Lötschberg, Frutigen.

SAVIOZ & MARTI SA, Sion   VS58 C

The SIG list of references gives two ETB 70 locomotives delivered to this company, which has several branches in the area.

H.R. SCHMALZ SA, Ardon   VS59 C

later: Batigroup (merger of Schmalz and Stuag)
later: Implenia (merger of Batigroup and Zschokke)

A civil engineering company with headquarters in Bern. Locomotives were stored here between contracts.

Gauge: 600 mm   Locn: 273 $ 5875/1172   Date: 09.2008

361		4wDM	Bdm	O&K	?	19xx		s/s
464		4wDM	Bdm	unknown	?	19xx		s/s
466		4wBE	Ba	SSW EL 8	6071	1959	(c)	s/s
468	RITA	4wBE	Ba	SSW EL 8	6092	1961	(c)	s/s
901		4wBE	Ba	TIBB	0117	1959	(c)	s/s
902		4wBE	Ba	SSW	?	19xx	(f)	s/s
903		4wBE	Ba	SSW	?	19xx	(f)	s/s
904		4wBE	Ba	SSW EL 8	6094	1961	(c)	s/s
		4wBE	Ba	unknown	?	19xx	(i)	s/s

(c)   by 1970 stored at Ardon between contracts.
(f)   plates of Gebrüder Meier AG, Zollikofen, Berne.
(i)   is this identical to one of 902 or 903?

SEBA APROZ SA NENDAZ, Aproz   VS60

To connect the Migros mineral water source with the CFF a 2.4 kilometre long branch from Ardon CFF station was inaugurated on 4 July ?1973. It includes a 133 metre long bridge over the R. Rhône. There are no locomotives present.

Gauge: 1435 mm   Locn: 273 $ 5689/1167   Date: 01.2008

SOCIÉTÉ POUR L'AMÉLIORATION DU CHARBON VALAISANS (SAPAV), Grône   VS61

Coal mines exploited 1916-57. Anthracite was transported by an aerial ropeway to the station of Granges bound for von Roll, Gerlafingen, the cement works at Reuchenette and Wildegg, the +GF + steelworks at Schaffhausen and Italy.

Gauge: 500 mm   Locn: ?   Date: 01.2000

           4wBE   Ba1   Stadler   15   1945   new   c1957 scr

SYNGENTA, Monthey VS62

formerly: Novartis AG (1996-2000)
            CIBA-GEIGY AG (1970-1996)
            CIBA SA (1946-1970)
            Gesellschaft für Chemische Industrie (19xx-1946)
            Basler Chemische Fabrik (1904-19xx)
            Société des Usines de produits chimiques (1896-1904)
            sucrerie Helvétia (1892-1897)

"491" Henschel 31209 of 1966 (type DHG 240B) returns from the CFF exchange sidings across the road into the chemical works in Monthey on 30 September 2008.

On liquidation of the sugar factory a chemical works was installed on the same site. From 1997 the logistics (including locomotives) have been transferred to Compagnie Industrielle de Monthey SA (CIMO).

Gauge: 1435 mm   Locn: 272 $ 5631/1226   Date: 01.2003

	4wBE	Ba	unknown,BBC	? ,1302	1919	new	s/s
	4wDM	Bdm	Kronenberg DT 90/120	131	1954	new	1984 scr
1	0-4-0DM	Bdm	SLM Em 2/2	3464	1930	(c)	(3)
491	4wDH	Bdh	Henschel DHG 240B	31209	1966	(d)	
492	4wDH	Bdh	Henschel DHG 240B	31554	1971	new	(5)
493	4wDH	Bdh	Henschel DHG 300B	32474	1984	(f)	
	4wDM R/R	Bdm	Mercedes MB Trac	?	xxxx		

(c)   ex CIBA, Basel /19xx.
(d)   ex Henschel /1970.
(f)   ex Henschel /1985.
(3)   1970 to Silo Olten AG, Olten (SOAG).
(5)   30/10/2003 to NEWAG; Alstom, Bangkok [Thailand]; 200x Kualar Lumpur [Malaysia].

TAMOIL SA, Collombey VS63

formerly: RSO Services SA, Collombey (? till 21/12/1998)
            Raffinerie du Sud-Ouest SA (RSO) (at latest 1967 till ?)
            Raffineries du Rhône, SA, Collombey (1963 till at latest 1967)

This oil refinery was first operational in 1963. The offices are located in the western part of the site in Collombey while the railway installations are all across the R. Rhône to the east and so in Vaud. They are connected to the CFF station of Saint-Triphon, see under Canton VD.

TAUCHERVERSUCH, Le Bouveret　　　　　　　　　　　　　　　　VS64

While it was operating on Lac Léman the public-carrying bathoscape "J. PICCARD" was taken to and from the water on a trolley mounted on a section of standard gauge track. This locomotive was incorporated into the system as a counterweight.

Gauge: 1435 mm　　　　　　　Locn: 262 $ 5533/1374　　　　　　　Date: 07.1966

Eb 3/5 5834　　　　2-6-2T　　　1C1h2t　SLM Eb 3/5　　　　2556　　1916　　(a)　　1966 s/s
(a)　　ex CFF "5834" /1963.

TROTTET RÉCUPÉRATION SA, Collombey-Muraz　　　　VS65

A recycling firm that commenced operations in 1990.

Schöma 2168 of 1959 (type CHL 20GR) awaits work on 21 May 1993.

Gauge: 1435 mm　　　　　　　Locn: 274 $ 5615/1260　　　　　　　Date: 03.2003

　　　　　　　　　　4wDH　　Bdh　　Schöma CHL 20GR　　2168　　1960　　(a)　(1)
Tm^I 472　　　　　4wDM　　Bdm　　RACO 85 LA7　　　　1678　　1963　　(b)
(a)　new to Alfred Teves GmbH, Gifhorn [D] (as CHL 30R); reconditioned by Schöma; ex LSB /1990.
(b)　ex CFF "Tm^I 472" /2006.
(1)　2003-4 scrapped.

TRUCHETET & BESSON, Sembrancher　　　　　　　　　　VS66 C

A civil engineering firm (based in Dijon [F] and Lausanne) involved in the construction of the Martigny-Orsières railway (1908-1910). See also the entry under Truchetet in Canton Vaud.

Gauge: 750 mm　　　　　　　Locn: ?　　　　　　　　　　　　　　Date: 08.1995

　　　　　　　　　　0-4-0T　　Bn2t　　O&K 20 PS　　　　　2895　　1908　　new　s/s

UNKNOWN CONCERN, Isérables   VS67 C
Civil engineering concern?

Gauge: 600 mm   Locn: ?   Date: 08.2002

		4wDM	Bdm	O&K MD 2b	25133	1951	(a)

(a) built by Schöma as their 1241; recorded as 700/750 mm; new, via MBA, Dübendorf; apparently seen 8/2002.

UNKNOWN CONCERN, Unknown location   VS68 C
A tunnel engineering concern.

Gauge: 900 mm   Locn: ?   Date: 04.1999

		4wBE	Ba	SIG ATM 800	800 820	1980	(1)

(1) 19xx to Furka-Oberalp "Ta 4982"; 1994 into service for shunting at Glisergrund.

UNKNOWN CONCERN, Zeusier   VS69 C
A construction concern involved in the Zeusier hydroelectric dam scheme.

Gauge: ? mm   Locn: ?   Date: 06.2002

		4wBE	Ba	SSW EL 9	?	195x	(a)	s/s
		4wDM	Bdm	O&K MD 2b	25209	1951	(b)	s/s

(a) ex MBA, Dübendorf /1954.
(b) ex MBA, Dübendorf /1955 (750 mm gauge).

UTAS, Brig   VS70 C
Probably a civil engineering firm. It has also been seen referred to as UTALS, and no records of either have so far been found.

Gauge: 600 mm   Locn: ?   Date: 08.2002

		4wDM	Bdm	O&K MV0	25102	1951	new	(1)

(1) 1953 to M. H. Bezzola AG, Biel.

VALMETAL SA, Martigny   VS71
formerly: Ugine-Kuhlmann (1965-1975)
          Société des Produits Azotes (SPA) (1906-1965)

A chemical factory built in 1906 and rail connected from 1908. The rail connection had a gradient of 3.3%. The firm was absorbed by Ugine-Kuhlmann in 1965 and later closed. The disused installations were taken over in 1975 by the scrap merchant Valmetal and finally demolished in 1990.

Gauge: 1435 mm   Locn: 282 $ 5728/1063   Date: 07.1990

1		4wBE	Ba2	SWS,MFO	?	1919	new	
	rebuilt	4wDE	Bde			19xx		1989 scr
2		4wPM	Bpm	Breuer III	?	1931	(b)	(2)
1		4wDM	Bdm	RACO 80 SA3	1388	1949	(c)	(3)

(b) ex CFF "Tm 462" /1961.
(c) 1979 ex Migros, Schönbühl.

(2) 1979 withdrawn; 1989 scrapped.
(3) 1990 to Cablofer, Bex.

VAPAROID AG, Turtmann   VS72
formerly: Vaparoid Stia AG.
A company manufacturing insulating and waterproofing materials.

Gauge: 1435 mm		Locn: 273 or 274 $ 6199/1287				Date: 04.2005
237 834-5	4wDM	Bdm	RACO 85 LA7	1605	1961	(a)
(a)	ex CFF "Tm^I 427" /1996.					

VEUTHEY & CIE SA, Martigny   VS73
Metal dealer and hardware store set up in 1882 and rail connected to the Martigny-Orsières line in 1954.

Gauge: 1435 mm			Locn: 282 $ 5731/1062				Date: 10.1995
TM	511	4wDM	Bdm	RACO 45 PS	1248	1937	(a) (1)
	5	4wDH	Bdh	KHD KG245 B ex	56848	1958	(b) 2005 scr

(a)   new to SBB "Tm 325"; rebuilt /1941 and /1968; ex MO "TM 511" /1983.
(b)   ex Bayer AG, Dormagen [D] "5" /1991 via WLH, Hattingen [D]; also referred to as type A8M517 R.
(1)   1991 to Asper; 1997 to Swisstrain.

VISENTINI & BILLIEUX, Martigny   VS74 C
A consortium of the civil engineering firms Visentini of Martigny and Billieux of Saint-Maurice.

Gauge: 600 mm		Locn: ?				Date: 08.2002
	4wDM	Bdm	O&K MV0	25161	1952	(a) s/s

(a)   new via MBA, Dübendorf.

ZSCHOKKE AG, Gondo   VS77 C
A tunnel construction project.

Gauge: 600 mm		Locn: ?				Date: 01.2009
	4wDM	Bdm	R&H *LBU*	*386623*	*1955*	(a) (1)

(a)   a Ruston & Hornsby locomotive was seen in use during the approximate period 1950-1956. Details were not recorded at the time of observation. The locomotive indicated appears to be the most likely candidate of those delivered new via RACO. The hypothesis is based on the facts that this one was delivered as a 600 mm gauge locomotive on 29/4/1955 and had to be supplied with operating instructions in French if possible. If the locomotive here was, in fact, a second-hand one then the locomotive data does not apply.
(1)   to P. Schlatter, Münchwilen; by 2008 at Montan- und Werksbahnmuseum, Graz, Graz [A].

Zug   ZG

LOGISTIKBASIS DER ARMEE (LBA)   ZG1

formerly: Bundesamt für Betriebe des Heeres (BABHE) (till 31.12.2003)
            Ozo-Lager, Oberkriegskommissat (OKK)
            AG für Petroleum-Industrie (IPSA)

Fuel stores (BEBECO-Tankstelle).

Gauge: 1435 mm   Locn: 235 $ 6759/2219   Date: 06.1991

| | 2 | 4wDM | Bdm | Kronenberg DT120 | 117 | 1950 | (a) | (1) |
| | 11 | 6wDE | Cde | Stadler | 116 | 1963 | (b) | |

(a)   new to IPSA.
(b)   new to OKK.
(1)   1961 to Kaufmann AG, Thörishaus (a scrap dealer).

F. LUSSER & CO, Baar   ZG2 C

A public works company.

Gauge: 750 mm   Locn: ?   Date: 04.2002

ALBIS	0-4-0T	Bn2t	Krauss-S IV cd	2647	1893	(a)	(1)
SIHL	0-4-0T	Bn2t	Krauss-S IV cd	2648	1893	(a)	(2)
BAAR	0-4-0T	Bn2t	Krauss-S IV fg	2817	1893	(a)	(3)

(a)   new for construction of the Thalwil-Zug railway.
(1)   to Mirra & Cie.
(2)   1896-1908 to Favetto & Catella, Brunnen.
(3)   1896-1900 to Fritz Marti, Winterthur.

NESTLÉ AND ANGLO-SWISS CONDENSED MILK CO, Cham   ZG3

formerly: Anglo-Swiss Condensed Milk Co (1866-1905)

The Anglo-Swiss Condensed Milk Co produced both condensed milk and milk powder. Rail connection was provided from 1879 and by 1910 there were 952 metres of track. The plant closed in 1934.

Gauge: 1435 mm   Locn: 235 $ 6775/2260   Date: 06.1991

| | 4wBE | Ba2 | Rastatt,MFO | -,21 | 1910 | new | (1) |

(1)   c1934 to Cellulosefabrik Attisholz (see Riedholz).

PAPIERFABIKEN CHAM TENERO AG, Cham   ZG4

formerly: Papierfabrik AG (till ?1978)

A paper factory founded in 1658. It was rail connected from 1920 by a line from the Nestlé factory.

Gauge: 1435 mm   Locn: 235 $ 6773/2266   Date: 10.1996

Ta	1	4wBE	Ba2	SIG,MFO	?	1919	new	
E 2/2 No. 2		0-4-0T	Bn2t	Henschel Riebeck	20593	1925	(b)	(2)
		4wDH	Bdh	KHD A8L614 R	57216	1961	(c)	(3)

(b)   purchased from Papierfabrik Perlen "3" /1962.
(c)   hired from Asper /1990.

(2) 1973 withdrawn; 1981 privately preserved; 1988 to VVT, St. Sulpice NE.
(3) 1990 returned to Asper.

C. VANOLI AG, Rotkreuz   ZG5 C

see under: C. Vanoli AG, Samstagern.

V-ZUG AG, Zug   ZG6
formerly: Verzinkerei Zug AG
This factory produces household appliances and is connected to the SBB via a U-shaped siding crossing some streets of the town. Address: Industriestrasse 66, 6300 Zug.

Gauge: 1435 mm   Locn: 235 $ 6821/2260   Date: 02.1993
              0-4-0DM   Bdm   KHD KS55 B   57552   1962   (a)
(a)   ex Heinr. Aug. Schulte, Eisen AG, Düsseldorf-Grafenberg [D] /1991; into service 1992.

Untergrundbahn

Running from the Hauptbahnhof this line is now dismantled and replaced by an automatic conveyor system.

Gauge: 600 mm Locn: ? Date: 01.2008

	4wWE	Beg	MFO		?	1938	new	s/s
	4wWE	Beg	MFO		?	1938	new	s/s

(b) ex SBB "Em2/2 101" /1933.
(f) ex Die Schweizerische Post, Bern "8" for storage by 3/2007.
(g) ex Die Schweizerische Post, Luzern "4" /1999-2000.
(h) ex PTT, Zürich-Sihlpost, Zürich "6" /1985.
(i) ex Die Schweizerische Post, Bern "8" by 3/2007.
(k) ex Die Schweizerische Post, Luzern "14" /2000.

(1) 1966 to PTT Lausanne.
(2) 1952 to SERSA for construction of the Sembrancher-Le Châble line; later MO "Tm 2/2 11".
(3) 1965 to PTT Lausanne.
(4) 1985 to PTT Luzern.
(5) 1985 to PTT Zürich-Mülligen, Zürich-Altstetten.
(6) 2008 to RM workshops, Oberburg.
(8) 1999 to Die Schweizerische Post, Härkingen.
(9) 2000 to Die Schweizerische Post, Bern.
(10) 2007 to Die Schweizerische Post, Bern.

see: Revue PTT n°12.1952 pages 481-484

DIETIKER METALLHANDEL AG, Regensdorf ZH037

The company imports, exports and trades in raw iron, metals, scrap metal and other materials. It lies at the end of a siding through the streets from the western end of Regensorf-Watt station. Address: Althardstrasse 345, 8105 Regensdorf.

Gauge: 1435 mm Locn: 215 $ 6769/2551 Date: 11.1994

	1	4wDH	Bdh	Jung?	?	19xx	(a)	(1)
		4wDM R/R	Bdm	Mercedes Unimog U400	?	8/2002	(b)	

(a) ex Thommen AG, Kaiseraugst /1986.
(b) Zagro conversion; new via Robert Aebi & Co, Zürich.

(1) 1995 withdrawn; 1996-97 s/s.

DIETIKER, Zürich-Seebach ZH038 C

A public works firm.

Gauge: 750 mm Locn: ? Date: 03.2002

5	IDA	0-4-0T	Bn2t	Hanomag	8009	1922	(a)	(1)
9	MARTHA	0-4-0T	Bn2t	Märkische 50 PS	296	1898	(a)	(2)

(a) purchased from Kibag AG, Bäch SZ /1960-62.

(1) 1964-70 to J. Heusser, Schlieren.
(2) fitted with boiler O&K 344 of 1899; 1965 displayed in a procession in Kreuzlingen; 19xx s/s.

GEBRÜDER DÜBENDORFER, Bassersdorf ZH039

formerly: Heinrich Dübendorfer-Keller (till 1930)
later: Dübendorfer AG (from 2001)

A company founded in 1901 to provide timber transport. It opened its first of several gravel pits in 1926.

Gauge: 600 mm		Locn: ?				Date: 05.2002		
	4wDM	Bdm	O&K(M) MD 1		7750	1937	(a)	(1)

(a) purchased via O&K/MBA, Dübendorf.
(1) 19xx to Scheifele, Dänikon ZH.

EISENBAHN TRANSPORTMITTEL AG (ETRA), Zürich   ZH040 C

A company with railway wagons for hire, it also provided locomotives for hire.
Address: Angererstrasse 6 (offices).

Gauge: 1435 mm			Locn: 225 $ 6827/2466			Date: 08.1990		
Em 3/3 260 747-9	0-6-0DH	Cdh	Henschel V 60	30038	1959	(a)	(1)
Em 3/3 260 355-3	0-6-0DH	Cdh	Jung V 60	12485	1957	(b)	(2)
Em 3/3 12	0-6-0DH	Cdh	MaK V 60	600026	1956	(c)	(3)

(a) ex DB "260 747-9" /1990.
(b) ex DB "260 355-3" /1990.
(c) ex DB "260 106-0" /1989; leased to and renumbered by Sensetalbahn (STB) "12" /1989.

(1) 5/2000 sold to Mittelweserbahn [D] "V661".
(2) 1990 sold to Kieswerk Steinigand AG (Kiestag), Wimmis.
(3) 3/2000 sold to Karsdorfer Eisenbahngesellschaft mbH (KEG) [D] "0652".

EMBRAPORT, Embrach   ZH041

A duty free port and a transhipment centre that commenced activities in 1976 and was rail connected to Embrach-Rorbas station. It is owned by Zürcher Freilager AG. Rail traffic had ceased by 2007.

Gauge: 1435 mm			Locn: 215 $ 6869/2638			Date: 04.1988		
	4wDM R/R	Bdm	Mercedes Unimog		?	19xx		?
2	4wDE	Bde	Stadler,BBC	143,	?	1974	(b)	(2)

(b) into service 09/12/1974.
(2) 1982 to Maggi AG, Kemptthal "2"; 2006 to Thommen-Furler AG, Rüti bei Büren.

ENTSORGUNG + RECYCLING ZÜRICH (ERZ),
Wallisellen/Aubrugg   ZH042

formerly: EWZ Elektritätwerke der Stadt Zürich (till 1996)

A six kilometre long underground line, following the steam pipes between the Aubrugg remote heating power station and the Zürich Polytechnik (Wässerwiese site). The line was constructed using other battery locomotives.

Gauge: 600 mm			Locn: 225 $ 6858/2519			Date: 09.1998
	4wBE	Ba1	Jung EZ10	13810	1970	(a)
	4wBE	Ba1	Jung EZ10	13879	1970	(a)

(a) purchased second-hand from Jung /1970.

ENTSORGUNGSZENTRUM RICHI WEININGEN AG, Weiningen ZH
ZH043

formerly: Richi AG

A concrete and gravel works with a biomass electrical power station. A Garrett road-roller is on display. Address: Im Riesentobel, 8104 Weiningen ZH.

Gauge: 1435 mm			Locn: 225 $ 6740/2520					Date: 02.2002
E3592	0-4-0WT	Bn2t	Krauss-S XXVII au		5564	1906	(a)	(1)
2	4wPM	Bpm	RACO RA11		1260	1940	(b)	

(a) ex Sulzer-Escher Wyss, Zürich /1986 via display at Kindergarten, Oetwil an der Limmat.
(b) ex Shell AG, La Renfile, Genève /1985.
(1) 2002 to a private collector and stored at Bülachguss AG, Bülach.

FAVRE & CO, Zürich   ZH044

alternative: Asphalt-, Cement- & Betonbau-Gesellschaft.
This was a company dealing in asphalt, cement and concrete. The exact location is not known.

Gauge: 1435 mm		Locn: ?				Date: 06.1988
	0-4-0PM	Bbm	Deutz C XIV R 22/25 PS	3922	1920	(a) s/s

(a) new, the data comes from the Deutz delivery list.

FELDMAIER & CIE, Wildberg   ZH045 C

Gauge: 700 mm		Locn: ?				Date: 05.2008
	0-4-0T	Bn2t	Heilbronn II ur	12	1870	new (1)

(1) 18xx to Grubitz & Ziegler, Solothurn.

FELD- UND WERKBAHN FREUNDE (FWF), Otelfingen   ZH046 M

This organisation was founded 7 December 1986. Individuals or groups of individuals own a number of the locomotives, particularly those in the store. Locomotives from members may be present from time to time. Conversely some locomotives are stored or maintained elsewhere and may be rotated with those at Otelfingen. Some locomotives may be loaned either for long or short terms. The principle store is not in the immediate area and visits are not offered. A "Festbahn" is installed at various locations in the area for specific periods: one or more working locomotives are in use on such lines. The group also loans locomotives to the Verein Bergwerk Riedhof at the Götschihof in the Aeugstertal.

Standing by the entrance is an overhead shovel provided by Aktiengesellschaft für Bergbau- und Hüttenbedarf, Salzgitter [D] as their 8544 of 1950 (type H1 300/5)

The plate on the locomotive listed here as O&K 25166; but which could be read as O&K 25170.

The two Ruston and Hornsby locomotives, 285352 of 1950 (type 30 DLU) in front of 256188 of 1948 (type 30 DL), which has been given a steam outline. Both have frame extensions to enable them to be used on 750 mm track though neither is using the facility here.

			Locn: 225 $ 6714/2254			Date: 06.2007
Gauge: 750 mm						
	4wDM	Bdm	O&K MV2a	25323	1954	(a)
	4wDM	Bdm	O&K MV4a	26245	1963	(a)
	1ABE	1Aa	Stadler,BBC	17, ?	1945	(c)

			Locn: 225 $ 6714/2254			Date: 06.2007
Gauge: 600 mm						
No 8456	4wDM	Bdm	Ammann	-	19xx	(aa)
	4wDMSO	Bdm	Ammann	-	19xx	(ab)
	4wDM	Bdm	Ammann	-	19xx	(ac)
	4wPM	Bbm	Austro-Daimler	?	19xx	(ad)
	4wBE	Ba	BBA Metallist	?	19xx	(ae)
	4wBE	Ba	BBA B360	Nr 277	9/1982	(ae)
	Bo+Bo BE	BBa	BBA B660	Nr 774	8/1988	(ae)
	4wDM	Bdm	Brun FL 12	34	19xx	(ah)
	4wDM	Bdm	Brun 2HK65	43	19xx	(ai)
	4wDM	Bdm	Brun 2 DS 90E	67	19xx	(aj)
Magdalena	4wDM	Bdm	Deutz OME117 F	10286	1936	(ak)
	4wDM	Bdm	Deutz OMZ117 F	10876	1933	(al)
Inv No. 282/159	0-4-0DM	Bdm	Deutz MLH322 F	21529	1938	(am)
6 SUSI	4wDM	Bdm	Diema DS28	2702	1964	(an)
	4wDM	Bdm	Gmeinder 20/24PS	1557	1936	(ao)
	4wDM	Bdm	JW JW8 -PONY	051	1950	(ap)
	4wDM	Bdm	JW JW8 -PONY	225	1954	(aq)
	4wDM	Bdm	Jung MS 131	?	1929-33	(ar)
2	4wDM	Bdm	Jung MS 131	4704	1929	(as)
	4wDM	Bdm	Jung MS 131	4930	1930	(at)
KANTON AARGAU 95x	4wDM	Bdm	Jung EL 105	5799	1934	(au)
	4wDM	Bdm	Jung EL 105	7372	1937	(av)
	4wDM	Bdm	Jung ZL 233	8457	1939	(aw) (49)
	4wBE	Ba	Jung Ez 21	12906	1958	(ax)
	0-4-0DM	Bdm	KHD F2L514 F	55827	1954	(ay)
	0-4-0DM	Bdm	LKM Ns2f	262050	1958	(az)
	0-4-0DM	Bdm	LKM Ns2f	262057	1959	(ba) (53)
917	0-4-0DM	Bdm	LKM Ns2f	262211	1959	(bb)
	4wDM	Bdm	MBA MD 1	2008	c1950	(bc)
	4wDM	Bdm	O&K MD 2b	?	19xx	(bd)
	0-4-0DMSO	Bdm	O&K(M)	780	1914	(bf)
281.21.101 No 4	4wDM	Bdm	O&K(M) S5	1543	192x	(bg)
781 KIESEL	4wDM	Bdm	O&K(M) H 1	2063	192x	(bh)
	4wDM	Bdm	O&K(M) RL 1a	*4021*	*1930*	(be)
	4wDM	Bdm	O&K(M) RL 1a	4210	1930	(bi)
	4wDM	Bdm	O&K(M) RL 1a	4332	1930	(bj)
	4wDM	Bdm	O&K(M) RL 1a	4453	1931	(bk)
	4wDM	Bdm	O&K(M) RL 1a	4593	1931	(bl)
	4wDM	Bdm	O&K(M) RL 1a	4690	1932	(bm)
	4wDM	Bdm	O&K(M) RL 1a	4899	1933	(be) (66)
	4wDM	Bdm	O&K(M) MD 1	7750	1935	(bo)
281.21.106 00117	4wDM	Bdm	O&K(M) LD 2	8191	1937	(as)
	4wDM	Bdm	O&K(M) RL 1c	8371	1937	(bq)
281.21.103	4wDM	Bdm	O&K(M) RL 1c	8456	1937	(br)
281.21.105 No. 5	4wDM	Bdm	O&K(M) LD 2	8587	1937	(bg)
	4wDM	Bdm	O&K RL 1c-16/1	25017	1950	(bt)
LAUFEN	4wDM	Bdm	O&K MV0	25046	1951	(bu)
	4wDM	Bdm	O&K MV0	25072	1951	(bv)
	4wDM	Bdm	O&K MV0	25074	1951	(bw)
	4wDM	Bdm	O&K 3 D	21504	1939	(bx) (50)
	4wDM	Bdm	O&K MV1	25140	1952	(bx) (50)
	4wDM	Bdm	O&K MV0	25166	1952	(bz)
KBB	4wDM	Bdm	O&K MD 2b	25232	1952	(ca)
281.21.102	4wDM	Bdm	O&K(S) MD 2b	1518	194x	(as)
	4wDM	Bdm	RACO	?	19xx	(cc)
	4wDM	Bdm	RACO	?	19xx	(cd)
	4wDM	Bdm	RACO	?	19xx	(ce)

		4wDM	Bdm	RACO 16PS	1264	1944	(cf)
	rebuilt	4w	2			19xx	
		4wDM	Bdm	RACO 12/16PS	1301	1945	(cg)
Fanny		4wDM	Bdm	RACO 12/16PS	1320	194x	(ch)
		4wDM	Bdm	RACO 12/16PS	1333	194x	(ci)
		4wDM	Bdm	RACO 12/16PS	1366	1949	(cj)
		4wDMSO| Bdm	R&H 30DL	256188	1948	(ck)
		4wDM	Bdm	R&H 30DLU	285352	1950	(cl)
		4wDM	Bdm	Schöma KDL 8	3295	1970	(cm)
		4wBE	Ba	Schalke,SSW EL 9	?	19xx	(cn)
		4wBE	Ba	Schalke,SSW EL 9	57610,6067	1960	(co)
		0-4-0DM	Bdm	Spoorijzer A2L514	60001	1960	(cp)
		4wBE	Ba2	Stadler	36	8/1948	(cq)

Gauge: 500 mm   Locn: 225 $ 6714/2254   Date: 06.2007

| | | 4wDM | Bdm | R&H 30DLU | 285351 | 1950 | (da) | (105) |

(a) 750 mm; ex K. Hürlimann Söhne AG, Brunnen /1994.
(c) rebuilt from a Deutz 21PS locomotive, probably C XIV F 1583 1915; 750 mm; ex K. Hürlimann Söhne AG, Brunnen /1994.
(aa) ex Jean Bollini, Baulmes /2006; Gerate Nr 8456, probably Jung ZM 233 8456 1939.
(ab) used at Tourbières de la Rougève, Semsales, ex Momect SA, Collombey /1997.
(ac) ex Gips Union SA, Bex by /1999.
(ad) ex T. R. Unternehmung by /1997.
(ae) ex P. Schlatter, Münchenstein by /1997.
(ah) ex Eisenbergwerk Gonzen AG, Sargans (500 mm) by /2006; Fabr. No. 93913, PS10-12.
(ai) ex Manzini Construction SA, Aigle /2008.
(aj) ex Auf der Maur Bau Company, Steinen by /2003; Fabr. No. 26483, PS 16.
(ak) new to Robert Aebi & Co AG; at Camp d'éclaireurs, Prahins /1970; ex ? /19xx.
(al) new to Robert Aebi & Co AG, Zürich; ex Meier & Jäggi AG, Breitenbach /19xx.
(am) new to Robert Aebi & Co AG, Zürich; plates of Hans F Würgler Ing. Bureau, Zürich-Albisrieden; ex ? by /2003; F2L612 F motor 2135031/032.
(an) new to J. Schmidheiny & Co. AG, Heerbrugg (500 mm); used at Zürcher Ziegeleien AG, Ziegelei Giesshübel (600 mm); ex Oliver Weder, Diepoldsau by /2003 via Belloli, Grono and other private owners.
(ao) ex Meier & Jäggi AG, Breitenbach by /2006.
(ap) ?new to Ziegelei Gebrüder Fink, Riedtwil (500 mm); ex Herr A. Schwarz, Sumiswald (400 mm) /2008.
(aq) new to Ciments de Saint-Ursanne, Saint-Ursanne ; ex Herr A. Bauer, Lüsslingen /1996; motor "3780"
(ar) ex Valli AG Strassenbau, Aarau /1988.
(as) ex Toneatti AG, Bilten /2006.
(at) new to Robert Aebi & Co AG, Zürich ; ex Dampf Express Alpenhof, Bülach by/1996.
(au) ex Wasserbauamt Frick (500 mm) by /1999; KANTON AARGAU 950 or 951 painted over by /2000.
(av) new via U. Ammann A.G., Langenthal to ?; Brun plates; ex ? by /2006; steam outline removed /2008.
(aw) ex Oliver Weder, Diepoldsau /1997-98.
(ax) ex Oliver Weder, Diepoldsau by /1996.
(ay) ex Fritz Wyss AG, Kieswerk Leuzigen, Leuzigen /1993 after a period on display in Leuzigen.
(az) new to VEB Gießereisandwerke Nudersdorf [DDR]; ex Quarzsand GmbH Nudersdorf [D] "3" /1990.
(ba) new to VEB Ziegelwerke Halle, Betriebsteil Wansleben am See [DDR]; ex privately owned [D] /1990.
(bb) new to VEB Quarzsandwerk Nudersdorf, Werk Pödelwitz [DDR]: ex Quarzsand GmbH, Nudersdorf [D] /19xx.
(bc) used at Ziegelei Erhai AG, Frauenfeld; ex Plättli-Zoo, Frauenfeld /19xx.
(bd) ex Ziegelei Lufingen /198x.
(be) new as 4wDM Bdm; ex ? by /2006.
(bf) ex Toneatti AG, Bilten /1995.
(bg) new as 4wPM Bpm; used by A. Kiesel, Tiefbau AG, Winterthur; ex Pfenninger Bau AG, Schlieren /1992.
(bh) ex Société Jurassienne de Matériaux de Construction SA, Delémont by /2006.
(bi) ex Rudolf Jenzer AG, Frutigen /2006.
(bj) ex Bauunternemung Kühnis AG, Oberriet by /2006
(bk) ex Brandenberg Bau AG, Flaach /2002.
(bl) ex Wilh. Wälti, Giswil by /1996.
(bm) ex Jean Bollini, Baulmes by /2006.
(bo) ex Scheifele, Dänikon ZH by /2006.
(bq) ex Gantenbein, Buchs /1994; repaired with parts of RL1a 4560 1932; which frame is present is not known.

(br)	ex Toneatti AG, Bilten /1995, Toneatti quotes type RL1a.
(bt)	O&K records indicate built as MD 2b by Schöma as 1157 and supplied to P. Nolden, Euskirchen [D]; this conflicts with the plate carried; ex ? by /1996.
(bu)	ex Keramik Laufen, Laufen.
(bv)	ex Salanfe SA, Vernayaz (800 mm) by /1999 via Oliver Weder, Diepoldsau.
(bw)	ex Ziegelei Landquart, Landquart (750 mm) by /1996.
(bx)	used by Schafer & Mugglin, Liestal; ex Ed. Schäfer & Co AG, Aarau /1994.
(bz)	ex Payot, Villeneuve-Clarens by /1999; plate damaged or reused - can be read as 25170.
(ca)	built as Schöma 1309; new to "Eichi", Alpnachdorf (750 mm); ex Asper /1994; the "KBB" text may refer to Kantonal-Bernischer Baumeisterverband; plate appears to read MD 20.
(cc)	used by AG Heinrich Hatt-Haller AG (HHH); ex Hess AG, Wald ZH /19xx.
(cd)	ex Herr A. Schwarz, Sumiswald (400 mm) /1996.
(ce)	ex ? by /1997; SMM motor, Aebi gearbox, no RACO supports, exhaust cleaner by cab.
(cf)	used by Bauunternehmung Bless AG, Dübendorf; ex Widmer AG, Grämigen /1994; engine removed and fitted in "Fanny"; crane mounted.
(cg)	ex Reifler & Guggisberg, Ing AG/SA, Pieterlen, Biel-Bienne "Nr 70/03" by /1997.
(ch)	used by AG Heinrich Hatt-Haller AG (HHH); ex Hess AG, Wald ZH /19xx; 199x name added "FANNY"; 199x "Fanny".
(ci)	ex Eternit AG, Niederurnen by /2006.
(cj)	used by Nicollier; ex Schaffner & Huldi AG, Affoltern am Albis by /1997.
(ck)	used at Tourbières de La Rougève, Semsales; ex Momect SA, Collombey /1997.
(cl)	ex F.lli, Somaini, Grono /19xx.
(cm)	ex Theiler & Kalbermatter AG, Buchrain by 6/1999.
(cn)	used by ?; ex MBA by /2003.
(co)	used by Frutiger Söhne AG, Thun; ex MBA by /2003.
(cp)	ex P. Schlatter, Münchwilen AG /19xx; engine 275444/45; Spoorijzer licence plate.
(cq)	ex Seeburger & Jordi AG, Frutigen /1995.
(da)	ex von Roll AG, Gerlafingen "166" /1996.
(49)	by 1999 moved elsewhere.
(50)	by 1996 s/s.
(53)	2005 to Herr A. Schwarz, Sumiswald (400 mm).
(66)	2004 to FIXIT SA, Bex for display.
(105)	by 6/1999 to Dampf Express Alpenhof, Bülach.

FERRIFF AG BAUUNTERNEHMER, Zürich   ZH047 C

A construction company.

Gauge: 600 mm   Locn: ?   Date: 03.2002

	0-6-0T	Cn2t	Henschel	15851	1917	(a)	(1)

(a)   purchased from Hans Rohrer, Zürich /1928.
(1)   1928 sold to Anton Bless & Co, Dübendorf.

FIETZ & LEUTHOLD AG, Zürich   ZH048 C

A construction company.

Gauge: 600 mm   Locn: ?   Date: 04.2002

LIMMAT	0-4-0T	Bn2t	Maffei 20/25 PS	2954	1910	(a)	1937 s/s
	0-4-0T	Bn2t	Zobel	551	1907	(b)	(2)
	4wDM	Bdm	O&K(M) MD 2	12056	194x	(c)	
	4wDM	Bdm	O&K(M) MD 2	12057	194x	(c)	
	4wDM	Bdm	O&K RL 2	20177	1931	(e)	
8	4wBE	Ba1	Stadler	32	1948	(f)	(6)

Gauge: 750 mm   Locn: ?   Date: 03.2002

SANTERNO	0-4-0T	Bn2t	Heilbronn II	227	1886	(g)	(7)

(a)   purchased from Robert Aebi & Co, Zürich /1919.
(b)   purchased from Ackermann, Bärtsch & Cie, Mels (or Fritz Marti AG, Bern) /1919.
(c)   gauge unknown; purchased via MBA, Dübendorf.

(e) purchased via MBA, Dübendorf.
(f) purchased from Hew et Co, Chur.
(g) purchased from RTU Rickentunnel-Unternehmung AG /1918.
(2) 1960 removed from boiler register.
(6) 1953 re-engineered by Stadler for Arbeitsgemeinschaft Töss, Bäretswil.
(7) 1936 removed from boiler register.

FISCHER & SCHMUTZIGER, Zürich   ZH049 C
A public works concern

Gauge: 750 mm   Locn: ?   Date: 04.2002

IMPAVIDA	0-4-0T	Bn2t	Krauss-S IV pp	2098	1889	new	(1)
	0-4-0T	Bn2t	Krauss-S IV ww	2375	1890	new	s/s
SPERANZA	0-4-0T	Bn2t	Krauss-S IV tu	3190	1895	new	(3)

(1) 1889-1908 sold to Buchser & Broggi, Degersheim.
(3) 1895-1905 sold to Gribi & Hassler, Burgdorf.

GASWERK DER STADT WINTERTHUR, Winterthur   ZH052
A gasworks opened in 1859, rail connected from 1899 and closed in 1969.

Gauge: 1435 mm   Locn: 216 $ 6957/2612   Date: 03.1988

    0-4-0Fic   Bf2   SLM   1141   1898   new   (1)

(1) 1976 sold to S. Lory, Winterthur; 1980 to Technorama, Winterthur; 2007 loan to Martin Horath, Goldau.

GASWERK DER STADT ZÜRICH (GWZ), Schlieren   ZH053
A gasworks built in 1898, enlarged 1928-33 and closed in 1974. After closure the site was used by more than one concern for locomotive repair. Later some buildings were preserved and the rest of the site reused.

One of the two coke cars built by Hübscher with Sulzer motors seen parked out of use shortly after closure on 4 October 1976.

Gauge: 1435 mm			Locn: 225 $ 6773/2507				Date: 03.1996	
	1	0-4-0T	Bn2t	Krauss-S XXXV vv	3775	1898	new	(1)
	2	0-6-0WT	Cn2t	Krauss-S XLV ac	5331	1905	new	(2)
	1	0-4-0F	Bf2	SLM	3566	1932	new	c1970 str
	2	2-6-0T	1Cn2t	SLM Ed 3/4	1799	1907	(d)	(4)
	3	0-6-0WT	Cn2t	SLM	1902	1908	new	(5)
	2	0-6-0WT	Cn2t	SLM E 3/3	1400	1901	(f)	1965 scr
		4wPM	Bbm	RACO 40 PS	11	1930	new	(7)
	4	0-6-0DH	Cdh	Henschel DH 500Ca	31079	1964	(h)	(8)
	1	Bo2D	B2d	Hübscher	-	1952	(i)	s/s
	2	Bo2D	B2d	Hübscher	-	1952	(i)	s/s

(d) purchased from SMB "Ed 3/4 2" /1932.
(f) new to JS "861"; purchased from SBB "E 3/3 8435" /1947.
(h) new as type DH 500ex; this locomotive is fitted with a reduced height cab.
(i) self propelling quench cars with Sülzer involvement, perhaps for motors, and full railway buffers and drawgear. Details above not verified.

(1) 1909-10 sold to Schweizerische Wagons- und Aufzügefabrik AG (SWS), Schlieren.
(2) 1932 sold to Kieswerk Hardwald, Dietikon.
(4) 1946 to HOVAG, Ems.
(5) 1996 to Swisstrain, but not moved till ?2007 to Le Locle.
(7) 19xx to Cartiere di Locarno, Tenero.
(8) 1983 to Geleisegenossenschaft Ristet, Birmensdorf ZH "2 Marianne".

GELEISEGENOSSENSCHAFT RISTET-BERGERMOOS, Birmensdorf ZH ZH054

formerly: Bauunternehmung Heinrich Stutz, Zürich

"Marianne" Henschel 31079 of 1964 (type DH500C) leaves her shed on 3 June 2008 preparatory to shunting the sidings of the industrial estate. The locomotive still has the unusual low roofed cab from its days with the Gaswerk der Stadt Zürich at Schlieren.

Originally there was an industrial branch serving the gravel-pits of Heinrich Stutz. With the arrival of other industry in the area a combine has been formed to provide services to all, including Scheller & Cie AG, a company whose rail-connection was changed from Dietikon to this branch.

Gauge: 1435 mm		Locn: 225 $ 6747/2465					Date: 09.1996
1 Em 837 901-8 Monica							
	0-6-0DH	Cdh	Henschel DH 500Ca	30511	1962	(a)	
2 Em 837 900-0 Marianne							
	0-6-0DH	Cdh	Henschel DH 500Ca	31079	1964	(b)	

(a) new, "1" till /1995; hired to SERSA /1996; and returned.
(b) ex Gaswerk der Stadt der Zürich, Schlieren "4" /1983, "2" till /1995.

GEMEINDE ADLISWIL, Schule Werd   ZH055 M

Photograph 1 June 2008.
The locomotive is located in a playground on the Kanalweg.

Gauge: 1435 mm		Locn: 225 $ 6823/2411				Date: 02.2008
4	0-6-0WT	Cn2t	SLM E 3/3	1016	1897	(a)

(a) ex Sihltalbahn (SiTB) "4" /1965.

GEMEINDE SCHLIEREN, Dietikon-Au   ZH056 M

The locomotive was placed on display on the edge of a housing development.

Gauge: 1435 mm		Locn: ?				Date: 06.2008	
327	0-6-0T	Cn2t	Hohenzollern Leverkusen	4267	1922	(a)	(1)

(a) new to Badische Anilin und Sodafabrik (BASF), Ludwigshafen [D]; fitted with boiler Hohenzollern 4268 (BASF "60") /19xx; ex Rheinische Braunkohlenwerke AG (RBW) [D] "327" 12/4/1973.
(1) 1981 to private ownership .

GEMEINDE HORGEN, Schule Tannenbach   ZH057 M

The locomotive was placed in a playground on the Fuchsenweg.

Gauge: 1435 mm		Locn: 225 $ 6865/2358				Date: 03.2008	
3	0-6-0WT	Cn2t	SLM E 3/3	1015	1897	(a)	1988 scr

(a) ex Sihltalbahn (SiTB) "3" /1956.

GIPS-UNION AG, Zürich   ZH058

This is the location of the head office of this company which operates many quarries and plaster works throughout Switzerland. The locomotives were almost certainly used elsewhere than Zürich.

Gauge: ? mm   Locn: ?   Date: 05.2002

	4wDM	Bdm	O&K(S) RL 1a	1081	19xx	(a)	(1)

Gauge: 600 mm   Locn: ?   Date: 05.2002

	4wDM	Bdm	O&K(M) RL 1a	5363	1934	(b)	(1)
	4wDM	Bdm	O&K(M) RL 1c	7759	1937	(c)	(1)
	4wDM	Bdm	O&K(M) RL 1c	8559	1937	(d)	(1)
	4wDM	Bdm	O&K MV0	25175	1952	(e)	(5)

(a) purchased from MBA, Dübendorf /1951, O&K(S) was not specified, but in view of the type, year of transaction, and the fact that 1082 was recorded as O&K(S) this appears to be a reasonable assumption.
(b) purchased from Fürstliches Bauamt Liechtenstein, Vaduz /1951 via MBA, Dübendorf.
(c) new to Stuag.-Öst. Straßenbauunternehmung AG, Wien [A]; purchased from MBA, Dübendorf /1953.
(d) purchased from MBA, Dübendorf /1950.
(e) purchased new via Wander-Wendel/MBA, Dübendorf; either it is the locomotive described under Ziegelei Lüchinger, Oberriet or a mistake for one of the others in the same batch (25173-25176).

(1) 19xx to ?
(5) 19xx to Ziegelei Lüchinger, Oberriet,SG (Ziegelei Schmidheiny, Oberriet).

GLAUS-NÄGELI, Zürich-Altstetten   ZH059 C

A public works firm?

Gauge: ? mm   Locn: ?   Date: 05.2002

	4wPM	Bpm	O&K(M) S10	1890	192x	(a)	s/s

(a) purchased via O&K/MBA, Dübendorf; type S1 is given in the MBA records.

A. GOSSWEILER, Zürich   ZH060 C

A building concern (Bauunternehmer in some versions of the name).

Gauge: 800/600 mm   Locn: ?   Date: 04.2002

	0-4-0T	Bn2t	Jung 40 PS	206	1895	(a)	s/s

Gauge: 750 mm   Locn: ?   Date: 05.2008

	0-4-0T	Bn2t	Märkische 30 PS	37	1895	new	s/s

(a) purchased from F. Marti, Winterthur /1900, only the maker's number is known for certain, the builder, year of construction and gauge are based on the correct identification of the locomotive.

ING. GREUTER AG, Hochfelden   ZH061 C

A civil engineering company with special interest in tunnel linings and water control. The company often uses sidings at Bülach for its purposes. Office address: Langmattstrasse 8, 8182 Hochfelden.

Gauge: 1435 mm   Locn: 215 $ 6824/2647   Date: 02.2008

	4wDH	Bdh	KHD A6M517 R	55126	1952	(a)	derelict
Tm 237 906-3	0-4-0DH	Bdh	Henschel DH 240B	30594	1964	(b)	

(a) new to Gußstahlwerk Wittmann AG, Hagen-Haspe [D]; ex Schlatter Peter AG, Rheinsulz /200x; the oval Deutz plate from this locomotive is now affixed to KHD 56810 (see: Stöckli AG, Sursee).
(b) ex Gotthard Schnyder AG, Emmen /2003.

HÄUTE- & FETTWERK AG (HFZ), Zürich-Altstetten   ZH062
later: Centravo AG

Located within the abattoirs this plant treated skins and fat. The railway siding ceased to be used from about 1990-1992 and the tracks were removed in 1995.

Gauge: 1435 mm   Locn: 225 $ 6797/2492   Date: 10.1995

8	4wDH	Bdh	ABL IV NHT	4702	1956	(a)	(1)
	4wDM	Bdm	RACO VL82PS	1490	1957	(b)	(2)

(a)   ex the HFZ plant in Bellinzona (Centrale Pelli & Fonditoio Grassi SA) /19xx; Breuer style with HydroTitan transmission.
(b)   ex Shell, Birsfelden Hafen /1979 via Asper; Breuer style.
(1)   1986 to DVZO; 1997 to IG Lokdepot Winterthur, Winterthur; 2001 to Historisches Bahnbetriebswerk Arnstadt, Arnstadt [D].
(2)   1994-95 s/s.

HEFTI AG, Zürich-Altstetten   ZH063

This chemical factory was rail connected to the SBB at Altstetten station. It closed 1995-1996.

Gauge: 1435 mm   Locn: 225 $ 6796/2492   Date: 06.1994

4wDH	Bdh	Diema DVL15/2	3228	1971	(a)	(1)

(a)   new on 7/1/1972 with 12kW motor; purchased via Asper; later a new motor fitted.
(1)   1996 to ?, Bodio; perhaps via an intermediary. Current owner and location unknown.

AG HEINRICH HATT-HALLER (HHH), Zürich   ZH064 C
alternative: Zürich-Binz

A building firm (Hoch-& Tiefbau in some versions of the name). In 1982 it was absorbed into Zschokke AG and since then the works area has been redeveloped. The SIG list of references gives four ET 20L5 locomotives delivered here.

Gauge: 750 mm   Locn: ?   Date: 03.2002

	0-4-0T	Bn2t	Hanomag	8009	1922	(a)	(1)
	0-6-0T	Cn2t	Henschel Brigadelok	5672	1901	(b)	(2)
	0-4-0T	Bn2t	Jung 50 PS	1684	1911	(c)	(3)
	0-4-0T	Bn2t	Krauss-S IV w	1292	1883	(d)	(4)
HOHENLOHE	0-4-0T	Bn2t	Märkische 50 PS	192	1896	(e)	(4)
	0-4-0T	Bn2t	O&K 50 PS	1136	1903	(f)	(4)

Gauge: 600 mm (mainly)   Locn: ?   Date: 03.2002

4wPM	Bbm	O&K(M) S10	1747	192x	(g)	s/s
4wDM	Bdm	O&K(M) RL 1a	4452	1931	(h)	s/s
4wDM	Bdm	O&K(M) RL 1a	4901	1933	(h)	s/s
4wDM	Bdm	O&K(M) LD 2	8341	1937	(j)	(10)
4wDM	Bdm	O&K(M) RL 1c	11425	1939	(h)	s/s
4wDM	Bdm	RACO	?	19xx		(12)
4wDM	Bdm	RACO 12/16PS	1320	194x		(12)

(a)   hired from RUBAG /1920; purchased /1920-29.
(b)   purchased from O&K, Zürich /1923; used by Hilti & Züblin at Schaan during the construction of the Etzel dam /1936-42.
(c)   purchased from Mineral AG, Brig /1936-42; used at Braunkohlenwerk Zell LU /194x.
(d)   purchased from Mineral AG, Brig /1941.
(e)   purchased from Entreprise pour la construction du tunnel des Forces du Rhône, Chippis /1908-41.
(f)   purchased from Robert Aebi & Co, Zürich /1939-43.
(g)   unknown gauge; purchased from O&K, Zürich.
(h)   purchased from O&K, Lager Zürich.
(j)   purchased from Baukonsortium Zweilütschinen.

(1) 1931 sold to Kibag AG, Zürich.
(2) 1948 sold to Brun & Cie, Nebikon.
(3) 1966 donated to Technorama, Winterthur for preservation.
(4) 1955 removed from boiler register.
(10) 19xx sold to Hoch-& Tiefbau, Interlaken.
(12) 19xx to AG Hess AG, Wald ZH; later FWF, Otelfingen.

HESS AG, Wald ZH ZH065 C

A civil engineering company.

Gauge: 600 mm   Locn: ?   Date: 06.1991

4wDM	Bdm	RACO 12/16PS	1320	194x	(a)	(1)
4wDM	Bdm	RACO	?	19xx	(a)	(1)

(a) ex AG Heinrich Hatt-Haller (HHH), Zürich-Binz.
(1) 19xx OOU; 1993 donated to FWF, Otelfingen.

J. HEUSSER, Schlieren ZH066 M

The locomotive is part of a private collection and is stored on a wagon within the secure compound of the gasworks at Schlieren.

Gauge: 750 mm   Locn: 225 $ 6775/2507   Date: 03.2002

0-4-0T	Bn2t	Hanomag	8009	1922	(a)

(a) new to Rollmaterial AG, Zürich (900 mm); purchased from Dietiker, Zürich-Seebach /1964-1970.

HOLCIM KIES UND BETON AG, Werk Hüntwangen ZH067

formerly: Holderbank Ciments et Bétons (HCB) (till 2001)
Kieswerk Hüntwangen AG

A gravel works with a branch line from Hüntwangen-Wil station to the works. These are situated within and to one side of a large pit that reaches towards the German border (which here lies to the west). A spur from that line leads to a second pit. Address: Bahnhofstrasse, CH-8194 Hüntwangen.

Gauge: 1435 mm   Locn: 215 $ 6803/2713   Date: 08.2001

1	4wDH	Bdh	O&K MB 9 N	26656	1968	(a)	(1)
	4wDH	Bdh	O&K MB 170 N	26741	1972	(b)	(2)
	0-6-0DH	Cdh	Cockerill	4224	1972	(c)	
EM 837 910-9 GUSTI	6wDH	Cdh	MaK G 763 C	700100	1991	(d)	
FRIEDA	0-8-0DH	Ddh	MaK 1200 D	1000250	1963	(e)	(5)
	4wCE	Bel	Vollert	?	19xx	(f)	

(a) ex Oranje Nassau Mijn, Heerlen [NL] via WBB /197x.
(b) hired from Vanoli /198x-2001.
(c) ex steelworks (unknown) [B] via Locorem/Asper /1986.
(d) ex Holderbank Cement und Beton (HCB), Rekingen AG /1998.
(e) hired from LSB /1990 and again /1991.
(f) ex BEMO /19xx; also recorded as a Vollert and a Windhoff – clarification required.
(1) by 2004 to HOLCIM, Untervaz; 17/7/2006 to Shunter [NL] "304 MYRTHE" via Unirail [D].
(2) 1987 and 2006 returned to Vanoli "V 2".
(5) 1991 returned to LSB; 1993 returned to SERSA as "FRIEDA Bm 847 959-4".

HOLLIGER IMBALLEGNO AG, Rafz ZH068

formerly: HIAG Imballegno AG
Hiag Paletten und Verpackungen AG (till ?1998)

CH 365 ZH

Holzindustrie AG (HIAG)

A timber yard. The transfer of the locomotive from the adjacent SIG-Holzwerk is not confirmed.
Address: Rüdlingerstrasse 790, 8197 Rafz.

Gauge: 1435 mm		Locn: 215 $ 6833/2732				Date: 01.2008
	4wDH	Bdh	Schöma CFL 30DCR	2682	1964	(a)
(a)	ex SIG-Holzwerk, Rafz /19xx.					

ERWIN HUBER GETRÄNKE, Zürich   ZH069

A dealer in drinks located in buildings of Locher & Cie AG. The Zürcher Museumsbahn (ZMB) store locomotives here. Rail traffic had ceased by 2008. Address: Allmendstrasse 91, 8041 Zürich.

Gauge: 1435 mm		Locn: 225 $ 6817/2442				Date: 03.1991
	0-4-0DM	Bdm	O&K RL 4	20077	1930	(a)
(a)	200x on loan from Locher & Cie AG, Zürich-Manegg.					

M.F. HÜGLER AG, Dübendorf   ZH070

A recycling company. Address: Usterstrasse 99, 8600 Dübendorf.

Gauge: 1435 mm			Locn: 225 $ 6900/2502			Date: 02.2003	
		4wDM	Bdm	RACO 60 LA7	1497	1957	(a) (1)
Tm	237 900-6	4wDM	Bdm	Schöma CFL 150DBR	2292	1960	(b) ?s/s
		4wDM R/R	Bdm	Mercedes Unimog	?	19xx	
(a)	new to SBB "Tmˡ 302"; ex SBB "Tmˡ 402" /1995.						
(b)	new to Rohstoffbetriebe Oker, Oker [D]; ex Florin AG, Muttenz 3/1996 via LSB.						
(1)	1996 to LSB.						

HUNZIKER & CO, Zürich   ZH071 C

Gauge: 750 mm		Locn: ?				Date: 03.2002
	0-4-0T	Bn2t	O&K 40 PS	1086	1903	(a) (1)
Gauge: ? mm		Locn: ?				Date: 05.2002
	4wDM	Bdm	O&K(M) RL 1a	4416	1931	(b) s/s
(a)	purchased from Minder & Galli, ?Engelberg /19xx; Engelsberg appears in the O&K delivery lists.					
(b)	new to Italy as a 600 mm locomotive; purchased via O&K/MBA, Dübendorf /19xx.					
(1)	19xx-31 sold to Kies AG Bollenberg, Tuggen.					

KELLER AG ZIEGELEIEN   ZH072

formerly: Ziegelei Keller & Cie

Pfungen

The brickworks, built in 1889, later became the headquarters of a company with works in many areas. This site is situated on the northern side of the railway at the station. The narrow gauge rail system had closed by 1976 though traces of the incline on which the wagons were hauled by chain could still be seen in 2005. Address: Ziegeleistrasse, 8422 Pfungen.

Gauge: 1435 mm		Locn: 216 $ 6912/2636				Date: 04.1988
No. 5800	4wPM	Bbm	Breuer		642	192x	(a) a1989 s/s

Gauge: 500 mm		Locn: 216 $ 6912/2636				Date: 02.1989	
	0-4-0T	B?2t	Maffei	3505	1909	(b)	(2)
	0-4-0T	Bn2t	Couillet	?	1xxx	(c)	s/s
	4wDM	Bdm	O&K(S) RL 1 s	1550	1947	(d)	(4)
	4wDM	Bdm	O&K(M) RL 1a	4004	1930	(e)	

The following locomotives were delivered by MBA, Dübendorf to the order of the company at Pfungen, but they may have been used at their other brickworks.

Gauge: 600 mm		Locn: ?				Date: 05.2002	
	4wDM	Bdm	O&K(M) RL 1a	3948	1930	(e)	s/s
	4wDM	Bdm	O&K(M) RL 1c	7812	1937	(e)	s/s
	4wDM	Bdm	O&K(M) MD 1	8202	1937	(e)	s/s

Gauge: ? mm		Locn: ?				Date: 05.2002	
	4wDM	Bdm	MBA MD 1	2004	c1950		
	4wPM	Bpm	O&K(M) H 1	2064	192x	new	s/s

Dättnau-Töss

formerly: Keller & Cie Pfungen, Werk Dättnau

This brickworks in the southern outskirts of Winterthur also had a railway on which the wagons were hauled by chain. This too was derelict by 1976.

Gauge: 500 mm		Locn: 216 $ 6949/2593				Date: 02.1989	
	4wDM	Bdm	O&K(S) RL 1 s	*1474*	194x		(11)

Schloß Teufen, Stat. Embrach

Gauge: 500 mm		Locn: ?				Date: 04.2009	
	4wDM	Bdm	O&K(S) RL 1a	4322	1930	(l)	s/s

(a) mentioned in the Breuer list of references of 1928, but not the one of 1926.
(b) purchased from Cementfabrik Rüthi, Rüthi SG /1922.
(c) building date unsure; purchased from ? /1922.
(d) originally 600 mm; purchased from the Schlatt brickworks ca. /1960. Other sources quote an MBA locomotive; this may be the same locomotive or an additional one.
(e) purchased from O&K/MBA, Dübendorf.
(l) new to Herm. Keller, Ziegelei, Schloß Teufen, Stat. Embrach.
(2) 1922 sold to IRR Rorschach via RUBAG.
(4) 1974 OOU; 19xx returned to Keller AG Ziegeleien, Schlatt for the company's museum.
(11) seen derelict in 1976; the plate could not be read properly and, in consequence, the number given may be incorrect.

REINHARD KERN STRASSENBAU AG, Bülach   ZH073 M

The founder of this company (Reinhard Kern) maintains a private collection on the premises. There is a short street-side railway for operating trains. Address: Solistrasse 86/88, Bülach.

The older locomotive from Kalkfabrik Spühler AG, Rekingen AG seen on 12 August 2005. It is O&K 2196 of c1926 (type S 10 a), but proudly displays the fact that is has been rebuilt with a Deutz A4L514 motor.

Gauge: 600 mm   Locn: 225 $ 6829/2649   Date: 08.2005

1	4wDM	Bdm	O&K(M) S 10 a	2196	c1926	(a)	
2	4wDM	Bdm	O&K(M) MD 2r	12002	1944	(b)	
3	0-4-0DM	Bdm	LKM Ns2f	48432	1953	(c)	
Diesellok Nr 4	0-4-0DM	Bdm	LKM Ns2f	?	195x	(d)	(4)

(a) ex Kalkfabrik Spühler AG, Rekingen AG /1989-90.
(b) ex Bauunternehmung Bless AG, Zürich "281.22.202".
(c) new to VEB Vereinigte Sodawerke Bernburg, Staßfurt [D] "4"; ex Tagesförderbahn am Bergbaumuseum Ottiliaeschacht, Clausthal-Zellerfeld [D] /1989-90.
(d) ex ? [D] /1989-90.
(4) 2007 to Parkbahn Letten, Rümlang.

A. KIESEL TIEFBAU AG, Winterthur   ZH074 C

A public works company.

Gauge: ? mm   Locn: ?   Date: 05.2002

4wDM	Bdm	O&K(M) H 1	2063	192x	(a)	(1)
4wPM	Bpm	O&K(M) RL 1a	2181	192x	(b)	s/s
4wDM	Bdm	O&K(M) RL 1	3645	1929	(c)	(3)

(a) new as 4wPM Bpm; purchased via O&K/MBA, Dübendorf.
(b) purchased via O&K/MBA, Dübendorf.
(c) gauge 600 mm; purchased via O&K/MBA, Dübendorf.
(1) 19xx to Pfenninger Bau AG, Schlieren; if correctly identified later to FWF, Otelfingen.
(3) 19xx sold to Kühnis & Lüchinger, Oberriet.

KIESWERK HOLBERG, Kloten — ZH073

A gravel pit with a railway providing the gravel used for enlarging the Kloten airport runway 1959-60.

Gauge: 750 mm   Locn: ?   Date: 03.2002

	0-4-0T	Bn2t	Märkische 50 PS	302	1898	(a) (1)

(a) made available by VEBA AG, Zürich /1959.
(1) 1960 returned to VEBA AG, Zürich.

KIESWERK TIEFENBRUNNEN AG, Zürich — ZH078

A company providing sand and gravel.

Gauge: ? mm   Locn: ?   Date: 05.2002

4wDM	Bdm	O&K(S) RL 1c	1506	194x	(a)

(a) purchased via O&K/MBA, Dübendorf.

KINDERGARTEN, Oetwil an der Limmat — ZH079 M

Gauge: 1435 mm   Locn: 225 $ 6723/2526   Date: 01.2008

E3592	0-4-0WT	Bn2t	Krauss-S XXVII au	5564	1906	(a) (1)

(a) new to Sulzer & Wyss; ex Sulzer-Escher Wyss, Zürich /1972.
(1) 1986 to Richi AG, Weiningen ZH.

KUMMLER + MATTER AG, Zürich — ZH080 C

This railway overhead construction company owns metre gauge locomotives that spend most of their time on the RhB. It has also operated a standard gauge locomotive. For more details see under Kantons AR and GR. Address: Hohlstrasse 176, 8026 Zürich.

LAGERHAUS KLOTEN AG, Kloten — ZH081

formerly: Löhne & Kern AG (1899-1953)

Originally this was a metal construction workshops, particularly for the building industry. Since closure the buildings have been used as warehouses.

Gauge: 1435 mm   Locn: 215 $ 6866/2561   Date: 05.1990

4wPM	Bbm	Breuer	?	c1930	(a)	s/s
4wPM	Bbm	Breuer III	?	1928	(b)	(2)
4wPM	Bbm	Breuer III	1159	1929	(c)	(2)

(a) mentioned in the Breuer list of references of about 1930, but not the one of 1928.
(b) ex SBB "438" /1960.
(c) ex SBB "899" /1965.
(2) stored for a long time; 1990 to VVT, St. Sulpice NE; later scrapped.

LIMBERG & GRAENER, Winterthur — ZH082

The business of this concern is not known.

Gauge: 750 mm   Locn: ?   Date: 05.2008

	0-4-0T	Bn2t	Heilbronn III	38	1874	new s/s

LOCHER & CIE AG, Zürich-Manegg — ZH083

A civil engineering company founded in 1830. As this workshop was rail connected only from 1908, the standard gauge steam engine could not have been used for traction purposes. The concern transferred to Baden in 1998. The SIG list of references gives two ET 40L and five ETB 70 locomotives delivered here. Address: Allmendstrasse 91, 8041 Zürich.

Gauge: 1435 mm   Locn: 225 $ 6812/2444   Date: 03.1991

	0-4-0T	Bn2t	Jung 80 PS	100	1891	(a)	(1)
Tm 2	0-4-0DM	Bdm	O&K RL 4	20077	1930	new	(2)

Gauge: 600 mm   Locn: ?   Date: 05.2008

0-4-0T	Bn2t	Hagans	230	1890	new	s/s
4wDM	Bdm	O&K RL 2	2774	1928	new	s/s
4wDM	Bdm	O&K RL 1a	4171	1930	(e)	s/s
4wDM	Bdm	O&K RL 1a	4211	1930	(f)	s/s
4wDM	Bdm	O&K RL 1a	4212	1930	(f)	s/s
4wDM	Bdm	O&K(M) LD 2	7385	1937	(h)	s/s
4wDM	Bdm	O&K(M) LD 2	7465	1937	new	(9)
4wDM	Bdm	O&K(M) RL 1c	7880	1937	new	s/s
4wDM	Bdm	O&K RL 2	20130	1931	new	s/s
4wDM	Bdm	O&K RL 2	20131	1931	new	s/s

(a) probably from construction of Sihltalbahn (SiTB).
(e) imported via Lagerhausgesellschaft, Basel.
(f) new via O&K, Zürich.
(h) ex Th. Moser & Co, Bienne /19xx.
(1) 1894 to Chemie Uetikon AG, Uetikon.
(2) 200x on loan to Erwin Huber Getränke, Zürich-Manegg.
(9) 19xx to Reifler & Guggisberger, Bienne.

LOCHER HAUSER AG, Schlieren — ZH084

formerly: Adolf Locher AG (till 199x)

An iron and steel wholesale company with a rail connection to the branch line serving Kieswerk Hardwald. In 199x the buildings were sold to AW Demontage AG and the rail connection abandoned.

Gauge: 1435 mm   Locn: 225 $ 6747/2506   Date: 07.1993

4wPM	Bbm	RACO 35 PS	881	1927	(a)	a1971 s/s
4wDM	Bdm	KHD KK140 B	57487	1962	(b)	(2)

(a) ex CFF '839' 9/1960.
(b) ex Domaniale Mijn Maatsch., Kerkrade [NL] /1973 via Asper.
(2) 1993-2001 s/s.

LOKSERVICE BURKHARDT AG (LSB), Rüti ZH — ZH085 C

This company repairs, supplies and hires locomotives. The details of the known longer-term hires are detailed under the firms hiring the locomotives. The following are locomotives known to have been owned by the company. Other locomotives may be present for overhaul from time to time. Address: Weierstrasse 50a, Rüti (offices) and J. Müller, Hinwil (workshops).

Gauge: 1435 mm   Locn: 236 $ 7085/2345   Date: 03.1996

	4wDH	Bdh	Diema DVL 120	2620	1962	(a)	
	6wDH	Cdh	Esslingen Neuhof II	5164	1956	(b)	1997 s/s
	6wDH	Cdh	Esslingen Guthof	5224	1956	(c)	1993 scr
	4wDH	Bdh	Gmeinder Köf II	4675	1951	(d)	(4)
	4wDH	Bdh	Gmeinder Köf II	5008	1957	(e)	1989 scr
	4wDH	Bdh	Gmeinder V12-16	5369	1955	(f)	(6)
03	0-4-0DH	Bdh	Henschel DH 240B	29705	1958	(g)	(7)

	0-4-0DH	Bdh	Henschel DH 240B	29968	1959	(h)	(8)
	0-6-0DH	Bdh	Henschel DH 500Ca	30515	1963	(i)	(9)
04	BBDH	BBdh	Henschel V 100	30541	1962	(k)	(11)
	0-4-0DH	Bdh	Henschel DH 240B	30594	1964	(l)	(12)
	4wDH	Bdh	Henschel DHG 240Bex	30864	1964	(m)	
	4wDH	Bdh	Henschel DHG 240Bex	31091	1964	(n)	(14)
	4wDH	Bdh	Henschel DHG 275B	31098	1968	(o)	(15)
	6wDH	Cdh	Henschel DHG 500C	31111	1965	(p)	(16)
Em 837 901-9, ex 05	6wDH	Cdh	Henschel DHG 500C	31184	1967	(q)	
	6wDH	Cdh	Henschel DHG 500C	31236	1968	(r)	
	4wDH	Bdh	Henschel DHG 300B	32473	1984	(s)	(19)
	0-6-0DH	Cdh	Jung R 40 C	12041	1961	(t)	
Em 837 903-4	0-6-0DH	Cdh	Jung R 42 C	12351	1956	(u)	2001 scr
Em 837 904-2 JOSEFA	0-6-0DH	Cdh	Jung R 42 C	13290	1961	(v)	(22)
KROKODIL	4wDH	Bdh	KHD A6M517 R	46994	1949	(w)	(23)
	4wDH	Bdh	KHD A6M517 R	55126	1952	(x)	(24)
	0-4-0DH	Bdh	KHD A4L514 R	56488	1957	(y)	
	0-4-0DH	Bdh	KHD A4L514 R	56810	1957	(z)	(26)
	6wDH	Bdh	KHD MG 530C	57562	1963	(aa)	(27)
	4wDH	Bdh	KHD A6M617 R	57914	1965	(ab)	(28)
	4wDH	Bdh	KHD KG275 B	57831	1965	(ac)	(29)
	4wDH	Bdh	KHD KG230 B	57869	1965	(ad)	(30)
Tm 237 859-4	4wDH	Bdh	KrMa ML 225 B	18701	1960	(ae)	(31)
	6wDH	Cdh	KrMa ML 700 C	18850	1962	(af)	
	6wDH	Cdh	KrMa M 350 C ex	19399	1969	(ag)	(33)
	0-6-0DH	Cdh	Krupp 1W1C	3781	1961	(ah)	(34)
847 956-0 MANUELA	BBDH	BBdh	Krupp V 100	4381	1962	(ai)	(35)
	0-4-0DH	Bdh	MaK 240 B	220055	1960	(aj)	1997 scr
	0-8-0DH	Ddh	MaK 650 D	500020	1956	(ak)	19xx scr
	0-8-0DH	Ddh	MaK 650 D	600412	1962	(al)	1993 scr
	0-8-0DH	Ddh	MaK 1200 D	1000250	1963	(am)	(39)
	4wDH	Bdh	O&K MB 5 N	26203	1963	(an)	
	4wDH	Bdh	O&K MB 5 N	26583	1966	(ao)	(41)
	4wDH	Bdh	O&K MB 5 N	26600	1966	(ap)	(42)
	4wDH	Bdh	O&K MB 5 N	26602	1968	(aq)	
3	4wDH	Bdh	RACO 80 ST4	1408	1951	(ar)	(44)
	4wDM	Bdm	RACO 60 LA7	1497	1957	(as)	(45)
	4wDH	Bdh	RACO 70 DA4 H	1817	1974	(at)	(46)
	4wDH	Bdh	Schöma CHL 20GR	2168	1960	(au)	(47)
	4wDH	Bdh	Schöma CFL 150DBR	2292	1960	(av)	(48)
1	4wDE	Bde	Stadler	140	1972	(aw)	
Em 847 901-6	BBDH	BBdh	U23A LDH370	23879	1979	(ax)	
Tm 1	4wDE	Bde2	ACMV,BBC	1259, ?	1959	(ay)	
	BBDH	BBdh	U23A LDH45	24122	1980	(az)	
Am 847 961-0	BBDH	BBdh	KHD V 100	57361	1962	(ba)	(53)
Em 847 904-0	BBDH	BBdh	Henschel DHG 1200 BB	31575	1973	(bb)	
Tm 237 954-3	4wDH	Bdh	KrMa M 250 B ex	19279	1965	(bc)	
	6wDH	Cdh	KrMa M 700 C	19675	1973	(bd)	
	6wDH	Cdh	KrMa M 700 C	19687	1973	(be)	

Gauge: 1000 mm   Locn: ?   Date: 10.1998

	4wDH	Bdh	Schöma CFL 150DCL	4807	1965	(ca)	(301)

Gauge: 900 mm   Locn: ?   Date: 03.1991

	0-4-0DH	Bdh	O&K MV9	25833	1958	(da)	(401)
	0-4-0DH	Bdh	O&K MV9	25892	1959	(da)	(402)

Gauge: 785 mm   Locn: 236 $ 7058/2398   Date: 07.1994

10	4wDH	Bdh	KHD KG230 BS	57850	1965	(ea)	1994 scr
11	4wDH	Bdh	KHD KG230 BS	58110	1966	(eb)	1994 scr
12	4wDH	Bdh	KHD KG230 BS	58111	1966	(ec)	1994 scr
19	4wDH	Bdh	KHD KG230 BS	57894	1966	(ed)	(504)

(a)   ex Lübecker Hafenbahn [D] /1990 via OnRail; rebuilt by LSB /1994.

(b) new to Hüttenwerke Oberhausen AG (HOAG), Oberhausen [D] "250"; ex Eisenbahn und Häfen, Dortmund [D] "236" /1991 via OnRail.
(c) new to Hüttenwerke Oberhausen AG (HOAG), Oberhausen [D] "303"; ex Eisenbahn und Häfen, Dortmund [D] "382" /1991 via OnRail.
(d) new to DB "Köf 6126"; ex DB "323 940-7" /1987 via OnRail.
(e) new to DB "Köf 6308"; ex DB "323 619-7" /1988 via OnRail.
(f) new to Lever-Sunlicht, Mannheim-Rheinau [D] "10"; ex Lonza AG, Waldshut [D] "444" /1994.
(g) new to Rheinelbe Bergbau AG, Gelsenkirchen [D] "Bd 2"; ex Ruhrkohle AG [D] "471" /1990 via OnRail.
(h) ex Preußen Elektra AG, Kraftwerk Kassel "3" /1991 via OnRail.
(i) ex Rheinisch-Westfälische Elektrizitätswerke AG (RWE), Kraftwerk Frimmersdorf [D] "4"
(k) new to DB "V100 1192"; later "211 192-0"; ex OnRail /1991.
(l) new to Deutsche Edelstahlwerke AG (DEW), Krefeld [D] "8" (as type DH 240 according to the Henschel information); ex EH "410" /1995 via OnRail.
(m) new to Union-Kraftstoff (UK), Speyer [D] "653"; ex PACTON Eisenbahnservice + Spezialtransporte Kay Winkler, Radevormwald [D] /1999.
(n) new to Basler Rheinschiffahrt AG (BRAG), Tankleger Birsfelden Hafen BL; ex Van Ommeren-Bragtank AG, Birsfelden-Hafen /1992.
(o) new to E. Merck AG, Darmstadt [D] "2"; ex Merck KGaA, Darmstadt [D] "2" /1999.
(p) new to Steinkohlen-Elektrizitäts-AG (STEAG), Kraftwerk Herne [D] "2"; ex ? [F] /1994.
(q) new to Hibernia AG, Zechenbahn- und Hafenbetriebe Ruhr-Mitte, Gladbeck [D] "41-500"; ex Deutsche Steinkohle AG, RAG Bahn- und Hafenbetriebe, Gladbeck [D] "454" /2003 via VFST.
(r) new to Hibernia AG, Zechenbahn- und Hafenbetriebe Ruhr-Mitte, Gladbeck [D] "30-500"; ex Deutsche Steinkohle AG, RAG Bahn- und Hafenbetriebe, Gladbeck [D] "451" /2003.
(s) new to Bayerische Elektrizitäts-Lieferungs AG (BELG), Kraftwerk Arzberg [D]; ex E.ON Kraftwerke GmbH, Arzberg [D] /2004.
(t) ex Neukölln-Mittenwalder Eisenbahn Gesellschaft AG, Berlin [D] "ML 00602" /1991 via Jung/MaK.
(u) new to Farbwerke Hoechst AG vorm. Meister Lucius & Brüning, Werk Gersthofen [D] "5"; ex Stora Enso Baienfurt GmbH & Co. KG, Baienfurt [D] "2" (and "839") /2001.
(v) new to Farbwerke Hoechst AG vorm. Meister Lucius & Brüning, Werk Gersthofen [D] "6 Judy"; ex Clariant Produkte (Deutschland) GmbH, Werk Gersthofen [D] "6" /1997.
(w) new to DB "Köf 6113"; later DB "323 993-6"; ex Schmolz & Bickenbach, Düsseldorf [D] /1989 via OnRail.
(x) new to Gußstahlwerk Wittmann AG, Hagen-Haspe [D]; ex KG Schüssler, Gevelsberg [D] /1988 via OnRail.
(y) new to Kronprinz AG, Solingen-Ohligs, Werk Immigrath [D]; ex Mannesmann Edelstahlrohr GmbH, Langenfeld [D] /1990 via OnRail.
(z) new to Berkenhoff & Drebes, Drahtwerke, Asslar/Wetzlar [D] "1"; ex Thyssen Draht, Asslar/Wetzlar [D] "1" /1990; has acquired oval plates from Deutz 55126, see above.
(aa) new to Dortmund-Hörder-Hütten-Union AG (DHHU), Dortmund-Hörde, Werk Hörde [D] "173"; ex Raffineriegesellschaft Vohburg/Ingolstadt mbH (RVI), Vohburg a. d. Donau, Ingolstadt [D] /1991 via OnRail.
(ab) new to DB "Köf 6814"; later DB "323 334-3"; ex Migros Volketswil, Schwerzenbach /1988.
(ac) new to Lokalbahn Lam-Kötzting (LLK) [D] "L 02"; ex Lonza G+T AG, Bodio /1996.
(ad) new to Vereinigte Tanklager und Transportmittel GmbH, Hamburg [D] "AE 3005"; ex Monteforno Acciaierie e Laminatori SA, Bodio, Bodio /19xx via Asper.
(ae) new to Westfälische Union AG, Hamm [D]; ex Thyssen Draht, Hamm [D] "2" /1991 via OnRail.
(af) new to Hoesch Hüttenwerke AG, Dortmund [D] "5"; ex Mannesmann Hoesch Präzisrohr GmbH, Hamm [D] 5/1998.
(ag) new to Lonza AG, Waldshut [D] "289"; ex Lonza AG, Waldshut [D] "443" before /1994.
(ah) new to Fried. Krupp Maschinenfabriken - Lokbau, Essen [D] "M 1", ex MaK, Werkstatt Moers [D] /1991 (another source gives via OnRail).
(ai) new to DB "V100 1271"; ex DB "211 271-2" /199x via Layeritz.
(aj) new to Deutsche BP AG, Raffinerie Hünxe [D] "3" ; ex Deutsche BP AG, Mainz-Gustavsburg [D] "2054" /1991 via OnRail.
(ak) new to Hüttenwerke Ilsede-Peine AG, Peine [D] "90"; ex Gribskovbanens Driftsselskab (GDS), Hillerød [DK] "L2" /1991 via OnRail.
(al) new to Hillerød-Frederiksværk-Hundested Jernbane (HFHJ), Hillerød [DK] "M 11"; ex MaK /1990.
(am) ex Wanne-Bochum-Herner Eisenbahn (WBHE) [D] "V 15" /1989 via OnRail.
(an) new to John Deere Lanz, Mannheim [D] "222"; ex Hafenverwaltung Kehl, Kehl [D] "488" by /2001.
(ao) new to Dynamit Nobel AG, Werk Rheinfelden [D] "253"; ex Karl Kaufmann, Thörishaus /1993 and from Hafenverwaltung Kehl, Kehl [D] /2005.
(ap) new to Dortmunder Union-Brauerei AG (DAB), Dortmund [D] "1"; ex Thévenaz-Leduc, Ecublens VD "3" /1979.

CH 372 ZH

Gauge: 600 mm			Locn: ?			Date: 05.2002
	4wDM	Bdm	O&K(M) RL 2	2698	1927	new (1)

(1) 19xx sold to Mangold & Cie, Zürich.

SCHÄRER, SCHWEITER, METTLER AG (SSM), Horgen-Oberdorf
ZH121

formerly: Maschinenfabrik Schweiter AG (1893-1989)

A construction company, particularly of textile industry machinery, with a siding alongside a loading dock adjacent to the station. Production had ceased by 2007.

Gauge: 1435 mm			Locn: 225 $ 6869/2349			Date: 12.1997
	A1BE	A1a	unknown	?	19xx	(1)
	4wBE	Ba1	Stadler,BBC	40, ?	1950	new (2)

(1) c1950 probably scrapped.
(2) 1997 to Hans Moser, Biberist for scrap.

SCHAFFNER & HULDI AG, Affoltern am Albis   ZH122 C
The company deals in industrial wastes.

Gauge: 600 mm			Locn: 225 $ 6761/2362			Date: 02.1988
	4wDM	Bdm	RACO 12/16PS	1366	1949	(a) (1)

(a) purchased from Nicollier? /198x.
(1) by 1997 to FWF, Otelfingen.

SCHAFIR & MUGGLIN, Zürich   ZH123 C

The Zürich-based subsidiary of this public works concern. The SIG list of references gives one ET 20L5 locomotive delivered here.

Gauge: 750 mm			Locn: ?			Date: 03.2002	
BIRS	0-4-0T	B?2t	Henschel Monta	17674	1920	(a)	(1)
BUBI	0-4-0T	Bn2t	Jung 10 PS	1349	1909	(b)	(2)
MUTTENZ	0-4-0T	Bn2t	Jung 20/25 PS	3620	1924	(b)	(2)
SPERANZA	0-4-0T	Bn2t	Krauss-S XXVII vv	3254	1895	(b)	(2)

LICHTENSTEIG	0-4-0T	Bn2t	Krauss-S XXVII bg	6173	1909	(b)	(2)	
BASEL	0-4-0T	Bn2t	O&K 80 PS	1622	1905	(f)	(6)	
VEVEY	0-4-0T	Bn2t	O&K 50 PS	2907	1908	(f)	(7)	
	0-4-0WT	Bn2t	O&K 90 PS	3216	1908	(h)	(8)	
TÄUFFELEN	0-4-0T	Bn2t	O&K 50 PS	3702	1910	(i)	(1)	
ZÜRICH	0-4-0T	B?2t	O&K 70 PS	4375	1911	(j)	(10)	
MURI	0-4-0T	B?2t	O&K 70 PS	6020	1914	(k)	(11)	
BERNA	0-4-0T	B?2t	O&K 70 PS	6572	1913	(f)	(12)	
LIMMAT	0-4-0T	B?2t	O&K 70 PS	6573	1913	(k)	(11)	
GWATT	0-4-0T	B?2t	O&K 50 PS	7520	1916	(k)	(10)	

(a) purchased from Jardini, Basel /1930; used in the construction of the hydroelectric power stations at Klingnau from /1931 and Rupperswil-Auenstein from /1944.
(b) transferred from the Muri AG subsidiary /1930 used in the construction of the hydroelectric power station at Klingnau from /1931.
(f) transferred from the Muri AG subsidiary /1933-39; used in the construction of the hydroelectric power station at Klingnau from /1931.
(h) purchased from Hatt-Haller & Züblin, Zürich /1922-39.
(i) purchased from O&K, Zürich /1932; used in the construction of the hydroelectric power station at Klingnau from /1931.
(j) purchased from Frutiger & Lanzrein, Bern /1926, used in the construction of the hydroelectric power station at Rupperswil-Auenstein from /1944.
(k) transferred from the Muri AG subsidiary /1933-39; used in the construction of the hydroelectric power stations at Klingnau from /1931 and Rupperswil-Auenstein from /1944.

(1) 1953 removed from boiler register.
(2) 1939-53 transferred to the Liestal subsidiary; 1953 removed from boiler register.
(6) 1939-53 transferred to the Liestal subsidiary; 1960 removed from boiler register.
(7) 1939-53 transferred to the Liestal subsidiary.
(8) 1955 sold to Kieswerk Hardwald, Dietikon (gauge 750 mm or 900 mm?).
(10) 1944-53 transferred to the Liestal subsidiary.
(11) 1944-59 transferred to the Liestal subsidiary.
(12) 1966 removed from boiler register.

SCHEIFELE CONTAINERS AG, Dänikon ZH   ZH124

formerly: Scheifele

This firm originally provided carpentry services. Later it added container rental. Address: Unterdorfstrasse 21, 8114 Dänikon.

Gauge: 600 mm   Locn: ?   Date: 04.2009

4wDM	Bdm	O&K(M) MD 1	7750	1937	(a)	(1)

(a) ex Gebrüder Dübendorfer, Bassersdorf /19xx.
(1) by 2006 to FWF, Otelfingen.

EMIL SCHELLER & CIE AG, Dietikon   ZH127

A factory, rail connected to Zürich-Altstetten from 1896. The connection was reorganised to connect with Birmensdorf-Ristet ca. 1980.

Gauge: 1435 mm   Locn: 225 $ 6730/2508   Date: 03.1988

	1ABE	1Aa1	BBC Akku-Fahrz.	1098	1917	new
rebuilt	4wD?M	Bd?m	RACO 50 ST	1421	1952	(1)

(1) 1975 to Vanoli for use on the Rangierbahnhof Limmattal (RBL) project.

SCHLACHTHOF ZÜRICH, Zürich-Altstetten   ZH128

The slaughterhouse complex was built 1906-09 and rail connected from 1904 with a line through the streets from Altstetten SBB. The rails were removed in 1995.

SIG-HOLZWERK, Rafz　　　　　　　　　　　　　　　　　　ZH138

formerly:　　Schweizerische Industrie-Gesellschaft (SIG), Holzlager und Sägewerk (from 1917)
The joiners shop of SIG whose main site is at Neuhausen am Rheinfall.

Gauge: 1435 mm　　　　　　　Locn: 215 $ 6831/2730　　　　　　　　Date: 04.1995

	xxBE	xa	SIG,MFO	?	1897	(a)	s/s
	4wDH	Bdh	Schöma CFL 30DCR	2682	1964	new	(2)

(a)　　ex SIG, Neuhausen am Rheinfall as a conversion from a battery locomotive; it is not clear if only one or both axles were driven.
(2)　　19xx to HIAG Imballegno AG, Rafz. ZH.

A. SPALTENSTEIN, Zürich　　　　　　　　　　　　　　　　ZH139 C

later:　　Spaltenstein Hoch+Tiefbau AG (from 2008)
A public works firm. The continued existence of the locomotive is not confirmed.

Gauge: 600 mm　　　　　　　Locn: ?　　　　　　　　　　　　　　Date: 05.2002

	4wDM	Bdm	O&K(M) RL 1a	4871	1933	(a)	?

(a)　　purchased via O&K/MBA, Dübendorf.

SPINNEREI STREIFF AG, Aathal-Seegräben　　　　　　　ZH140

A weaving factory founded in 1825, for which a new building was built in 1962. Production had ceased by 2008.

Gauge: 1435 mm　　　　　　　Locn: 226 $ 7004/2436　　　　　　　　Date: 04.1988

	4wPM	Bbm	Breuer III	?	1928	(a)	?1982 s/s

(a)　　ex SBB "Tm 439" /1962.

STADLER WINTERTHUR, Winterthur　　　　　　　　　　　ZH141

formerly:　　Sulzer-Winpro (till 2006, though Stadler took control in 2005)
　　　　　　Schweizerische Lokomotiv- und Maschinenfabrik (SLM) (1873-1998)

SLM was formed and rail connected in 1873. The firm was integrated into the Gebrüder Sulzer AG group in the 1970s and the buildings were taken over by Stadler in 2005.

Gauge: 1435 mm　　　　　　　Locn: 216 $ 6960/2603　　　　　　　　Date: 06.1993

	0-4-0T	Bn2t	SLM	89	1876	new	(1)
	0-4-0T	Bn2t	SLM	388	1884	new	(2)
1	0-4-0T	Bn2t	SLM	2090	1910	new	(3)
	A1BE	A1a	MFO	?	1918	new	(4)
	0-4-0DM	Bdm	SLM Tm	2853	1923	new	(5)
	4wDH	Bdh	SLM Tm^{IV}	5069	1975	(f)	(6)

(f)　　new; MTU motor.

(1)　　18xx sold to ?, La Spezia [I].
(2)　　1911 sold to Papierfabrik Landquart, Landquart.
(3)　　200x sold to Ralph Schorno, Winterthur.
(4)　　19xx either to Gebrüder Sulzer, Bülach or scrapped.
(5)　　rebuilt as SLM 3608/36 and sold to Saïs AG, Horn.
(6)　　1987 to SBB "Tm^{IV} 8797".

STADT DIETIKON, Dietikon   ZH142 M

The locomotive is displayed (with other railay equipment) outside the premises of the Modell Bahn Club Dietikon (MBCD), itself adjacent to the SBB station. It is intended to transfer the locomotive to the Historische Seethalbahn, Bremgarten (West) and to make it operational.

Gauge: 1435 mm   Locn: 225 $ 672910,251185   Date: 03.2007

      0-6-0WT Cn2t SLM E 3/3   900 1894 (a) display

(a) new to NOB "256"; later SBB "8554"; ex Von Moos, Emmenbrücke "5" /1986 via private ownership.

STAUB & CO AG, Männedorf   ZH143

formerly: Gerberei Staub & Co

A tannery started in 1866 and rail connected from 1894. Nowadays it produces plastic elements for machines.

Gauge: 1435 mm   Locn: 226 $ 6943/2348   Date: 07.1988

     1APM 1Abm SAFIR   - 1908-9 (a) 1913 wdn
     4wPM Bbm   O&K(M) ?U5a ? 1913   (2)

(a) built by the SAFIR motorcar firm /1908-09.
(2) early 1950s scrapped.

STAUFFER SCHIENEN-UND SPEZIALFAHRZEUGE, Schlieren
ZH144 C

see: Stauffer Schienen-und Spezialfahrzeuge, Frauenfeld

GUSTAV STEINER AG, Embrach   ZH145 C

later: Steiner Teufen AG (in liquidation 2007)

This company was a gravel supplier. The source data (MBA records) quotes "Steiner, Embrach". This has been tentatively identified as above, but that is not confirmed.

Gauge: 600 mm   Locn: ?   Date: 05.2002

      4wDM Bdm   O&K(M) RL 1 3829 1929 (a) s/s?

(a) purchased via O&K/MBA, Dübendorf.

STEINER & CIE, Zürich   ZH146 C

A builder's merchant (Baugeschäft).

Gauge: 600 mm   Locn: ?   Date: 05.2002

      0-4-0T Bn2t O&K 20 PS   2459 1908 (a) (1)

(a) purchased from O & K, Zürich /1920.
(1) 1937 removed from boiler register.

STEINER & CIE, Zürich   ZH147 C

An office for civil engineers (Ingenieurbureau).

Gauge: 750 mm   Locn: ?   Date: 03.2002

WEISSENSTEIN Bn2t 0-4-0T Krauss-S IV bh 5295 1905 (a) (1)

(a) purchased from Robert Aebi & Co, Zürich /1914-23.

(1)   1923-36 sold to Bauunternehmung Dünnernkorrektion Los D, Olten.

STRASSENBAULEITUNG ZÜRICH, Zürich   ZH148 C
The cantonal road building directorate.

Gauge: 600 mm   Locn: ?   Date: 05.2002

| | 4wDM | Bdm | O&K(M) RL 1c | 9555 | 1938 | (a) | s/s |

(a)   purchased new via O&K Wien [A].

SULZER AG   ZH149
formerly: Gebrüder Sulzer (GS) (from 1834)

Werk Winterthur

A foundry, with a diesel motor and machinery plant, rail connected in 1860. In 1907, the company opened an additional factory at Oberwinterthur. Latterly Werk Winterthur locomotives were sometimes present at Werk Oberwinterthur. The small steam locomotive operated at both works. As it was constructed at the same time as Werk Oberwinterthur, it could be considered as belonging there. Gradually all manufacturing was transferred to Werk Oberwinterthur and industrial activity at Werk Winterthur ceased in 1990. The apparently missing "2" from the standard gauge list was, in fact, an unknown narrow gauge locomotive. The Schweizerische Lokomotiv- und Maschinenfabrik (SLM) was integrated into Gebrüder Sulzer AG group in the 1970s, but is listed in this Handbook separately.

Gauge: 1435 mm   Locn: 216 $ 6966/2626   Date: 05.1988

1	0-4-0T	Bn2t	SLM E 2/2	917	1895	new	(1)
4	0-6-0WT	Cn2t	SLM E 3/3	1333	1901	(b)	(2)
5	0-6-0WT	Cn2t	SLM E 3/3	3610	1936	(c)	(3)
	2-6-0T	1Cn2t	SLM Ed 3/4	1488	1903	(d)	1957 scr
	xxBE	xa	ACMV,MFO	?	1926		s/s
	A1BE	A1a	unknown,MFO	?	19xx		s/s
	A1BE	A1a	unknown,MFO	?	19xx		s/s

Gauge: 720 mm   Locn: 216 $ 6966/2626   Date: 05.1988

1	0-4-0T	Bn2t	SLM	1068	1897	new	(8)
	4wBE	Ba	unknown	?	19xx	(i)	s/s
	4wBE	Ba	unknown	?	19xx	(i)	s/s

Werk Oberwinterthur

A foundry and machine shop initially constructed in 1907 and rail connected from 1909. Locomotives from Werk Winterthur have been stored here from time to time.

Gauge: 1435 mm   Locn: 216 $ 6995/2622   Date: 07.1997

	xBE	xa	unknown	?	1912	(k)	s/s
3	0-4-0T	Bn2t	SLM	1836	1907	(l)	(12)
4	0-6-0T	Cn2t	SLM E 3/3	1333	1901	(m)	1962 scr
2	2-6-0T	1Ch2t	SLM Ed 3/4	1489	1903	(n)	(14)
	0-6-0DH	Cdh	Krupp,Sulzer 350 PS	3324,345	1954	new	(15)
	4wDH	Bdh	SLM Tm^{IV}	4810	1969	new	(16)

(b)   new to GTB "4"; purchased from BLS "73" /1910.
(c)   purchased from HWB "5" /1947.
(d)   new to RSG "1"; purchased from RPB "1" /1952.
(i)   purchased via RACO.
(k)   a Platformwagen, with either one or both axles driven; OFT records of 1918 state that it was put into service in 1915.
(l)   new; on loan several times to Sulzer AG, Werk Winterthur.
(m) new to Gürbetalbahn "4"; ex Sulzer AG, Werk Winterthur /1945.
(n)   new to RSG "2", ex RPB "2" /1949.

CH 393 ZH

(1) 1953 sold to Monteforno, Bodio "2".
(2) 1945 to Sulzer AG, Werk Oberwinterthur.
(3) 1976 withdrawn; 1981 to VHS, Luzern.
(5) 1972-198x to Sulzer AG, Werk Oberwinterthur.
(8) 1956 placed on display in the "Park im Grüne", Rüschlikon; later scrapped.
(12) 1972 donated to DVZO; 1978-97 on display at the SBB Rangierbahnhof Limmattal (RBL).
(14) 1972 to DVZO "2 HINWIL".
(15) 1993 to EUROVAPOR, Sektion Sulgen/Rorschach, Sulgen "Em 3/3".
(16) 1994 to SBB "Tm 8797".

SULZER RÜTI, Rüti ZH ZH152

formerly: Maschinenfabrik Rüti AG (MSR) (1886-1982)
Caspar Honneger (1847-1886)

A factory that wove materials and also turned them into finished products. It was constructed in 1847 and rail connected to Rüti SBB station by a siding 2.25 kilometres long. A section near the station was rack equipped. The original layout was replaced in the 1960s with the result that the rack section was then on a curving concrete viaduct. Production ceased in 1997.

Gauge: 1435 mm + Strub rack   Locn: 226 $ 7071/2356   Date: 12.2002

	1		2-2-0RT	1Aan2t	Aarau	11	1876	new	
	1	rebuilt	2-2-0RT	1Aan2t	SLM	-	1893		
	1	rebuilt	0-4-0RT	Ban2t	SLM		1925		(1)
Eh 2/2	2		0-4-0RT	Ban2t	SLM Eh 2/2	2126	1910	new	(2)
	3		0-4-0RT	Ban2t	SLM	4046	1951	new	(3)
Tmh	4	CHAPPI	4wRDE	Bbde	SLM,BBC Tmh,Thm	4400,6724	1962	new	(4)

(1) 1964 withdrawn; by 9/1992 to Technorama, Winterthur; by 2006 on loan to Serge Bourginet, Seewen SZ.
(2) 1962 donated to Knie Kinderzoo, Rapperswil and placed on display.
(3) 1997 donated to EUROVAPOR, Sektion Sulgen/Rorschach, Heiden "ROSA" for use on the Rorschach-Heiden-Bergbahn (RHB) (see AR section of this Handbook).
(4) 2002 hired to Kummler & Matter, Zürich for use on the Rorschach-Heiden-Bahn (RHB) "237 916"; 2006 sold to Rorschach-Heiden-Bergbahn (RHB).

SWISSMILL, Zürich-Industriequartier ZH153

formerly: Swissmill Zürich-Rivaz (from 1998)
Mühlegenossenschaft Schweiz Konsumverein (from 1912)
Coop Mühle, Zürich (from 1896)

A long siding runs from near the Escher Wyss Platz along the street to warehouses by the R. Limmat and the site can be seen from trains running via Wipkingen. The line was electrified 1898-1982. The history of this flourmill goes back to 1780 and it remains a subsidiary concern of the Coop. Address: Division der Coop. Basel, Sihlquai 306, 8037 Zürich.

(a) hired from LSB /1990.
(b) new to DB "V100 1124", ex DB "211 124-3" /1990 via OnRail/LSB.
(c) new to DB "Köf 6814"; later DB "323 334-3"; ex LSB /1991; initially stationed at Rotkreuz.
(d) new to Gelsenkirchener Bergwerks AG, Rheinelbe, Gelsenkirchen [D] "Cd 5"; ex Ruhrkohle AG (RAG) [D] "V425" /1992 via OnRail/LSB.
(e) new to Geldner Rheinlager AG, Birsfelden; purchased from Schlachthof Zürich /1991 after a period of hire.
(f) ex Varde-Nørre Nebel Jernbaneselskab (VNJ) [DK] "DL14" /2002 via PACTON/LSB.
(g) new to Hjorring Privatbaner (HP) [DK] "DL16" /ex Stadtwerk Düsseldörf [D]/LSB /2002-4.
(h) new to DB "Köf 6712"; ex DB "323 782-3" /1984; from Oswald Steam (OSS) with the site.
(i) new to DB "Köf 6744", later "324 814-4"; ex SOB /1985; initially stationed at Rothenburg.
(j) new to DB "V100 1047", ex DB "211 047-6" /1990 via OnRail/LSB.
(k) new to Oranje Nassau Mijnen, Heerlen [NL]; ex HOLCIM, Hüntwangen /1987.
(l) new via MBA as "V2"; hired out for various contracts.
(m) purchased from Emil Scheller AG, Dietikon /1976; used during the construction of the SBB marshalling yards, Zürich-Altstetten.
(n) new to ARBED, Burbach, Saarbrücken [D]; ex Stahlwerke Röchling-Burbach, Werk Burbachhütte [D] "2" /1983 via Asper; also recorded as KG 200B.

(1) 1990 returned to LSB.
(2) 2008 sold to LSB via Unirail.
(3) 1999 to Kummler + Matter AG, Zürich; 25/09/2006 to Unirail.
(8) 1993 sold to VVT, St. Sulpice NE.
(9) 7/11/2006 to Unirail "4".
(10) 2007 to ? via Stauffer/Unirail.
(11) by 10/1996 to HOLCIM, Hüntwangen; then Untervaz; 17/7/2006 to Shunter [NL] "304 MYRTHE" via Unirail.
(14) 1986 transferred to Langenthal and stored; 1986 donated to Dampfbahn Furka-Bergstrecke (DFB) "HGm 2/2 51".

see: www.vanoli-ag.ch

VEBA VEREINIGTE BAUUNTERNEHMER AG, Zürich   ZH161 C

alternative: Veba AG

This building company has ceased to operate. The SIG list of references gives one ET 40L locomotive delivered here.

Gauge: 750 mm   Locn: ?   Date: 03.2002

		0-4-0T	Bn2t	Freudenstein	210	1905	(a)	(1)
		0-4-0T	Bn2t	Märkische 60 PS	22	1893	(b)	(2)
7		0-4-0T	Bn2t	Märkische 50 PS	302	1898	(c)	(3)
AUENSTEIN		0-4-0T	Bn2t	SLM	3833	1944	(d)	(4)
WILDEGG		0-4-0T	Bn2t	SLM	3834	1944	(d)	(5)
RUPPERSWIL		0-4-0T	Bn2t	SLM	3835	1944	(d)	(6)

(a) used during the construction of the RhB; purchased from ? /1946 and used for Kloten airport construction and hydroelectric power station contracts.
(b) property of Losinger & Cie, Bern; used by VEBA /1953.
(c) purchased from J. Seeberger, Frutigen /1939-57; used during the construction of the water courses for a hydroelectric power station /1957.
(d) purchased from Kraftwerkbau Rupperswil-Auenstein /1943-53.

(1) 1957 removed from boiler register.
(2) 19xx returned to Losinger & Cie, Bern.
(3) 1959-60 used at the Holberg gravel pits in connection with building the runway at Kloten; to store at the Niederhasli subsidiary; 1964 removed from boiler register: 197x sold to K. Keller, Rorschach.
(4) 1959-60 used at the Holberg gravel pits in connection with building the runway at Kloten; 1960-73 sold to Messerli, Kaufdorf.
(5) 1959-60 used at the Holberg gravel pits in connection with building the runway at Kloten; 1967 placed on display at Turgi Bahnhof (under the care of Schuljugend Turgi); 1994 to Schinznacher Baumschulbahn (SchBB), Schinznach Dorf.
(6) 1960 used as a stationary boiler at a bridge construction site at Nydegg, Bern; 1960-73 sold to Messerli, Kaufdorf.

VERBAND OSTSCHWEIZERISCHER LANDWIRTSCHAFTLICHER GENOSSENSCHAFTEN (VOLG), Winterthur ZH162

One site of the agrarian cooperative of eastern Switzerland. These mills and warehouses were erected in 1898. The sidings were electrified till around the 1960s.

Gauge: 1435 mm   Locn: 216 $ 6971/2614   Date: 08.1988

    4wWE Bg unknown   ? 19xx s/s

VEREIN BERGWERK RIEDHOF, Aeugstertal ZH163 M

This group has set up an underground museum devoted to the coal mine that existed to the west of the Götschihof. There is also a surface railway; locomotives for this are on loan from the Feld- und Werkbahn Freude (FWF), Otelfingen.

Gauge: 600 mm   Locn: 225 $ 6792/2374   Date: 06.2006

  see: www.bergwerk-riedhof.ch/museum.html

VEREINIGTE FÄRBEREIEN & APPRETUR AG (VFAAG), Thalwil ZH164

formerly: A. Weidmann, Seidenfärberei (till 1933)
alternative: Färberei Thalwil

A (cloth) dyeing and finishing works, rail connected in 1896. The locomotive also performed shunting operations in Thalwil station under contract to the SBB. A. Weidmann died in 1928 and the new title was adopted in 1933. After further detail name changes production ceased in 2004.

Gauge: 1435 mm   Locn: ?   Date: 03.2008

  4 0-4-0WT Bn2t SLM E 2/2 725 1892 (a) (1)

(a) new to Sihltalbahn (SiTB) "2", renumbered "4" /1893; purchased from Sihltalbahn (SiTB) /1896.
(1) 1937 sold for scrap; ca. /1944 scrapped.

VEREIN TRAM-MUSEUM ZÜRICH ZH165 M

Wartau

A collection of tramway vehicles was initially established within this tram depot; one industrial locomotive was added.

Gauge: 1435 mm   Locn: 225 $ 6795/2507   Date: 12.2007

    4wWE Bg MFO   ? 1898 (a) (1)

Burgwies

On 26 May 2007 a new location was opened in the converted depot at Burgwies on the VBZ route used by the Forchbahn trains. Address: Forchstrasse 260, 8008 Zürich.

(a) ex SWISSMILL, Sihlquai, Zürich /1997.
(1) ?2007 to private collector.

  see. www.tram-museum.ch/frameset.html

GEBRÜDER VOLKART, Winterthur ZH166

This concern, founded in 1851, had offices in Winterthur and Bombay (India). Locomotives ordered from Winterthur and exported to India are excluded from this Handbook.

WEIACHER KIES AG, Weiach ZH167

Gravel pits belonging to the Haniel group, rail connected to the east of Zweideln SBB station and started in 1962. Address: Im Hard, 8187 Weiach.

Gauge: 1435 mm   Locn: 215 $ 6763/2689   Date: 09.2002

	4wDH	Bdh	RACO 100 SA7 HT	1595	1961	new	(1)
Tm 2/2 Nr 1 Julia	4wDE	Bde1	Moyse BN40E260D	1301	1975	(b)	OOU
Tm 2	4wDE	Bde1	Moyse BN34E210B	1410	1977	(c)	(3)
Tm 2/2 237923-8 Britta	4wDE	Bde1	Moyse BN40E260D	1264	1973	(d)	
237926-1 Veronika	4wDH	Bdh	Schöma CHL 350GR	5712	2002	(e)	

(b) new via Asper.
(c) ex Casam, Coutances [F] /1985 via Desbrugères/Asper.
(d) ex Krupp-Klöckner AG, Osnabrück [D] "2" /1989 via OnRail/LSB.
(e) new; another source gives the type as CFL 350GR.
(1) 1975 to Migros, Cadenazzo via Asper.
(3) 2001 to LSB.

WERKZEUGMASCHINENFABRIK BÜHRLE-OERLIKON (WBO), Zürich-Seebach ZH168

formerly: Schweizer Werkzeugmaschinenfabrik Oerlikon (SWO) (from 1906)

This factory makes machine tools, armaments and railway brakes. It was rail connected to Seebach station in 1907.

Gauge: 1435 mm   Locn: 225 $ 6829/2516   Date: 03.1988

	1APM	1Abm	Orion	-	1916	(a)	s/s
	A11ABE	A11Aa2	SIG,MFO	?	1944	new	(2)

(a) rebuild of a railcar built by Orion of Zürich and tested on the UeBB "20".
(2) 1980 sold to Baumgartner AG, Buchs ZH; 1984 scrapped at Stadler, Bussnang.

WETZIKON-MEILEN BAHN, Station, Grüningen ZH169 M

A motor coach from the Trogenerbahn has been restored and placed on display near the old WMB main station.

Gauge: 1000 mm   Locn: 226 $ 7000/2377   Date: 03.2008

	4w-4wWERC	SIG,MFO	?	1903	(a)	display

(a) new to TB "CFe 4/4 1"; converted to TB "Xe 4/4 23"; restored as WMB "3".

WOLF & GRAFS, Zürich ZH170 C

A contractor?

Gauge: 796 mm   Locn: ?   Date: 03.2002

BERTHA	0-4-0T	Bn2t	Heilbronn II alt	30	1874	new	(1)
CLARA	0-4-0T	Bn2t	Heilbronn II alt	31	1874	new	(1)

Gauge: 750 mm   Locn: ?   Date: 03.2002

RHEINFELDEN	0-4-0T	Bn2t	Heilbronn II alt	28	1873	new	(3)

(1)   1xxx sold to Grubitz & Ziegler, Solothurn.
(3)   18xx sold to Vicarino & Curty, Pratteln for Bötzbergbahn construction.

WOLF & WEISS, Zürich ZH171 C
Either a dealer in equipment for building sites or a public works firm.
Gauge: 600/750 mm Locn: ? Date: 03.2002
 0-4-0T Bn2t Jung 40 PS 98 1891 (a) s/s
(a) noted during the building of the RhB Albula line /1900.

ZEHNDER & CO, Zürich ZH172 C
A public works firm?
Gauge: 600 mm Locn: ? Date: 08.2002
 4wDM Bdm O&K MV0 25163 1952 (a) s/s
(a) purchased new via MBA, Dübendorf.

ZIEGELEI LUFINGEN, Lufingen ZH173
A brickworks.
Gauge: 600 mm Locn: ? Date: 06.1991
 4wDM Bdm O&K MD 2b ? 19xx (a) (1)
 4wDM Bdm O&K(M) RL 1a 4899 1933 (a) (1)
(a) ex ? /19xx.
(1) 198x donated to FWF, Otelfingen.

ZIEGELEI ROBERT MEIER, Riedikon ZH174
A brickworks.
Gauge: ? mm Locn: ? Date: 05.2002
 4wPM Bbm O&K(M) S5 1588 192x (a) s/s
(a) purchased via O&K/MBA, Dübendorf.

ZIEGLER-HUBER & CIE, Zürich ZH177
The following locomotive was hired, probably for use as a stationary boiler.
Gauge: 1000 mm Locn: ? Date: 02.2008
 0-4-0T Bn2t Jung 50 PS 920 1905 (a) (1)
(a) hired from Robert Aebi & Co AG /1938.
(1) returned to Robert Aebi & Co AG.

AG CONRAD ZSCHOKKE, Zürich ZH178 C
later: Zschokke-Holding
 Implenia AG (from 1 January 2006)
Conradin Zschokke founded this firm in 1909, which later became a world wide construction concern, till its merger with Batigroup in 2005-6. The SIG list of references gives four ETM 50 locomotives delivered here.

ED. ZÜBLIN & CIE AG, Zürich   ZH179 C

A public works company.

Gauge: ? mm   Locn: ?   Date: 03.2002

HELVETIA		0-4-0T	Bn2t	Heilbronn III	269	1891	(a) (1)

(a)   new to Messing, Laufenburg (900 mm); purchased from Baumann & Stiefenhofer, Wädenswil /1937.
(1)   1942 sold to ? [D] via King, Zürich-Seebach.

ZÜRCHER FREILAGER AG, Zürich-Albisrieden   ZH180

Warehouses and a duty free area. The steam locomotives are stored here for a private individual.

Gauge: 1435 mm   Locn: 225 $ 6793/2482   Date: 06.2007

	4wPM	Bbm	Breuer	?	c1930	(a)	s/s
	4wDH	Bdh	Jung ZN 233	13061	1958	new	(2)
Tm 2/2	4wDE	Bde	Stadler	139	1972	(c)	
GWZ 2	0-6-0WT	Cn2t	Krauss-S XLV ac	5331	1905	(d)	
3	0-6-0WT	Cn2t	SLM E 3/3	1359	1901	(e)	

(a)   mentioned in the Breuer list of references of about 1930, but not the one of 1928.
(c)   new as "2".
(d)   new to Gaswerk Zürich, Werk Schlieren "2"; ex Schweizerische Reederei AG, Basel Kleinhüningen /1973; in private ownership.
(e)   new to SCB "41"; later SBB "8410"; ex von Moos Stahl, Emmenbrücke "3" /1973; in private ownership.
(2)   1975 to SOB "32"; later Schweizerisches Militärmuseum (SMM), Full.

ZÜRCHER MUSEUMSBAHN (ZMB), Sihlwald   ZH181 M

A museum railway operating on the erstwhile Sihltalbahn (SiTB). Not all the stock is kept at Sihlwald. Address: Rämistrasse 7, 8024 Zürich.

Gauge: 1435 mm   Locn: 225 $ 6847/2359   Date: 12.2008

2	0-6-0WT	Cn2t	SLM E 3/3	795	1893	(a)	
	0-6-0WT	Cn2t	SLM E 3/3	1221	1900	(b)	
Tm 9 MAX	4wDM	Bdm	Kronenberg DT 90/120	126	1952	(c)	
Tm 2/2 10	4wDM	Bdm	RACO 45 PS	1380	1949	(d)	
De 3/4 41	Bo1AWE		SIG,MFO	?	1925(31)	(e)	
FCe 2/4 84	1AA1WERC		SWS,MFO	?	1924	(f)	
WM BDe 2/4 3	1AA1WERC		SWS,SAAS	?	1938	(g)	
SZU 576 592	BoBoWERC		SWS,MFO	?	1968	(h)	

(a)   new to Sihltalbahn (SiTB) "2"; 1924 to Gaswerk Basel "GAS 2"; ex Rive Bleu Express, Le Bouveret "HANSLI" /2005.
(b)   new to Sihltalbahn (SiTB) "5".
(c)   ex OKK, Niedergesteln "Tm 7" /1999 via OKK, Oberburg.
(d)   new to SBB "Tm 536" as 4wPM Bbm; ex SZU "10" (second SZU tractor with this number).
(e)   new to Sihltalbahn (SiTB) "De 3/3 1"; rebuilt as "De 3/4 2" /1931; renumbered "De 3/4 41" /1973.
(f)   new to Sihltalbahn (SiTB).
(g)   new to Sensetalbahn (STB) "101"; to be disposed of end/2009.
(h)   new to SZU "BDe 4/4"; to be transferred third quarter /2009.

see: www.museumsbahn.ch

ZÜRCHER PAPIERFABRIK AN DER SIHL AG, Zürich-Manegg
ZH182

formerly: Mechanische Papierfabrik an der Sihl
trade name :   "Sihl"
later: Sihl-Eika Papier AG

A paper factory rail connected to the Sihltalbahn (SiTB) in 1893. Since 1991 a road/rail locomotive has performed shunting (details unknown). Address: Giesshuebelstrasse 15, 8045 Zürich.

Gauge: 1435 mm   Locn: 225 $ 6812/2443   Date: 05.1988

4wDM	Bdm	JW DM50V10	51.105	1952	(a)	(1)

(a)   new to Zellulosewerk Sankt Michael [A]; ex ? /1980.
(1)   1991 sold to Asper but stored at Manegg till 1992.

ZÜRCHER ZIEGELEIEN AG (ZZ)   ZH183

Ziegelei Rafz

formerly: Ziegelei Schmidheiny
later: ZZ Wancor

This brickworks lies about a kilometre from the village centre; the railway system has been abandoned. It connected the quarry with the brickworks and also the works with the SBB station. Address: Landstrasse 75, 8197 Rafz.

Part of the railway system in Ziegelei Rafz seen on 23 March 1968.

Gauge: 600 mm   Locn: 215 $ 6836/2745   Date: 02.1989

0-6-0T	Cn2t	Maffei	3508	1909	(a)	(1)
4wDM	Bdm	O&K(M) RL 1	?	19xx		
4wDM	Bdm	O&K(M) MD 2	?	19xx		
4wDM	Bdm	O&K(S) RL 1c	1501	1947	(d)	
4wDM	Bdm	Diema DS28	2581	1962	(d)	(5)

Ziegelei Brunau, Zürich-Giesshübel

Ziegelei Albishof-Heurieth AG was formed in 1907 by the grouping of Ziegelei Albishof and Ziegelei Heurieth. Then in 1912 the fusion of Ziegelei Albishof-Heurieth AG and Mechanische Backsteinfabrik, Zürich led to the formation of Zürcher Ziegeleien AG. The railway system was abandoned ca. 197x.

Built under Deutz licence, Spoorijzer 60001 of 1960 (type A2L514) stands in retirement with the FWF, Otelfingen collection on 29 April 2000. The diamond-shaped licence plate is just visible below the radiator.

Gauge: 600 mm　　　　　　Locn: 225 $ 6810/2463　　　　　　　　　　Date: 02.1989

1	?	?	O&K(M)?	?	19xx		s/s
4	4wDM	Bdm	O&K(S) MD 2	1520	1947	(d)	s/s
14	0-4-0DM	Bdm	Spoorijzer A2L514	60001	1960	(h)	(8)
(28)	4wDM	Bdm	Diema DS28	2580	1962	new	(9)
	4wDM	Bdm	Diema DS28	2581	1962	(j)	(10)
78	4wDM	Bdm	Diema DS28	2702	1964	(k)	(11)

Ziegelei Tiergarten, Zürich Giesshübel

Gauge: 600 mm　　　　　　Locn: ?　　　　　　　　　　　　　　　Date: 05.2008

3	4wDM	Bdm	O&K(M)	?	19xx		s/s
5	4wDM	Bdm	unknown	?	19xx		s/s
6	4wDM	Bdm	MBA?	?	19xx		s/s
11	4wDM	Bdm	KHD A2L514 F	55764	1954	new	s/s
12	4wDM	Bdm	KHD A2L514 F	56807	1957	(p)	s/s
	4wDM	Bdm	Diema DTL90/1.3	3199	1971	(q)	(17)
	4wDM	Bdm	Diema DTL90/1.3	3200	1971	(q)	(17)

Zürich

This section is reserved for locomotives delivered to the company, but for which no record of the place of use has been found.

Gauge: 600 mm Locn: ? Date: 04.2009

4wDM	Bdm	O&K(M) RL 1	3644	1929	(s)	s/s
4wDM	Bdm	O&K(M) RL 1a	4651	1932	(s)	s/s
4wDM	Bdm	O&K(M) RL 1a	4686	1932	(s)	s/s
4wDM	Bdm	O&K(M) RL 1a	4880	1933	(s)	s/s
4wDM	Bdm	O&K(M) RL 1a	5364	1934	(s)	s/s
4wDM	Bdm	O&K(M) RL 1c	8423	1937	(s)	s/s

(a) see note for this locomotive under Robert Aebi & Cie AG, Regensdorf, new as 750 mm; ?hired from Robert Aebi & Co, Zürich /1917.
(d) purchased via MBA.
(h) new via P. Schlatter, Münchwilen AG; engine 275444/45.
(j) transferred from Ziegelei Rafz /1975; not used at Giesshübel.
(m) new to Schmidheiny & Co, Ziegelei, Heerbrugg (500 mm); ex ? /197x; not used at Giesshübel.
(p) imported via Lagerhausgesellschaft, Basel; also reported as "Tm 2/2 1".
(q) new via Asper.
(s) new via O&K, Zürich.
(1) 1920 returned to Robert Aebi & Co, Zürich.
(5) 1975 transferred to Ziegelei Giesshübel; but not used there.
(8) 19xx to P. Schlatter, Münchwilen AG; later FWF, Otelfingen.
(9) by 1969 sold to Eternit AG, Niederurnen.
(10) 19xx to Vereinigte Ziegelwerke, Schwenningen [D]; 1983 to Baustoffwerke Mühlacker, Mühlacker [D]; later Heddesheimer Feldbahn, Guldental [D].
(11) 19xx to Belloli, Grono; later Oliver Weder, Diepoldsau; later FWF, Otelfingen.
(17) 1977 sold to ? [Venezuela] via Asper.

Miscellaneous

This section lists two types of concerns. The first is those that have operations in more than one canton and for which the locomotives are for one reason or another peripatetic so that a listing by canton is likely to be out-of-date by compilation date. These locomotive listings are not complete; please refer to the appropriate web-sites for more details. The second type is for unknown concerns or locations.

CLASSIC RAIL   MI01 M

This is an association of groups with heritage material. Individual items may be in store, hired to other concerns or operational. Some locomotives are hired out for extended periods. There are several locations, for instance Fleurier, Glarus, Holderbank, Münchenstein, Oensingen, Prattlen, Romanshorn, Sulgen and Winterthur. Administrative Address: Baarerstrasse 73, 6302 Zug.

Gauge: 1435 mm   Locn: various   Date: 03.2008

Tm^1 419	4wDM	Bdm	CMR Tm^l	'Tm^l 319'	1961	(a)	
RFe 4/4 603	BoBoWE	BBw4	SLM,BBC,MFO RFe 4/4	3695, ?	1938	(b)	
MThB Re 416 625-2	BoBoWE	BBw4	SLM,MFO Re 4/4^l	3878, ?	1944	(c)	
MThB Re 416 626-0	BoBoWE	BBw4	SLM,SAAS Re 4/4^l	3890,5650-3	1945	(d)	(4)
MThB Re 416 627-8	BoBoWE	BBw4	SLM,MFO Re 4/4^l	3897, ?	1945	(e)	
MThB Re 416 628-6	BoBoWE	BBw4	SLM,MFO Re 4/4^l	4018, ?	1950	(f)	
Re 4/4^l 10042	BoBoWE	BBw4	SLM,MFO Re 4/4^l	4021, ?	1950	(g)	
Re 4/4^l 10046	BoBoWE	BBw4	SLM,SAAS Re 4/4^l	4025, ?	1950	(h)	
Ae 3/6^l 10639	2Co1WE	2C1w3	SLM,BBC Ae 3/6^l	2982,2083	1924,25		
Ae 4/7 10908	2Do1WE	2D1w4	SLM,BBC Ae 4/7	3179,2528	1927		
Ae 4/7 10914	2Do1WE	2D1w4	SLM,BBC Ae 4/7	3238,2818	1927		
Ae 4/7 10922	2Do1WE	2D1w4	SLM,MFO Ae 4/7	3246, ?	1927		8/2008 scr
Ae 4/7 10943	2Do1WE	2D1w4	SLM,SAAS Ae 4/7	3425,1700-1	1930		
Ae 4/7 10948	2Do1WE	2D1w4	SLM,SAAS Ae 4/7	3445,1700-6	1930		
Ae 4/7 10950	2Do1WE	2D1w4	SLM,SAAS Ae 4/7	3451,1700-8	1930		(4)
Ae 4/7 10951	2Do1WE	2D1w4	SLM,SAAS Ae 4/7	3456,1700-9	1930		
Ae 4/7 10961	2Do1WE	2D1w4	SLM,BBC Ae 4/7	3442,3355	1930		
Ae 4/7 10999	2Do1WE	2D1w4	SLM,MFO Ae 4/7	3541, ?	1932		
Ae 4/7 11001	2Do1WE	2D1w4	SLM,MFO Ae 4/7	3546, ?	1932		
Ae 4/7 11002	2Do1WE	2D1w4	SLM,MFO Ae 4/7	3547, ?	1932		
Ae 4/7 11010	2Do1WE	2D1w4	SLM,SAAS Ae 4/7	3488,2052-2	1931		(4)
Ae 4/7 11015	2Do1WE	2D1w4	SLM,SAAS Ae 4/7	3540,2525-3	1932		
Ae 4/7 11022	2Do1WE	2D1w4	SLM,BBC Ae 4/7	3537,3719	1932		
BLS De 4/5 796	A1ABoWE	A1ABw4	SLM,SAAS CFe 4/5	3290, ?	1928	(x)	(24)
BLS Ae 6/8 206	1CoCo1WE	1CC1w12	SLM,SAAS Ae 6/8	3679, ?	1939		
EBT Be 4/4 104	BoBoWE	BBw4	SLM,SAAS Be 4/4	3554, ?	1932		
EBT Be 4/4 105	BoBoWE	BBw4	SLM,SAAS Be 4/4	3555, ?	1932		
BLS Ee 3/3 401	0-6-0WE	Cw1	SLM,SAAS Ee 3/3	3806,4840	1942	(28)	
Ee 3/3 16315	0-6-0WE	Cw1	SLM,BBC Ee 3/3	3192,2537	1926,7		
Ee 3/3 16316	0-6-0WE	Cw1	SLM,BBC Ee 3/3	3193,2538	1926,7		
Ee 3/3 16325	0-6-0WE	Cw1	SLM,BBC Ee 3/3	3232,2815	1928		scr?
Ee 3/3 16326	0-6-0WE	Cw1	SLM,BBC Ee 3/3	3233,2816	1928		

(a)   new to SBB "Tm^l 319".
(b)   into service /1940; sold to SOB "21" /1944; withdrawn 12/1996.
(c)   new to SBB "402"; renumbered "Re 4/4^l 10002" 27/11/1959.
(d)   new to SBB "409"; renumbered "Re 4/4^l 10009" 17/07/1959.
(e)   new to SBB "416"; renumbered "Re 4/4^l 10016" 2/06/1959.
(f)   new to SBB "439"; renumbered "Re 4/4^l 10039" 24/12/1959.
(g)   new to SBB "442"; renumbered "Re 4/4^l 10042" 8/07/1960.
(h)   new to SBB "446"; renumbered "Re 4/4^l 10046" 6/09/1962.
(x)   new to BLS "CFe 4/5 726".

(4)   2007 hired to Rail4Chem via Taurino Traction.
(24) 2008 offered for scrap.
(28) 8/2008 scrapped at Kaiseraugst.

see: www.privat-bahn.de/Classic_Rail.html
see: www.elektrolok.de/Museum/museum_sbb-nostalgie.htm

L. FAVRE   MI02

This railway construction concern and agent took delivery of the following locomotives, before starting on the Gotthard tunnel (1871). The concern had offices in Genève and Lausanne, but where these locomotives were employed is not known.

Gauge: 1435 mm   Locn: ?   Date: 05.2008

0-4-0T	Bn2t	Anjubault	66	1862	new	
0-6-0T	Cn2t	Schneider 42	799	1859	new	
0-6-0T	Cn2t	Schneider 42	800	1859	new	
0-6-0T	Cn2t	Schneider 64	1113	1867	(d)	

Gauge: 1000 mm   Locn: ?   Date: 05.2008

0-4-0T	Bn2t	Schneider	1055	1867	(d)

(d)   new, locomotive recorded as delivered to Favre, Genève.

LINDER & FAVETTO, Unknown location   MI03 C

A public works concern, probably connected with Favetto & Catolla, Brunnen.

Gauge: ? mm   Locn: ?   Date: 03.2002

0-4-0T	Bn2t	O&K 50 PS	391	1900	(a)	(1)

(a)   new via O&K Strasbourg [F].
(1)   1901-06 to Favetto & Catolla, Brunnen.

MAZOT & SIMARD, Unknown location   MI04 C

An entry exists in the SLM lists for a locomotive that was also suitable for metre gauge.

Gauge: 1435 mm   Locn: ?   Date: 02.2008

0-4-0T	Bn2t	SLM	182	1880	new	s/s

MINDER & GALLI, Unknown location   MI05 C

formerly: Galli & Cie
            Ing. Galli & Co

A public works concern with several branches in Switzerland.

Gauge: 750 mm   Locn: ?   Date: 03.2002

ANGELO	0-4-0T	Bn2t	Heilbronn II	274	1891	(a)	(1)
ALBIS	0-4-0T	Bn2t	Krauss-S IV cd	2647	1893	(b)	(2)
	0-4-0T	Bn2t	Krauss-S IV kl	2960	1893	(c)	(3)
EOLO	0-4-0T	Bn2t	Krauss-S IV tu	3190	1895	(d)	(4)
	0-4-0T	Bn2t	O&K 30 PS	375	1899	new	(5)
	0-4-0T	Bn2t	O&K 50 PS	529	1900	new	s/s
	0-4-0T	Bn2t	O&K 30 PS	700	1900	(g)	(7)
	0-4-0T	Bn2t	O&K 30 PS	985	1902	(h)	(8)
	0-4-0T	Bn2t	O&K 40 PS	1086	1903	(i)	(9)

(a)   ex Aebli, Rossi & Krieger, Schaffhausen /1902 via Minder & Galli, Seftigen.
(b)   new to Galli.
(c)   ex Vogel & Frei, Widnau to Galli & Cie for the building of the Fussacher Durchstich (IRR).

(d) delivered new to Immensee for either Ing. Galli, Willisau or Galli & Cie Bauunternehmer, Meggen probably for the construction of the Luzern-Immensee railway line.
(g) new to Ing. Galli & Co via O&K Strasbourg [F].
(h) new via O&K, Strasbourg, [F].
(i) new to Minder & Galli, Engelsberg [possibly Engelberg].
(1) 1902-20 to Dr. G. Lüscher Ing. Bauunternehmer, Aarau.
(2) before 1908 sold for construction of the St. Gallen-Wattwil section of the Bodensee-Toggenburg railway.
(3) before 1897-1900 to St. Rossi, St Gallen.
(4) 1897 sold to ? & Cie, Ing., Grosshöchstetten.
(5) 1899-1925 to Frutiger & Lanzrein, Bern.
(7) 1900-1910 for construction of the St. Gallen-Wattwil railway.
(8) 1923 to O&K, Zürich.
(9) 19xx to Hunziker & Co, Zürich.

A. RENNER, Unknown location   MI06 C
A public works concern?

Gauge: 730 mm   Locn: ?   Date: 03.2002

	0-4-0T	Bn2t	O&K 50 PS		726	1900	new

SBB-HISTORIC   MI07 M

Formed in 1999 when the SBB became a commercial operation. Locomotives may be found almost anywhere on the system. The stock is increasing. A base and workshops were established in Olten. Other work is carried out in Brugg AG, Erstfeld, Winterthur and Delémont. Some items are on long or short-term loan to the VHS, Luzern while others are in store. Some items may be loaned to groups at Rapperswil, Rorschach, Balsthal and elsewhere. Operational items may be located as and when required for a period. SBB Historic plan to be shortly acquiring additional Ae6/6. These locos have not yet been confirmed but could include some or all of 11411, 11425 and 11490.

Gauge: 1435 mm   Locn: various   Date: 11.2008

SCB Ec 2/5	28 GENF	0-4+6	B3n2	Esslingen Engerth	396	1858		(V)
5469	(JS 35)	4-4-0T	2Bnt	Esslingen	2498	1892		
Ed 2*2/2	196	0-4-4-0T 4cc BBn4vt		Maffei Duplex-Compound	1710	1893	(c)	(V)
UeBB CZm 1/2	31	2-2-0TRC	A1	Esslingen	- 1902(07,80)			
A 3/5	705	4-6-0 4cc	2Cn4v	SLM A 3/5	1932	1908	(e)	
Eb 3/5	5819	2-6-2T	1C1h2t	SLM Eb 3/5	2220	1912		
B 3/4	1367	2-6-0	1Ch2	SLM B 3/4	2557	1916		(V)
C 5/6	2965	2-10-0 4cc	1Eh4v	SLM C 5/6	2518	1916		(V)
C 5/6	2978	2-10-0 4cc	1Eh4v	SLM C 5/6	2612	1917		(D)
E 3/3	5	0-6-0T	Ch2t	SLM E 3/3	3610	1936	(j)	(E)
D1/3	1 LIMMAT	4-2-0	2An2	SLM C 1/3	3937	1946	(k)	(V)
SW Ce 4/4	1	0-4-4-0WE	BBw2	SLM,MFO Ce 4/4	1524, 1	1903	(l)	(V)
SW Ce 4/4	2	0-4-4-0WE	BBw2	SLM,MFO, ?CIEM Ce 4/4 1665, 2		1904	(m)	
Tm^I	475	4wDM	Bdm	RACO 85 LA7	1688	1964		
Ta	971	4wBE	Ba	Olten	?	1927		(V)
RAe 2/4	1001	Bo2WERC	B2w2	SLM,SBB,BBC,MFO,SAAS CLe 2/4 3581,3983 1935,36			(p)	
RBe 2/4	1003	Bo2WERC	B2w2	SLM,SBB,BBC,MFO,SAAS CLe 2/4 3604,3985		1936	(q)	(V)
RAe 4/8	1021	2 carWERC		SLM,SWS,MFO,SAAS RAe 4/8 3689, ?		1939		
RAe^{II}	1053	6 carWERC		SIG,MFO	?	1961	(s)	

Class	Number	Wheel arr.		Builder	Works no.	Year		Notes
BDe 4/4	1643	BoBoWERC	BBw4	SLM,MFO CFe 4/4	4105,5441	1954	(t)	(W)
BDe 4/4	1646	BoBoWERC	BBw4	SLM,SAAS CFe 4/4	4108,5444	1954	(u)	(E)
De 4/4	1679	BoBoWE	BBw4	SAAS,SWS	?	1928		
Re 4/4^I	10001	BoBoWE	BBw4	SLM,MFO Re 4/4^I	3877, ?	1944	(w)	(O)
Re 4/4^I	10044	BoBoWE	BBw4	SLM,SAAS Re 4/4^I	4023, ?	1950	(x)	(R)
Ae 3/5	10217	1Co1WE	1C1w6	SLM,SAAS Ae 3/5	2948,600003	1924		(B)
Ae 3/6^{II}	10264	2Co1WE	2C1w6	SLM,SAAS Ae 3/6^{III}	3061, ?	1925		
Ae 3/6^{III}	10439	4-6-2WE	2C1w2	SLM,MFO Ae 3/6^{III}	3092, ?	1925	(aa)	(O)
Ae 3/6^I	10650	2Co1WE	2C1w3	SLM,BBC Ae 3/6^I	3049,2111	1925		(28)
Ae 3/6^I	10664	2Co1WE	2C1w3	SLM,BBC Ae 3/6^I	3088,2125	1925		(Z)
Ae 3/6^I	10700	2Co1WE	2C1w3	SLM,MFO Ae 3/6^I	3159, ?	1927		
Ae 4/7	10905	2Do1WE	2D1w4	SLM,BBC Ae 4/7	3176,2525	1927		
Ae 4/7	10949	2Do1WE	2D1w4	SLM,SAAS Ae 4/7	3449,1700-7	1930		(A)
Ae 4/7	10976	2Do1WE	2D1w4	SLM,MFO Ae 4/7	3427, ?	1930		
Ae 6/6	11402 URI	CoCoWE	CCw6	SLM,BBC Ae 6/6	4051,5430	1951		(E)
Ae 6/6	11403 SCHWYZ	CoCoWE	CCw6	SLM,BBC Ae 6/6	4138,6004	1955		(L)
Ae 6/6	11407 AARGAU	CoCoWE	CCw6	SLM,BBC Ae 6/6	4142,6008	1955		(O)
Ae 6/6	11411 ZUG	CoCoWE	CCw6	SLM,BBC Ae 6/6	4146,6012	1955	(aj)	(E)
Ae 6/6	11416 GLARUS	CoCoWE	CCw6	SLM,BBC Ae 6/6	4234,6051	1957		(28)
Ae 6/6	11425 ZÜRICH	CoCoWE	CCw6	SLM,BBC Ae 6/6	4244,6060	1957		(O)
Ae 8/14	11801	(1A)A1A(A1)+(1A)A1A(A1)WE 2x(1B1B1)w8		SLM,BBC Ae 8/14	3501,3360	1931		(E)
Ae 8/14	11852	(1A)A1A(A1)+(1A)A1A(A1)WE 1B1B1+1B1B1w8		SLM,MFO Ae 8/14	3685, ?	1938		(V)
Be 4/6	12320	2-4-4-2WE	1BB1w2	SLM,BBC Be 4/6	2762,1592	1920,21		(W)
Be 4/6	12332	2-4-4-2WE	1BB1w2	SLM,BBC Be 4/6	2809,1804	1922		(E)
Be 4/7	12504	1Bo1Bo1WE	1B1B1w8	SLM,SAAS Be 4/7	2716,94004	1920		
Be 6/8^{II}	13254	2-6-6-2WE	1CC1w2	SLM,MFO Ce6/8^{II}	2674, ?	1919	(aq)	(V)
Ce 6/8^I	14201	2-6-6-2WE	1CC1w4	SLM,MFO Be 6/8^I	2649,1205	1918,19		(E)
Ce 6/8^{II}	14253	2-6-6-2WE	1CC1w2	SLM,MFO Ce 6/8^{II}	2673, ?	1919		(E)
Ce 6/8^{III}	14305	2-6-6-2WE	1CC1w2	SLM,MFO Ce 6/8^{III}	3076, ?	1925		
Bm 4/4^{II}	18451	BoBoDE	BBDE	SLM,BBC Am 4/4	3691, ?	1939	(au)	(O)
Fe 4/4	18518	BoBoWE	BoBoDE	SAAS,SWS Fe 4/4	999019, ?	1928		(V)
Tem^{II}	281	0-4-0W+DEBw2		Tuchschmid,MFO Tem^{II}	?	1967	(aw)	
Tm^I	480	4wDM	Bdm	RACO 85 LA7	1698	1964	(aw)	
Tm^I	453	4wDM	Bdm	RACO 85 LA7	1639	1962	(av)	(U)

Gauge: 1000 mm   Locn: ?   Date: 11.2007

| Deh 4/6 | 914 | Bo2BoRWE | B2bBw6 | SLM,SAAS Fhe 4/6 | 3731, ? | 1941 | (ba) | |

(c)   ex SBB "7696".
(e)   frame of "A3/5 778", boiler of "A3/5 739" (SLM 1754 1906) and plates of "A3/5 705" (SLM 1550 1904).
(j)   new to HWB "5"; ex Sulzer AG, Winterthur /1981.
(k)   replica of Kessler 78 of 1847 for the centenary celebrations in 1947; this is sometimes quoted as the building date.
(l)   locomotive for experimental electrification Seebach-Wettingen.
(m)   locomotive for experimental electrification Seebach-Wettingen; carries an MFO plate, but also reported to have CIEM participation.
(p)   new to SBB "CLe 2/4 201"; SLM used this works number twice.
(q)   new to SBB "RCe 2/4 203"; "603" /1947-58; ex SBB "RBe 2/4 1003".
(s)   includes power car from "1055" from 16/12/1986 and replacement driving trailer "1053/1" from /1963.

(t) BBC allocated a number, but the electrical part was provided by MFO; new to SBB "CFe 4/4 863"; reclassified "BDe 4/4 1643" 4/1959.
(u) BBC allocated a number, but the electrical part was provided by SAAS; new to SBB "CFe 4/4 866"; reclassified "BDe 4/4 1646" 3/1961.
(w) new to SBB "Re 4/4 401"; renumbered "Re 4/4^I 10001" 2/11/1959.
(x) new to SBB "Re 4/4 444"; renumbered "Re 4/4^I 10044" 19/12/1962.
(aa) plates of "Ae 3/6^{II} 10439" (SLM 3013 1934) mounted on "Ae 3/6^{III} 10452".
(aj) new to SBB "Ae 6/6 11411"; ex 30/4/2009.
(aq) ex "Ce6/8^{II} 14254" /1944.
(au) new to SBB "Am 4/4 1001"; renumbered "18451" /1959; reclassified "Bm^{II} 4/4 18451" /1963.
(aw) ex SBB "Tm^I 480"; withdrawn 11/2008.
(ay) new to SBB "Tm^I 353"; ex SBB "Tm^I 453"; withdrawn 4/2009.
(ba) new to SBB "Feh 4/6 916"; entered service /1942.
(28) to be cannibalised for spares.
(A) long term loan to Bahnpark, Augsburg [D].
(B) based in Bern in 2009.
(D) based in Delémont in 2008.
(E) based in Erstfeld in 2008.
(L) stored in Bellinzona in 2009.
(O) based in Olten in 2008.
(S) on loan to Serge Bourginet, Schwyz for restoration.
(U) based at Vallorbe in 2009.
(V) long term loan to Verkehrshaus der Schweiz (VHS), Luzern.
(W) based in Winterthur in 2008.
(Z) long term loan to Eisenbahnfreunde Zürichsee rechtes Ufer, Rapperswil.

see: www.sbbhistoric.ch/index.cfm?43581BFF2B351571EFE10393ED551018
see: www.elektrolok.de/Museum/museum_sbb-nostalgie.htm
see: www.ae47.ch/ae66/html/03_projekt.htm

SWISSTRAIN MI08 M

An association formed in 1998 by Christoph Bachmann with the objective of preserving and repairing small railway locomotives. It absorbed the collection of Betriebsgesellschaft Historischer Schienenfahrzeuge (BHS) and was part of Classic Rail. Initially the locomotives were collected at Liesberg; from 1988 most moved to Bodio and then, starting in 2001, to Le Locle. Others were located in, for instance, Murten and Holderbank. In 2008 the group became independent, purchased the CFF depot in Payerne and hires the one in Le Locle. Address: Case Postale 75, 2400 Le Locle.

Gauge: 1435 mm   Locn: various   Date: 06.2008

M-O TM 511		4wDM	Bdm	RACO 45 PS	1248	1937	(a)	(L)
OKK	4	4wDM	Bdm	RACO RA11	1387	1949	(b)	(L)
BTA	3	4wDH	Bdh	RACO 80 ST4	1408	1951	(c)	(L)
	8	4wDH	Bdh	RACO 80 ST4	1422	1953	(d)	
		4wDM	Bdm	RACO 60 LA7	1497	1957	(e)	(5)
Tm^{II}	777	4wDM	Bdm	RACO 95 SA3	1433	1953	(g)	
	3	4wPM	Bbm	Kronenberg DT 60	116	1950	(h)	(L)
	2	4wDM	Bdm	Kronenberg DT 60/90P	121	1951	(i)	(L)
		4wDM	Bdm	Kronenberg DT 90	145	1962	(j)	(10)
		4wDM	Bdm	Breuer III	? 1931(69)		(k)	(11)
		4wPM	Bbm	Breuer V	3054	1952	(l)	(L)
	2	4wDM	Bdm	Gebus Lokomotor	565	1957	(m)	(13)
GWZ	3	0-6-0WT	Cn2t	SLM	1902	1908	(n)	(L)
PTT	7	0-6-0WE	Cw1	SLM,BBC Ee 3/3	3188,2533	1926,27	(o)	(P)
		0-4-0DM	Bdm	SLM Em 2/2	3412	1930	(p)	
	13	4wPM	Bbm	SLM Tm	3579	1934	(q)	
"MAX"	3	0-4-0DM	Bdm	SLM Tm	3690	1938	(r)	
		4wBE	Ba	Olten?,MFO	?	c1932	(s)	(21)
		1AWE	1Ag1	ACMV,BBC	? ,1338	1921	(t)	(L)
Te^{II}	221	0-4-0WE	Bw1	SLM/SAAS Te 2/2	3186/818001	1926	(u)	(P)
RFe 4,4	601	BoBoWE	BBw4	SLM,BBC,MFO RFe 4/4	3693, ?	1938	(v)	(P)

BDe 4/4 1632	BoBoWERC BBw4	SLM,MFO CFe 4/4	4063,4592	1952	(w)	(B)		
Ae 3/6^{II} 10448	4-6-2WE	2C1w2	SLM,MFO Ae 3/6^{II}	3022,?	1924			(BL)
Ae 3/6^{I} 10601	2Co1WE	2C1w3	SLM,BBC Ae 3/6^{I}	2740,1541	1920,21	(y)	(L)
Ae 3/6^{I} 10639	2Co1WE	2C1w3	SLM,BBC Ae 3/6^{I}	2982,2083	1924,25		(P)
Ae 3/6^{I} 10693	2Co1WE	2C1w3	SLM,MFO Ae 3/6^{I}	3152,?	1927		(P)
Ae 4/7 10902	2Do1WE	2D1w4	SLM,BBC Ae 4/7	3133,2522	1926,27		(P)
Ae 4/7 10987	2Do1WE	2D1w4	SLM,MFO Ae 4/7	3485,?	1931		(L)
Ae 4/7 11000	2Do1WE	2D1w4	SLM,MFO Ae 4/7	3543,?	1932		(L)
Ee 3/3 16313	0-6-0WE	Cw1	SLM,BBC Ee 3/3	3190,2535	1926,7		(H)
Ee 3/3 16332	0-6-0WE	Cw1	SLM,BBC Ee 3/3	3396,3315	1930		(L)
SBB Ee 6/6 16801	0-6-6-0WE	CCw2	SLM,SAAS Ee 6/6	4038,5427	1950	(af)	(L)
BLS Ae 6/8 208	1CoCo1WE 1CC1w12		SLM,SAAS Ae 6/8	3797,?	1942	(ag)	(L)
BLS Ce 4/4 311	0-4-4-0WE	BBw2	SLM,BBC	2693,1319	1919,20	(ah)	(M)
BLS Ce 4/4 315	0-4-4-0WE	BBw2	SLM,BBC	2843,1588	1922	(ai)	(L)
BLS Ce 4/4 316	0-4-4-0WE	BBw2	SLM,BBC	2844,1589	1922	(aj)	(10)
BT Be 4/4 12	BoBoWE	BBw4	SLM,SAAS Be 4/4	3510,?	1931		(L)
OeBB RBe 2/4 202	Bo2WERC	B2w2	SLM CLe 2/4	3634	1937	(al)	(B)
SMB Be 4/4 171	BoBoWE	BBw4	SLM,SAAS	3557,?	1932	(am)	(D)
EBT Be 4/4 101	BoBoWE	BBw4	SLM,SAAS Be 4/4	3551,?	1932		(M)
Tm^{III} 9526	4wDH	Bdh	RACO 225 SV4 H	1931	1986	(ao)	(P)

(a)　new to SBB "Tm 325"; to MO "TM 511" 8/1964; ex Asper /1997.
(b)　ex OKK, Altdorf UR /1999.
(c)　new to OKK; Zollikofen "3"; loan from LSB /1997.
(d)　ex BTA, La Plaine /1999.
(e)　new to SBB "Tm^{I} 302"; ex LSB /1995 (not confirmed that Swisstrain was involved).
(g)　ex SBB "Tm^{II} 777" by /2000.
(h)　ex Eidgenössische Alkoholverwaltung Romanshorn /2000.
(i)　ex Chemische Fabrik Schweizerhall (CFS), Basel St-Johann /1995.
(j)　new to Firestone, Pratteln "1"; ex Borner AG, Trimbach /1995 via BHS.
(k)　ex COOP, Olten /1995.
(l)　new to Zollfreilager, St. Margrethen; ex VVT, St. Sulpice NE /1996.
(m)　ex Migros, Birsfelden /1995 via HEG.
(n)　ex Gaswerk der Stadt Zürich, Schlieren; ZH /1995.
(o)　new to SBB "E 3/3 16311"; ex Die Schweizerische Post, Däniken SO "7" /1997.
(p)　new to PTT, Zürich "1"; ex PTT, Olten /1997.
(q)　new to CFF "Tm 214"; ex BTR-Prébéton SA, Avenches /1998.
(r)　ex Novartis Services AG, Basel Kleinhüningen /1999.
(s)　ex von Roll, Bern /1997.
(t)　ex voie industrielle de la Guinguette, Vevey by /2007.
(u)　ex SBB "Te 221" /1998.
(v)　into service /1940; sold to BT "25" /1944; sold to SZU "51" /1977; withdrawn and sold to OeBB /1994.
(w)　new to SBB "CFe 4/4 852"; reclassified "BDe4/4 1632" 7/1960; BBC allocated a number, but the electrical part was provided by MFO.
(y)　new to SBB "10301"; renumbered "Ae 3/6^{I} 10601" /1924; ex SBB /?2008 via Classic Rail.
(af)　BBC allocated a number, but the electrical part was provided by SAAS.
(ag)　into service /1943.
(ah)　new to GBS "Ce 4/6 311".
(ai)　new to BN "Ce 4/6 315".
(aj)　new to BN "Ce 4/6 316".
(al)　new to SBB "CLe 2/4 207" /1938; reclassified "RCe 2/4 607" /1949; "RBe 2/4 607" /1956; renumbered "RBe 2/4 1007" /1959; ex SBB /1974.
(am)　new to EBT "107"; ex RM.
(ao)　new to SBB "Tm^{III} 9526"; received from Rail Arena, Brittnau-Wikon /2009.

(5)　1998 to private ownership; 2002 scrapped.
(11)　by 10/2007 to Bahn Museum Kerzers (BMK), Kerzers.
(10)　8/2008 scrapped at Kaiseraugst.
(13)　1997 to Veluwsche Stoomtrein Maatschappij (VSM), Beekbergen [NL].
(21)　by 2008 to Rail Arena, Brittnau-Wikon.

(B)　at Balsthal in 2008.
(BL)　at Biel in 2008.

(D) on loan to Dampf Bahn Bern, Burgdorf /2008.
(H) at Holderbank 2008.
(L) at Le Locle 2007-8.
(M) at Le Locle 2007-8 as a spare parts bank.
(P) at Payerne 2008-9.

see: www.swisstrain.ch

UNKNOWN OPERATOR, Unknown location   MI09

Gauge: 600 mm   Locn: ?   Date: 04.2009

4wDM	Bdm	JW JW8	8094	1951	new	
4wDM	Bdm	JW JW8	8257	1955	new	

Gauge: 750 mm   Locn: ?   Date: 05.2009

4wDH	Bdh	Schöma CFL 200DCL	4763	1984	(c)	(3)

(c) new to Alfred Kunz, München for Tunnel Makkah-Taif [Saudi Arabia]; ex Schöma 1992.
(3) returned to Schöma by /2001.

ZÜRCHER ZIEGELEIEN AG, Unknown location   MI10

The locomotive was delivered to one of the brickworks of this large industrial group.

Gauge: 600 mm   Locn: ?   Date: 08.2002

4wDM	Bdm	O&K MD 2b	25126	1951	(a)	s/s

(a) built by Schöma as their number 1232; new via MBA, Dübendorf.

UNLOCATED AGENT'S LOCOMOTIVES   MI99 C

The following are some of the locomotives delivered to agents in Switzerland but for which there is no other data. Some may have been hire locomotives, but the majority was delivered to an unknown customer. A few locomotives have been found in the nearby Aarlberg region of Austria. Perhaps some of those listed were also exported. Those locomotives suspected of being used outside Switzerland are omitted.

ROBERT AEBI & CIE AG, Regensdorf

For details see the entry for this firm under Regensdorf.

Gauge: 750 mm   Locn: ?   Date: 02.2008

4wPM	Bbm	Ruhrthaler 12/16 PS	484	1921	new	
4wDM	Bdm	R&H 48DL	296043	1950	(b)	
4wDM	Bdm	R&H 48DL	296045	1950	(b)	
4wDM	Bdm	R&H 48DL	296046	1950	(b)	

Gauge: 750/600 mm   Locn: ?   Date: 02.2008

0-4-0DM	Bdm	Deutz OMZ117 F	11002	1933	new	
4wDM	Bdm	R&H 30DLU	285353	1950	new	
4wDM	Bdm	R&H 30DLU	285354	1950	new	
4wDM	Bdm	R&H 30DLU	285355	1950	new	
4wDM	Bdm	Ruhrthaler 30 DML/S2	757	1927	new	
4wDM	Bdm	Ruhrthaler 30 DML/S2	781	1927	new	
4wDM	Bdm	Ruhrthaler 15 DML/S1	1052	1929	new	(11)
4wDM	Bdm	Ruhrthaler 22 DML/S2	1053	1929	new	

Gauge: 600 mm   Locn: ?   Date: 02.2008

0-4-0DM	Bdm	Deutz PME117 F	9947	1931	new
0-4-0DM	Bdm	Deutz PME117 F	10205	1931	new

0-4-0DM	Bdm	Deutz PME117 F	10268	1931	new
0-4-0DM	Bdm	Deutz PME117 F	10274	1931	new
0-4-0DM	Bdm	Deutz PME117 F	10275	1931	new
0-4-0DM	Bdm	Deutz PME117 F	10276	1931	new
0-4-0DM	Bdm	Deutz PME117 F	10277	1931	new
0-4-0DM	Bdm	Deutz PME117 F	10278	1931	new
0-4-0DM	Bdm	Deutz PME117 F	10279	1931	new
0-4-0DM	Bdm	Deutz PME117 F	10281	1931	new
0-4-0DM	Bdm	Deutz OME117 F	10287	1931	new
0-4-0DM	Bdm	Deutz OME117 F	10288	1932	new
0-4-0DM	Bdm	Deutz OME117 F	10295	1932	new
0-4-0DM	Bdm	Deutz MLH332 F	10744	1932	new
0-4-0DM	Bdm	Deutz OME117 F	10846	1932	new
0-4-0DM	Bdm	Deutz OME117 F	10859	1933	new
0-4-0DM	Bdm	Deutz OME117 F	10865	1933	new
0-4-0DM	Bdm	Deutz OME117 F	10866	1933	new
0-4-0DM	Bdm	Deutz OME117 F	10873	1933	new
0-4-0DM	Bdm	Deutz OME117 F	10884	1933	new
0-4-0DM	Bdm	Deutz OME117 F	11009	1933	new
0-4-0DM	Bdm	Deutz MLH514 F	11010	1933	new
0-4-0DM	Bdm	Deutz OMZ117 F	11604	1933	new
0-4-0DM	Bdm	Deutz OME117 F	11908	1934	new
0-4-0DM	Bdm	Deutz OME117 F	11979	1934	new
0-4-0DM	Bdm	Deutz MLH514 F	13481	1935	new
0-4-0DM	Bdm	Deutz OME117 F	15467	1936	new
0-4-0DM	Bdm	Deutz OME117 F	15490	1936	new
0-4-0DM	Bdm	Deutz OME117 F	15491	1936	new
0-4-0DM	Bdm	Deutz OMZ117 F	15492	1936	new
0-4-0DM	Bdm	Deutz OMZ117 F	17038	1936	new
0-4-0DM	Bdm	Deutz OMZ117 F	19519	1937	new
4wDM	Bdm	Jung MS 13	3893	1927	new
4wDM	Bdm	Jung MS Platform	4026	1927	new
4wDM	Bdm	Jung MSZ13	4070	1927	new
4wDM	Bdm	Jung MS 131	4138	1928	new
4wDM	Bdm	Jung MS 131	4391	1929	new
4wDM	Bdm	Jung MS 131	4392	1929	new
4wDM	Bdm	Jung MS 131	4457	1929	new
4wDM	Bdm	Jung MS 130	4487	1929	new
4wDM	Bdm	Jung MS 131	4503	1929	new
4wDM	Bdm	Jung MS 131	4549	1929	new
4wDM	Bdm	Jung MS 131	4690	1929	new
4wDM	Bdm	Jung MS 131	4709	1930	new
4wDM	Bdm	Jung MS 131	4980	1930	new
4wDM	Bdm	Jung MS 131	5101	1930	new
4wDM	Bdm	Jung MS 131	5138	1930	new
4wDM	Bdm	Jung MS 131	5139	1930	new
4wDM	Bdm	Jung MS 131	5165	1931	new
0-4-0Tm	B2ntk	Maffei TL – 30 PS	2894	1908	new
0-4-0Tm	B2ntk	Maffei TL – 30 PS	2895	1909	new
4wPM	Bbm	Ruhrthaler 10/13 PS	485	1921	new
4wPM	Bbm	Ruhrthaler 10/13 PS	489	1921	new
4wPM	Bbm	Ruhrthaler 10/13 PS	491	1921	new
4wPM	Bbm	Ruhrthaler 12/16 PS	505	1921	new
4wPM	Bbm	Ruhrthaler 12/16 PS	528	1921	new
4wPM	Bbm	Ruhrthaler 10/13 PS	531	1921	new
4wPM	Bbm	Ruhrthaler 10/13 PS	532	1921	new
4wPM	Bbm	Ruhrthaler 11/13 PS	533	1922	new
4wPM	Bbm	Ruhrthaler 14/16 PS	538	1922	new
4wPM	Bbm	Ruhrthaler 11/13 PS	541	1922	new
4wPM	Bbm	Ruhrthaler 11/13 PS	549	1922	new
4wPM	Bbm	Ruhrthaler 11 PS	563	1922	new
4wPM	Bbm	Ruhrthaler 14 PS	567	1922	new
4wPM	Bbm	Ruhrthaler 14 PS	574	1922	(dj)
4wPM	Bbm	Ruhrthaler 8 MT	616	1925	new

4wPM	Bbm	Ruhrthaler 8 MT	623	1925	new	
4wPM	Bbm	Ruhrthaler 8 MT	624	1925	new	
4wPM	Bbm	Ruhrthaler 8 MT	625	1925	new	
4wPM	Bbm	Ruhrthaler 8 PS	685	1925	new	
4wDM	Bdm	R&H 16/20 HP	218003	1946	(dp)	
4wDM	Bdm	R&H 13DL	247165	1947	new	
4wDM	Bdm	R&H 13DL	247166	1947	new)	
4wDM	Bdm	R&H 13DL	247167	1947	new	
4wDM	Bdm	R&H 13DL	247168	1947	new	
4wDM	Bdm	R&H 13DL	247169	1947	new	
4wDM	Bdm	R&H 13DL	247170	1947	new	
4wDM	Bdm	R&H 13DL	247172	1947	new	
4wDM	Bdm	R&H 13DL	247173	1947	new	
4wDM	Bdm	R&H 25/30 HP	252086	1947	new	
4wDM	Bdm	R&H 20DL	252848	1947	new	
4wDM	Bdm	R&H 20DL	256800	1948	new	
4wDM	Bdm	R&H LBT	398112	1956	new	

Gauge: 600/500 mm   Locn: ?   Date: 02.2008

4wDM	Bdm	Jung MS 131	4995	1930	new
4wPM	Bbm	Ruhrthaler 8 MT	615	1924	new
4wDM	Bdm	Ruhrthaler 8 DML	636	1925	new
4wDM	Bdm	Ruhrthaler 8 DML	647	1925	new
4wPM	Bbm	Ruhrthaler 8 MT	656	1925	new

Gauge: 500 mm   Locn: ?   Date: 02.2008

0-4-0DM	Bdm	Deutz MLH322 F	10605	1931	new
4wDM	Bdm	Jung MS 13 Platform	4112	1928	new
4wDM	Bdm	Jung MS 13 Platform	4113	1928	new
4wDM	Bdm	Jung MS 13 Platform	4237	1928	new
4wDM	Bdm	Jung MS 13 Platform	4238	1928	new
4wDM	Bdm	Jung MS 13 Platform	4334	1928	new
4wDM	Bdm	Jung MS 131	5105	1929	new

Gauge: ? mm   Locn: ?   Date: 03.2008

0-4-0DM	Bdm	Deutz OMZ117 F	19643	1938	new

In the period 1944-49 the company built for stock several batches of narrow gauge diesel locomotives. The order numbers and the size of batch are known, but not the builder's numbers. A probable allocation of builder's number can be derived, but it may be faulty. Reference to the locomotive list in this Handbook indicates some locomotives that might also fall into the "lost" category. This list is adapted from data calculated by S. Jarne.

Batch	Type	Year	Number	Likely Builder's numbers	In this Handbook
1448	16 PS	1944	6	1262-1267	1264
1449	16 PS	1944	6	1268-1275	-
1450	16 PS	1944	6	1276-1283	-
1452	16 PS	1945	6	1285-1291	-
1453	16 PS	1945	11	1292-1302	-
1454	16 PS	1945	10	1304-1312	1304
?	50 PS	1946	4	1339-1342	-
1462	16 PS	1948	10	1344-1353	1349
1463	16 PS	1948	10	1354-1363	-
1466	16 PS	1949	10	1366-1375	1366

(b) initially supplied without motors, suitable motors followed later.
(dj) reconditioned Heeresverwaltung locomotive (Ruhrthaler 278 1915).
(dp) reconditioned Ministry of Supply locomotive.

(11) returned; reconditioned as Ruhrthaler 1118.

ARMEMENT SEEGMULLER SA, Basel

This shipping firm had dockside warehouses and silos in Strasbourg [F] and Kehl [D]; the location of a similar establishment in Basel has not been determined. Two locomotives are recorded as being delivered to them, but whether for use or for forwarding elsewhere is not clear.

Gauge: 600 mm   Locn: ?   Date: 04.2009

4wDM	Bdm	Diema LR	626	1931	(a)	s/s
4wDM	Bdm	Diema LR	627	1931	(a)	s/s

(a)   new via Comessa Schiltigheim, Strasbourg [F].

BRUN & CIE AG, Nebikon

Gauge: 600 mm   Locn: ?   Date: 04.2009

4wDM	Bdm	Diema B 10	641	1931	new
4wDM	Bdm	Diema B 10	661	1932	new
4wDM	Bdm	Diema DS12	664	1933	new

MASCHINEN-UND BAHNBEDARF (MBA), Dübendorf

alternative: Orenstein & Koppel, Zürich

For details of the concern see under MBA, Dübendorf. Some O&K locomotives were imported via Wander-Wendel.

Gauge: 750 mm   Locn: ?   Date: 03.2008

4wDM	Bdm	O&K(M) RL 1a	4466	1931	new
4wDM	Bdm	O&K MD 2b	25203	1951	(b)
4wDM	Bdm	O&K MD 2b	25471	1952	(c)
4wBE	Ba	SSW EL 7	6068	1959	new
4wBE	Ba	SSW EL 7	6072	1959	new
4wBE	Ba	SSW EL 7	6073	1959	new
4wBE	Ba	SSW EL 7	6189	1964	new

Gauge: 700 mm   Locn: ?   Date: 03.2008

4wDM	Bdm	O&K MV2	25330	1955	(aa)

Gauge: 600/750 mm   Locn: ?   Date: 12.2008

?4wBE	?Ba	SSW EL 8	5872	1957	new
?4wBE	?Ba	SSW EL 8	5873	1957	new
?4wBE	?Ba	SSW EL 8	5886	1957	new
?4wBE	?Ba	SSW EL 8	5902	1957	new
?4wBE	?Ba	SSW EL 8	5965	1958	new
?4wBE	?Ba	SSW EL 8	5976	1958	new
?4wBE	?Ba	SSW EL 8	6070	1959	new
?4wBE	?Ba	SSW EL 8	6074	1960	new
?4wBE	?Ba	SSW EL 8	6075	1960	new
?4wBE	?Ba	SSW EL 8	6086	1960	new
?4wBE	?Ba	SSW EL 8	6087	1960	new
?4wBE	?Ba	SSW EL 8	6088	1960	new
?4wBE	?Ba	SSW EL 8	6089	1961	new
4wBE	Ba	SSW EL 7	6090	1961	new
4wBE	Ba	SSW EL 7	6091	1961	new
?4wBE	?Ba	SSW EL 8	6093	1961	new
?4wBE	?Ba	SSW EL 8	6095	1961	new

Gauge: 600 mm   Locn: ?   Date: 12.2008

4wDM	Bdm	O&K(M) RL 1a	4897	1933	new
4wDM	Bdm	O&K(M) RL 1a	4898	1933	new
4wDM	Bdm	O&K(M) RL 1c	7708	1937	new
4wDM	Bdm	O&K(M) RL 1c	11048	1939	new
4wDM	Bdm	O&K(M) RL 1c	11433	1940	new

4wDM	Bdm	O&K RL 2	20136	1931	new
4wDM	Bdm	O&K MV 0	25174	1952	(cf)
4wDM	Bdm	O&K MD 2b	25471	1952	(cg)
4wBE	Ba	SSW EL 9	5407	1952	(ch)
4wBE	Ba	SSW EL 9	5497	1953	new
4wBE	Ba	SSW EL 9	5590	1954	new
4wBE	Ba	SSW EL 9	5591	1954	new
4wBE	Ba	SSW EL 9	5598	1954	new
4wBE	Ba	SSW EL 9	5660	1955	new
?4wBE	?Ba	SSW EL 8	5661	1955	new
4w+4wBE	B+Ba	SSW EL 9 Doppellok	5662	1955	new
4w+4wBE	B+Ba	SSW EL 9 Doppellok	5663	1955	new
?4wBE	?Ba	SSW EL 8	5666	1955	new
?4wBE	?Ba	SSW EL 8	5667	1955	new
?4wBE	?Ba	SSW EL 8	5668	1955	new
?4wBE	?Ba	SSW EL 8	5682	1955	new
?4wBE	?Ba	SSW EL 8	5683	1955	new
?4wBE	?Ba	SSW EL 8	5690	1955	new
?4wBE	?Ba	SSW EL 8	5691	1955	new
?4wBE	?Ba	SSW EL 8	5801	1956	new
?4wBE	?Ba	SSW EL 8	5858	1958	new
4wBE	Ba	SSW EL 9	5884	1959	new
4wBE	Ba	SSW EL 9	5885	1959	new
4wBE	Ba	SSW EL 9	5895	1957	new
4wBE	Ba	SSW EL 9	5896	1957	new
4wBE	Ba	SSW EL 9	5910	1957	new
4wBE	Ba	SSW EL 9	5911	1957	new
4wBE	Ba	SSW EL 9	5919	1957	new
4wBE	Ba	SSW EL 9	5920	1957	new
4wBE	Ba	SSW EL 9	5964	1958	new
4wBE	Ba	SSW EL 9	5970	1958	new
4wBE	Ba	SSW EL 9	5991	1958	new
4wBE	Ba	SSW EL 9	5996	1958	new
?4wBE	?Ba	SSW EL 8	6069	1959	new
?4wBE	?Ba	SSW EL 8	6185	1964	new

(b) built by Schöma as their number 1272; imported via Wander-Wendel.
(c) built by Schöma as their number 1393; despatched to Zürich (no further details).
(aa) ex O&K, Amsterdam [NL] /1962.
(cf) imported via Wander-Wendel.
(cg) built by Schöma as their number 1394; despatched to Switzerland (no further details).
(ch) imported via FKG.

FRITZ MARTI, Winterthur and Bern

A number of the locomotives imported by this agent were for use in Italy or by public railways. Many locomotives from the list of those delivered to the order of Fritz Marti have been eliminated on the grounds of falling into these two classes. However some may still listed below.

Gauge: 1445/800 mm   Locn: ?   Date: 04.2009

0-4-0T	Bn2t	Jung 80 PS	66	1889	
0-4-0T	Bn2t	Jung 80 PS	67	1889	

Gauge: 1435 mm   Locn: ?   Date: 04.2009

0-4-0T	Bn2t	Maffei bay D VI	1369	1884	

Gauge: 1435/750 mm   Locn: ?   Date: 04.2009

0-4-0T	Bt	Jung 40 PS	159	1893	

Gauge: 1000/800 mm   Locn: ?   Date: 04.2009

0-4-0T	Bn2t	Jung 40 PS	15	1886	
0-4-0T	Bn2t	Jung 40 PS	26	1887	

Gauge: 1000/700 mm			Locn: ?			Date: 04.2009
	0-4-0T	Bt	Jung 20 PS	94	1890	

Gauge: 800 mm			Locn: ?			Date: 04.2009
	0-4-0T	Bn2t	Jung	13	1886	
	0-4-0T	Bn2t	Jung 50 PS	21	1886	
	0-4-0T	Bn2t	Jung 50 PS	22	1886	
	0-4-0T	Bn2t	Jung 50 PS	52	1888	
	0-4-0T	Bn2t	Jung 50 PS	64	1889	
	0-4-0T	Bn2t	Jung 40 PS	65	1889	
	0-4-0T	Bn2t	Jung 50 PS	101	1891	
	0-4-0T	Bt	Jung 60 PS	143	1893	
	0-4-0T	Bt	Jung 60 PS	144	1893	
	0-4-0T	Bt	Jung 120 PS	145	1893	
	0-4-0T	Bt	Jung 120 PS	146	1893	
	0-4-0T	Bt	Jung 120 PS	147	1893	
	0-4-0T	Bt	Jung 30 PS	2245	1914	

Gauge: 800/600 mm			Locn: ?			Date: 04.2009
	0-4-0T	Bt	Jung 40 PS	154	1894	

Gauge: 775 mm			Locn: ?			Date: 04.2009
	0-4-0T	Bn2t	Jung 25 PS	27	1887	
	0-4-0T	Bn2t	Jung 25 PS	40	1888	

Gauge: 750 mm			Locn: ?			Date: 04.2009
	0-4-0T	Bn2t	Jung 25 PS	14	1886	
	0-4-0T	Bn2t	Jung	18	1886	
	0-4-0T	Bn2t	Jung 50 PS	19	1886	
	0-4-0T	Bn2t	Jung 30 PS	29	1887	
	0-4-0T	Bn2t	Jung 20 PS	58	1889	
	0-4-0T	Bt	Jung 50 PS	158	1893	
	0-4-0T	Bt	Jung 50 PS	224	1895	
	0-4-0T	Bt	Jung 50 PS	225	1895	

Gauge: 750/600 mm			Locn: ?			Date: 04.2009
	0-4-0T	Bn2t	Jung 40 PS	97	1891	
	0-4-0T	Bt	Jung 50 PS	1411	1911	

Gauge: 650 mm			Locn: ?			Date: 04.2009
	0-4-0T	Bn2t	Jung 50 PS	75	1889	
	0-4-0T	Bn2t	Jung 50 PS	76	1889	
	0-4-0T	Bn2t	Jung 50 PS	116	1891	
	0-4-0T	Bn2t	Jung 50 PS	117	1891	

Gauge: 600 mm			Locn: ?			Date: 04.2009
L'ALLIER	0-4-0T	Bn2t	Corpet	425	1884	(ka) s/s
	0-4-0T	Bn2t	Jung 40 PS	625	1903	
	0-4-0T	Bn2t	Jung 40 PS	626	1903	
	0-4-0T	Bn2t	Jung 30 PS	1049	1907	
	0-4-0T	Bn2t	Jung Helikon	1396	1909	
	0-4-0T	Bn2t	Jung Helikon	1397	1909	
	0-4-0T	Bn2t	Jung Helikon	1398	1909	

Gauge: 600/500 mm			Locn: ?			Date: 04.2009
	0-4-0T	Bn2t	Jung 10 PS	91	1891	

Gauge: 500 mm			Locn: ?			Date: 04.2009
	0-4-0T	Bn2t	Jung	16	1886	
	0-4-0T	Bn2t	Jung 20PS	24	1891	

Gauge: 488 mm			Locn: ?			Date: 04.2009
	0-4-0T	Bn2t	Jung 15PS	34	1887	

(ka) new to Poulangeon & fils; ex ? /1901. The Corpet locomotive index lists two locomotives from 1884 neither of which is fully defined – perhaps they are one and the same.

ROLLMATERIAL UND BAUMASCHINEN AG (RUBAG), Zürich

The company is a supplier of construction equipment. The offices are now (2009) in Birsfelden, but would appear to have been in Seebach (later Zürich-Seebach) when they were importing Diema locomotives.

Gauge: 700 mm Locn: ? Date: 04.2009

4wDM	Bdm	Diema LR	399	1928	new	

Gauge: 600 mm Locn: ? Date: 04.2009

4wDM	Bdm	Diema 8-10PS	165	1925	new	
4wDM	Bdm	Diema 8-10PS	187	1926	(bb)	
4wDM	Bdm	Diema 8-10PS	188	1926	(bc)	
4wDM	Bdm	Diema 8-10PS	202	1926	(bd)	
4wDM	Bdm	Diema 8-10PS	203	1926	(bd)	
4wDM	Bdm	Diema 8-10PS	205	1926	(bd)	
4wDM	Bdm	Diema 8-10PS	208	1926	(bg)	
4wDM	Bdm	Diema 8-10PS	214	1926	(bh)	
4wDM	Bdm	Diema 8-10PS	220	1926	(bh)	
4wDM	Bdm	Diema 8-10PS	226	1926	(bg)	
4wDM	Bdm	Diema 8-10PS	235	1926	(bg)	
4wDM	Bdm	Diema 8-10PS	251	1927	(bh)	
4wDM	Bdm	Diema 8-10PS	261	1927	(bg)	
4wDM	Bdm	Diema LR	354	1927	new	
4wDM	Bdm	Diema LR	365	1928	new	
4wDM	Bdm	Diema LR	367	1928	(bg)	
4wDM	Bdm	Diema LR	368	1928	(bg)	
4wDM	Bdm	Diema LR	369	1928	(bg)	
4wDM	Bdm	Diema LR	476	1929	(bs)	
4wDM	Bdm	Diema LR	482	1929	(bt)	
4wDM	Bdm	Diema LR	484	1929	(bs)	
4wDM	Bdm	Diema LR	486	1929	(bs)	
4wDM	Bdm	Diema A	544	1929	(bs)	
0-4-0?DM	Bdm	Ruhrthaler 10 DKL/S1	1178	1931	new	
0-4-0?DM	Bdm	Ruhrthaler 10 DKL/S1	1190	1931	new	
0-4-0?DM	Bdm	Ruhrthaler 10 DKL/S1	1210	1931	(bu)	(46)

Gauge: 500 mm Locn: ? Date: 04.2009

4wDM	Bdm	Diema LR	401	1928	(ca)
4wDM	Bdm	Diema LR	479	1929	(cb)

(bb) new, Basel Badischer Bahnhof (as (bh)?).
(bc) new, Schaffhausen (for onward transmission?).
(bd) new, Station Basel Badischer Bahnhof (as (bh)?).
(bg) new, Seebach.
(bh) new, Welltifurrer Int. Transport AG, Basel Badischer Bahnhof (for onward transmission).
(bs) new, Zürich-Seebach.
(bt) new, Zürich-Seebach, for Basel-St.Johann.
(bu) new, but may never have reached Switzerland.
(ca) new, Basel. A possible destination is Ziegelwerk AG Gettnau.
(cb) new, Zürich-Seebach for Gisikon. A likely destination is Ziegelei Körblingen.

(46) ?returned; reconditioned as Ruhrthaler 1338 1934.

SCHWEIZERISCHE ELEKTRIZITÄTS- UND VERKEHRSGESELLSCHAFT (SEVG), Basel

alternative: Société Suisse d'Electricité et de Traction
             Suiselectra

A financing company founded in 1924. The location where the locomotive was used and its operator are not known.

Gauge: 1435 mm		Locn: ?				Date: 04.2009
	4wDM	Bdm	O&K RL 4	20170	1931	new s/s

SCHWEIZERISCHE LOKOMOTIV- UND MASCHINENFABRIK (SLM), Winterthur

Gauge: 1435 mm		Locn: ?				Date: 04.2009
	0-4-0T	Bn2t	Maffei bay D VI	1374	1884	(a)

(a) ordered by Bayerische Staatsbahn [D], but delivered as shown.

HANS F WÜRGLER ING. BUREAU, Zürich-Albisrieden

This concern acted as an agent for several locomotives. It is not known if this one remained their property or was delivered to a customer.

Gauge: 600 mm		Locn: ?				Date: 04.2009
	0-4-0DM	Bdm	Deutz OMZ117 F	10860	1933	(a)

(a) imported via Lagerhausgesellschaft, Basel; fitted with motor type A2L514 in /1963.

Number   Year Type   Other   Number Reference(s)

Locomotive index

Where there is more than one not fully defined locomotives at the same concern column five is used to indicate this. (*2 meaning two similar locomotives, etc).

Internationale Fabrik für Bergbahnen, Aarau   Aarau
In the literature (Hefti and Messerschmidt) different versions of the data are given. The builder's numbers quoted here must be treated with caution.

10	1876		BE035, BE102, LU39, SO37
11	1876		SZ02, ZH152, ZH155
13	1878		BL24

Antonio Badoni spA, Lecco [I]   ABL
ABL built some locomotives to Breuer design under licence (the type IV below)

?	19xx		TI03
279	1962 1L		TI35
4702	1956 IV NHT		SO37, TI16, ZH034, ZH062
4707	1957 IV NHT		SO45
5232	1969 V-A		TI03
6171	1980 VI/C		GE17

Ateliers de Constructions Mécaniques de Vevey SA, Vevey   ACMV

?	1921	BBC	1338	VD46, MI08
?	1926	MFO	?	ZH149
?	194x	Spälti	-	VD47
1259	1959	BBC	?	BE071, ZH085
1362	1961	BBC	?	VD03

see: RACO   Aebi

Allgemeine Elektricitätsgesellschaft, Berlin [D] (1889-1945)   AEG

| 4561 | 1930 Ks | | BS08 |

AB Gävle Vagnverkstad, Uppsalaleden [S] (AGV) (to ABB 1989)   Ageve
each type has its own number series

?	19xx		AG089
?	19xx		SZ19
x779	19xx DMD		LU03, LU33
?	1978		GR36
707	1972		AG089
779	1976 D12		AG089
780	1976		AG089
811	1978		GR36
813	1976		SZ19
865	1980		SZ19
951	1984 D12		AG089
1201	1994 D12		AG089
1202	1994 D12		AG089

Ammann Schweiz AG, Eisenbahnstrasse 25, Langenthal   Ammann
formerly: Ulrich Ammann Baumaschinen AG, Langenthal
the majority of locomotivesappear to be modified Jung locomotives.

-	19xx			FR13, VS52, ZH046
-	19*39 ZM 233*	Jung	8456	VD12, ZH046
-	194x			VS54
-	1949			VD41, ZH046

Anjubault, Paris [F]   Anjubault
predecessors of Corpet-Louvet

Number	Year Type	Other	Number	Reference(s)
11	1856			VD36
18	1857			VD36
19	1857			VD36
20	1857			VD36
21	1857			VD36
22	1857			VD36
66	1862			MI02
75	1862			ZH029

Allmänna Svenska Elektriska Aktiebolaget, Stockholm [S]   ASEA

merged with BBC in 1988

?	19xx			AG089

Victor Asper, Küsnacht ZH   Asper

398	1948			SG34
402	1949 type 14			SG34
?	1953			UR09

Austro-Daimler, Wiener Neustadt [A]   Austro-Daimler

IRS convention   A-D

?	19xx			ZH046
?	19xx			BL20
?	1930			AG006
102	19xx PS 6			GR20
259	19xx PS 6			SG38, SG43
541	19xx PS 6			SG38, SG68

Brissoneau & Lotz, Creil [F]   B+L

?	19xx			SO37
?	1960 040DE600			VD64
?	1961 040DE600			VD47

Heinrich Bartz KG, Dortmund-Körne [D]   Bartz

963	1952			VS22
966	1952			VS22
1114	1957			VS22
1503	1959 GA 01a			ZH017

Betrieb für Bergbauausrüstungen Aue BBA (SDAG Wismut) [D]   BBA

?	19xx B660			ZH017
?	19xx Metallist			ZH046
277	9/1982 B360			ZH046
774	8/1988 B660			ZH046

AG Brown Boveri & Cie, Baden   BBC

Kollektivgesellschaft Brown, Boveri & Cie. (1891-1900)

Merged with Allmänna Svenska Elektriska Aktiebolaget (ASEA) (1988)   ABB

Up to number 3999 the locomotives were built in any one of company's works in Mannheim [D], CEM [F], TIBB [I] or Baden, Zürich-Oerlikon or Münchenstein. From number 4000 onwards blocks of numbers were allocated to each works; though irregularities have been noticed.

The "Transportwagen" and "Platformwagen" with only one driven axle built by this firm appear to have been all built with this axle and cab adjacent; but it is not clear if this should be considered the front or rear end. In this Handbook the BBC convention is used.

Generally BBC provided electrical gear to fit into locomotives mechanically constructed elsewhere. It is IRS custom to list the mechanical part first and order such a list by the mechanical firm. However, in this case not all such builders have been determined. For this manufacturer therefore as many numbers as possible are listed below.

Another complication arises with a number of preserved locomotives that makes itself particularly obvious when dealing with this firm. It is common practice to list all the

Number	Year	Type	Other	Number	Reference(s)

manufacturers involved with a class. However each individual locomotive has often been built by only a limited number of these. Cases are known of BBC allocating a number and then passing the work to another firm to complete.

To make this list more useful known BBC numbers used in locomotives from other manufacturers are included here.

Number	Year	Type	Other	Number	Reference(s)
727	1913	Ge 2/4	SLM	2308	GR15, ZH154
729	1913	Ge 2/4	SLM	2310	LU39
845	1912	Transportwagen			AG005
860	1917	Transportwagen			AG005
879	1914	Akku-Fahrz.			BE052, FR01
968	1914	BCFhe 4/4	SWS	?	VD10
1008	1914	Transportwagen			AG005
1010	c1914	Akku-Fahrz.	Spälti	-	AG005, BL04
1054	1916(29)	Ge 4/4	SLM	3295	VD10
1098	1917	Akku-Fahrz.			ZH127
1099	1917	Akku-Fahrz.	Oehler		SO13
1205	1918	Be 6/8^I	BBC	2649	MI07
1209	1917	Akku-Fahrz.			BL18
1210	1917	Akku-Fahrz.			BL04
1214	1918	Akku-Fahrz.			BE023
1232	1919	Akku-Fahrz.			GE15
1319	1920/19	ex Ce 4/6	SLM	2693	MI08
1320	1920/19	ex Ce 4/6	SLM	2694	TI17
1321	1920/19	ex Ce 4/6	SLM	2695	BE018
1338	1921	Rangierlok	ACMV	-	VD46
1339	1921	Akku-Fahrz.			GE15
1425	1920	Akku-Fahrz.			AG005
1426	1920	Akku-Fahrz.			AG005
1541	1921/20	Ae 3/6^I	SLM	2740	AG090, MI08
1546	1921/20	Ge 6/6	SLM	2754	LU39
1550	1921/20	Ge 6/6	SLM	2758	FR01
1580	1920	Akku-Fahrz.			NE05
1588	1922	ex Ce 4/6	SLM	2843	MI08
1589	1922	ex Ce 4/6	SLM	2844	MI08
1592	1921/20	Be 4/6	SLM	2762	MI07
1804	1922	Be 4/6	SLM	2809	AG090, MI07
1886	1922	Ge 6/6	SLM	2839	GR44
1895	1924	Akku-Fahrz.			ZH027
1942	1923	Ta	Olten	?	VD34, VD47, VS39
1945	1923	Ta	Olten	?	FR01, VD40, VS39
1968	1923-4	BCFe 4/4	CEM	?	TI34
2083	1925/24	Ae 3/6^I	SLM	2982	MI08
2111	1925	Ae 3/6I	SLM	3049	MI07
2125	1925	Ae 3/6^I	SLM	3088	MI07, SG17
2140	1926	De 6/6	SLM	3056	SO29
2242	1925	Ge 6/6	SLM	3045	GR38
2522	1927/26	Ae 4/7	SLM	3133	MI08
2525	1927	Ae 4/7	SLM	3176	MI07
2528	1927	Ae 4/7	SLM	3179	MI01
2533	1927/26	Ee 3/3	SLM	3188	SO11, MI08
2535	1927/26	Ee 3/3	SLM	3190	MI08
2537	1927/26	Ee 3/3	SLM	3192	MI01
2538	1927/26	Ee 3/3	SLM	3193	MI01
2808	1928	Ee 3/3	SLM	3225	TG10
2815	1928	Ee 3/3	SLM	3232	MI01
2816	1928	Ee 3/3	SLM	3233	MI01
2818	1927	Ae 4/7	SLM	3238	MI01
2835	1928/27		HStP	1576	JU06
2967	1929	Ge 6/6	SLM	3297	GR38
2968	1929	Ge 6/6	SLM	3298	GR38
3315	1930	Ee 3/3	SLM	3396	MI08
3355	1930	Ae 4/7	SLM	3442	MI01

Number	Year	Type	Other	Number	Reference(s)
3360	1931	Ae 8/14	SLM	3501	MI07
3366	1930	BCFeh 4/4	SLM,SIG	3463	AR1
3703	1932	Gepäck-Tw	SIG	?	VD10
3719	1932	Ae 4/7	SLM	3537	MI01
3723	1933/32	Ae 4/7	SLM	3547	AG104, MI01
3983	1936/35	CLe 2/4	SLM,MFO,SAAS,SBB	3581	MI07
3985	1936	CLe 2/4	SLM,MFO,SAAS,SBB	3604	MI07
4451	1941	BCFeh 2/4	SLM	3765	VS48
4507	1944	Ae 4/4	SLM	3883	BE010
4529	1945	Ee 3/3	SLM	3899	BE084
4535	1946	Che 2/4	SLM	3949	VS37
5429	1951	Ae 6/6	SLM	4050	TI17
5430	1951	Ae 6/6	SLM	4051	MI07
5459	1953	Tm^{III}	SLM	4112	AG005, AG053
5464	1953	HGe 4/4	SLM	4079	LU38
6003	1954	Tm^{III}	SLM	4129	AG005, AG068, BL04
6004	1955	Ae 6/6	SLM	4138	MI07
6007	1955	Ae 6/6	SLM	4141	OW5
6008	1955	Ae 6/6	SLM	4142	MI07
6012	1955	Ae 6/6	SLM	4146	MI07
6014	1955	Ae 6/6	SLM	4148	LU39
6017	1955	Ae 4/4	SLM	4128	LU39
6026	1956	Tm^{III}	SLM	4158	LU16, VD66, TG20, ZH006
6051	1957	Ae 6/6	SLM	4234	MI07
6053	1957	Ae 6/6	SLM	4236	LU10
6060	1957	Ae 6/6	SLM	4244	MI07
6088	1960	FRITZ	SIG	?	AG002, VD49, ZH136
6089	1959	EMIL	SIG	?	AG005, ZH019
6117	1959(79)		SIG	?	LU37
6130	1960	Antonio	SIG	?	AG092, ZH136
6147	1959	Ee 3/3	SLM	4360	LU08, ZH036
6149	*1960*	MUTZ	SIG	?	BE034
?	1961		ACMV	*1362*	VD03
?	1961	ERICH	SIG	?	ZH136
6179	1962	Tm^{III}	SLM	4438	BS06
6180	1962	Tm^{III}	SLM	4439	BS06, TG19
6183	1962	Tm^{III}	SLM	4440	AG057
6184	1962	Thm,Tmh	SLM	4400	SG03, ZH152
6405	1964	WALTER	SIG	?	AG005, ZH019
6406	1964	LISI	SIG	?	LU28
6407	1963	Ae 8/8	SLM	4443	BE010
?	1965		SWS,MFO	?	TG10
7516	1965	Tm^{III}	SLM	4564	BL36, TG20
7531	1965	Tm^{III}	SLM	4579	TG20
7539	1966	Tm^{III}	SLM	4586	BL36, BS06, TG20
7540	1966	Tm^{III}	SLM	4587	BS06, TG12
7639	1968	Madeleine	SIG	?	AG009
7640	1984		Stadler	162	AG046
7701	*1967*	FRIDOLIN	SIG	?	AG023

Bedia Maschinenfabrik GmbH & Co. KG, Bonn [D]   Bedia

| 313 | 1995 | D105/17B - Umbau | | | BE096 |

Bell & Cie, Kriens   Bell

Bern-Lötschberg-Simplon, Spiez   BLS

| - | 1962 | | | | BE010 |

Baldwin Locomotive Works, Eddystone [USA]   BLW

This title applies from 1902

| 4xxxx | c1917 | VM | | | VS20, VS56 |

Number	Year	Type	Other	Number	Reference(s)
72381	1945	141.R			SH010

Berliner-Maschinenbau-Actien-Gesellschaft vormals L. Schwartzkopff, Berlin [D]
BMAG

Number	Year	Type	Other	Number	Reference(s)
12226	1943	BR 52	Škoda	1584	NE15, ZH105

Société Anonyme Boilot-Pétolat (anciennement A. Pétolat), Dijon, Côte d'Or [F]
Boilot-Pétolat

Number	Year	Type	Other	Number	Reference(s)
10183	19xx				JU01

Borsig Lokomotiv-Werke GmbH, Berlin-Hennigsdorf [D]   Borsig

Number	Year	Type	Other	Number	Reference(s)
5683	1906				ZH099
6250	1907				BE008, BE060, BE061, UR16, ZH010
6485	1908				BE024, SO19, SZ10, ZH117
6486	1908				BE024
6487	1908				BE024
6488	1908				BE024
6849	1908				SO05
6850	1909				SO05
6977	1909				AG047
6989	1908				BE024
6990	1908				BE024, ZH117
6991	1908				BE024, SG34, ZH117
6992	1908				BE024, SG34, ZH117
7258	1910				AG047
7263	1910				BL06
7442	1910				AG013
7605	1910				GR28
7842	1911				AG011, BE009, ZH002
7885	1911	DrL			VD08
7897	1911	DrL			VD08
7898	1911	DrL			VD08
7899	1911	DrL			VD08
7900	1911	DrL			VD08
7901	1911	DrL			VD08
8169	1911	DrL			VD08
8226	1912				VD08
8376	1912	DrL			SO05
8377	1912	DrL			SO05
8378	1912	DrL			SO05
8379	1912	DrL			SO05
8380	1912	DrL			SO05

Breuer-Werke GmbH, Höchst, Frankfurt am Main [D]   Breuer
Maschinen- und Armaturenfabrik, vorm. H. Breuer & Co., Frankfurt/Main-Höchst [D]
1969 to Krauss-Maffei

Number	Year	Type	Other	Number	Reference(s)
?	19xx				AG009
?	19xx				AG031
?	19xx				BL14
?	19xx				BS19
?	19xx				UR12
?	19xx				ZH024
?	19xx				ZH029
?	1915				VD26
?	192x I, II or III		Stadler	21	AG020
?	192x I, II or III				SO30
?	192x				BE087
?	192x				SO18
?	192x				TG06
?	192x				VD28
631	*1925* II				BE118
642	192x				ZH072

Number	Year	Type	Other	Number	Reference(s)
?	1927	II or III			BE103
?	1928	III			BS01
?	1928	III			NE15
?	1928	III			ZH081
?	1928	III			ZH140
746	1928				VD26
1096	192x	III			BL10, JU10
1159	1929	III			NE15, ZH081
1201	1928	III			SZ06
?	1929	III			AG109
?	1929	III			BS01
?	1929	III			SG16
?	1929	III			SO41
?	1929	III			TI40
1235	c1929	40 PS			TI21
?	c1930	II, III or IV			LU16
?	c1930				SO40
?	c1930				ZH081
?	c1930				ZH156
?	c1930				ZH180
?	1930				BL32
?	c1930	III			BS02
?	c1930	III			FR09
?	1930	III			SG16
?	1930	III			SG16
?	1930	III			TG18
?	1930	IV			AG094
?	1931(69)	III			FR01, SO10, MI08
?	1931	III			BS01
?	1931	III			LU39
?	1931	III			VS71
?	1931	IV			BE030
?	1931	IV			BE107
?	1931	IV			BE123
?	c1943				BL34, NE09
3033	1951	V			VD26, VD34, VS39
3039	1951	V			BE112, SO40
3049	1952	V	RACO	1424	JU12, NE15, SO44, SO45
3054	1952	V			NE15, SG32, SG63, MI08
3067	1953	VL			AG038, AG080, AG097
3088	1955	VL			BL34, FR01, ZH110
3093	1955	VL			BL43

Brookville Locomotive Company, Brookville, Pennsylvania [USA]   Brookville

Number	Year	Type			Reference(s)
3517	1949	BMD-8			VS56
3518	1949	BMD-8			VS56

Brun & Cie AG, Nebikon   Brun

Plates on the locomotive indicate a two-digit locomotive number (Masch. No.) and a five-digit manufacturing number (Fabr. No.). Both are given when known.

Number	Year	Type		Number	Reference(s)
?	19xx				BE085
?	19xx				LU40
?	*1938*	FD 16			SG49
?	1939				SG18
?	1940	2 DS 90E			SG18
33	1938				SG18
34	1938	FL 12		93913	SG18, ZH046
35	1940				SG18
36	1943	FL 12			BE032, LU30
41	19xx	B12		41983	LU30
43	19xx	2HK65			VD52, ZH046
48	19xx	B20/2HK65		58603	AG054
67	19xx	2 DS 90E		26483	SZ01, ZH046

Number	Year Type	Other	Number	Reference(s)
78	1946			SG34
86	1951 16		71783	VD31, VD47
?	1965			BS07

Canada Works, Birkenhead [GB]　　　　　　　　　　　　　　　　　　　Canada

114	1863			TI32

Compagnie Électro-Mécanique, Paris [F]　　　　　　　　　　　　　　　CEM

French subsidiary of BBC

?	1923-4	BBC	1968	TI34

Compagnie de Chemins de Fer Départementaux　　　　　　　　　　　CFD

Locomotive construction is the responsibility of CFD Bagnères, 65202 Bagnères de Bigorre [F]

702	1997 BB700HR			VD47
851	2000 BB800			ZH004, TG20
152-001	2004			VD65

Chemins de Fer Fédéraux, Yverdon　　　　　　　　　　　　　　　　　CFF (Yverdon)

?	1906	MFO	?	FR16

Pierwsza Fabryka Lokomotyw w Polsce (FABLOK), Chrzanów　　　　　Chrzanów

IRS convention　　　　　　　　　　　　　　　　　　　　　　　　　　　　　　Chrz

Although Chrzanów is now in Poland it has - as Krenau - been in Germany. Many other variations of the name are known - but are not relevant here

2667	1952 Slask			NE15
4778	1956 TKt48			NE15

Compagnie de l'Industrie Électrique et Mécanique, Genève　　　　　CIEM

later SAAS

?	1898			VD21

Českomoravská Kolben-Daněk (ČKD), Prag [CZ]　　　　　　　　　　　ČKD

4686	1959 T 211.0			GR29
?	1994 T 239.1			GR18

see: Kaelble-Gmeinder　　　　　　　　　　　　　　　　　　　　　　　　　CKG

Canadian Locomotive Company, Kingston, Ontario [CAN]　　　　　　CLC

2399	1947 141.R			ZH130

Constructions mécaniques SA, Renens VD　　　　　　　　　　　　　CMR

The plates of some CFF Tm' display a running number in place of a works number on the builder's plate; these are quoted in the text as -for example- 'Tm' 319' so as to be able to distinguish the locomotives.

'Tm' 318'	1960 Tm'			TI35
'Tm' 319'	1961 Tm'			MI01
'Tm' 320'	1960 Tm'			VD27
'Tm' 365'	1962 Tm'			AG046
'Tm' 421'	1961 Tm'			NE04

Cockerill Mechanical Industries, Seraing [B]　　　　　　　　　　　　　CMI

S.A. Cockerill-Ougrée, Seraing [B] (1955-66)
Société pour l'exploitation des Etablissements John Cockerill, Seraing [B](1842-1955)
John Cockerill, Seraing [B]

2104	1898 II			AG005, AG027
2951	1920 IV			NE15
4143	1964			AG074, ZH097
4224	1972			ZH067

| Number | Year | Type | Other | Number | Reference(s) |

Société pour l'exploitation des Etablissements John Cockerill Cockerill
IRS convention Cock
see: CMI

Corpet, Louvet & Compagnie, La Courneuve, Seine Saint Denis [F] Corpet
Vve L Corpet et L Louvet, Paris [F]
Société Corpet Louvet et Cie. (from 1867; as successor to Anjubault)
(1952 merged into Fives Lille)

?	1884			NE13	
425	*1884*			*MI99*	
1195	1908			VS08	

SA des Usines Métallurgiques du Hainaut, Couillet [B] Couillet

| ? | 1xxx | | | ZH072 | |

Société Nouvelle des Établissements Decauville Ainé, Petit-Bourg, Corbeil [F]
Decauville
IRS convention Dcv
Société Anonyme Decauville, Corbeil [F]

| 91 | 1890 | | | NE08 | |

Deutsche Maschinenbau-Aktiengesellschaft, Duisburg [D] Demag
founded 1910

?	191x			VS16	
2937	*1941*			AG089	

Motorenfabrik Deutz AG, Köln [D] (1896-1930) Deutz
Humboldt-Deutz-Motoren AG, Köln [D] (1930-38)
Klöckner-Humboldt-Deutz AG (KHD) (from 5/11/1938) KHD
The firm of Motorenfabrik Deutz AG was reorganised to become Klöckner-Humboldt-Deutz AG on 5 November 1938. Where possible the abbreviation Deutz and KHD is used as appropriate.

For locomotives built before about 1960 the type, as given in the firm's documentation, almost always has a terminal letter or group of letters. These letters do not appear on the plates usually present in the cab of the locomotives. The codes include F (Feldbahnlok), R (Rangierlok), Gr (Grubenlok) and Str (Strassenbahnlok).

Also on the "builder's plates" in the cab are the engine type and engine number. This engine type is sometimes very similar or identical to the locomotive type, so that there may be inconsistencies when the engine is changed. The locomotive numbers have a maximum of five digits. The engine numbers typically consist of a seven-digit number and a two-digit one (for example 1643544/47); understood to be for the motor and gearbox respectively.

The Deutz type for the standard DR (later DB) Köf II is A6M517 R.

From 1954 a new type code was introduced in parallel with the existing one which was not finally dropped till the early 1960s. Some locomotives may be described in either way and the new code has been applied retroactively to some locomotives built before 1954. Where appropriate both types are given. The new code is more easily converted into the "wheel arrangement" than the old one. For instance KS55 B = Kleinlok, mit Stangen, 55PS, two-axled = 0-4-0D. For more details
see: http://www.werkbahn.de/eisenbahn/lokbau/typen/typen_V.htm

When KHD ceased to handle them, the mining locomotive and spare parts programs were shared amongst Bedia AG and Rensmann AG.
see: www.werkbahn.de/eisenbahn/lokbau/khd.htm

?	19xx			ZH012	
32	1901 C I F			VS12	
983	1913 C XIV F			SG34	

Number	Year	Type	Other Number	Reference(s)
1235	1913	X IV R		VD47
1462	1914	C V F		VD69
1463	1914	C V F		VD69
1477	1914	C XIII		VD74
1583	1915	C XIV F		SZ06
1633	1915	R IV		LU32
?	1917	R III		TG08
2760	1918	R III		LU32, ZH096
?	1920			SG18
3922	1920	C XIV R		ZH044
?	1921			SG18
4079	1921	C XIV F		LU41
6454	1923	ML128 F		BE121
8402	1928	MLH222 G		SG18, SG74
9279	1929	PME117 F		AG047
9318	1929	MLH228 F		AG092
9718	1931	MLH332 F		SZ09
9947	1931	PME117 F		MI99
10003	1931	MLH322 F		BE121
10205	1931	PME117 F		MI99
10207	1931	PME117 F		SO38
10226	1931	PMZ117 F		ZH114
10268	1931	PME117 F		MI99
10274	1931	PME117 F		MI99
10275	1931	PME117 F		MI99
10276	1931	PME117 F		MI99
10277	1931	PME117 F		MI99
10278	1931	PME117 F		MI99
10279	1931	PME117 F		MI99
10281	1931	PME117 F		MI99
10282	1932	OME117 F		LU30
10286	1932	OME117 F		VD15, VD58, ZH046
10287	1931	OME117 F		MI99
10288	1932	OME117 F		MI99
10295	1932	OME117 F		MI99
10296	1932	OME117 F		SO37
10605	1931	MLH322 F		MI99
10744	1932	MLH332 F		MI99
10811	1932	OME117 F		GE24
10830	1932	OMZ117		SZ06
10846	1932	OME117 F		MI99
10859	1933	OME117 F		MI99
10860	1933	OMZ117 F		MI99
10865	1933	OME117 F		MI99
10866	1933	OME117 F		MI99
10873	1933	OME117 F		MI99
10876	1933	OMZ117 F		SO23, ZH046
10884	1933	OME117 F		MI99
11002	1933	OMZ117 F		MI99
11008	1933	OME117 F		SG54
11009	1933	OME117 F		MI99
11010	1933	MLH514 F		MI99
11069	1933	OME117 F		SO37
11604	1933	OMZ117 F		MI99
11908	1934	OME117 F		MI99
11979	1934	OME117 F		MI99
13481	1935	MLH514 F		MI99
15467	1936	OME117 F		MI99
15490	1936	OME117 F		MI99
15491	1936	OME117 F		MI99
15492	1936	OMZ117 F		MI99
17038	1936	OMZ117 F		MI99
?	1937	OME117		BE092

Number	Year	Type	Other	Number	Reference(s)
19519	1937	OMZ117 F			MI99
19643	1938	OMZ117 F			MI99
21114	1937	MLH714 F			SO37
21529	1938	MLH322 F			ZH046
26099	1940	A6M420 R			AG009
39663	1940	OMZ122 F			AG088
46994	1949	A6M517 R			ZH085, ZH160
47169	1951	A4M517 G			AG089, SG18, SG49, SG64
47304	1944	A6M517 R			NE11, NE15
55125	1952	A6M517 R			AG057
55126	1952	A6M517 R			SG31, ZH001, ZH061, ZH085
55182	1952	A4L514 R	KS55 B		BE088
55442	1952	A4M517 G			GR35, SG18, SG64
55754	1955	A6M617 R			BL08, VD40
55764	1954	A2L514 F			ZH183
55827	1954	A2L514 F			BE121, ZH046
56209	1955	A6M517 G			AG089, SG18, SG64
56281	1956	A2L514 F			SG85
56282	1956	A12L614 R			BS02, BS17, BS24
56464	1956	T4M625 R			BL05, BL36, BS17, BS24
56488	1957	A4L514 R			ZH085
56506	1957	A2M517 G			SG18, SG49
56507	1957	A2M517 G			SG18, SG49
56511	1957	A8L614 R			AG016, ZH029
56794	1957	A8L614 R	KK130 B		AG080
56807	1957	A2L514 F			ZH183
56810	1957	A4L514 R			LU27, ZH085
56837	1958	A8L614 R			BS14, BS17
56848	1958	KG245 B ex	A8M517 R		VS73
56893	1959	A4L514 R			SG32
56900	1958	A8L614 R			SH09, TI72, ZH024
57069	1960	A12L714 R			AG016, ZH029
57181	1961	KG230 BS			UR09, ZH160
57198	1961	A12L714 R	KS230 B		AG078
57204	1961	MS430 C			AG004, BE059, ZH006
57216	1961	A8L614 R			BL52, BL53, ZG4, ZH006
57309	1960	A6M617 R			BE033
57338	1960	A6M617 R			BE033, TG20
57355	1961	V 100			ZH136
57361	1962	V 100			BL02, ZH160
57487	1962	KK140B			ZH084
57552	1962	KS55 B			ZG6
57562	1963	MG 530C			ZH085
57667	1964	KG230 B			AG094, ZH109
57668	1964	KG230 B			BE048
57687	1963	KG125 BS			AG009
57688	1963	KG125 BS			AG009
57703	1964	KK135 B ex			VD71
57831	1965	KG275 B			TI72, ZH085
57850	1965	KG230 BS			ZH085
57869	1965	KG230 B			TI64, ZH006, ZH085
57894	1966	KG230 BS			ZH085
57900	1965	KK140 B			BL09, BL32, BL52
57906	1965	A6M617 R			TI14
57914	1965	A6M617 R			ZH085, ZH094, ZH160
57938	1965	KG230 B			BE105
58110	1966	KG230 BS			ZH085
58111	1966	KG230 BS			ZH085
58164	1967	MG600 C			BS17, BS24
58175	1967	KG230 B ex			SZ06
58228	1968	KS55 B			VD39, VD66

Diepholzer Maschinenfabrik (Fr. Schöttler GmbH), Diepholz [D] — Diema

Number	Year	Type	Other	Number Reference(s)
?	19xx			AG098, ZH093
?	19xx			GR10
?	19xx	DS30		GL10
165	1925	8-10 PS		MI99
187	1926	8-10 PS		MI99
188	1926	8-10 PS		MI99
202	1926	8-10 PS		MI99
203	1926	8-10 PS		MI99
205	1926	8-10 PS		MI99
208	1926	8-10 PS		MI99
214	1926	8-10 PS		MI99
220	1926	8-10 PS		MI99
226	1926	8-10 PS		MI99
235	1926	8-10 PS		MI99
251	1927	8-10 PS		MI99
261	1927	8-10 PS		MI99
354	1927	LR		MI99
365	1928	LR		MI99
367	1928	LR		MI99
368	1928	LR		MI99
369	1928	LR		MI99
399	1928	LR		MI99
401	1928	LR		LU40
476	1929	LR		MI99
479	1929	LR		LU30
482	1929	LR		MI99
484	1929	LR		MI99
486	1929	LR		MI99
544	1929	A		MI99
626	1931	LR		MI99
627	1931	LR		MI99
641	1931	B 10		MI99
661	1932	B 10		MI99
664	1923	DS12		MI99
1280	1948	DS12		ZH017
1797	1955	DL6		LU30
1883	1956	DS30		GE14
2054	1957	DS30		AG012, GL10
2072	1957	DL8		VD55
2360	1960	DFL 90/0.2		BE004, GE14, JU02, VS03
2391	1960	DS40		GE14, TI08, UR13
2527	1962	DL8		GE24, VD55
2556	1962	DS28		UR13
2580	1962	DS28		GL01, GL07, SG08, ZH183
2581	1962	DS28		ZH183
2590	1963	DS30		GE14
2620	1962	DVL 120		ZH085
2626	1963	DL8		GE24, VD55
2638	1963	DS20		LU30
2702	1964	DS28		SG79, ZH046, ZH183
2715	*1964*	*DFL 90/0.2*		*BE004*
2729	1964	DFL 90/0.2		BE004, GR35, VS03
2733	1964	DS20		BL30
2790	1965	DS11/2		SZ29
2803	1965	DFL 60/11		LU30
2987	1958	DS90/1		AG088, AG114
3154	1970	DVL90/1.1		SG06
3199	1971	DTL90/1.3		ZH183
3200	1971	DTL90/1.3		ZH183
3228	1971	DVL15/2		ZH063
3229	1971	DVL15/3		TI35, UR10
3238	1972	DVL150/1.2		TI03, TI40

Number	Year	Type	Other	Number	Reference(s)
3328	1973	DVL60/1.3			SG42
3516	1974	DVL90/1.1			GR06, GR10
4315	1979	DFL40/1.7			GR46
?	1981	DFL 90/0.2			GR35
4525	1981	DFL60/1.6			SZ06
5146	1991	DFL150/1.2			SZ06
5147	1991	DFL150/1.2			SZ06
5239	1993	DVL200/2			AG057

Elektrische Fahrzeuge AG, Zürich   EFAG

Number	Year	Type	Other	Number	Reference(s)
?	1931				BE054
2121	19xx				VS40
2123	19xx				VS40

Elektricitäts-Gesellschaft Alioth, Münchenstein   EGA
to AG Brown Boveri & Cie (1913)

Emam srl, 59, V. Xxv Aprile, 22070, Guanzate [I]   EMAM

Number	Year	Type	Other	Number	Reference(s)
?	19xx	Th00D			UR13
030-3025	1996	T250D	Belloli	?	GR02
1/CM 102	19xx	Th00D			UR13
4/CM 102	19xx	Th00D			UR13
5/CM 102	19xx	Th00D			UR13

Maschinenfabrik Esslingen, Esslingen am Neckar [D]   Esslingen
IRS convention   Essl
VRS convention   ME

Number	Year	Type	Other	Number	Reference(s)
396	1858	Engerth			MI07
481	1859	Engerth			BE013
2224	1887				ZH029, ZH034
2498	1892				MI07
?-	1902(07,80)				MI07
4183	1927	BR 99.1			VD10
4227	1929				UR09
5164	1956	Neuhof II			ZH085
5224	1956	Guthof			ZH085

Fauvet et Girel, Lille [F]   Fauvet-Girel

Number	Year	Type	Other	Number	Reference(s)
?	1968				GR23

Compagnie Fives-Lille, Fives-Lille [F]   Fives-Lille
Société Fives-Lille-Cail, Fives-Lille et Denain [F] (1958-61)
Compagnie Fives-Lille pour Constructions Mécaniques et Entreprises, Fives bei Lille [F] (1865-1958)

Number	Year	Type	Other	Number	Reference(s)
3587	1909				VD10
4714	1931				AG105, ZH105
?	1965				VS55

Société anglo-franco-belge des Ateliers de La Croyère, Seneffe et Godarville [B]   Franco-Belge
Title valid 1927-53

Number	Year	Type	Other	Number	Reference(s)
2686	1953				AG088

Stahlbahnwerke Freudenstein & Co. AG, Berlin [D]   Freudenstein
IRS convention   Freud
to Orenstein &Koppel 1905

Number	Year	Type	Other	Number	Reference(s)
93	1901				ZH088
106	1901				AG067, BL22, BS16, SG30, VS29, ZH088, ZH117
157	1904				SO05

Number	Year	Type	Other	Number	Reference(s)
31575	1973	DHG 1200 BB			SO37, VD70, ZH085
31948	1977	DHG 1200			ZH160
31949	1977	DHG 1200			ZH160
31988	1978	DHG 300B			SG12
32473	1984	DHG 300B			ZH085
32474	1984	DHG 300B			VS62
32475	1982	DHG 300B			VS01
32479	1981	DHG 700C			VS01

Aktiengesellschaft für Lokomotivbau Hohenzollern, Düsseldorf-Grafenburg [D]
Hohenzollern

IRS convention — Hohen

Number	Year	Type	Other	Number	Reference(s)
120	1879				SG21
462	1888				AG042, ZH087
2015	1911				SO17
4267	1922	Leverkusen	Hohenzollern	4268	SH09, ZH056, ZH105

Homemade
IRS convention — HM homemade

This section covers some of those locomotives produced locally; others are listed under the operators.

Number	Year	Reference(s)
-	xxxx	AG010
-	197x	VD78
-	1915	VD03
-	1919	VD03
-	1976	VD68
-	1981	VD68
-	1985	VD68

Herm. Hüscher Söhne, Eisenbau-Werkstätte, Schaffhausen — Hübscher
Hübscher & Co., Maschinen- und Stahlbau, Schaffhausen

Number	Year	Reference(s)
-	1952	ZH053

Forges, Usines et Fonderies de Haine, Saint-Pierre [B] (FUF) — HStP
Société Anonyme des Forges, Usines et Fonderies, Haine-Saint-Pierre

Number	Year	Other	Number	Reference(s)
1576	1927/28	BBC	2835	JU06

Maschinenbau-Anstalt Humboldt, Köln-Kalk [D] — Humboldt

Number	Year	Reference(s)
?	1906	SG34

Internationale Rheinregulierung — IRR

Number	Year	Other	Number	Reference(s)
-	1950	Heilbronn	228	SG34

Jäger — Jäger

Number	Year	Other	Number	Reference(s)
?	1916	BBC	1010	AG005, BL04

Jura-Cement-Fabriken — JCF

Number	Year	Other	Number	Reference(s)
?	1xxx	BBC	?	AG057

Jonneret SA, Genève — Jonneret
In liquidation by 2008

Number	Year	Reference(s)
?	19xx	ZH104, ZH131

Arn. Jung Lokomotivfabrik GmbH, Jungenthal [D] — Jung

Number	Year	Type	Reference(s)
13	1886		MI99
14	1886	25 PS	MI99
15	1886	40 PS	MI99
16	1886		MI99
18	1886		MI99
19	1886	50 PS	MI99

Number	Year	Type	Other Number	Reference(s)
21	1886	50 PS		MI99
22	1886	50 PS		MI99
24	1891	20 PS		MI99
26	1887	40 PS		MI99
27	1887	25 PS		MI99
29	1887	30 PS		MI99
34	1887	15 PS		MI99
40	1888	25 PS		MI99
52	1888	50 PS		MI99
58	1889	20 PS		MI99
59	1889	50 PS		TI29, TI66, TI73
60	1889	50 PS		TI29
64	1889	50 PS		MI99
65	1889	40 PS		MI99
66	1889	80 PS		MI99
67	1889	80 PS		MI99
69	1889	30 PS		SO19, SZ10
75	1889	50 PS		MI99
76	1889	50 PS		MI99
91	1891	10 PS		MI99
94	1890	20 PS		MI99
97	1891	40 PS		MI99
98	1891	40 PS		GR38, SG34, SG53, ZH171
100	1891	80 PS		ZH029, ZH083
101	1891	50 PS		MI99
104	1891	50 PS		TI04, TI20
105	1891	50 PS		TI49, ZH113
116	1891	50 PS		MI99
117	1891	50 PS		MI99
143	1893	60 PS		MI99
144	1893	60 PS		MI99
145	1893	120 PS		MI99
146	1893	120 PS		MI99
147	1893	120 PS		MI99
154	1894	40 PS		MI99
158	1893	50 PS		MI99
159	1893	40 PS		MI99
178	1893	50 PS		BE060, TI29
206	1895	40 PS		ZH060, ZH087
224	1895	50 PS		MI99
225	1895	50 PS		MI99
?	19xx			ZH037
625	1903	40 PS		MI99
626	1903	40 PS		MI99
668	1903			AG033
785	1904	125 PS		SO09, ZH088
911	1905	80 PS		AG049, LU19, OW4, TG04, TG17
920	1905	50 PS		AG007, TI14, TI30, TI33, VD02, ZH002, ZH177
1049	1907	30 PS		MI99
1310	1908	50 PS		BE060, TI29
1349	1909	10 PS		AG086, BE061, BE086, BL35, VD60, ZH123
1396	1909	Helikon		MI99
1397	1909	Helikon		MI99
1398	1909	Helikon		MI99
1411	1911	50 PS		MI99
1427	1910	Helikon		BL22, BS16
1515	1910	80 PS		AG086, BL35, SG34
1516	1910	80 PS		AG067, SG34
1603	1910	50 PS		VD33
1622	1911	80 PS		AG067, AG086, BE061, BL35, SG34
1682	1911	80 PS		AG067, AG086, BE061, BL35, SG34
1683	1911	80 PS		AG067, AG086, BE061, BL35, SG34

Number	Year Type	Other	Number	Reference(s)
1684	1911 50 PS			BE061, BE110, LU06, VS49, ZH064, ZH155
1693	1911 20 PS			AG088, BE061, BE109, VD05, VD58, VS29, VS54, ZH002, ZH020, ZH088
2245	1914 30 PS			MI99
3024	1919 Helikon			AG067, ZH002
3025	1919 Helikon			VS31, ZH002
3620	1924 20/25 PS			AG086, BL35, ZH014, ZH123
3621	1924 20/25 PS			ZH014
3893	1927 MS 13			MI99
4026	1927 MS Platform			MI99
4039	1927 MS13x Platform			ZH002
4070	1927 MSZ12			MI99
4112	1928 MS 13 Platform			MI99
4113	1928 MS 13 Platform			MI99
4138	1928 MS 131			MI99
4237	1928 MS 13 Platform			MI99
4238	1928 MS 13 Platform			MI99
4334	1928 MS 13 Platform			MI99
?	1929-33 MS 131			AG095, ZH046
4391	1929 MS 131			MI99
4392	1929 MS 131			MI99
4457	1929 MS 131			MI99
4487	1929 MS 130			MI99
4503	1929 MS 131			MI99
4549	1929 MS 131			MI99
4690	1929 MS 131			MI99
4704	1929 MS 131			GL09, ZH046
4709	1930 MS 131			MI99
4930	1930 MS 131			ZH032, ZH046
4935	1930 MS 131			AG087
4936	1930 MS 131			AG087
4940	1930 MS 131			AG087
4941	1930 MS 131			AG087
4980	1930 MS 131			MI99
4995	1930 MS 131			MI99
5060	1930 Pm			ZH112
5061	1930 Pm			ZH112
5076	1931 Pm			ZH112
5101	1930 MS 131			MI99
5105	1929 MS 131			MI99
5138	1930 MS 131			MI99
5139	1930 MS 131			MI99
5165	1931 MS 131			MI99
5799	1934 EL 105			AG106, ZH046
5810	1934 ZL 105			AG088, LU41
6880	1936 EL 105			SG79
7372	1937 EL 105			ZH046
7863	1938 ZN 133			SG81
7974	1938 EL 110			LU30
8208	1938 ZL 105			AG088, LU41, SG79
8457	1939 ZL 233			SG79, ZH046
8812	1940 VN 234			AG097
8860	1939 ZL 114			SG79
9268	1941 BR 64			BE113
9515	1941 ZL 105			AG088, LU41
9519	1941 ZL 105			AG088, LU41
10966	1948 EL 105			FR01, ZH029
11024	1952 ZL 105			FR12
11509	1956 RK 12 B			SO21
11795	1953 Flink			SG77
12041	1961 R 40 C			NE09, UR03, VS01, ZH085
12126	1954 Flink			SO08
12348	1956 R 40 C			TG20

Number	Year Type	Other	Number	Reference(s)
12351	1956 R 42 C			ZH085
12485	1957 V 60			BE053, ZH040
12749	1956 RK 15 B			TG09
12837	1957 R 40 C			GR23
12883	1957 R 30 B			SG16
12906	1958 Ez 21			VS27, ZH046
12907	1958 Ez 21			VS27
12994	1959 R 30 B			SG31
12998	1960 R 30 C			BL36
12999	1958 Flink			SO08, SO26
13008	1958 Freia			SO08
13061	1958 ZN 233			AG091, ZH180
13150	1959 Köf II			NE15, ZH105, ZH160
13156	1960 Köf II			BS12
13171	1960 Köf II			BS18
13172	1960 Köf II			BS12
13174	1960 Köf II			BS17, BS24
13178	1960 Köf II			SO08
13182	1960 Köf II			ZH160
13272	1960 RK 8 B			BL28, ZH031
13288	1961 R 42 C			ZH136
13290	1961 R 42 C			LU29, ZH085
13372	1961 RK 20 B			GE16, GR42
13372	1961 RK 20 B			VD19
13397	1961 RK 15 B			BL46, BS03
13399	1961 RK 15 B			BL46, BS29
13435	1962 RK 11 B			TG08, TG12
13458	1962 V 100			ZH136
13620	1962 R 30 C			GR23
13634	1963 RK 20 B			AG072
13691	1964 RK 8 B			BE071, BL13
13810	1970 EZ10			ZH042
13869	1965 RC 24 B			SO08
13874	1965 Flink			SO08
13879	1970 EZ10			ZH042
13951	1966 Unilok-2			TG21
14025	1969 RC 24 B			BL05, LU10
14038	1969 RK 11 B			SG67
14127	1971 RK 8 B			AG028
14129	1973 RK 8 B			AG097, BL29
14136	1972 RC 24 B			SO08
14137	1972 RK 24 B			BS03
14154	1973 RC 24 B			BL46
14163	1974 RC 43 C			LU29

Jenbachwerke AG, Jenbach [A] JW

Each class has its own number series. This is always shown on the builder's plate affixed to the locomotive, though it may be combined with other digits. The number recorded in the manufacturer's records is formed of the number in the series with the class indicated in different positions within this number. In the list which follows the left hand number is that which appears on the locomotive and the right hand one that in the manufacturer's records.

?	19xx JW8			JU03
051	1950 JW8 - PONY		8051	BE089, BE127, BE128, ZH046
?	1950			AG088
77.08	1950 DM50F-10		8077	GR10
94	1951 JW8		8094	MI09
100	1951 JW8		8100	MI99
101	1951 JW8		8101	MI99
225	1954 JW8 - PONY		8225	BE002, JU03, ZH046
238	1954 JW8		8238	OW6
257	1955 JW8		8257	MI09
319	1957 JW10a		8319	OW6

Number	Year	Type	Other	Number	Reference(s)
539	1983	JW10a		8539	OW6
2019	1950	JW20		20019	SG34
2379	1962	DM20/1		20379	JU12
2450	1965	DM20/1		20450	BE089, JU12, SO37
51.105	1952	DM50V10		50105	ZH006, ZH182
3.453.31	1967	DM500.8.10		100231	AG089
3.453.36	1970	DM100.6.10		100236	AG089
3.684.080	1982	DH660C54		600080	AG103

Besucherbergwerk Käpfnach, Horgen   Käpfnach

Number	Year	Type	Other	Number	Reference(s)
1	1989				ZH017
?	2006				ZH017

Maschinenfabrik Karlsruhe AG, Karlsruhe/Baden [D]   Karlsruhe

Maschinenbau-Gesellschaft Karlsruhe [D]
The spelling of Carlsruhe was used when the earlier locomotives were constructed.

Number	Year	Type	Other	Number	Reference(s)
284	1866				SG64
?	1894				BE061
1369	1894				BE061, GR38, VD23, VD56
2051	1918				VD10

Klöckner-Humboldt-Deutz AG   KHD

see: Deutz

Elsässische Maschinenbaugesellschaft Andreas Köchlin & Cie. Mulhouse [F]
    Köchlin

1872 merged to form SACM

Number	Year	Type	Other	Number	Reference(s)
1289	1870	68bis			BE046
1290	1870	68bis			BE046

Lokomotivfabrik Krauss & Co AG, Marsfeld, München [D]   Krauss-M
Lokomotivfabrik Krauss & Co AG, Sendling, München [D]   Krauss-S
Lokomotivfabrik Krauss & Co AG, Linz [A]   Krauss-L

Locomotive production commenced in 1867 at Marsfeld, but from 1875 an additional factory was started at Sendling. In 1882 locomotive production also started in Linz [A]. A common number list was maintained for all the steam locomotives produced. Some works plates give the address as "München + Linz", while others indicate either "Linz" or "München". The location of manufacture has no known impact on this Handbook. In the last ten years of production (1921-30) Linz maintained its own separate list and one of that series appears in this Handbook.

Number	Year	Type	Other	Reference(s)
?	18xx		?	SG09
71	1869	IV f	M	AG011
188	1872	XVIII a	M	LU16, SO28
290	1874	XIII d	S	JU04, SO39
330	1874	IV m	S	BS22
400	1874	IV n	S	BS22
401	1874	IV n	S	BS22
425	1874	IV o	S	AG011, SO22
430	1874	IV o	S	AG057, SG64, SO19, ZH088
434	1874	XXVII	S	AG055
485	1876	XVIII f	S	SG24
487	1875	XIII f	S	VD37
559	1876	XXVII b	S	BS09
799	1879	IV q	S	TI12
1049	1882	XVIII m	S	BS30
1150	1882	XXIV a	M	AG015, AG046
1190	1882	LXV	S	SG34
1284	1883	LXXI	S	SG34
1285	1883	LXXI	S	SG34
1292	1883	IV w	S	GE08, SG64, ZH064

Number	Year	Type	Other	Number	Reference(s)
1430	1883	LXXI a	S		SG34
1719	1886	LVIII i	S		SO37
1720	1886	LVIII i	S		VS01
2098	1889	IV pp	S		SG09, SG11, ZH049
2375	1890	IV ww	S		ZH049
2605	1891	XVIII bb	S		AG103
2615	1892	XIV u	S		SG56
2647	1893	IV cd	S		SG09, ZG2, ZH010, ZH087, MI05
2648	1893	IV cd	S		SG09, SO05, ZG2
2808	1893	IV fg	S		SZ16, ZH115
2811	1893	IV fg	S		BE117, SH01, ZH010
2817	1893	IV fg	S		ZG2, ZH087
2836	1895	LXXII l	S		LU35
2936B	1894	Zwilling	S		GE18, GL06, VS31
2960	1893	IV kl	S		AG057, GR38, SG09, SG19, SG34, SG78, SO19, ZH010, ZH087, ZH115, MI05
3048	1894	LXV s	S		SG53
3052A	1894	Zwilling f	S		VS31
3185	1895	IV tu	S		SG09, SZ10, ZH115, ZH117
3190	1895	IV tu	S		SG11, SO19, ZH049, MI05
3254	1895	XXVII vv	S		AG086, BE036, BL35, SG09, ZH002, ZH123
3562	1896	LXXII f	S		SG34
3566	1897	XXXV nn	S		SG34
3589	1897	LXXII f	S		SG34
3683	1897	XXXV qq	S		SG34
3684	1897	XXXV qq	S		SG34
3699	1897	XXXVV ss	S		SG34
3754	1897	LXXII f	S		SG34
3775	1898	XXXV vv	S		ZH053, ZH133
3813	1898	XXIV g	M		TI72
3816	1898	AV m	L		BL16
3839	1898	VXXII g	S		BE124, SG34
3878	1897	LXXII g	S		SG34
4048	1899	XIV mm	S		NE08, VD60, VS29, VS30, ZH002
4132A	1899	Zwilling n	S		BE060, BE069, SO38
4278	1900	XXXV lt	S		VD10
4589	1901	IV zs	S		BE028, GR38, SG53, SG55, SO19, ZH002
4642	1902	IV zu	S		GR38
5295	1905	IV bh	S		AG067, AG102, BS04, SG13, SO02, SO09, SO43, ZH002, ZH147
5331	1905	XLV ac	S		BS23, ZH053, ZH093, ZH180
5492	1905	LXV n	S		BS04, SG13
5564	1906	XXVII au	S		TG10, ZH024, ZH079, ZH043, ZH118
5666	1907	XXVIII p	S		AG033
5757	1908	IV bv	S		SG09, SG11, ZH010
6173	1909	XXVII bg	S		AG086, BE086, BL35, SG09, ZH002, ZH123
6426	1911	XIV am	S		BE061
6427	1911	XIV am	S		UR07, ZH002, ZH005, ZH014
6502	1913	XIV anBBC	S		ZH002, ZH014
7349	1917	LXXXI c	M		AG088
7759	1920	VXI de	S		BE006
7760	1920	VXI de	S		BE006
7884	1921	VXI dl	S		BE006
7889	1921	VXI dl	S		BE006
7899	1921	LXI dm	S		AG016, ZH029, ZH118

Lokomotivfabrik Krauss & Co AG, Linz [A]   Krauss(L)

Locomotives produced at Linz 1882-1921 are included in the common list under Krauss above. The 351 locomotives built 1921-30 are in a separate number series (1171-1521). The one locomotive from that list in this Handbook is:

Number	Year	Type			Reference(s)
1430	1927	BBÖ 629			FL2, FL3

Number	Year	Type	Other	Number	Reference(s)

Krauss-Maffei GmbH, München-Allach [D]   KrMa
IRS convention   KM
Krauss-Maffei AG, München-Allach [D] (till 1996)
Founded in 1931 by the two named constituents.

Number	Year	Type	Reference(s)
16388	1945	40 PS	NE15
17627	1949	KDL 10	BE027
18320	1958	ML 500 C	SZ09
18356	1957	ML 500 C	FR03
18701	1960	ML 225 B	BL36, ZH085
18850	1962	ML 700 C	ZH085
18916	1962	V 100.10	ZH136
19089	1963	ML 700 C	AG033, ZH136
19279	1965	M 250 B ex	ZH085
19399	1969	M 350 C ex	BL19, ZH085
19405	1968	M 500 C ex	LU16
19675	1973	M 700 C	ZH085
19687	1973	M 700 C	ZH085

Rudolf Kröhnke, Buxtehude [D]   Kröhnke

Number	Year	Type	Reference(s)
298	1957	Lorenknecht	LU30
300	1958	Lorenknecht	LU30
313	1959	Lorenknecht	SG79

KROMAG Metallindustrie- Gesellschaft mbH, Hirtenberg [A]   Kromag

Number	Year	Type	Reference(s)
?	19xx		SG85

S. Kronenberg, Fabrique de Machines, Lucerne   Kronenberg
IRS convention   KRO
S. Kronenberg, Maschinenbau, Lucerne
Kronenberger Söhne & Co, Lucerne

Number	Year	Type	Reference(s)
101	1938	DT 50	LU16
102	1940	DT 60	SG18, SG36
103	1940	DT 60P	BE023
104	1941	DT 45P	LU02
110	1947	DT 90	BS14, TI62
112	1948	DT 60P	AG022, JU01, LU22
113	1948	DT 90	BS29
115	1949	DT 90	BL52, TI64
116	1950	DT 60	TG02, MI08
117	1950	DT 120	BE048, ZG1
118	1951	DT 60	BE037, SG14
119	1950	DT 50	SO26
120	1950	DT 90	BE072, SZ13, UR12
121	1951	DT 60/90P	BS03, MI08
122	1951	DT 50/90	BL28
123	1952	DT 90/120	BE059, BE016, BE072
124	1952	DT 90/120	BE059, VS44
125	1952	DT 90/120	NE02, TG21
126	1952	DT 90/120	BE059, UR12, VS14, ZH181
127	1952	DT 90/120	BE016, BE072
128	1952	DT 90/120	*BE072*, TI55
129	1952	DT 180/250	VS36, VS38
130	1953	DT 180/250	GR06, TI64
131	1954	DT 90/120	VS62
141	1956	DT 180/250	AG015, AG048, LU29
142	1957	DT 100/220	BE059, BE072, SZ13
143	1958	D 250/300	AG103
144	1961	3 L 600	AG015, AG048, BS17
145	1962	DT 90	BL17, SO07, MI08
146	1963	DT 120	*BE059*, UR12

Number	Year	Type	Other	Number	Reference(s)
147	1964	DL 200			AG097, SG14

Fried. Krupp AG, Essen [D]   Krupp

Number	Year	Type	Other	Number	Reference(s)
3324	1954	350 PS	Sulzer	345	TG03, ZH149
3340	1956	225 PS			BL43
3464	1955	200 PS			TI21, VD30
3465	1955	200 PS			TI21, VD30
3487	1957	B 330			ZH136
3781	1961	1W1C			ZH035, ZH085, ZH136
4359	1962	V 100			ZH136
4381	1962	V 100			ZH085

Société Anonyme des Ateliers de Construction de la Meuse, Sclessin [B]   La Meuse

Number	Year	Reference(s)
2252	1910	VD33
2400	1910	VD33

Levahn Mekaniske Verksted Aktieselskap, Skøyen [N]   Levahn

Number	Year	Reference(s)
?	19xx	AG089

Lokomotivbau-Elektrotechnische Werke 'Hans Beimler', Hennigsdorf [DDR]   LEW

to AEG in 1992

Number	Year	Type	Reference(s)
12246	1968	V60	FR03, VD59

Lima Locomotive Works, Lima [USA]   Lima

title valid 1916-1947

Number	Year	Type	Reference(s)
8939	1945	141.R	ZH130

LOWA Lokomotivbau Karl Marx, Babelsberg [DDR]   LKM

ex O&K, Werk Babelsberg from 18/03/1948

Number	Year	Type	Reference(s)
?	195x	Ns2f	ZH073, ZH106
48432	1953	Ns2f	ZH073
248665	1955	Ns2f	SG79
262005	1958	Ns2f	AG088
262050	1958	Ns2f	ZH046
262057	1959	Ns2f	BE089, ZH046
262211	1959	Ns2f	ZH046
265023	1970	V23^l	TI71

J. A. Maffei AG, Lokomotiv & Maschinenfabrik, München [D]   Maffei

Number	Year	Type	Reference(s)
1369	1884	bay D IV	MI99
1374	1884	bay D IV	MI99
1709	1910	Duplex-Compound	MI07
2833	1908	40 PS	BL22, BS16, LU18
2894	1908	TL – 30 PS	MI99
2895	1909	TL – 30 PS	MI99
2920	1909	40 PS	AG047, GR13, TI09, TI52
2954	1910	20/25 PS	SG15, ZH002, ZH048, ZH093
2983	1909		SO29, SO44
3126	1910		TI17
3129	1910		AR2
3505	1909		AG067, SG15, SG64, ZH072, ZH117
3508	1909		AG067, SG72, TG22, TI46, ZH002, ZH183
3512	1911		ZH002
3513	1911		BE009, BE061, BE095, ZH002
3521	1911		ZH002
3564	1911		BL44, LU34
3581	1911		ZH002, ZH015

Number	Year	Type	Other Number	Reference(s)
1491	192x	S5		BE086
1520	1922	L308		BE020, BE068, FR01
1543	192x	S5		GL09, ZH046
1545	192x	S5		BE122
1588	192x	S5		ZH174
1613	192x	S5		ZH134
1735	1923			GR12, GR34
1747	192x	S10		ZH064
1751	192x	S5		BE080
1890	192x	S10		ZH059
1891	1924	S10		SG85
1969	c1924	S10		TI49
2063	192x	H 1		ZH046, ZH074, ZH111
2064	192x	H 1		ZH072
2181	192x	RL 1a		ZH074
2182	192x	H 1		SG85
2196	c1926	S 10 a		AG060, ZH073
2231	1926	M		VS30
2308	1926	H 1		GL04
2346	1926	S 10 a		AG057
2385	1927	H 2		BE001, BE071, FR01
2572	1927	S 10 a		AG057
2614	1927	RL 2		SG06
2680	1927	M		AG048
2698	1927	RL 2		BE086, BL35, ZH086, ZH120
2743	1928	M		BE099, SG06
2774	1928	RL 2		ZH083
2858	1928	RL 2		SG06
2899	1928	H 1		SG85
2973	1928	M		ZH131
3011	1928	RL 1		SG48
3012	1928	RL 1		SG55, ZH114
3356	1929	RL 1		BL25
3466	1929	RL 1		BL25
3474	1929	RL 1		AG020, AG112
3490	1929	RL 1		SO38, TG24
3491	1929	RL 1		SO38, TG05
3615	1929	RL 1		SZ09
3616	1929	RL 1		TI54
3644	1929	RL 1		ZH074
3645	1929	RL 1		SG07, SG84, ZH074
3660	1929	RL 1		BL30
3661	1929	RL 1		SG07, SG79, SG85
3709	1929	RL 1		ZH014
3710	1929	RL 1		SZ09
3763	1929	RL 1		TI54
3816	1929	RL 1		AG020
3826	1929	RL 1		BS13
3829	1929	RL 1		ZH145
3839	1929	RL 1		SZ09
3840	1929	RL 1		SG22
3907	1929	RL 1		SG85
3908	1929	RL 1		GR13
3945	1930	RL 1a		ZH158
3946	1930	RL 1a		TI52
3947	1930	RL 1a		TI52
3948	1930	RL 1a		ZH072
4003	1930	RL 1a		AG010, AG020, AG061
4004	1930	RL 1a		ZH072
4005	1930	RL 1a		AG056, AG089, AG110
4020	1930	RL 1a		GR13
4021	1930	RL 1a		JU05, ZH046
4022	1930	RL 1a		ZH014, ZH088

Number	Year	Type	Other	Number	Reference(s)
4023	1930	RL 1a			SG85
4109	1930	RL 1a			TI41
4170	1930	RL 1a			BS07
4171	1930	RL 1a			ZH083
4209	1930	RL 1a			SG35, ZH098
4210	1930	RL 1a			BE045, BL21, ZH046, ZH088
4211	1930	RL 1a			ZH083
4212	1930	RL 1a			ZH083
4291	1930	RL 1a			SG23, SG85
4307	1930	RL 1a			BE060, VS21
4322	1930	RL 1a			ZH072
4329	1930	RL 1a			ZH114
4331	1930	RL 1a			SO15
4332	1930	RL 1a			SG07, ZH046
4333	1930	RL 1a			ZH114
4360	1931	RL 1a			ZH158
4361	1931	RL 1a			FR04
4376	1931	RL 1a			BE080
4377	1931	RL 1a			FL1
4378	1931	RL 1a			TI57
4398	1931	RL 1a			AG048
4405	1931	RL 1a			BE130
4416	1931	RL 1a			ZH071
4439	1931	RL 1a			AG010, AG020
4440	1931	RL 1a			BE091
4452	1931	RL 1a			ZH064
4453	1931	RL 1a			ZH021, ZH046
4465	1931	RL 1a			GR36
4466	1931	RL 1a			MI99
4551	1931	RL 1a			GE09
4593	1931	RL 1a			OW7, ZH046
4602	1931	RL 1a			BL25
4609	1932	LD 16			GL09, *SG79*, ZH113
4610	1932	LD 16			TG13, ZH114
4612	1931	RL 1a			AG083
4614	1931	RL 1a			BL01
4615	1931	RL 1a			AG048, BL10
4627	1931	RL 1a			SG10
4639	1931	RL 1a			AG048, ZH014
4640	1931	RL 1a			BL21
?	1932	RL 1a			AG010
4650	1932	RL 1a			SG22
4651	1932	RL 1a			ZH183
4652	1932	RL 1a			SG04
4684	1932	LD 16			TG27
4685	1933	LD 16			AG011
4686	1932	RL 1a			ZH183
4690	1932	RL 1a			VD12, ZH014, ZH046
4704	1932	RL 1a			*ZH014*
4705	1932	RL 1a			*ZH014*
4706	1932	RL 1a			VD12
4826	1932	RL 1a			GL04
4834	1933	RL 1a			GR30
4835	1933	RL 1a			AG056, BS07
4871	1933	RL 1a			ZH139
4880	1933	RL 1a			ZH183
4881	1933	RL 1a			TG27
4883	1933	LD 2			SG85
4897	1933	RL 1a			MI99
4898	1933	RL 1a			MI99
4899	1933	RL 1a			VD41, ZH046, ZH173
4900	1933	RL 1a			AG030
4901	1933	RL 1a			ZH064

Number	Year	Type	Other	Number	Reference(s)
4911	1933	LD 2			BE130
5310	1934	RL 1a			AG020
5363	1934	RL 1a			FL1, ZH058
5364	1934	RL 1a			ZH183
5642	1934	RL 1a			GL04
5643	1934	RL 1a			FL4
5644	1934	RL 1a			SG40, SG59
5645	1934	MD 1			AG059
5680	1934	RL 1a			TI54, VD35
5681	1934	RL 1a			AG056
5691	1934	LD 2			TG27
5763	*1934*	LD 2			ZH114
5994	1935	RL 1a			GL02
6141	1935	LD 2			BE080
6404	1935	LD 2			VD29
7127	1936	LD 2			BE091
7385	1937	LD 2			BE066, ZH083
7452	1937	MD 2			GL02
7465	1937	LD 2			BE080, ZH083
7576	1937	RL 1c			BE041, BE085
7708	1937	RL 1c			MI99
7750	1935	MD 1			ZH039, ZH046, ZH124
7759	1937	RL 1c			ZH058
7806	1937	MD 2			AG089
7809	1937	RL 1c			ZH086
7810	1937	RL 1c			FL1
7811	1937	RL 1c			AG030
7812	1937	RL 1c			ZH072
7858	1937	RL 1a			TI56
7859	1937	RL 1c			*BS07*, FL1
7879	1937	RL 1c			BS07
7880	1937	RL 1c			ZH083
7901	1937	MD 1			AG071
7902	1937	MD 1			BS27
8102	1937	RL 1c			OW7
8103	1937	RL 1c			AG048
8191	1937	LD 2			GL09, SG55, ZH046, ZH113
8202	1937	MD 1			ZH072
8282	1937	MD 1			ZH022
8293	1937	LD 2			ZH027
8340	1937	RL 1a			ZH014
8341	1937	LD 2			BE007, BE041, ZH064
8371	1937	RL 1c			SG22, ZH046
8372	1937	RL 1c			SZ12
8400	1937	LD 2			AR5, FL1
8422	1937	RL 1c			AR5, SZ19, ZH112
8423	1937	RL 1c			ZH183
8456	1937	RL 1c			GL09, ZH046
8457	1937	RL 1c			GR38
8476	1940	MD 2			SG34
8558	1937	RL 1c			AG029
8559	1937	RL 1c			ZH058
8585	1937	LD 2			BE077
8586	1937	LD 2			SG49, SG82
8587	1937	LD 2			GL09, ZH046
8588	1937	LD 2			AG077
8605	1938	MD 1			ZH027
8606	1938	MD 1			BE089, ZH027
8767	1938	MD 1			AG059
9400	1939	MD 2			SG34
9555	1938	RL 1c			ZH148
10502	1940	MD 1			BL33
10503	1940	MD 1			BE060, SZ14

Number	Year	Type	Other	Number	Reference(s)
10677	1939	LD 2			VD52
10678	1939	LD 2			SG49, SG82
10685	1941	MD 1			ZH018
10701	1940	MD 2			AG089
11048	1939	RL 1c			MI99
11049	1939	RL 1c			GE24
11366	1940	MD 1			GE14
11367	1940	MD 1			GE14
11413	1939	MD 1			NW1
11414	1939	MD 1			GR16, VD57
11415	1939	MD 1			BL33
11425	1939	RL 1c			ZH064
11426	1939	RL 1c			AG011
11427	1939	RL 1c			GL09
11428	1939	RL 1c			SZ12
11433	1940	RL 1c			MI99
11434	1940	RL 1c			GR08
11435	1940	RL 1c			GR08
11438	1940	MD 1			GR36
11562	1940	MD 1			VD14
11563	1940	MD 1			GE08
11713	1943	MD 3			AG089, SG34
11714	1947	MD 3			ZH014
11715	1947	MD 3			ZH014
12002	1944	MD 2r			ZH014, ZH073
12004	1944	MD 2r			SG28, SG80, ZH014
12005	1944	MD 2			ZH114
12006	c1944	MD 2			ZH022
12008	1944	MD 2r			ZH014
12009	194x	MD 2			BE040
12030	194x	MD 2			AG060
12039	194x	MD 2			BE091
12040	c1943	MD 3			BE060
12043	194x	RL 2			SG48
12056	194x	MD 2			ZH048
12057	194x	MD 2			ZH048
12058	194x	MD 2			BE091
12231	1950	MD 2			GR32, GR36, ZH088
12234	1952	MD 2r			GR09

Orenstein & Koppel, Sestao San Giovanni [I]   O&K(S)

similar to the construction performed by MBA in Zürich this works built its own locomotives ca. 1943-50.

Number	Year	Type	Other	Number	Reference(s)
?	19xx				TG26
1081	19xx	RL 1a			ZH058
1082	19xx	RL 1a			ZH108
1433	c1943	RL 1c			BE060
1473	19xx	MD 2			BE060
1474	194x	RL 1 s			ZH072
1475	194x	MD 2s			BE080
1501	1947	RL 1c			VD14, ZH183
1502	1947	RL 1c			SG52
1503	194x	RL 1c			BE129
1504	194x	RL 1c			LU41
1505	194x	RL 1c			AG111
1506	194x	RL 1c			ZH078
1517	194x	MD 2			TI28
1518	194x	MD 2b			GL09, GR09, ZH046
1519	194x	MD 2s			ZH112
1520	1947	MD 2			ZH183
1545	194x	MD 2s			GR22
1548	194x	MD 2s			GR36, ZH022

Number	Year	Type	Other	Number	Reference(s)
1550	1947	RL 1 s			TG06, ZH072
1552	194x	RL 1c			BL25
1554	1947	RL 1c			AG052, GR18, ZH105
1605	194x	L 200			VS21
1639	1948	RL 1c			SG06
1640	194x	RL 1c			AG048
1641	194x	RL 1c			BE080

Motorenfabrik Oberursel AG, Oberursel [D] — Oberursel

Number	Year	Type	Other	Number	Reference(s)
?	19xx				BE024, VD22
?	1911				SZ22
?	c1915				VD10, VD18, VD32, VS28

Roheisenwerke Oehler & Co, Aarau — Oehler
Oehler & Cie, Aarau

Number	Year	Type	Other	Number	Reference(s)
?	19xx				GL03
?	1912		BBC	?	LU19, SG77
?	1917	Akku-Fahrz.	BBC	1099	SO13
413	1941	LO 1541			TI70
?	1942(7x)				VD68
492	1942	1042			TI24
582	1944				BE091
701	1946	ELG1042			BE128
?	1950				BE091
779	1950	G2049			TI22
832	1950	G3050			AG011
868	1950	G3050			AG011
958	1951	G2050			TI27
998	1954	G2050			TI53
?	1955	G5055			AG011
1025	1955	G3050			AG011
1100	1957	G3050			AG011
1101	1957	G3050			AG011
1136	1958	G3053			TI10
1163	1958	D1158			TI58
?	c1960				AG083
1215	1960	G3050			AG011
2127	1987				VS32
2139	1988				TI58

Hauptwerkstätte SCB, Olten — Olten
Hauptwerkstatt SBB (Olten)

Number	Year	Type	Other	Number	Reference(s)
7	1863				BE036
8	1863				BE036
20	1870				BE070, BE102, JU11, LU39, SO35
?	1923		BBC	1942	VD34, VD47, VS39
?	1923		BBC	1945	FR01, VD40, VS39
?	1927				MI07
?	c1932		MFO	?	AG081, BE116, MI08

Orion SA, Zürich — Orion
road motor vehicles

Number	Year	Type	Other	Number	Reference(s)
-	1916				ZH168

Etablissement A. Pinguely, Lyon [F] — Pinguely

Number	Year	Type	Other	Number	Reference(s)
354	1919				VS08

Plasser & Theurer, Johannesgasse 3, 1010 Wien [A] — Plasser

Number	Year	Type	Other	Number	Reference(s)
?	1967				VD64

Plus AG, Dornacherstrasse 110, 4147 Aesch — Plus
Providers of electrical equipment

Number	Year	Type	Other	Number	Reference(s)

Plymouth Ironworks Corporation, Plymouth, Ohio [USA]   Plymouth

Number	Year	Type	Other	Number	Reference(s)
7305	19xx				AG089
7571	19xx				AG089
7731	19xx				AG089

Robert Aebi & Cie AG, Zürich   RACO

Robert Aebi AG (from 1996)

After about 1950 the locomotive type was given as two or three character groups, where the first group represented the power (in PS) and the second the "type". Before that the two were recorded separately and, particularly for the tractors, both parts have not always been recorded.

Number	Year	Type	Other	Reference(s)
?	19xx			BE058, ZH046
?	19xx			GL07
?	19xx			SG85
?	19xx			VS34
?	19xx			BE089, ZH017, ZH046
?	19xx			ZH046, ZH064, ZH065
?	1923			SG85
11	1930	40 PS		TI14, ZH053
103	1933	RA7		BE072, SZ13
104	1933	RA7		BE063, BE090, FR01
881	1927	35 PS		ZH084
963	1930	40 PS		TG07, ZH029
977	1930	25 PS		VD73
982	1931	40 PS		AG006
993	1931	40 PS		LU38
1001	1931	*70 PS*	*Breuer*	ZH133
1066	1933	25 PS		AG053, AG081
1212	1933	45 PS		AG094, BE020
1215	1933	RA7		SO20
1219	1933	45 PS		ZH135
1224	1934	45 PS		BL14
1225	1934	45 PS		AG034, BL04
1232	1935	45 PS		AG022, AG078
1235	1935	45 PS		BS01, BS18
1237	1935	45 PS		AG016, ZH029
1243	1936	45 PS		BE062
1246	1936	45 PS		SG16
1247	1937	45 PS		AG022, BE062
1248	1937	45 PS		BE032, VS73, ZH006, ZH153, MI08
1250	1937	45 PS		SG16
1253	1938	45 PS		ZH029
1257	1939	RA11		TI39, VS01
1259	1940	30 PS		BE072, TI55
1260	1940	RA11		GE20, VD16, ZH043
?	1944			SG38
1264	1944	16PS		SG80, ZH014, ZH046
1301	1945	12/16PS		BE080, ZH046
1304	1945	12/16PS		SO37
?	194x	50 PS		VD58
1312	1945	50 PS		AG089, SG34
1313	1945	50 PS		FR05, SG34, SG58
1314	1945	50 PS		FR05
1315	1945	50 PS		FR05
1316	1945	50 PS		FR05
1317	1945	50 PS		FR05
1320	194x	12/16PS		ZH046, ZH064, ZH065
1321	1945	50 PS		FR05
1333	194x	12/16 PS		GL01, GL07, ZH046
1343	1947	80 SA 3		BE103
1349	1947	12/16PS		SO37
1364	1947	RA11		ZH029, AG016
1365	1948	RA11		BE114

Number	Year	Type	Other	Number	Reference(s)
1366	1949	12/16PS			ZH046, ZH122
1376	1948	Tm11			VD43, VS06
1377	1948	RA11			LU09
1378	1948	45 PS			UR09
1380	1949	45 PS			ZH181
1383	1949	TM11			VD71
1387	1949	RA11			MI08, UR12
1388	1949	80 SA3			BE065, BS28, VD17, VS71
1397	1950	80 SA3			AG091, SG62, TG17
1405	1951	80 SA3			SZ20
1408	1951	80 ST4			BE016, BE072, ZH085, MI08
1417	1953	80 SA3			SG00, TG20
1421	1952	50 ST			ZH127, ZH132, ZH160
1422	1953	80 ST4			GE04, MI08
1423	1953	80 ST4			OW3
1424	1952	V	Breuer	3049	JU12, SO44, SO45
1431	1953	95 SA3			TI17
1433	1953	95 SA3			MI08
1439	1955	95 SA3			TI17
?	19xx	95 SA3 RS			TI71
1451	1955	95 SA3 RS			AG097
1453	1955	95 SA3 RS			AG022, BL26
1455	1955	95 SA3 RS			TG10
1461	1955	95 SA3 RS			TG20
1474	1956	95 SA3 RS			BL12
1476	1956	95 SA3 RS			JU04
1481	1956	95 SA3 RS			AG022
1489	1957	95 SA3 RS			SG65
1490	1957	VL82PS	Breuer		BL34, ZH062
1497	1957	60 LA7			ZH070, ZH085, MI08
1498	1958	60 LA7			VD62
1500	1958	60 LA7			AG113
1505	1958	95 SA3 RS			JU04
1512	1958	95 SA3 RS			AG017
1520	1958	95 SA3 RS			SG62
1521	1958	95 SA3 RS			BE055
1523	1958	95 SA3 RS			ZH029
1533	1958	95 SA3 RS			TI74
1546	1959	65 SA3 HT			UR09
1547	1959	65 SA3 HT			UR09
1549	1958	30 VA/HT			NE12, SG31
1552	1959	95 SA3			AG065
1562	1959	95 SA3			AG032
1569	1959	100 SA4 HT			BE059
?	19xx	85 LA7			BE072
1574	1960	85 LA7			UR15
1576	1960	85 LA7			SO08
1577	1960	85 LA7			BL27, LU22
1578	1960	85 LA7			BE062
1588	1960	95 SA3 RS			VS44
1595	1961	100 SA7 HT			TI62, ZH167
1596	1961	85 LA7			VD13
1599	1961	85 LA7			TI17
1601	1961	85 LA7			FR09
1603	1961	85 LA7			AG018
1605	1961	85 LA7			VS72
1608	1961	85 LA7			NE04, SG66
1610	1961	85 LA7			BE052
1615	1961	95 SA3 RS			ZH034
1620	1961	95 SA3 RS			AG105
1629	1962	140 SA3			UR09
1630	1962	140 SA3			UR09
1634	1964	300 PS HT			FR16, SO37

Number	Year	Type	Other	Number	Reference(s)
1635	1964	300 MA7 HT			JU12, SO44
1639	1962	85 LA7			MI07
1641	1962	85 LA7			SG66
1649	1962	85 LA7			VS06
1653	1963	85 LA7			TG30
1654	1963	85 LA7			GE07
1655	1963	85 LA7			TG16
1657	1963	85 LA7			BE106
1659	1963	85 LA7			TI62
1660	1963	85 LA7			BE106
1662	1963	95 SA3 RS			TI17
1664	1963	95 SA3 RS			TG20
1667	1963	95 SA3 RS			LU09
1676	1963	85 LA7			SO07
1678	1963	85 LA7			VS65
1682	1964	85 LA7			VS14
1688	1964	85 LA7			MI07
1690	1964	85 LA7			BE019
1695	1964	85 LA7			BE106
1697	1964	85 LA7			SO36
1698	1964	85 LA7			MI07
1699	1966	200 PS HT			SO37, SO44
1707	1964	95 SA3 RS			SH05
1708	1964	95 SA3 RS			BE107
1723	1965	85 LA7			TI14
1727	1965	85 LA7			JU11
1731	1965	85 LA7			VD17
1732	1965	85 LA7			SG70
1737	*1965*	*95 SA3 RS*			*TI71*
1740	1965	95 SA3 RS			ZH029
1744	1965	95 SA3 RS			AG104
1766	1968	40 MA4 H			AG018, ZH094
1783	1967	95 SA3 RS			TG20
1790	1968	95 SA3 RS			BE055
1798	1971	95 MA3 RS			AG036, TG020
1811	1972	85 DA7			JU12, SO44
1817	1974	70 DA4 H			SG31, ZH028, ZH085
1818	1975	100 DA4 H			SH09, ZH034
1820	1976	210 SV4 H			VD19
1826	1975	100 DA3 H			TG20
1838	1978	225 SV4 H			TI71
184x	1980				SG73, TG20
1867	1981	225 SV4 H			SZ20
1870	1981	225 SV4 H			BE033
1873	1982	225 SV4 H			TI71
1876	1982	225 SV4 H			TI71
1877	1982	225 SV4 H			ZH034
1881	1982	225 SV4 H			BE033
1928	1986	225 SV4 H			BE033
1931	1986	225 SV4 H			AG081, MI08
1932	1986	225 SV4 H			BE113
1936	1987	225 SV4 H			ZH034
2023	1995				BE059
3021	1949	RA11			BE103, FR01, ZH027

Waggonfabrik Rastatt GmbH, Rastatt [D]   Rastatt
later: BWR Waggonreparatur GmbH, Rastatt [D]

-	1910		MFO	21	BL10, SO08, ZG3

Ruston & Hornsby Ltd, Lincoln [GB]   R&H
IRS convention   RH
In general the maker's plates read "Ruston & Hornsby".

Number	Year	Type	Other	Number	Reference(s)
218003	1946	16/20 HP			MI99
247165	1947	13DL			MI99
247166	1947	13DL			MI99
247167	1947	13DL			MI99
247168	1947	13DL			MI99
247169	1947	13DL			MI99
247170	1947	13DL			MI99
247172	1947	13DL			MI99
247173	1947	13DL			MI99
249517	1947	48DL			AG089
252086	1947	25/30 HP			MI99
252840	1949	88DS			GE16, VD16, VD19
252848	1947	20DL			MI99
256188	1948	30DL			VS52, ZH046
256189	1948	30DLU			SG34
256190	1948	30DL			FR13, GR46, SG18, SG49
256800	1948	20DL			MI99
285351	1950	30DLU			SO37, ZH032, ZH046
285352	1950	30DLU			GR43, ZH046
285353	1950	30DLU			MI99
285354	1950	30DLU			MI99
285355	1950	30DLU			MI99
285356	1950	30DLU			SO37
296043	1950	48DL			MI99
296044	1952	48DL			BE082, VS54
296045	1950	48DL			MI99
296046	1950	48DL			MI99
321726	1951	88DS			VD17, VS01
323589	1952	30DL			SO37
323590	1952	30DL			SO37
323591	1952	30DL			SO37
371380	1954	30DL			SO37
386623	1955	LBU			AG82, VS77
386851	1955	48DL			SZ09
398112	1956	LBT			MI99
476114	1962	LBT			SO37
497713	1963	LBT			SO37
497743	1963	LSSH			VD70
497744	1963	LSSH			VD70
497745	1963	LSSH			VD70
497746	1963	LSSH			UR06, VD70, VS55
518188	1965	LFT			SO37
518189	1965	LFT			SO37

Ringhoffer, Praha [CZ] — Ringhoffer

Number	Year	Type	Other	Number	Reference(s)
?	1909		Rieter	?	TI37

Robel Bahnbaumaschinen GmbH, Industriestraße 31, Freilassing [D] — Robel

Number	Year	Type	Other	Number	Reference(s)
?	1931				AG006
2353	1994	Bamow AG 54			ZH136

Ruhrthaler Maschinenfabrik H. Schwarz & Co. Mülheim (Ruhr) [D] — Ruhrthaler Ruhr

IRS convention

Number	Year	Type	Other	Number	Reference(s)
210	1913	16 PS			UR11
484	1921	12/16 PS			MI99
485	1921	10/13 PS			MI99
489	1921	10/13 PS			MI99
491	1921	10/13 PS			MI99
505	1921	12/16 PS			MI99
528	1921	12/16 PS			MI99
531	1921	10/13 PS			MI99
532	1921	10/13 PS			MI99
533	1922	11/13 PS			MI99

Number	Year	Type	Other	Number	Reference(s)
538	1922	14/16 PS			MI99
541	1922	11/13 PS			MI99
543	1922	11/13 PS			SG82
549	1922	11/13 PS			MI99
563	1922	11 PS			MI99
567	1922	14 PS			MI99
574	1922	14 PS			MI99
615	1924	8 MT			MI99
616	1925	8 MT			MI99
623	1925	8 MT			MI99
624	1925	8 MT			MI99
625	1925	8 MT			MI99
636	1925	8 DML			MI99
647	1925	8 DML			MI99
656	1925	8 MT			MI99
685	1925	8 PS			MI99
757	1927	30 DML/S2			MI99
781	1927	30 DML/S2			MI99
1052	1929	15 DML/S1			MI99
1053	1929	22 DML/S2			MI99
1092	1930	22 DML/S2			BE060
1178	1931	10 DKL/S1			MI99
1190	1931	10 DKL/S1			MI99
1210	1931	10 DKL/S1			MI99
3257	1954	G22Z			BE039
3278	1955	G42Z			AG088
3286	1955	G22Z			BE039
3291	1955	G42Z			AG088
3325	1955	D54Z			AG088
3530	1957	NO1206V			AG094
3943	1970	G100HVG			GR10

SA des Ateliers de Sécheron, Genève — SAAS

Number	Year	Type	Other	Number	Reference(s)
?	1xxx				GE17
94004	1920	Be 4/7	SLM	2716	MI07
600003	1924	Ae 3/5	SLM	2948	MI07
818001	1926	Te^{II}	SLM	3186	MI08
999019	1928	Fe 4/4	SWS	?	MI07
1700-1	1930	Ae 4/7	SLM	3425	MI01
1700-6	1930	Ae 4/7	SLM	3445	MI01
1700-7	1930	Ae 4/7	SLM	3449	MI07
1700-8	1930	Ae 4/7	SLM	3451	ZH130, MI01
1700-9	1930	Ae 4/7	SLM	3456	MI01
2080-1	1931	Be 4/4	SLM	3509	AR6
2525-3	1932	Ae 4/7	SLM	3540	MI01
4840	1942	Ee 3/3	SLM	3806	MI01
5650-3	1945	Re 4/4^{I}	SLM	3890	MI01
5427	1950	Ee 6/6	SLM	4038	MI08
5444	1954	CFe 4/4	SLM	4108	MI07

Société Alsacienne de Constructions Mécaniques, Usine de Grafenstaden, Ht. Rhin [F] — SACM-G

Maschinenfabrik Grafenstaden (till 1919)

Number	Year	Type	Other	Number	Reference(s)
4172	1890				VD10, VS08
7916	1943	BR 52			ZH033

Société Alsacienne de Constructions Mécaniques, Usine de Mulhouse, Ht. Rhin [F] — SACM-M

Number	Year	Type	Other	Number	Reference(s)
3963	1886				BS06

Safir Engineering, Ltd [GB] — Safir

road motor vehicles

Number	Year	Type	Other	Number	Reference(s)
-	1908-9				ZH143

Société d'Applications des Moteurs à Huile Lourde, Paris [F]   SAMHUL

Number	Year	Type	Other	Number	Reference(s)
?	19xx				BE114

S.A. de Transbordement et Manutention, Basel   SATRAM

Number	Year	Type	Other	Number	Reference(s)
-	c1955				BS19
-	1972				BS17, BS19

Schweizerische Bundesbahnen   SBB(Biel)

Number	Year	Type	Other	Number	Reference(s)
?	1919		CIEM	?	VS13

Gew. Schalker, Eisenhütte Maschinenfabrik GmbH, Gelsenkirchen [D]   Schalke

Number	Year	Type	Other	Number	Reference(s)
?	19xx	EL 9	SSW	?	ZH046
57610	1960	EL 9	SSW	6067	BE032, ZH046, ZH088
57623	1959	Abraumlok	BBC	5737	TI21
57624	1959	Abraumlok	BBC	5738	TI21, VD30
57625	1959	Abraumlok	BBC	5739	TI21, VD30
57626	1959	Abraumlok	BBC	5740	TI21, VD30
67040	1967	3A16	SSW	6301	AG089

Schneider & Cie.   Schneider

Société des Forges et Ateliers du Creusot (Usines Schneider), Le Creusot, S & L [F]
Société des Forges et Ateliers du Creusot, Le Creusot [F]
Le Matériel de Traction Electrique, Le Creusot, Jeumont, Lyon [F] (from 1954)

Number	Year	Type	Other	Number	Reference(s)
?	18xx				VD21
799	1859	42			MI02
800	1859	42			MI02
1055	1867	50			MI02
1113	1867	64			MI02
1559	1873	50 bis			TI32
1560	1873	50 bis			TI32
1715	1874	50 ter			TI32
1716	1874	50 ter			TI32
1863	1876	50 ter			TI32
1864	1876	50 ter			TI32
2000	1879	50 ter			TI32
2001	1879	50 ter			TI32
4932	1951	195			NE15
6174	1982	BB800H			ZH004, TG20

Christoph Schöttler Maschinenfabrik GmbH, Diepholz [D]   Schöma

Number	Year	Type	Other	Number	Reference(s)
1177	1950				GR31
1211	1951	MD 2b	O&K	25064	ZH112
1230	1951	MD 2b	O&K	25125	BE017
1232	1951	MD 2b	O&K	25126	MI10
1240	1951	MD 2b	O&K	25132	AG083
1241	1951	MD 2b	O&K	25133	VS67
1272	1951	MD 2b	O&K	25203	MI99
1273	1951	MD 2b	O&K	25204	BE032
1274	1951	MD 2b	O&K	25205	BE032, LU30
1279	1951	MD 2b	O&K	25208	TI24
1280	1951	MD 2b	O&K	25209	VS69
1308	1951	MD 2b	O&K	25231	OW2, ZH008
1309	1951	MD 2b	O&K	25232	OW2, ZH046
1393	1952	MD 2b	O&K	25471	MI99
1394	1952	MD 2b	O&K	25472	MI99
1499	1953	CDL 15			LU30
1533	1954	CDL 10			SH02
1642	1955	CDL 15			VS23
2069	1957	CDL 28			AG040
2151	1958	CFL 80DR			BE021, SG31

Number	Year	Type	Other	Number Reference(s)
2168	1960	CHL 20GR		SG27, VS65, ZH085
2292	1960	CFL 150DBR		BL19, ZH070, ZH085
2381	1960	CFL 80DR		SG31
2399	1961	CFL 120DBR		AG097
2630	1963	CFL 150DBR		GE15
2682	1964	CFL 30DCR		ZH068, ZH138
2683	1964	CDL 20		LU
2724	1964	CFL 80DBR		ZH118, ZH136
2801	1964	CFL 40DR		BS01, FR11
3136	1969	CFL 60DR		AG035, BS01
3295	1970	KDL 8		LU33, ZH046
3325	1972	CFL 200DCL		SH04
3686	1973	CFL 60DZ		BE090, FR01
4418	1981	CFL 180DCL		GR02, GR03
4419	1981	CFL 180DCL		GR02, GR03
4451	1981	CHL 20G		LU30
4452	1994	CFL 250DVR		AG108
4495	1981	CFL 180DCL		GR04
4504	1982	CFL 180DCL		SZ12, VS05
4529	1982	CFL 60DZR		ZH153
4649	*198x*	*CFL 60DZR*		*ZH006*
4665	1983	CFL 200DCL		AG089, JU02
4666	1983	CFL 200DCL		JU02
4667	1983	CFL 200DCL		AG089
4668	1983	CFL 200DCL		JU02
4669	1983	CFL 200DCL		AG089
4670	1983	CFL 200DCL		AG089
4762	1984	CFL 200DCL		JU02
4763	*1984*	*CFL 200DCL*		*MI09*
4807	1985	CFL 150DCL		BE033, ZH085
4816	*1985*	*CHL 20G*		*ZH006*
4942	1987	CFL 350DCLR		SO08
4999	1989	CFL 180DCL		BE005, TI08, UR13
5005	1989(93)	CFL 180DCL		VS05
5006	1989	CFL 180DCL		VS19
5014	1989	CFL 180DCL		VS19
5022	1989	D60		GR02, GR03
5060	*1989*	*CFL 180DCL*		*VS05*
5061	1989	CFL 180DCL		VS19
5062	1989	CFL 180DCL		GR02
5064	1989	CFL 180DCL		GR02
5089	*1990*	*CFL 200DCL*		*VS05*
5106	1990	CFL 150DCLR		TG08
5134	1990	CFL 180DCL		GR02, UR02
5140	1990	CHL 30G		LU30
5147	1990	CFL 150DCL		GR04
5149	1990	CFL 150DCL		VS19
5159	1989	CFL 180DCL		GR02
5170	1991	CFL 180DCL		GR02, GR03, GR17
5171	1991	CFL 180DCL		GR03, GR17
5172	1991	CFL 180DCL		GR02, GR03, GR17
5176	1990	CFL 180DCL		BE096, SZ12, UR04
5177	1990	CFL 180DCL		SZ12, UR04
5189	1991	CFL 180DCL		GR01, TI08, UR13
5190	1991	CFL 180DCL		GR01, TI08, UR13
5192	1991	CFL 180DCL		BE096
5193	1991	CFL 180DCL		BE096
5194	1991	CFL 180DCL		BE096
5196	1991	CFL 180DCL		BE096
5197	1991	CFL 100DCL		BE003
5198	1991	CFL 100DCL		BE003
5201	1991	CFL 150DCL		GR04
5241	1991	CFL 100DCL		LU33

Number	Year	Type	Other	Number	Reference(s)
5242	1991	CFL 100DCL			AG064, LU03, LU33
5254	1991	CFL 200DCL			VS05
5255	1991	CFL 200DCL			VS05
5273	1992	CFL 200DCL			VS05
5275	1992	CFL 180DCL			UR02
5276	1992	CFL 180DCL			UR02
5301	1993	CHL 60G			GR10
5311	*1993*	*CFL 200DCL*			*VS05*
5316	*1993*	*CFL 200DCL*			*VS05*
5362	1994	CFL 200DCLR			SO21
5368	1993	CFL 200DCL			GR03, GR05
5369	1993	CFL 200DCL			GR03, GR05
5370	1993	CFL 200DCL			GR03, GR05
5374	1993	CFL 180DCL			GR04
5375	1993	CFL 180DCL			GR04
5376	1993	CFL 180DCL			GR04
5386	1993	CFL 180DCL			GR47
5387	1993	CFL 180DCL			GR47
5388	1994	CFL 180DCL			VS19
5389	1994	CFL 180DCL			VS19
5390	1994	CFL 180DCL			VS19
5417	1995	CFL 200DCL			GR03
5418	1995	CFL 200DCL			GR03
5419	1995	CFL 200DCL			GR03, GR04
5420	1995	CFL 200DCL			GR03, GR04
5424	1995	CFL 180DCL			GR04
5435	1995	CFL 180DCL			VS19
5530	1997	CFL 180DCL			TI31
5531	1997	CFL 180DCL			TI31
5624	1999	CFL 200DCL			GR02
5647	2000	CHL 60G			LU04
5648	2000	CHL 60Tandem			LU04
5712	2002	CHL 350GR			ZH167
5730	2002	CFL 200DCL			TI23
5731	2002	CFL 200DCL			TI23
5732	2002	CFL 200DCL			TI23
5733	2002	CFL 200DCL			TI23
5734	2002	CFL 350DCL			TI23
5735	2002	CFL 350DCL			TI23
5736	2002	CFL 350DCL			TI23
5737	2002	CFL 350DCL			TI23
5738	2002	CFL 350DCL			TI23
5739	2002	CFL 350DCL			TI23
5740	2002	CFL 350DCL			TI23
5741	2002	CFL 350DCL			TI23
5742	2002	CFL 350DCL			TI23
5743	2002	CFL 350DCL			TI23
5744	2002	CFL 350DCL			TI23
5745	2002	CFL 350DCL			TI23
5766	2003	D60-20			TI23
5767	2003	D60-20			TI23
5787	2003	CFL 350DCL			TI23
5788	2003	CFL 350DCL			TI23
5789	2003	CFL 350DCL			TI23
5790	2003	CFL 350DCL			TI23
5791	2003	CFL 350DCL			TI23
5792	2003	CFL 350DCL			TI23
5793	2003	CFL 350DCL			TI23
5794	2003	CFL 350DCL			TI23
5795	2003	CFL 350DCL			TI23
5796	2003	CFL 350DCL			TI23
5797	2003	CFL 350DCL			TI23
5798	2003	CFL 350DCL			TI23

Number	Year	Type	Other	Number	Reference(s)
5799	2003	CFL 200DCL			TI23
5800	2003	CFL 200DCL			TI23
5801	2003	CFL 200DCL			TI23
5802	2003	D60-20			UR02
5803	2003	CFL 180DCL			UR02
5804	2003	CFL 180DCL			UR02
5805	2003	CFL 180DCL			UR02
5806	2003	CFL 180DCL			UR02
5807	2003	CFL 180DCL			UR02
5808	2003	CFL 180DCL			UR02
5809	2003	CFL 180DCL			UR02
5810	2003	CFL 180DCL			UR02
5811	2003	CFL 180DCL			UR02
5812	2003	CFL 180DCL			UR02
5813	2003	CHL 60G			GR47
5837	2003	CFL 200DCL			GR02
5838	2003	CFL 200DCL			GR02
5839	*2003*	*CFL 200DCL*			*GR02*
5840	2003	CFL 180DCL			GR02
5841	2003	CFL 180DCL			GR02
5842	2003	CFL 180DCL			GR02
5843	2003	CFL 180DCL			GR02
5844	2003	CFL 180DCL			GR02
5845	2004	CFL 180DCL			GR02
5846	2004	CFL 180DCL			GR02
5853	2004	CFL 180DCL			GR02
5854	2004	CFL 180DCL			GR02
5855	2004	CFL 180DCL			GR02
5856	2004	CFL 180DCL			GR02
5871	2004	CFL 180DCL			GR02
5872	2004	CFL 180DCL			GR02
5873	2004	CFL 180DCL			GR02
5874	2004	CFL 180DCL			GR02
5876	2004	CFL 350DCL			TI23
5877	2004	CFL 350DCL			TI23
5878	2004	CFL 350DCL			TI23
5879	2004	CFL 350DCL			TI23
5880	2004	CFL 350DCL			TI23
5881	2004	CFL 350DCL			TI23
5882	2004	CFL 200DCL			TI23
5883	2004	CFL 200DCL			TI23
5910	2004	CHL 60G			GR47
5925	2004	CFL 180DCL			GR02
5926	2004	CFL 180DCL			GR02
5927	2004	CFL 180DCL			GR02
5933	2004	CFL 180DCL			UR02
5934	2004	CFL 180DCL			UR02
5935	2004	CFL 180DCL			UR02
5966	2005	CFL 180DCL			UR02
5967	2005	CFL 180DCL			UR02
5968	2005	CFL 180DCL			UR02
5969	2005	CFL 180DCL			UR02
6015	2005	CFL 180DCL			GR02
6016	2005	CFL 180DCL			GR02
6038	2005	CFL 350DCL			TI23
6039	2005	CFL 350DCL			TI23
6040	2005	CFL 350DCL			TI23
6041	2005	CFL 350DCL			TI23
6068	2005	CFL 350DCL			TI23
6070	2006	CFL 200G			TI23
6071	2006	CFL 200G			TI23
6183	2007	CFL 350DCL			TI23
6184	2007	CFL 350DCL			TI23

Number	Year	Type	Other	Number	Reference(s)
6185	2007	CFL 350DCL			TI23
6186	2007	CFL 200DCL			TI23
6187	2007	CFL 200DCL			TI23
6253	2008	CFL 180DCL			UR01
6254	2008	CFL 180DCL			UR01
6255	2008	CFL 180DCL			UR01
6256	2008	CFL 180DCL			UR01
6257	2008	CFL 180DCL			UR01
6258	2008	CFL 180DCL			UR01
6259	2008	CFL 180DCL			UR01
6260	2008	CFL 180DCL			UR01
6261	2008	CFL 180DCL			UR01
6321	2008	CFL 350DCL			TI23
6322	2008	CFL 350DCL			TI23
6323	2008	CFL 350DCL			TI23
6324	2008	CFL 350DCL			TI23
6325	2008	CFL 350DCL			TI23
6326	2008	CFL 350DCL			TI23
6327	2008	CFL 350DCL			TI23

Société Française de Locotracteurs [F]   **SFL**

Founded in 1980 by Fauvet-Girel and others.

Number	Year	Type	Other	Number	Reference(s)
300	1994	385DH44NIND			VD70
301	1994	385DH44NIND			VD70

Schweizerische Industriegesellschaft, Neuhausen am Rheinfall   **SIG**

Number	Year	Type	Other	Number	Reference(s)
N7798	1894		SAAS	?	LU39
?	1897		BBC	?	LU39
?	1897		MFO	?	SH08, ZH138
?	1897		MFO	?	VD10
?	1899		EGA	?	BL04
?	19xx			*3	AG089
?	19xx				BE080
?	19xx				GE05
?	19xx			*2	GE06, GE24
?	19xx			*2	SZ19
?	19xx				TI21
?	19xx			*5	UR05
?	19xx			*6	VS04
?	19xx ATS 100			*2	GR35
?	19xx ATS 100			*3	UR13
?	19xx EIM 100			*2	LU12
?	19xx ET 15KL				BE54
?	19xx ET 15KL				BS16
?	19xx ET 15KL				GR28a
?	19xx ET 15KL			*2	LU33
?	19xx ET 20L5				BE040
?	19xx ET 20L5			*4	GR36
?	19xx ET 40L				ZH083
?	19xx ET 40L				ZH161
?	19xx ET 70				GR35
?	19xx ETB 50				BE80
?	19xx ETB 50				GE06, GE24
?	19xx ETB 50				GR32
?	19xx ETB 50			*4	GR34b
?	19xx ETB 50				GR35
?	19xx ETB 50				LU33
?	19xx ETB 50				VS11
?	19xx ETB 70			*2	AG083
?	19xx ETB 70				BE087a
?	19xx ETB 70				GR10
?	19xx ETB 70			*2	GR35
?	19xx ETB 70			*5	GR43a

Number	Year	Type	Other	Number	Reference(s)
?	19xx	ETB 70		*11	LU12
?	19xx	ETB 70		*5	SG37a
?	19xx	ETB 70		*15	TI30a
?	19xx	ETB 70		*2	UR06a
?	19xx	ETB 70		*9	UR13
?	19xx	ETB 70			VD53a
?	19xx	ETB 70		*2	VS58
?	19xx	ETB 70			ZH009
?	19xx	ETB 70			ZH013
?	19xx	ETB 70			ZH023
?	19xx	ETB 70		*4	ZH112
?	19xx	ETE 70		*9	GL10
?	19xx	ETE 70		*2	BE040
?	19xx	ETM 50			GE06, GE24
?	19xx	ETM 50			GE06
?	19xx	ETM 50		*5	VS10
?	19xx	ETM 50		*4	ZH178
?	19xx	ETM 50		*2	VS35
?	19xx	ETM 70		*2	UR13
?	19xx	ETM 100			BE060
?	19xx	ETR 70			GR10
?	19xx	ETS 25		*3	GR28a
?	19xx	ETS 35			BE040
?	19xx	ETS 35			TI54a
?	19xx	ETS 35			VS43a
?	19xx	ETS 35			TI02
?	19xx	ETS 50			BS16
?	19xx	ETS 50		*3	UR08a
?	19xx	ETS 100		*4	BE087a
?	19xx	ETS 100			GR39a
?	19xx	ETS 100		*4	LU12
?	1903		MFO	?	ZH168
?	1905		EGA	?	VD10
?	1908(46,47)		SAAS	?	GR38
?	1909		SLM, EGA	?	VD10
?	1911(53)		SAAS	?	GR38
?	1911		SLM, EGA	?	AI1
?	1911		EGA	?	LU39
?	1912		MFO	?	SO39
?	1912		MFO	22	BL37, ZH102
?	1913		BBC	?	SO42
?	1915		MFO	?	SH04
?	1917		BBC	?	VD10
?	1918		MFO	?	ZH019
?	1919		MFO	?	ZG4
?	1920		MFO	?	SH08, ZH137
?	1921		MFO	?	SH04
?	1924		MFO	?	SG77
?	1925(31)		MFO	?	ZH181
?	1929		MFO	?	AR2
?	1930(37,61)		MFO (,SAAS)	?	SH09
?	1931		BBC	?	FR01, FR02
1165	19xx	ES 50			BE080, BE082, VS54
2939	1958				GE24
3844	1932(81)		SWS, BBC	?	BE018
3848	1932(81)		SWS, BBC	?	BE018
3850	1932(81)		SWS, BBC	?	BE018
3851	1932		SWS, BBC	?	BE018
?	1933		MFO	?	AR2
?	1933		MFO	?	TI37
?	1933		MFO	?	TI37
?	1933	Gepäck-Tw	BBC	3703	VD10
?	1935		MFO	?	LU39

Number	Year	Type	Other	Number	Reference(s)
?	1938		SAAS	?	AR2
?	1939		SAAS	?	LU39
?	1939		MFO	?	SG24, ZH102
?	1942		MFO	?	BE017
?	1943		SAAS	?	BE054, LU39
?	1944		MFO	?	ZH011, ZH168
?	1945	BCFe 4/8	SAAS	?	FR01
?	1948		SAAS	?	BE054
?	1959	EMIL	BBC	6089	AG005, ZH019
?	1959(79)		BBC	6117	LU37
?	1960	FRITZ	BBC	6088	AG002, VD49, ZH136
?	1960	Antonio	BBC	6130	AG092, ZH136
?	*1960*	MUTZ	BBC	6149	BE034
?	1961	ERICH	BBC	?	ZH136
?	1961		MFO	?	MI07
?	1964	WALTER	BBC	6405	AG005, ZH019
?	1965		SWS, SAAS, BBC, MFO	?	AG022
?	1964	LISI	BBC	6406	LU28
?	1968	Madeleine	BBC	7639	AG009
?	1966		SWS, SAAS, BBC, MFO	?	AG022
?	1966		BBC	?	LU37
?	1966		BBC	?	TI17
?	*1967*	FRIDOLIN	BBC	7701	AG023
?	1970				NE08
?	1959		BBC	?	LU13
?	1978		BBC	?	LU13
?	1979		BBC	?	LU13
520 501	1953				VD47
563 507	19xx				SO33
578 301	1957	ETB 70			AG083
588 101	1991	ETB 70			UR13
588 102	1958	ETB 70			LU33
618 211	1961	ETB 50			LU33
619 523	1963	ETM 30			UR13
619 524	1963				UR13
619 525	19xx				UR13
649 115	19xx				UR13
649 211	1965	ETB 70			LU33
669 214	1967	ETB 70			LU33
701 107	1973	ET 8			AG089
701 110	1973				AG089
705 409	1976	ETB 70			GR01
705 714	1975	ATS 610			AG089
707 728	1977				AG089
707 729	1977	ATS 100			UR13
707 730	1978	ATS 100			UR13
708 739	1978	ETR 70			GR10
709 446	1977				AG089
800 483	1980	ETB 70			UR13
800 820	1980	ATM 800			VS68
806 308	1986	BL 20 MG			TI54a
807 309	1987	BL 20 MG			TI54a

Škoda-Werke, Plzen [CZ]   Škoda

Number	Year	Type	Other	Number	Reference(s)
1584	1944	BR 52	BMAG	12226	NE15, ZH105

Schweizerische Lokomotiv- und Maschinen Fabrik, Winterthur   SLM

Number	Year	Type	Other	Number	Reference(s)
?	18xx				SG09
1	1873	H			LU20, LU39
6	1873	E 3/3			BS06
11	1873	E 3/3			BS06

Number	Year Type	Other	Number Reference(s)
35	1874		BE036
38	1874 E 3/3		BS02, SO37, SO44
40	1874		UR08
41	1874		ZH103
42	1874		BE036
45	1874		ZH103
46	1874		ZH103
47	1874		ZH103
49	1874		BE036
58	1874		GE13
61	1875		BE056, ZH103
62	1875		ZH103
64	1875		ZH103
66	1875		ZH103
67	1875		AG008
68	1880		TI59
72	1875		ZH103
74	1875		ZH103
76	1875		AG021
83	1875		AG044
84	1881		TI18
89	1876		ZH141
181	1880 G 2/2		SO05
182	1880		MI04
229	1881 Ed 3/3		LU39
236	1881 E 2/2		JU12, LU39, SO37, SO44
256	1882		BE074, BL41
316	1882		VD10
369	1884		GE12
370	1884 G 2/2		GR38
371	1884 G 2/2		VD56
377	1884		SO37
388	1884		GR42, ZH141
391	1884		ZH087
456	1887 G 3/3		BE061, BE119, TI34, TI68, VD56
462	1897 G 3/3		SG47
473	1887 G 3/3		BS04, GR38
474	1887 G 3/3		SH04
502	1888 HG 2/2		VD11, VD54
526	1888		GR38, SG19, ZH090
548	1888 G 3/3		BS04, GR38, VD56, ZH002
563	1889 Dampftriebwagen		LU39
575	1889		AG011, BE074
577	1899 G 3/4		GR38
604	1890 H		TI36
614	1890 HG 3/3		GR12, GR38
616	1890 HG 3/3		GR33, GR38
618	1890		VD10
619	1890		NE14
624	1890 G 3/3		SH04
629	1890 E 3/3		BE112, SO37, SO44
631	1890 E 3/3		BE112, FR01, NE15, SO37
635	1890 G 3/3		SH04
644	1890 G 3/3		BE057
679	1891 E 3/3		BL36
686	1891 E 2/2		SG24
716	1892 G 3/3		NE07
719	1892 H		BE098
725	1892 E 2/2		ZH164
726	1892 E 2/2		JU12, SO44
727	1892 E 3/3		BS02
748	1892		VS09
795	1893 E 3/3		BS02, VS55, ZH181

Number	Year Type	Other	Number	Reference(s)
863	1894 G 3/3			BE011, BE081, ZH155
897	1894 E 3/3			BS20, BS23
898	1894 E 3/3			BL36
899	1894 E 3/3			TI72
900	1894 E 3/3			LU32, VD24, ZH142
901	1894 E 3/3			BS06
917	1895 E 2/2			GR06, NE15, TI64, ZH149
922	1895 E 2/2			SO26
955	1896 E 3/3			SO37, SO44
958	1896 G 2/2 + 2/3			BE054
959	1896 G 2/2 + 2/3			BE054
1009	1896 E 3/3			BL36
1014	1896 E 3/3			LU32
1015	1897 E 3/3			ZH057
1016	1897 E 3/3			ZH055
1064	1897 E 3/3			GE22, VS39
1066	1898 He			VS37
1068	1897			ZH107, ZH149
1075	1897 Ec 3/3			SO44
1088	1898 E 3/3			JU06, JU12, NE08
1139	1898 HGe 2/2	BBC	?	LU39
1140	1898 HGe 2/2			AG093
1141	1898			SZ07, ZH052, ZH155
1194	1899 E 3/3			BE082, SO26
1197	1899 De 2/2	BBC	?	LU39
1201	1899			VS43
1202	1899			VS43
1219	1899 E 3/3			ZH029
1220	1899 E 3/3			SH09, SO29, SO37
1221	1900 E 3/3			ZH181
1267	1900			JU04, JU06, JU11, SO37
1292	1900 E 3/3			BS06
1293	1900 E 3/3			BS06
1308	1900			VS43
1326	1900			VS12
1327	1900			VS12
1328	1900			VS12
1329	1900			VS12
1330	1900			VS12
1331	1900			VS12
1332	1901 E 3/3			BE010, SO08
1333	1901 E 3/3			ZH149
1341	1901 G 3/3			BE081, VD10
1359	1900 E 3/3			LU32, ZH180
1361	1901 E 3/3			SO08
1362	1901 E 3/3			GE19
1377	1901 Ec 3/3			VS45
1378	1901 Ec 3/3			GR18
1387	1901 E 3/3			SG24, ZH034
1400	1901 E 3/3			ZH053
1410	1902 HG 2/3			GR18, GR41, UR09
1425	1902 Ed 3/4			AG092
1438	1902			BS02
1439	1902			BL11, BS06
1440	1902 G 3/3			BL16
1456	1902 E 3/3			TI64
1457	1902 E 3/3			SO08
1459	1902			VS12
1460	1902			VS12
1461	1900			VS12
1462	1900			VS12
1476	1902 G 3/4			GR38
1479	1902 G 3/4			AR2

Number	Year	Type	Other	Number	Reference(s)
1488	1903	Ed 3/4			ZH149
1489	1903	Ed 3/4			ZH034, ZH149
1511	1903	G 3/3			BE081
1524	1903	Ce 4/4	MFO	?	MI07
1532	1903	E 3/3			LU32
1533	1903	He	BBC	?	FR01
1534	1903	E 2/2			BE062, LU32
1568	1904	E 3/3			VS01
1623	1904	E 3/3			TI17, TI72
1665	1905	Ce 4/4	CIEM	?	MI07
1670	1905				JU04, JU06, JU11, JU12
1709	1906	G 4/5			GR38
1710	1906	G 4/5			GR38
1725	1906	HG 2/3			VS48
1726	1906	Ed 3/4			AG015, BE038, BE112
1798	1907	Ed 3/4			BS06, VS45, ZH034
1799	1907	Ed 3/4			BE020, GR18, ZH053
1805	1907	E 3/3			BS23, TG10
1807	1907	E 3/3			ZH029, ZH034
1810	1907	E 3/3			LU28
1836	1907				ZH034, ZH132, ZH149
1837	1907				BS02, BS23
1877	1907	E 3/3			AG033, BL36
1879	1907	E 3/3			LU32
1881	1907	E 3/3			GR18, JU04
1901	1908				BE034, BE112
1902	1908				ZH053, MI08
1904	1908	Ed 3/4			BE113, SO37
1932	1908	A 3/5			MI07
1967	1909	E 3/3			SG65
1971	1909	E 3/3			LU05, LU16
1972	1909	E 3/3			LU16
1974	1909	E 3/3			BE017, VD27, VS01
1993	1909	HG 3/3			LU39
2075	1910	E 3/3			LU32
2076	1910	E 3/3			GR18, TI17, ZH105
2079	1910	E 3/3			VS01
2086	1910	He 2/2	EGA	?	BL15
2090	1910	E 2/2			ZH130, ZH141
2095	1910	G 3/3			BE081, VD51
2096	1910				BS03
2097	1910				VD24
2126	1910	Eh 2/2			SG39, ZH152
2130	1910	E 3/3			VS01, VS17
2134	1911	E 3/3			GE19, NE15
2135	1911	E 3/3			LU39
2139	1911	E 3/3			ZH132
2149	1910	Xrotd			GR38
2160	1911	Ec 4/5			BE112
2168	1911	E 2/2			JU06, JU12, NE08, SO37
2211	1911	Eb 3/5			BE112
2212	1911	Eb 3/5			AG090, GL08
2220	1912	Eb 3/5			MI07
2239	1912	E 3/3			BS02
2263	1912	Ed 3/5			TG10
2276	1912	G 3/3			LU39
2298	1912	E 3/3			GE19, VD71, VS01
2299	1912	Xrotd			VD10
2304	1912	Be 5/7	MFO	?	LU39
2308	1913	Ge 2/4	BBC	727	GR15, ZH154
2310	1913	Ge 2/4	BBC	729	LU39
2315	1913	HG 3/4			UR09
2316	1913	HG 3/4			UR09

Number	Year	Type	Other	Number	Reference(s)
2317	1913	HG 3/4			VD10
2318	1913	HG 3/4			UR09
2327	1913	Ge 4/4	MFO	?	SH04, VD10
2328	1913	Ge 4/4	MFO	?	SH04
2341	1913	E 3/3			BE017, ZH034
2345	1913	E 3/3			LU28
2370	1913	HGe 3/3	MFO	?	FR01
2399	1913	Xrotd			UR09
2405	1913				VS16
2406	1913				VS16
2407	1913				VS16
2408	1913				VS16
2409	1913				VS16
2410	1913				VS16
2418	1914	HG 3/4			UR09
2419	1914	HG 3/4			UR09
2427	1914	Ed 4/5			BE112
2433	1914	Ge 4/6			GR38
2495	1914	C 5/6			SO35, TG03
2503	1914	E 3/3			VD27, VS01
2507	1914	E 3/3			BE064, GR34, SO44
2518	1916	C 5/6			UR14, MI07
2522	1916	C 5/6			TG03
2532	1915	E 2/2			AG103
2538	1915	E 3/3			VD27, VS01
2556	1916	Eb 3/5			VS64
2557	1916	B 3/4			MI07
2570	1916	E 3/3			GE19
2589	1917				VS45
2590	1917	Ed 2/2			BE017
2592	1917	E 2/2			BS06, SO26
2593	1917	T 2/2			VS45
2612	1917	C 5/6			MI07
2621	1917	E 4/4			VS01
2622	1917	Ta	MFO	?	BS15
2649	1918	Be 6/8^{I}	BBC	1205	MI07
2660	1919	Ta	MFO	?	BE072
2673	1919	Ce 6/8^{II}	MFO	?	MI07
2674	1919	Ce 6/8^{II}	MFO	?	MI07
2689	1919	Ce 4/6	MFO	?	BE018
2693	1919/20	ex Ce 4/6	BBC	1319	MI08
2694	1919/20	ex Ce 4/6	BBC	1320	TI17
2695	1919/20	ex Ce 4/6	BBC	1321	BE018
2706	1920	Ce 6/8^{II}	MFO	?	UR14
2716	1920	Be 4/7	SAAS	94004	MI07
2740	1920	Ae 3/6^{I}	BBC	1541	AG090, MI08
2754	1920	Ge 6/6	BBC	1546	LU39
2758	1921	Ge 6/6	BBC	1550	FR01
2762	1920	Be 4/6	BBC	1592	MI07
2773	1922	Ce 6/8^{II}	MFO	?	TI17
2809	1922	Be 4/6	BBC	1804	AG090, MI07
2839	1922	Ge 6/6	BBC	1886	GR44
2843	1922	ex Ce 4/6	BBC	1588	MI08
2844	1922	ex Ce 4/6	BBC	1589	MI08
2853	1923	Tm			ZH141
2871	1923	H			LU20
2940	1923				UR09
2948	1924	Ae 3/5	SAAS	600003	MI07
2982	1924,25	Ae 3/6^{I}	BBC	2083	MI08
2998	1925(56)	Te 2/3	MFO,BLS,SAAS	?	BE018
3022	1924	Ae 3/6^{II}	MFO	?	MI08
3035	1924	Em 2/2			VD47
3042	1925	Em 2/2			ZH093

Number	Year	Type	Other	Number	Reference(s)
3043	1925	H			LU20
3045	1925	Ge 6/6	BBC	2242	GR38
3049	1925	Ae 3/6^I	BBC	2111	MI07
3056	1926	De 6/6	BBC	2140	SO29
3061	1925	Ae 3/6^{III}	SAAS	?	MI07
3073	1925	Ce 6/8^{III}	MFO	?	ZH095
3076	1925	Ce 6/8^{III}	MFO	?	MI07
3088	1925	Ae 3/6^I	BBC	2125	MI07, SG17
3092	1925	Ae 3/6^{III}	MFO	?	MI07
3133	1926,27	Ae 4/7	BBC	2522	MI08
3152	1927	Ae 3/6^I	MFO	?	MI08
3159	1927	Ae 3/6^I	MFO	?	MI07
3176	1927	Ae 4/7	BBC	2525	MI07
3179	1927	Ae 4/7	BBC	2528	MI01
3186	1926	Te 2/2	SAAS	818001	MI08
3188	1926,27	Ee 3/3	BBC	2533	SO11, MI08
3190	1926,27	Ee 3/3	BBC	2535	MI08
3192	1926,27	Ee 3/3	BBC	2537	MI01
3193	1926,27	Ee 3/3	BBC	2538	MI01
3206	1927	BChm 2/2			UR09
3221	1927	Em 2/2			SG05, SG12
3225	1928	Ee 3/3	BBC	2808	TG10
3232	1928	Ee 3/3	BBC	2815	MI01
3233	1928	Ee 3/3	BBC	2816	MI01
3238	1927	Ae 4/7	BBC	2818	MI01
3246	1927	Ae 4/7	MFO	?	MI01
3260	1927	Em 2/2			ZH036, ZH136
3290	1928	CFe 4/5	SAAS	?	MI01
3295	1916(29)	Ge 4/4	BBC	1054	VD10
3297	1929	Ge 6/6	BBC	2967	GR38
3298	1929	Ge 6/6	BBC	2968	GR38
3299	1929				AG005, LU16
3340	1929	HGe 4/4	SWS,MFO	?	VS48
3396	1930	Ee 3/3	BBC	3315	MI08
3412	1930	Em 2/2			SG36, VD49, ZH036, MI08
3413	1930				UR09
3415	1930	Ee 2/2	SAAS	?	AG082
3425	1930	Ae 4/7	SAAS	1700-1	MI01
3427	1930	Ae 4/7	MFO	?	MI07
3442	1930	Ae 4/7	BBC	3355	MI01
3445	1930	Ae 4/7	SAAS	1700-6	MI01
3449	1930	Ae 4/7	SAAS	1700-7	MI07
3451	1930	Ae 4/7	SAAS	1700-8	ZH130, MI01
3456	1930	Ae 4/7	SAAS	1700-9	MI01
3459	1931(89)	BCFeh 4/4	BBC,SIG	?	JU06
3463	1930	BCFeh 4/4	BBC,SIG	3366	AR1
3464	1930	Em 2/2			BS15, JU04, SO36, VS62
3485	1931	Ae 4/7	MFO	?	MI08
3488	1931	Ae 4/7	SAAS	2050-2	ZH130, MI01
3501	1931	Ae 8/14	BBC	3360	MI07
3509	1931	Be 4/4	SAAS	2080-1	AR6
3510	1931	Be 4/4	SAAS	?	MI08
3511	1931	Be 4/4	SAAS	?	TG10, ZH034
3512	1931	Be 4/4	SAAS	?	TG03, ZH105
3513	1931	Be 4/4	SAAS	?	ZH034
3523	1931	E 2/2			GR42
3536	1932	Ae 4/7	MFO	?	SZ02
3537	1932	Ae 4/7	BBC	3719	MI01
3540	1932	Ae 4/7	SAAS	2525-3	MI01
3541	1932	Ae 4/7	MFO	?	MI01
3543	1932	Ae 4/7	MFO	?	MI08
3546	1932	Ae 4/7	MFO	?	MI01
3547	1932,33	Ae 4/7	BBC	3723	AG104, MI01

Number	Year	Type	Other	Number	Reference(s)
3551	1932	Be 4/4	SAAS	?	MI08
3552	1932	Be 4/4	SAAS	?	BE018
3554	1932	Be 4/4	SAAS	?	MI01
3555	1932	Be 4/4	SAAS	?	MI01
3557	1932	Be 4/4	SAAS	?	MI08
3566	1932				ZH053
3579	1934	Tm			VD62, MI08
3581	1934	Tm			BL29, BL42
3581	1935,36	CLe 2/4	BBC,MFO, SAAS,SBB	3983	MI07
3592	1935				NE15
3599	1935	Tm			FR09, SG36
3604	1936	CLe 2/4	BBC,MFO, SAAS,SBB	3985	MI07
3608	1936	Tm			SG61, TG08
3610	1936	E 3/3			ZH149, MI07
3634	1937	CLe 2/4			MI08
3677	1938	HGe 4/4^I	MFO	?	UR09
3678	1939	Ae 6/8	SAAS	?	BE010
3679	1939	Ae 6/8	SAAS	?	MI01
3685	1938	Ae 8/14	MFO	?	MI07
3689	1939	RAe 4/8	SWS,MFO,SAAS	?	MI07
3690	1938	Tm			BS15, MI08
3691	1939	Am 4/4	BBC	?	MI07
3693	1938	RFe 4/4	MFO	?	MI08
3695	1938	RFe 4/4	MFO	?	MI01
3731	1941	Fhe 4/6	SAAS	?	MI07
3732	1941	Fhe 4/6	SAAS	?	LU38
3740	1941	Fhe 4/6	SAAS	?	MI07
3758	1941	Te^I	MFO	?	LU38
3765	1941	BCFeh 2/4	BBC	4451	VS48
3770	1941	Te^I	MFO	?	LU38
3771	1941				AG092
3797	1942	Ae 6/8	SAAS	?	MI08
3806	1942	Ee 3/3	SAAS	4840	MI01
3833	1943				AG039, AG067, AG088, BE032, BE064, ZH161
3834	1943				AG067, AG088, AG090, BE032, ZH161
3835	1943				AG067, AG088, BE032, BE064, ZH161
3846	1944	Te^I	MFO	?	BE113
3865	1944	Te^I	MFO	?	ZH095
3877	1944	Re 4/4^I	MFO	?	MI07
3878	1944	Re 4/4^I	MFO	?	MI01
3883	1944	Ae 4/4	BBC	4507	BE010
3890	1945	Re 4/4^I	SAAS	5650-3	MI01
3897	1945	Re 4/4^I	MFO	?	MI01
3899	1945	Ee 3/3	BBC	4529	BE084
3927	1945	Te^I	MFO	?	BE113
3928	1945	Te^I	MFO	?	BE112
3930	1945	Te^I	MFO	?	BE112
3935	1946	Tm			BL09, BL32
3937	1946	C 1/3			MI07
3946	1946	Te^{III}	SAAS	?	AG022
3949	1946	Che 2/4	BBC	4535	VS37
3979	1948	Te	SAAS	?	FR16
3987	1949	Te^I			TI17
3990	1949	Te^I	MFO	?	ZH095
3991	1949	Te^I	MFO	?	LU21, TI42
4013	1949	Re 4/4^I	BBC	?	AG096
4018	1950	Re 4/4^I	MFO	?	MI01
4021	1950	Re 4/4^I	MFO	?	MI01
4023	1950	Re 4/4^I	BBC	?	MI07
4025	1950	Re 4/4^I	SAAS	?	MI01

Number	Year	Type	Other	Number	Reference(s)
4038	1950	Ee 6/6	SAAS	5427	MI08
4046	1951				SG03, TG03, ZH152
4047	1951	Te^{III}	SAAS	?	VD49, ZH036
4050	1951	Ae 6/6	BBC	5429	TI17
4051	1951	Ae 6/6	BBC	5430	MI07
4063	1952	CFe 4/4	MFO	4592	MI08
4079	1953	HGe 4/4	BBC	5464	LU38
4087	1952	HGe 4/4^{I}	MFO	?	UR09
4105	1954	CFe 4/4	MFO	5441	MI07
4108	1954	CFe 4/4	SAAS	5444	MI07
4112	1953	Tm^{III}	BBC	5459	AG005, AG053
4128	1955	Ae 4/4	BBC	6017	LU39
4129	1954	Tm^{III}	BBC	6003	AG005, AG068, BL04
4138	1955	Ae 6/6	BBC	6004	MI07
4141	1955	Ae 6/6	BBC	6007	OW5
4142	1955	Ae 6/6	BBC	6008	MI07
4146	1955	Ae 6/6	BBC	6012	MI07
4148	1955	Ae 6/6	BBC	6014	LU39
4158	1956	Tm^{III}	BBC	6026	LU16, VD66, TG20, ZH006
4199	1956	Tem^{III}	SAAS	?	AG067
4200	1956	Tem^{III}	SAAS	?	SH03
4218	1956	Tm^{III}			SO32
4234	1955	Ae 6/6	BBC	6051	MI07
4236	1957	Ae 6/6	BBC	6053	LU10
4244	1957	Ae 6/6	BBC	6060	MI07
4310	1961	Tem^{III}	SAAS	?	ZH034
4361	1959	Em 3/3	BBC	?	TI17
4400	1962	Tmh/Thm	BBC	6184	SG03, ZH152
4438	1962	Tm^{III}	BBC	6179	BS06
4439	1962	Tm^{III}	BBC	6180	BS06, TG19
4440	1962	Tm^{III}	BBC	6183	AG057
4443	1963	Ae 8/8	BBC	6407	BE010
4564	1965	Tm^{III}	BBC	7516	BL36, TG20
4579	1965	Tm^{III}	BBC	7531	TG20
4582	1965	Te^{III}	MFO	?	BE022, VD49
4583	1965	Te^{III}	MFO	?	JU09
4586	1965	Tm^{III}	BBC	7539	BL36, BS06, TG20
4587	1965	Tm^{III}	BBC	7540	BS06, TG12
4810	1969	Tm^{IV}			ZH149
4951	1973	Tm^{IV}			BE017
4952	1972	Tm^{IV}			SO37
4982	1973	Tm^{IV}			BE017
4983	1973	Tm^{IV}			BL07, BL43
4984	1973	Tm^{IV}			SO37
5069	1975	Tm^{IV}			ZH141
5286	1984	Ee 3/3	BBC	?	ZH036
5287	1985	Ee 3/3	BBC	?	BE022, ZH036
5288	1985	Ee 3/3	BBC	?	BE022, SO11
5289	1985	Ee 3/3	BBC	?	BE022
5467	1992	Ee 3/3	BBC	?	BE022, LU08, ZH036

NV Spoorijzer, Delft [NL]   Spoorijzer
IRS convention   Spoor

60001	1960	A2L514			AG089, ZH046, ZH183

Siemens-Schuckert Werke SSW, Berlin   SSW

?	19xx				SZ19, ZH046
?	19xx				VS59
?	19xx				VS59
?	19xx				ZH088
?	1922				SG18
?	1930				BL23
?	195x	EL 9			VS69

Number	Year	Type	Other	Number	Reference(s)
5277	1950	EL 9			GL10
5278	1950	EL 9			GL10
5301	1950	EL 9			GL05
5302	1950	EL 9			ZH112
5303	1950	EL 9			ZH112
5327	1951	EL 9			VD06
5375	1951	EL 9			VD05, ZH016
5376	1951	EL 9			VD05
5377	1952	EL 9			VD05
?	1952	EL 8			VS54
5395	1952	EL 9			VD05
5396	1952	EL 9			ZH016
5407	1952	EL 9			MI99
5408	1952	EL 9			VD05
?	1953	EL 9			VD05, VS34
5470	1953	EL 9			SG60
5497	1953	EL 9			MI99
5498	1953	EL 9			VD05
5510	1953	EL 9			VD05
5566	1954	EL 9			ZH112
5567	1954	EL 9			ZH016
5568	1954	EL 9			*TI02*
5590	1954	EL 9			MI99
5591	1954	EL 9			MI99
5598	1954	EL 9			MI99
5660	1955	EL 9			MI99
5661	1955	EL 8			MI99
5662	1955	EL 9 Doppelok			MI99
5663	1955	EL 9 Doppellok			MI99
5666	1955	EL 8			MI99
5667	1955	EL 8			MI99
5668	1955	EL 8			MI99
5682	1955	EL 8			MI99
5683	1955	EL 8			MI99
5690	1955	EL 8			MI99
5691	1955	EL 8			MI99
5722	1956		Henschel	28318	TI21
5725	1956		Henschel	28319	TI21
5801	1956	EL 8			MI99
5857	1957	*EL 8*	*EL 9*		VD20
5858	1958	EL 8			MI99
5867	1958	EL 9			BE060
5868	1957	EL 9			BE060
5869	1957	EL 9			BE060
5872	1957	EL 8			MI99
5873	1957	EL 8			MI99
5884	1957	EL 9			MI99
5885	1957	EL 9			MI99
5895	1957	EL 9			MI99
5896	1957	EL 9			MI99
5902	1957	EL 8			MI99
5910	1957	EL 9			MI99
5911	1957	EL 9			MI99
5919	1957	EL 9			MI99
5920	1957	EL 9			MI99
5964	1958	EL 9			MI99
5965	1958	EL 8			MI99
5970	1958	EL 9			MI99
5976	1958	EL 8			MI99
5991	1958	EL 9			MI99
5996	1958	EL 9			MI99
6067	1960	EL 9	Schalke	57610	BE032, ZH046, ZH088
6068	1959	EL 7			MI99

Number	Year	Type	Other	Number	Reference(s)
6069	1959	EL 8			MI99
6070	1959	EL 8			MI99
6071	1959	EL 8			VS59
6072	1959	EL 7			MI99
6073	1959	EL 7			MI99
6074	1960	EL 8			MI99
6075	1960	EL 8			MI99
6086	1960	EL 8			MI99
6087	1960	EL 8			MI99
6088	1960	EL 8			MI99
6089	1961	EL 8			MI99
6090	1961	EL 7			MI99
6091	1961	EL 7			MI99
6092	1961	EL 8			VD65, VS59
6093	1961	EL 8			MI99
6094	1961	EL 8			VS59
6095	1961	EL 8			MI99
6185	1964	EL 8			MI99
6189	1964	EL 7			MI99
6190	1966	EL 7			LU33

Spälti Fils et Cie, Vevey — Spälti

Electrical components

Ernst Stadler, Elektrische Fahrzeuge, Wädenswil *and* Bussnang — Stadler

Stadler Fahrzeuge AG, Bussnang
Stadler Bussnang AG, Bussnang

Number	Year	Type	Other	Number	Reference(s)
?	194x				GL10
?	1944		BBC	?	LU30
?	1944		BBC	?	SG85
10	1944				LU40
15	1945				VS61
17	1945		BBC	?	SZ06, ZH046
19	194x				AG042, GR45
20	1946		BBC	?	BL35, GR40
21	1947		Breuer		AG020
22	1946				SG34
23	1947				GR36
24	1947				BE040
25	1948				AG061
26	1948		BBC	?	AG020
27	1948				SZ09
28	1949				VS56
29	1949				VS56
-	1951				BE118
30	1951				BE047, BE078
31	1948				BE091, GR22
32	1948				BE091, GR22, ZH048
33	c1949				AG066, BL35
34	c1949				AG066, BL35
35	c1949				AG066, BL35
36	8/1948				BE040, BE092, ZH046
39	1950				SG82
40	1950		BBC	?	ZH121
43	1949				SG34
103	1953				BE023, BE112
104	1953				UR12
105	1953				SG34
-	1954				SG34
-	1956				SG34
106	1957				NE05
107	1958		BBC	*1422*	SO39
108	1958				SG34

Number	Year	Type	Other	Number	Reference(s)
109	1959				SG34
111	1960		BBC	?	AG082, BE118, SG05, TG19
112	1961				BE059, BE072, SZ13, UR12
113	1961				BE059, SG44
?	1963				BL23
116	1963				ZG1
117	1965				BE072, TI55, UR12
118	1965				FR11, LU24
119	1966				SG44, TG20
121	1967				BE059, SG44
124	1967				ZH089
131	1969				BL23, BS21, VD40
132	1970				AG048, BL10
136	1972				ZH130
137	1971				BL37
139	1972				ZH180
140	1972				ZH085, ZH109
141	1973				BE059
143	1974		BBC	?	BE106, ZH041, ZH102
144	1975				BE059
148	1976				BE059, UR12
155	1981				TG19
156	1979				BE059
157	1979				ZH027
162	1984		BBC	7640	AG046
241	1994		SIG	?	BE054
242	1994		SIG	?	BE054
243	1994		SIG	?	BE054
325	1998				BS06
522	1999				VD49
523	1999				SO11
524	1999				TG01
838	2003				BS06

Ferdinand Steck, Maschinenfabrik AG, 3533 Bowil — Steck

Number	Year	Type	Other	Number	Reference(s)
01	1993				AG009

Stemag GmbH, Worb, later Greffen — Stemag

Number	Year	Type	Other	Number	Reference(s)
?	19xx		von Roll	?	BS01
1	19xx	V			SG06

Gebrüder Sulzer, Winterthur + Oberwinterthur — Sulzer

Number	Year	Type	Other	Number	Reference(s)
MT 7	1922				LU39

Schweizerische Wagons- und Aufzügefabrik AG, Schlieren — SWS

Number	Year	Type	Other	Number	Reference(s)
?	19xx		MFO	?	ZH029
?	1903		EGA	?	VD10
?	1908		MFO	18	ZH128
?	1909		MFO	19	BE114, JU06
?	1911		MFO	?	LU28
?	1911		BBC	?	TI67
?	1913		BBC	?	VD02
?	1913		MFO	?	VD23
?	1917		MFO	?	BE090
?	1917		MFO	?	BE090
?	1917		MFO	?	ZH104
?	1919		MFO	?	VS71
?	1921		SAAS	?	BE104
?	1924		MFO	?	ZH181
?	1935		SAAS	?	VD51
?	1938		SAAS	?	ZH181
?	1939		MFO,BBC	?	GR38
?	1939		MFO,BBC,SAAS	?	SZ27

Number	Year	Type	Other	Number	Reference(s)	
?	1942		MFO	?	AG062	
?	1943		MFO	?	ZH028	
?	1955(73)		MFO	?	JU06	
?	1958		BBC	?	TI37	
?	1963		SAAS	?	TI37	
?	1965		MFO,BBC	?	TG10	
?	1968		MFO	?	ZH181	

Solothurn-Zollikofen-Bahn — SZB

Number	Year	Type	Other	Number	Reference(s)
-	1955				BE049
-	1958				BE049

Ferdinand Thébault, Marly-les-Valenciennes [F] — Thébault

Number	Year	Type	Other	Number	Reference(s)
?	1908				BE024

Technomasio Italiano Brown Boveri, Milano [I]-TIBB

Italian subsidiary of BBC

Number	Year	Type	Other	Number	Reference(s)
0117	1959				VD65, VS59

TransLok GmbH, Liebigstrasse 6a, 47608 Gelden [D] — TransLok

"Cargolok" production from 1999

Number	Year	Type	Other	Number	Reference(s)
180	2008	TL-DH440			BL34

Gebrüder Tuchschmid, Frauenfeld — Tuchschmid

Number	Year	Type	Other	Number	Reference(s)
?	1957	Tem^I	MFO	?	AG022
?	1957	Tem^I	MFO	?	TG03
?	1966	Tem^{II}	MFO	?	NE04
?	1966	Tem^{II}	MFO	?	TG09
?	1967	Tem^{II}	MFO	?	FR11
?	1967	Tem^{II}	MFO	?	MI07
?	1967	Tem^{II}	MFO	?	VD27

Uzinele 23 August, Bucureşti [RO] — U23A

Number	Year	Type	Other	Number	Reference(s)
21236	1971	LDH240			BS15
23879	1979	LDH370			AG092, AG103, ZH085
24122	1980	LDH45			ZH085

U.C.A., 30 Vaartkaai, 2170 Antwerpen (Merksen) [B] — UCA

Number	Year	Type	Other	Number	Reference(s)
?	2006	UCA-Trac			VD47
?	2008	UCA-Trac			VS44
?	2008	UCA-Trac			TI55

Unilok — Unilok

Founded by Hugo Aeckerle, Hamburg [D]. Production moved to Unilok Lokomotive Ltd., Oranmore, County Galway [IRL], though some locomotives have been built under licence elsewhere.

Number	Year	Type	Other	Number	Reference(s)
?	19xx	B6000S			BL13
A122	19xx	A2500			TI35
A208	1966	B6000S			TG21

Unknown builder — unknown

Number	Year	Type	Other	Number	Reference(s)
?	xxxx	R/R			BE108
?	19xx	R/R			AG002
?	19xx	R/R			BL29
?	19xx	R/R			LU14
?	19xx	R/R			SO32
?	19xx	R/R			TG08
?	xxxx				SG74
?	1xxx				AG042
?	1xxx				AG061
?	1xxx				AG079

Number	Year	Type	Other	Number	Reference(s)
?	1xxx				GL01
?	1xxx				SG09
?	1xxx				SG13
?	1xxx				SG30
?	1xxx				SG41
?	1xxx				SH06
?	1xxx		BBC	?	AG058
?	18xx				SG64
?	18xx				SO08
?	19xx				AG097
?	19xx				BE049
?	19xx				BL17
?	19xx				FR05
?	19xx				FR08
?	19xx			*2	GE24
?	19xx				GR32
?	19xx				GR46,
?	19xx				LU23
?	19xx				SG08
?	19xx				SG18
?	19xx				SG34
?	19xx				SO32
?	19xx				TI17
?	19xx				TI21
?	19xx				TI47
?	19xx				VD32
?	19xx				VD33
?	19xx				VS28
?	19xx				VS59
?	19xx				ZH121
?	19xx				ZH149
?	19xx				ZH162
?	19xx				ZH183
?	19xx		BBC	?	AG115
?	19xx		MFO	?	ZH149
?	1912				ZH149
?	1913				BE049
?	1916		MFO	?	SG69
?	1918		MFO	?	ZH024
?	1919		BBC	1302	VS62
?	1921		MFO	?	ZH001
1177	195x	CD33			GL10
?	1951				SG34

Sté des Ateliers de Construction du Val de Maizet, Ancien Etabts P. Drouville, Maizet (Calvados) [F]　　　　　　　　　　　　　　　　　　Val de Maizet

see: BLW

This company rebuilt World War I locomotives and re-sold them under their own name.

Ateliers de Constructions Mecaniques de Vevey SA　　　　　　　　Vevey

see: ACMV

Vollert GmbH + Co. KG, Weinsberg [D]　　　　　　　　　　　　　　Vollert

Vollert use the terms "Batterie, FFS" or "Diesel, FFS" to describe many of the appropriate Robots.

Number	Year	Type			Reference(s)
?	19xx				GR23
?	19xx				ZH067
?	1974	Robot			AG015
74/049 I	1974	V 7000			AG048
74/049 II	1974	V 7000			AG048
76/006	1976	DR 10 000			BS18
77/013	1977	Diesel, FFS			AG092

Number	Year	Type	Other	Number	Reference(s)	
81/009	1981	Kabeltrommel			SO31	
82/045	1982	Kabeltrommel			TG29	
83/092	1983	Kabeltrommel			TG29	
86/010	1986	Diesel, FFS			LU32	
01/023-1	2001	Kabeltrommel			BS28	
01/023-2	2001	Kabeltrommel			BS28	
02/003	2002	DER 3000			LU31	

Ludwig von Roll'schen Eisenwerke — von Roll

founded 1818 - slight changes in title till 1996
locomotives made at von Roll'sche Eisenwerke AG, Bern

Number	Year	Type	Other	Number	Reference(s)
-	19xx				AG057
-	19xx				LU32
-	1911		MFO	?	VD01
-	1928		SAAS	?	BE054
-	1946				SO32

see: MaK — **Vossloh**

Rheiner Maschinenfabrik Windhoff AG, Rheine [D] — Windhoff

IRS convention — Wind

Number	Year	Type			Reference(s)
?	1974				BE033
211	1931	6 II			AG084
1036	1948	Lg II			AG028
2253	1973				BE033
130426	1973	RW 50 E			GR23
260087	1992	RW 40 EM			BE105

AG der Lokfabrik Wiener Neustadt, Wiener Neustadt [A] — WN

formed 1873 from G. Sigl Locomotiv-Fabrikl
sold 1938 to Henschel & Sohn

Number	Year				Reference(s)
5597	1918				AG054

Zagro Bahn- und Baumaschinen AG, Mühlstrasse 11-15, 74906 Bad Rappenau-Grombach [D] — Zagro

Number	Year	Type			Reference(s)
80063	9.2007	Maxi Rangierer			BE029
90135	7.2005	MR 200 B			BS11

Zephir s.p.A, via Salvador Allende, 85 - 41100 Modena [I] — Zephir

Number	Year	Type			Reference(s)
?	1994	Lok 10.170			SO08
1440	1997	Lok 6.130			GR14

L. Zobel, Bromberg [D] (later Bydgostia [PL]) — Zobel

Production ceased in 1913 and was continued by Smoschewer & Co, Breslau [D] (later Wroclaw [PL]) with two locomotives in 1917 and full production in 1918.

Number	Year				Reference(s)
551	1907				SG01, ZH048
569	1909				AG102, SO38

Locations   Reference(s)

Location index

Aarau	AG047, AG057, AG070, AG079, AG083, AG085, AG099
Aarberg	BE046
Aathal-Seegräben	ZH140
Acquarossa	TI67
Adliswil	ZH055
Aeugstertal	ZH046, ZH163
Affoltern am Albis	ZH122, *see also* Zürich-Affoltern
Agno	TI38
Aigle	VD02, VD31, VD52, VD78
Airolo	TI11, TI21, TI24, TI32, TI40
Albisrieden	*see* Zürich-Albisrieden
Allschwil	BL01, BL33
Alpnach Dorf	OW2, OW4, OW6
Alpnachstad	LU38, OW5
Altdorf UR	UR07, UR08a, UR12, UR10, UR15
Altenrhein	*see* Staad SG
Altishausen	*see* Berg TG
Altishofen	LU10
Altstätten SG	SG52, SG72, SG85
Altstetten	*see* Zürich-Altstetten
Ambri	TI02, TI09
Amsteg	UR02, UR03, UR04, UR16
Andeer	GR34b
Andermatt	NW1, UR13
Appenzell	AI1, AR2
Aproz	VS60
Arbedo	*see* Castione-Arbedo
Arbon	TG04, TG17
Arch	BE063
Ardon	VS38, VS59
Arniberg	*see* Amsteg
Asp	AG083
Atisholz	*see* Riedholz
Attisholz	*see* Riedholz
Au ZH	ZH095
Aubrugg	*see* Wallisellen
Auenstein	AG039
Augst-Wyhlen	AG013
Auhafen	*see* Muttenz-Auhafen
Avenches	VD13, VD62
Baar	ZG2
Baden	AG005, AG011, AG034, AG090
Bäch SZ	SZ09
Bärengraben	*see* Deponie Bärengraben
Bäretswil	*see* Bauma
Bärschwil	SO41
Bâle Petite Hunigue	*see* Basel Kleinhüningen
Balerna	TI54
Balsthal	SO29, MI07, *see also* Oensingen
Baltschieder	VS03
Balzers	FL4
Barberine	*see* Le Châtelard VS
Bardonnex	GE24
Basel	BS04, BS05, BS07, BS09, BS16, BS22, BS27, BS30, MI99
Basel Badischer Bahnhof	BS01, BS08
Basel Dreispitz	BS06, BS11, BS12 *see also* Münchenstein, BL
Basel Kleinhüningen	BS02, BS13, BS14, BS15, BS17, BS18, BS19, BS20, BS23, BS24, BS28, BS29

Locations	Reference(s)
Basel St. Johann	BS03, BS21, BS29
Basel Wolf	BS01, see also Reinach BL
Basisstollen	SG18
Bassersdorf	ZH039
Baulmes	VD12, VD22
Bauma	ZH034
Bavois	VD60
Bavona	TI58
Beckenried	NW1
Bedretto	TI43
Bellerive JU	see Soyhières-Bellerive
Bellinzona	TI06, TI10, TI16, TI22, TI29, TI53, TI57, TI61, TI66, see also Osnoga
Berg TG	TG21
Bergün	GR44
Berikon	AG049
Bern	BE009, BE015, BE022, BE031, BE039, BE040, BE042, BE047, BE056, BE060, BE061, BE069, BE073, BE074, BE077, BE078, BE087a, BE091, BE094, BE104, BE109, BE112, BE116, BE119, MI99 see also Burgernziel and Wabern bei Bern
Bern-Bümpliz Süd	see Gwatt (Thun)
Bern-Riedbach	BE019
Bever	GR28
Bex	VD17, VD40, VD41, VD68, VS30
Beznau	see Döttingen-Klingnau
Biasca	TI49
Biberist	SO26
Biel	BE012, BE013, BE041, BE066, BE080, BE087, BE114, BE117, MI08
Biel-Bözingen	BE081, BE114
Biel-Mett	BE029, BE037
Bienne	see Biel
Bietschtal	VS03
Bilten	GL09
Bioley-Orjulaz	VD45
Birmensdorf ZH	ZH054
Birmenstorf AG	AG111
Birr	AG002, AG005, AG006, AG034, see also Baden
Birrfeld	AG002, AG003, AG006, AG034, AG082, see also Lupfig
Birsfelden	BL03, MI99
Birsfelden Hafen	BL07, BL13, BL34, BL49
Blonay	VD11
Blumenau	see Rapperswil-Blumenau
Bodio	TI55, TI15, TI27, TI64, TI72, MI08, see also Giornico
Böttstein	AG073
Bözingen	see Biel-Bözingen
Bonaduz	GR19
Boniswil	BE111
Bonstetten-Wettswil	see Filderen
Boudry	NE02
Boujean	see Biel-Bözingen
Bouveret	see Le Bouveret
Bowil	BE098
Bravuogn	see Bergün
Bregenz [A]	see Widnau
Breitenbach	SO23
Bremgarten West	AG046
Bremgarten AG	AG019
Brenzikofen-Herbligen	BE059
Brienz BE	BE028
Brig	BE024, VS12, VS16, VS32, VS43, VS48, VS49, VS70
Brittnau-Wikon	AG032, AG081, see also Walterswil SO
Broc	FR12
Bronschhofen	SG44

Locations	Reference(s)
Brugg AG	AG052, AG054, AG058, AG084, AG090, AG104, AG115, MI07, *see also* Villnachern
Bruggerhorn	*see* St. Margrethen-Bruggerhorn
Brunau	*see* Zürich-Brunau
Brunnen	SZ06, SZ20, SZ21
Buchmatt	BE055
Buchrain	LU33
Buchs AG	*see* Suhr
Buchs SG	FL2, SG06, SG22, SG65
Buchs ZH	ZH011
Buchs-Dällikon	ZH007
Bürglen UR	UR06a
Bütschwil	SG28
Bülach	ZH024, ZH032, ZH073, *see also* Hochfelden
Bümpliz Süd	*see* Bern-Bümpliz Süd
Bulle	FR04
Bure	JU09
Burgdorf	BE010, BE036, BE055, BE085, BE111, BE112, BE113, MI08, *see also* Buchmatt
Burgernziel	BE011
Burghof	BE084
Burgholz	BE067
Burgwies	ZH165
Buseno	GR16
Bussigny-près-Lausanne	VD34, VD64
Bussnang	TG19
Cadenazzo	TI18, TI69
Capolago	TI36
Carrières Bussien	*see* Le Bouveret
Carrières des Andonces	*see* St. Triphon
Carrières d'Arvel	*see* Villeneuve VD
Carrière de Ruan	*see* Le Châtelard VS
Carrières du Lessus	*see* Saint Triphon
Castagnola	TI52
Castione-Arbedo	TI37, TI39
Cham	ZG3, ZG4
Chamby	*see* Chaulin
Chandolan	FR14
Chandoline	VS38
Château d'Eau	VS54
Chaulin	VD10
Chavornay	VD30, VD59, VD61
Cheseaux-sur-Lausanne	VD58
Chiasso	TI41
Chippis	VS01, VS33, VS32, *see also* Oberems
Choindez	JU12
Chörbligen	*see* Gisikon
Chur	GR08, GR09, GR13, GR22, GR27, GR28a, GR29, GR32, GR34, GR36, GR41, GR45
Clarens	VD01, VD57
Claro	TI55
Col des Roches	*see* Le Locle-Col des Roches
Collombey	VS52, VS63, *see also* Saint-Triphon
Collombey-Muraz	VS20, VS65
Conthey	VS06
Corcapolo	TI24
Corcelles-près-Payerne	VD39, VD55
Cornaux NE	NE06, NE09, NE12
Cornavin	*see* Genève-Cornavin
Cossonay	VD03
Courtemaîche	*see* Bure
Couvet	NE05

Locations	Reference(s)
Cressier NE	see Cornaux NE
Crissier	VD04
Dällikon	see Buchs-Dällikon
Däniken SO	SO18
Därligen	BE099
Dättnau-Töss	ZH072
Dagmarsellen	LU18
Daillens	VD49, VD63
Dänikon ZH	ZH124
Davos	GR30
Davos-Islen	GR31
Davos-Laret	GR31
Degersheim	SG09, SG11
Deisswil	see Stettlen
Delémont	JU01, JU04, JU05, JU08, JU11, MI07
Denges	see Ecublens VD
Deponie Bärengraben	see Würenlingen
Derendingen	SO13, SO28, SO34
Diepoldsau	SG79
Dietikon	ZH109, ZH127, ZH142
Dietikon-Au	ZH056
Dixence	see Le Chargeur
Döttingen	AG052
Döttingen-Klingnau	AG078
Domat	see Ems
Domdidier	FR09
Dorénaz	see Vernayaz
Dornach	SO39
Dübendorf	ZH027, ZH030, ZH070, ZH088, MI99
Düdingen	FR02, FR11
Dulliken	SO20
Echallens	VD51
Eclépens	VD47
Ecublens VD	VD71
Effingen Bahnhof	AG065
Effretikon	ZH097
Egerkingen	see Härkingen
Eggerberg	VS24
Eglisau	ZH157, see also Zweidlen
Egnach	TG18
Eiken	AG096
Eifeld	see Wimmis
Einsiedeln	SZ14, SZ17, SZ27, SZ29
Elgg	ZH129
Embrach	ZH041, ZH145
Emmen	LU24, LU29
Emmenbrücke	LU15, LU19, LU32, LU35
Emosson	see Le Châtelard VS
Ems	GR18
Erlach	BE122
Erlen	TG09
Erstfeld	UR01, UR13, UR14, MI07
Escher Wyss Platz	see Zürich-Industriequartier
Eschlikon	TG26
Estavayer-le-Lac	FR03, see also Sévaz
Etzel	SZ05
Etzwilen	SH09
Evionnaz	VS53
Eyschachen	see Altdorf
Faido	TI08, TI49, TI59
Felsberg	GR21

Locations	Reference(s)
Felsenau	AG042
Ferpicloz	see Le Mouret
Fideris	GR24
Filderen	ZH004
Fionnay	VS18
Flaach	ZH021
Fleurier	NE11, MI01
Flums Hochwiese	SG74
Flüelen	UR06, UR08
Frauenfeld	TG01, TG05, TG13, TG15, TG20, TG23, TG27, TG28, TG29
Friedrich	see Steinbruch Friedrich
Frenkendorf	BL14, BL27, BL42
Fribourg	FR07, FR16
Frick	AG008, AG020, AG106
Frinvillier	BE082
Frutigen	BE024, BE045, BE092, BE095, BE096, BE110
Füllinsdorf	BL18
Full	AG016, AG091, AG105
Full-Reuenthal	see Full
Fussach [A]	see Widnau
Gais	AR1
Galerie 5	SG18
Galerie 12	SG18
Gampelen	see Witzwil
Gampel-Steg	see Niedergesteln and Steg VS
Gams	SG54
Gamsen	VS44
Genève	GE01, GE02, GE03, GE08, GE09, GE12, GE13, GE18, GE23, MI02
Genève-Cornavin	GE15
Genève-La Jonction	GE06
Genève-Le Praille	GE15, GE21
Genève-Le Renfile	GE20
Genève-Sécheron	GE17
Gerlafingen	SO12, SO37
Geroldswil	ZH020
Gesenk	SG18
Gettnau	LU13, LU40
Giesshübel	ZH183
Giornico	TI47, TI71
Gisikon	LU30
Gisikon-Root	LU16
Giswil	OW3, OW7
Giubiasco	TI35
Gland	VD16
Glarus	GL02, GL08, MI01
Glattbrugg	see Zürich-Altstetten
Goldach	SG33
Goldach-Rietli	SG24
Göschenen	UR12
Gösgen	see Olten-Gösgen
Goldau	SZ07, SZ08, SZ16, SZ23
Goldbach	see Lützelflüh-Goldbach
Gondiswil	BE008
Gondo	VS77
Goppenstein	BE081, VS57
Gossau SG	SG08, SG38, SG45, SG62, SG68, see also Lindenhof
Gotthard	MI02, MI99, see also Pollegio and Silenen
Grämigen	SG28, SG80
Grafstal ZH	SH03
Grande Dixence	VS38, see also Le Chargeur
Grellingen	BL52
Grône	VS61

Locations	Reference(s)
Grono	GR10, GR11, GR43
Grünenmatt	BE059
Grüningen	ZH169
Grüze	see Winterthur-Grüze
Guber	see Steinbruch Guber
Guntalingen	ZH114
Guttanen	BE054, BE097
Gwatt (Thun)	BE033
Haag	see Gams and Widnau
Härkingen	SO11
Haggen	see St. Gallen-Haggen
Hammer	see Olten-Hammer
Hardwald	see Schlieren
Hausen bei Brugg	AG082, see also Birrfeld
Haute-Nendaz	see Nendaz
Heerbrugg	SG85
Heiden	AR3, AR4
Heimberg	see Steffisburg
Heimiswil	see Burgdorf
Hemishofen	SH09
Herbligen	see Brenzikofen-Herbligen
Herblingertal	SH04
Hérémence	see Nendaz
Herisau	AR2, AR6
Herznach	AG010, AG056
Herzogenbuchsee	BE108
Hinteregg	ZH008
Hinterthal	see Muotathal
Hochdorf	LU41
Hochfelden	ZH061
Hochwiese	see Flums Hochwiese
Holderbank	MI01, MI08, see also Wildegg
Hondrich	BE004
Horgen	ZH095, ZH017, ZH057
Horgen-Oberdorf	ZH121
Horn	TG08, see also Goldach-Rietli
Horw	LU12, LU23
Hüntwangen	ZH067
Huttwil	BE059, BE016, BE018, BE020, BE113
Illsee	see Oberems
Immensee	SZ26, ZH160
Industriequartier	see Zürich-Industriequartier
Innertkirchen	BE054
Interlaken	BE041
Interlaken West	BE059, BE018, BE068
Intragna	see Corcapolo
Inwil	LU42
Iselle [I]	VS12, VS16, VS43
Isérables	VS67
Islen	see Davos-Islen
Joux Verte	VD65
Julier	GR40
Kaiseraugst	AG038, AG080, AG097
Kallnach	BE062, see also FR01
Kaltbrunn	SG53
Kandersteg	BE003
Kaufdorf	BE064
Kempten	ZH091
Kemptthal	ZH102
Kerenzerbergtunnel	see Weesen
Kerzers	FR01

Locations	Reference(s)
Klingnau	AG003, AG086, AG087, see also Döttingen
Klosters	GR03, GR05
Kloten	ZH073, ZH081
Klus	SO44
Koblach [A]	see Widnau
Koblenz	AG022, AG073
Kölliken	AG061
Körbligen	see Gisikon
Konolfingen	BE001, BE071
Kreissern	see Widnau
Kriens	LU03, LU05, LU08
Küblis	GR12
Küsnacht ZH	ZH006, see also Küssnacht am Rigi
Küssnacht am Rigi	SZ24
La Chaux-de-Fonds	NE04
La Comballaz	VD53a
La Croix-de-Rozon	see Bardonnex
La Fouly	VS54
La Jonction	see Genève-La Jonction
La Lienne	see Sion
La Paudèze	see Paudex
La Plaine	GE04
La Praille	see Genève-Le Praille
La Presta	see Travers
La Renfile	see Genève-Le Renfile
Lachen SZ	SZ12
Läufelfingen	BL21, see also Hauenstein
Landquart	GR20, GR38, GR42, GR46
Langenthal	BE057, BE124
Langnau-im-Emmental	BE030, BE107, BE120
Laret	see Davos-Laret
Laubegg	ZH018
Laufen	BL10, BL24, BL25
Laufenburg	AG024, AG027, AG029, AG043, AG062, AG073, AG095, AG107
Laupen BE	BE072, BE112
Lausanne	VD05, VD06, VD07, VD20, VD29, VD41a, VD42, VD44, VD49, MI02
Lausanne-Ouchy	VD36
Lausanne-Sébeillon	VD24
Lausen	BL48
Lavey	VD24
Lavorgo	TI01
Le Bouillet	see Bex
Le Bouveret	VS28, VS55, VS64
Le Chargeur	VS08
Le Châtelard VD	see Clarens
Le Châtelard VS	VS07, VS54
Le Day	VD73
Le Locle	NE07, MI08
Le Locle-Col des Roches	NE10
Le Mouret	FR15
Le Pont	VD27
Leibstadt	AG030, AG068
Lenzburg	AG011, AG021, AG044, AG045
Lerchenfeld	see Thun
Les Brenets	NE14
Les Clées	VD12
Les Planches VD	see Clarens
Letten	ZH106
Leuzigen	BE121
Liesberg	BL09, BL10, MI08
Liestal	BL35
Limmattal	ZH132

Locations	Reference(s)
Lindenhof	SG43
Locarno	TI20, TI28, TI46, TI58
Locarno-S. Antonio	TI34
Lochezen	see Walenstadt
Lochbach	see Oberburg
Lochfeld	see Laufen
Losone	TI04
Lüsslingen	BE002
Lützelflüh-Goldbach	BE052
Lufingen	ZH173
Lugano	TI07, TI30a, TI31, TI48, TI56, TI68
Lugano-Viganello	TI45
Lungern	OW1
Lupfig	AG002, AG006, AG036, AG082
Lustenau [A]	see Widnau
Lütisburg	SG80
Luterbach	see Riedholz
Luterbach-Attisholz	see Riedholz
Luzern	LU08, LU31, LU38, LU39, MI07, see also Kriens and Tunnel Sörenberg
Lyss	BE115
Männedorf	ZH143, see also Uetikon
Märstetten	TG07
Märwil	TG22
Magadino	TI12
Malerva	SG18
Malévaux	see Monthey
Manegg	see Zürich-Manegg
Marrogia-Melano	TI70
Martigny	VS71, VS73, VS74
Mauvoisin	VS34
Meggen	LU22
Meiringen	BE006
Melano	see Marrogia-Melano
Melide	TI13
Mels	SG01
Mendrisio	TI17, TI42, TI44, TI73
Menzonio	TI19
Mesocco	GR37
Mett	see Biel-Mett
Mex VS	VS21, VS29
Meyrin	GE05, GE11, see also ZIMEYSA
Möhlin	AG007, AG095, AG103
Mollis	GL04
Momont	see Pont-la-Ville
Montfaucon	see Pré-Petitjean
Monthey	VS23, VS36, VS62
Monthey	VD54
Montlingen	see Oberriet and Widnau
Mühleberg	see Bern
Mühlethal (I bis IV)	SH04
Müllheim	see Müllheim-Wigoltingen
Müllheim-Wigoltingen	TG24, TG30
Mülligen	see Zürich-Mülligen
Münchenbuchsee	see Zollikofen
Münchenstein	BL04, BL15, BL31, MI01
Münchenwiler	BE014
Münchwilen AG	AG089
Münstchemier	BE044
Muotathal	SZ03
Muraz (Collombey)	see Collombey-Muraz
Muri AG	AG001, AG063, AG109
Muri bei Bern	BE086

Locations	Reference(s)
Murten	MI08
Museumstrasse	see Zürich-Museumstrasse
Muttenz	BL02, BL19, BL26, BS16
Muttenz-Auhafen	BL05, BL08, BL28, BL38, BL43, BL46
Näfels	GL06, GL10
Naus	SG18
Nebikon	LU07, LU09, LU18, MI99
Nendaz	VS19
Netstal	GL03
Neuchâtel	NE01, NE03, NE13
Neuendorf	see Oberbuchsiten
Neuenegg	BE070, BE118
Neuhausen am Rheinfall	SH02, SH06, SH08
Niederbipp	BE105
Niederböttingen	see Bern-Riedbach
Niederer Erben	see Altstätten SG
Niedergesteln	VS14
Niederglatt ZH	ZH156, see also Niederhasli
Niederhasli	ZH031, ZH110
Niederschöntal-Frenkendorf	see Füllinsdorf
Niederurnen	GL01, GL07
Nuolen	SZ09
Nyon	VD56, VD72
Oberbuchsiten	SO24
Oberburg	BE059
Oberdiessbach	BE128
Oberdorf (ZH)	see Horgen-Oberdorf
Oberdorf SO	SO09
Oberems	VS40
Oberriet	SG07, SG40, SG71, SG84, SG85
Oberrüti	AG028
Obersaxen	see Tavansa
Oberwald	VS04
Oberwil	BL30
Oberwinterthur	ZH149
Obstalden	GL04
Oensingen	SO32, MI01
Oerlikon	see Zürich-Oerlikon
Oetwil an der Limmat	ZH079
Oey-Diemtigem	BE067
Olten	SO01, SO02, SO05, SO10, SO19, SO35, SO36, SO40, SO43, MI07, see also Trimbach
Olten-Gösgen	SO06, SO09, SO27
Olten-Hammer	SO31, SO38
Olten-Klingnau	SO03, see also Klingnau
Onnens	VD38
Orbe	VD21, VD66
Orjulaz	see Bioley-Orjulaz
Osogna	TI54a
Ostermundigen	BE021, BE035, BE072, BE083, BE102, BE123
Otelfingen	ZH046
Ouchy	see Lausanne-Ouchy
Paleu Sura	see Felsberg
Paudex	VD74
Paudez	see Paudex
Payerne	VD09, VD37, MI08
Perlen	see Gisikon-Root
Péry	BE017
Pfäffikon SZ	SZ15, SZ22, SZ28
Pfungen	ZH072
Pfyn	TG11

Locations	Reference(s)
Pieterlen	BE080
Pignia	see Andeer
Piotta	TI71
Pollegio	TI23
Ponte Chiasso [I]	TI03
Pont-de-la-Morge	VS35
Pont-la-Ville	FR05
Pontresina	GR38
Pradella	GR17
Prahins	VD15
Pratteln	BL12, BL17, BL20, BL23, BL36, BL37, BL44, BL45, BL47, TG20, MI01, see also Frauenfeld
Pré-Petitjean	JU06
Rafz	ZH068, ZH138, ZH183
Ramosch	GR01
Ramsen	SH05, SH09, SH10
Rapperswil SG	MI07, SG17, SG39, ZH095
Rapperswil-Blumenau	SG27, SG81
Raron	VS02, VS03, see also Steg VS
Rawyl	see Sion
Realp	UR05, UR09, UR11
Regensdorf	ZH002, ZH037, MI99
Reiden	LU01, LU14
Reinach BL	BL53
Rekingen AG	AG048, AG060, AG069, AG092
Renens VD	VD14, VD24, see also Crissier
Reuenthal	see Full
Rheinau	ZH159
Rheinfelden	AG014, AG033, AG112, see also Möhlin and Pratteln
Rheinsulz	AG089
Riburg	see Möhlin
Rickenbach-Attikon	ZH099
Riedbach	see Bern-Riedbach
Riedholz	SO08, SO17
Riedikon	ZH098, ZH174
Riedtwil	BE127
Rietli	see Goldach-Rietli
Ristet	see Birmensdorf, see also Frauenfeld
Rivaz	see Zürich-Rivaz
Rivera	TI74
Roche VD	VD47
Roggwil BE	BE130, see also Frauenfeld
Romanens	see Sâles (Gruyère)
Romanshorn	TG02, TG10, TG16, TG25, MI01
Rondchâtel	see Péry
Rondez	see Delémont
Root	see Gisikon-Root
Rorschach	SG03, SG14, SG29, SG37, TG03, MI07, see also Goldach, Sulgen and Widnau
Rossens FR	FR05
Rothenbrunnen	GR14
Rothenburg	LU17, LU19, LU21, LU36
Rothrist	AG035, AG037, AG041, AG094
Rotkreuz	ZG1, ZG5, see also Immensee
Rubigen	BE079
Rümlang	ZH136, see also Letten
Rüschlikon	ZH107
Rüthi SG	SG15
Rüti bei Büren	BE106
Rüti ZH	ZH085, ZH152
Rupperswil	AG067, AG113
Russin	GE10

Locations	Reference(s)
Rynächt Arsenal	*see* Altdorf
S. Antonino	TI62
S. Antonio	*see* Locarno-S. Antonio
Sagenwald	*see* Buchrain
Saint-Brais	JU07
Saint-Maurice	VS29, VS30, VS39, *see also* Mex VS
Saint-Prex	VD35, VD77
Saint-Triphon	VD16, VD19, VD70, VS63, *see also* Aigle
Saint-Ursanne	JU02, JU03
Sâles (Gruyère)	FR08
Samedan	GR15, GR32, GR38
Samnaun	GR07
Samstagern	ZH105, ZH160
San Bernardino	GR47
San Niclà	*see* Pradella
San Vittore	GR06
Sargans	SG18, SG36, SG49, SG65
Satigny	GE14, GE16, GE17
Saxon	VS31
Schaan	FL3
Schachen LU	LU02, LU10, *see also* Buchrain
Schaffhausen	SH01, SH03, SH04, SH010, ZH033
Schattdorf	*see* Altdorf
Schinznach Bad	AG006, AG110
Schinznach Dorf	AG012, AG088, AG114
Schlatt TG	TG06
Schlieren	ZH053, ZH066, ZH084, ZH092, ZH093, ZH111, ZH133, *see also* Airolo and Frauenfeld
Scuol	GR39a
Schönbühl	BE065
Schule Tannenbach	*see* Horgen
Schule Werd	*see* Adliswil
Schuls	see Scuol
Schwaderloch	AG071
Schwamendingen	ZH160
Schwanden GL	GL05
Schwarzenbach SG	SG05, SG31
Schwarzenburg	BE038
Schweizerhalle	BL11, BL29
Schwerzenbach	*see* Volketswil
Schwyz	SZ13, SZ18, *see also* Seewen SZ
Sébeillon	*see* Lausanne-Sébeillon
Sécheron	*see* Genève-Sécheron
Sedrun	GR02
Seebach	*see* Zürich-Seebach
Seegräben	*see* Aathal-Seegräben
Seewen SZ	SZ02, SZ09
Selfranga	*see* Klosters
Sembrancher	VS42, VS66
Semsales	FR13
Sennwald	SG02, SG04, *see also* Widnau
Sévaz	FR10
Sevelen	SG30
Siebnen	SZ19
Sierre	VS17
Siggenthal	*see* Würenlingen-Siggenthal
Sihlpost	*see* Zürich-Sihlpost
Sihlwald	ZH181
Silenen	UR02
Simplon	VS43
Sins	AG031
Sion	VS22, VS27, VS38, VS41, VS58

Locations	Reference(s)
Sissach	BL06, BL22, BL41 see also Tecknau
Sisseln AG	AG023
Sörenberg	see Tunnel Sörenberg
Solothurn	SO14, SO15, SO16, SO30, SO33, SO42
Sonceboz	BE005
Soyhières-Bellerive	JU10, see also Laufen
Spiez	BE010, BE018, BE112
St. Fiden	see St. Gallen-St. Fiden
St. Gallen	SG09, SG10, SG16, SG19, SG21, SG35, SG48, SG55, SG56, SG59, SG60, SG69, SG83
St. Gallen-Haggen	SG31
St. Gallen-St. Fiden	SG13
St. Gallen-Winkeln	SG57
St. Margrethen	SG32, SG42, SG61, SG63
St. Margrethen-Bruggerhorn	SG31
St. Moritz	GR35
St. Stephan	BE043
St. Sulpice NE	NE15
Staad SG	SG67
Stalden VS	VS37
Steffisburg	BE088
Steg VS	VS01, VS05
Stein AR	see Teufen AR
Stein am Rhein	SH07
Steinbruch Friedrich	see Laufen
Steinbruch Guber	see Alpnach Dorf
Steinbruch Lochezen	see Walenstadt
Steinbruch Zingel	see Seewen
Steinen	SZ01, SZ04
Stettlen	BE049
Strada en Engiadina	see Pradella
Subingen	SO45
Sugiez	FR06
Suhr	AG018, AG074
Sulgen	TG03, MI01
Sumiswald	BE089, see also Burghof
Sursee	LU11, LU18, LU27, LU34
Susch	GR04
Susten	VS45
Tavansa	GR34a
Taverne	TI30, TI60, TI62
Tecknau	BL06
Tenero	TI14, TI33
Territet	VD69
Teufen AR	AR5
Thalwil	ZH160, ZH164
Thörishaus	BE048
Thun	BE032, BE059, BE090, see also Uetendorf
Thusis	GR33, GR35, GR53a
Tiergarten	see Zürich-Tiergarten
Torrent de Saint-Barthélemy	see Mex VS
Töss	see Dättnau and Winterthur-Wülflingen
Travers	NE08, see also St. Sulpice NE
Triengen	LU28
Trimbach	SO07
Trimmis	GR40
Tuggen	SZ10
Tunnel Sörenberg	LU04
Turgi	AG017, AG090
Turtmann	VS46, VS72 see also Oberems
Uetendorf	BE032
Uetikon am See	ZH029

Locations	Reference(s)
Uetliberg	see Laubegg and Zürich
Unknown Galerie	SG18
Unknown location	SO04, TI09, TI65, VS68, MI03, MI04, MI05, MI06, MI10, MI99
Unterentfelden	AG040
Unterklein [A]	see Widnau
Unterterzen	SG82
Untervaz	GR23
Uznach	SG41
Uzwil	SG12
Vaduz	FL1
Val-de-Travers	see Travers
Vallée de Joux	see Le Pont
Vallorbe	VD08, VD33, VD48, VD53, see also Le Day
Vatryeret	see Sion
Vaulion	VD32
Vernayaz	VS13, VS45, VS56
Vétroz	see Conthey
Vernier	GE07, GE19, GE22
Vernier-Meyrin	see Meyrin, Vernier and ZIMEYSA
Vevey	VD23, VD26, VD46
Viganello	see Lugano-Viganello
Vild	SG18
Villars-sur-Glâne	FR14
Villeneuve VD	VD18
Villmergen	AG009
Villnachern	AG066
Visp	VS03, VS11, VS15, VS45, VS47
Vitznau	LU20
Volketswil	ZH094
Wabern bei Bern	BE034
Wädenswil	ZH010, ZH132, ZH135
Wäggital	SZ05
Wald ZH	ZH065
Waldenburg	BL16
Walenstadt	SG82
Walingen	see Rothenburg
Wallisellen	ZH042, ZH022, see also Winterthur
Waltenschwil	AG108
Walterswil SO	AG081
Wangen an der Aare	SO22
Wangen SZ	see Nuolen
Wartau	ZH165
Wasen im Emmental	BE093
Wasserauen	AI2
Wattwil	SG53, SG73
Wauwil	LU18
Weesen	SG37a
Weiach	ZH167
Weinfelden Süd	TG12
Weiningen ZH	ZH043
Werdenberg	SG23
Wettingen	AG093
Wettingen Dorf	AG098
Wettswil am Albis	see Filderen
Wetzikon ZH-Kempten	see Kempten
Widnau	SG34, SG58, SG64, SG77, SG78
Wikon	see Brittnau-Wikon
Wil SG	SG47, SG66, SG70
Wil ZH	see Hüntwangen
Wildberg	ZH045
Wildegg	AG015, AG057, AG102
Wimmis	BE023, BE053

Locations	Reference(s)
Winterthur	ZH033, ZH052, ZH074, ZH082, ZH087, ZH099, ZH130, ZH137, ZH141, ZH149, ZH153, ZH155, ZH158, ZH162, ZH166, MI01, MI07, MI99
Winterthur-Grüze	ZH135
Winterthur-Wülflingen	ZH089
Witzwil	BE103
Wohlen AG	AG055, AG059, *see also* Waltenschwil
Wolfsloch	SG18
Worblaufen	BE027, BE129
Wülflingen	*see* Winterthur-Wülflingen
Würenlingen	AG064
Würenlingen-Siggenthal	AG048
Wyhlen	*see* Augst-Wyhlen
Yverdon	VD28
Yvonand	VD43
Zell LU	LU06, LU13, LU37
Zermatt	VS09, VS10, VS37
Zeusier	VS69
Zingel	*see* Steinbruch Zingel
ZIMEYSA	GE16, GE17
Zofingen	AG053, AG072, AG077, AG081
Zollikofen	BE072
Zuchwil	SO21, *see also* Solothurn *and* Riedholz
Zürich	ZH005, ZH009, ZH012, ZH013, ZH014, ZH015, ZH016, ZH018, ZH023, ZH036, ZH040, ZH044, ZH047, ZH048, ZH049, ZH058, ZH060, ZH064, ZH069, ZH071, ZH078, ZH080, ZH086, ZH090, ZH103, ZH108, ZH112, ZH113, ZH115, ZH116, ZH117, ZH119, ZH120, ZH123, ZH134, ZH139, ZH146, ZH147, ZH148, ZH161, ZH165, ZH170, ZH171, ZH172, ZH177, ZH178, ZH179, MI99, *see also* Burgwies, Dübendorf, Regensdorf *and* Schwamendingen
Zürich-Affoltern	ZH028
Zürich-Albisrieden	MI99, ZH096, ZH180
Zürich-Altstetten	ZH035, ZH059, ZH062, ZH063, ZH128, *see also* Zürich-Mülligen
Zürich-Binz	ZH064
Zürich-Brunau	ZH003
Zürich-Hauptbahnhof	ZH132
Zürich-Industriequartier	ZH118, ZH153
Zürich-Manegg	ZH083, ZH182
Zürich-Mülligen	ZH036
Zürich-Museumstrasse	ZH132
Zürich-Oerlikon	*see* Zürich-Seebach
Zürich-Rivaz	*see* Zürich-Industriequartier
Zürich-Seebach	ZH001, ZH019, ZH038, ZH131, ZH168, MI99
Zürich-Sihlpost	ZH036
Zürich-Tiergarten	ZH183
Zug	ZG6, MI01
Zurzach	*see* Rekingen AG
Zweidlen	ZH104
Zweilütschinen	BE007
Zweisimmen	BE058
Zwingen	BL32

SÉCHERON SA, Zimeysa (GE17)
ABL (Badoni) 6171 1980 (type VI/C) takes a weekend break on 28 September 2008.

Andrea Gruber, Landquart (GR20)
Although out of use for half a century this Austro-Daimler petrol locomotive and its supporting railway equipment were photographed on 10 September 1996.

Jean-Claude Coquoz, Collombey-Muraz (VS20)
The World War I Baldwin Locomotive Works locomotive from the Carrières de Vongy [F] was seen in Collombey-Muraz on 30 September 2008

SALANFE SA, Vernayaz (VS56)
Modified Brookville 3517 of 1949 (a BMD-8) stands in the tunnels at Miéville on 22 October 2007.

SERSA AG, Rümlang (ZH136)

"ERICH" is one of a small series of diesel-electric locomotives built by SIG and BBC for industry and minor railways. It was photographed at Rümlang on 3 August 2005.

Ems-Chemie AG, Domat/Ems (GR18)

The ČKD locomotive (type T 239.1) overtakes a tamper owned by Parachini (part of the SERSA group) on 12 June 2007 while the standard gauge connection to the Stallinger Lager was being finalised. In common with a number of local shunting locomotives this one is equipped to handle both standard and metre gauge wagons.

6 (0-6-0WT SLM 2075 of 1910), a former SBB locomotive, was photographed by Andrew Smith on 27th July 1976.

Von Moos, Emmenbrücke (LU32)
11 (Moyse CN52EE500D 3525 1972) was photographed by Andrew Smith on 27th July 1976.

AVIA AG, Muttenz-Auhafen (BL05)
"RONNY" Tm 237 867-7" Gmeinder 5468 of 1970 (a D 35 B) pauses between duties outside the rail gate on 15 October 2007.

J. Heusser, Schlieren (ZH066)
Along with other railway material Hanomag 8009 of 1922 has been waiting (in Schlieren gasworks) for restoration for over forty years. It was photographed on 3 June 2008.

TENSOL RAIL SA (TI71)
Henschel 30506 of 1962 (a DH 500) is still labelled "Monteforno No 7" on 4 April 2005 as it brings a train of prepared permanent way material from Tensol Rail SA (its later owners) to Bodio FFS station.

EUROVAPOR, Worblaufen (BE027)
101 (Krauss Maffei 17627 of 1949) in steam at the SZB's Worblaufen depot on 25th July 1976.
(Andrew Smith)

Brenntag Schweizerhall, Basel St. Johann (BS03)
On disposal of the last conventional locomotive in 2003 two different Mercedes road-rail locomotives were acquired. The Unimog 1600 was seen at work on 17 October 2007.

Nordostschweizerische Kraftwerke (NOK), Döttingen-Klingnau (AG078)
"JANKA" KHD 57198 of 1961 (type A12L714 R) stands outside the workshop on 16 October 2007.

Giovanola Frères SA, Monthey (VS36)
Kronenberg 129 of 1952 (type DT 180/250 at work on 21 May 1993).

Von Roll, Delèmont
5 (SLM 1870 of 1905) was photographed in June 1974 in a scenic location at the foundry.
Rodney Weaver photograph, copyright IRS

Papierfabrik Perlen, Gisikon-Root (LU16)
3 (Henschel 20593 of 1925) at the paper factory in June 1974.
Rodney Weaver photograph, copyright IRS

Solvay (Suisse), Rekingen AG (AG092)
On 16 October 2007 the company's Moyse three-axled diesel-electric chain-driven 3537 of 1973 (type CN60EE500D) was on hire to Logistik- und Gewerbe-Zentrum Hochrhein AG. In this picture it is returning to that site having delivered a train to the SBB.

ULTRA-BRAG AG, Muttenz-Auhafen (BL46)
Orenstein & Koppel 26971 of 1981 (type MHB 101) positions some containers for despatch on 17 October 2007.

Vaparoid AG, Turtmann (VS72)
Many ex-SBB Tm^I are now in industrial service. This one was built by Robert Aebi & Co (RACO) as 1605 1061 (type 85LA 7). It was photographed on 2 October 2008.

ROUTORAIL SA, Zimeysa(GE16)
A Ruston and Hornsby 88DS locomotive (252840 of 1949) forms part of an equipment display. It was photographed on 28 September 2008.

Gaswerk der Stadt, Zürich (ZH048)
SLM 1141 of 1898 in steam in June 1974. Rodney Weaver photograph, copyright IRS

Raffinerie du Sud-Ouest SA (RSO), Saint-Triphon (VD70)
SFL locomotive, "2" SFL 301 of 1994 was the only working locomotive on 6 September 1996.

Stadler Bussnang AG, Bussnang (TG19)
The Stadler locomotive "Barry" used in the workshops and made from two previous locomotives. It was photographed on 29 April 2000.

Petroplus Tankstorage AG, Birsfelden Hafen (BL34)
TransLok 180 of 2008 (a cargolok TL-DH440) is being commissioned on 26 September 2008.

Lokservice Burkhardt AG (LSB), Rüti ZH (ZH085)
A Romanian (U23A) built locomotive is part of the LSB hire fleet. Here the LDH45 (24122 of 1980) is seen assisting Greuter at Biasca on 8 June 2008.

Von Roll, Choindez
9 (SLM 21698 of 1911) is seen at work near to the SBB station in June 1974
Rodney Weaver photograph, copyright IRS

HOLCIM Kies und Beton AG, Werk Hüntwangen (ZH067)
"GUSTI" MaK 700100 of 1991 (type G 763 C) was having its batteries charged on 8 September 2006.

Feldschlösschen Getränke AG, Depot Gurten, Solothurn (SO14)
The Windhoff "mule" was photographed on 14 May 1986. (Photograph: Anon).

Feldschlösschen-Getränke AG, Biel-Mett (BE029)
The Zagro Maxi Rangierer (80063 of 9.2007) was photographed on 14 November 2007.

Metallwerke, Dornach (SO39)
3 ZEPHIR (Krauss S 290 of 1874) was at Dornach in June 1974, although by this time it had been donated for preservation. Rodney Weaver photograph, copyright IRS

Feldschlösschen Brauerei (AG029)
One of the well kept locos at this often visited location (Krauss 5666) at work in June 1974.
Rodney Weaver photograph, copyright IRS